LEAVING
EARTH

Also by Robert Zimmerman

Genesis: The Story of Apollo 8
The Chronological Encyclopedia of Discoveries in Space

LEAVING EARTH

SPACE STATIONS, RIVAL SUPERPOWERS, AND THE QUEST FOR INTERPLANETARY TRAVEL

by

Robert Zimmerman

Joseph Henry Press
Washington, D.C.

Joseph Henry Press • 500 Fifth Street, N.W. • Washington, D.C. 20001

The Joseph Henry Press, an imprint of the National Academies Press, was created with the goal of making books on science, technology, and health more widely available to professionals and the public. Joseph Henry was one of the founders of the National Academy of Sciences and a leader in early American science.

Library of Congress Cataloging-in-Publication Data

Zimmerman, Robert, 1953-
  Leaving earth : space stations, rival superpowers, and the quest for interplanetary travel / by Robert Zimmerman.
     p. cm.
Includes bibliographical references and index.
  ISBN 0-309-08548-9 (Hardcover)
  1. Astronautics—History. 2. Outer space—Exploration—History. 3. Astronautics—Political aspects—History. I. Title.
  TL788.5.Z55 2003

                           2003007637

Cover: First two modules of the International Space Station. Photo by NASA/ Science Photo Library.

Printed in the United States of America.

To my wife Diane, who knows how to help me write.

Contents

Acknowledgments ix

Preface xi

1. Skyscrapers in the Sky 1

2. *Salyut*: "I Wanted Him to Come Home." 19

3. *Skylab*: A Glorious Forgotten Triumph 48

4. The Early *Salyuts*: "The Prize of All People" 81

5. *Salyut 6*: The End of Isolation 114

6. *Salyut 7*: Phoenix in Space 163

7. *Freedom*: "You've Got to Put on Your Management Hat . . ." 207

8. *Mir*: A Year in Space 227

9. *Mir*: The Road to Capitalism 270

10. *Mir*: The Joys of Freedom 303

11. *Mir*: Almost Touching   326

12. *Mir*: Culture Shock   375

13. *Mir*: Spin City   416

14. *International Space Station*: Ships Passing in the Night 446

Bibliography   467

Notes   483

Index   509

List of Illustrations

1. *Salyut* with approaching *Soyuz*, 28
2. *Skylab* with docked *Apollo* spacecraft and
   Salyut for scale, 52
3. *Salyut 3*, 87
4. *Salyut 4* with approaching *Soyuz*, 93
5. *Salyut 6*, 115
6. *Salyut 7* with transport-support module, 166
7. *Mir* core module, 230
8. *Mir* core with *Kvant*, 240
9. *Mir, Kvant, Kvant-2*, 274
10. *Mir, Kvant, Kvant-2, Kristall*, 284
11. *Mir, Kvant, Kvant-2, Kristall*, with *Sofora, Strela*,
    and docked *Soyuz-TM* and *Progress-M*, 312
12. *Mir, Kvant, Kvant-2, Kristall, Spektr*,
    with docked *Soyuz-TM*, 385
13. *Mir* complete, with *Kvant, Kvant-2, Kristall, Spektr, Priroda*,
    with docked *Soyuz-TM* and *Progress-M*, 407
14. *International Space Station*, as of December 2002, 450

Acknowledgments

No book can be written without the help and support of others. I must give special thanks to my interpreter, Andrew Vodostoy, and to all those who made my trip to Moscow possible, including Nina Doudouchava and her two children, Alice and Philip, Nicholai Mugue, Anatoli Artsebarski, Alexander Cherniavsky, and Galina Nechitailo. I must also thank the many cosmonauts, engineers, and scientists who gave me so much of their time in interviews when I met them in Russia. Authors Michael Cassutt and James Harford as well as Soviet space historians Asif Saddiqi, Bert Vis, and Charles Vic also deserve my gratitude for their advice about working in Moscow. Thanks must also go to David Harland and Michael Cassutt for reviewing my manuscript, Glen Swanson for helping me obtain Valeri Ryumin's diary, David S. Hamilton at Boeing for creating the International Space Station graphic, and Janet Ormes and the librarians at the Goddard Space Flight Center as well as Jane Odom, Colin Fries, and John Hargenrader and everyone else at the NASA History Office in Washington, D.C., for providing me more information than I imagined existed.

I also thank my editor, Jeff Robbins, for having faith in my writing talent, as well as all the talented people at the Joseph Henry Press for making my writing shine. This book would not exist without their effort.

Finally, I must recognize and praise the men and women, Russian and American, who risked their lives to fly into space and extend the range of human experience. It was their courage and dedication that actually wrote this history.

Preface

Societies change. Though humans have difficulty perceiving this fact during their lifetimes, the tide of change inexorably rolls forward, sometimes for better, sometimes for worse.

The story of the first space stations and the men and women who built and flew them is in most ways a story of the evolution of the Russian people. When they began their journey to the stars in 1957, they were an isolated, xenophobic, authoritarian culture ruled by an oppressive elite who believed that they had the right to dictate how everyone else should live their lives.

Forty years later, that same nation has become one of the world's newest democracies. Its borders are open, its people free, and its economy booming.

In the years between, driven by an inescapable, generations-old insecurity, Russia went out into space to prove itself to the world, and ended up taking the first real, long-term steps toward the colonization of the solar system. Cosmonauts, using equipment built by people only one generation removed from illiteracy, hung by their fingernails on the edge of space and learned how to make the first real interplanetary journeys. Sometimes men died. Sometimes they rose above their roots and did glorious and brave things. In the process, and most ironically, the space program that the communists supported and funded in their futile effort to reshape human nature helped wean Russia away from communism and dictatorship and toward freedom and capitalism.

Leaving Earth is my attempt to tell that story.

Nor is this book solely about how Russia changed in the late twentieth century. For Americans, this story carries its own lessons, lessons that some might find hard to take. For at the same time the Russians were pulling themselves out of tyranny as they lifted their eyes to the stars, the United States evolved from an innovative, free society to a culture that today seems bogged down with bureaucracy, centralization, and too much self-centeredness.

In the early 1970s, the United States had the tools, the abilities, the vision, the freedom, and the will to go to the stars. We had already explored the moon. Our rockets were the most powerful ever built. And we had launched the first successful space station, with capabilities so sophisticated that the Soviets took almost three decades of effort to finally match it. With only a little extra labor, that station could have been turned into a space vessel able to carry humans anywhere in the Solar System. The road was open before us, ours for the taking.

And then the will faded. For the next 30 years, the trail-blazing was taken up by others, as Americans chose to do less risky and possibly less noble tasks. More importantly, just as the bold Soviet space program helped teach the Russians to live openly and free, the top-heavy and timid American space program of the late twentieth century helped teach Americans to depend, not on freedom and decentralization, but on a centralized Soviet-style bureaucracy—to the detriment of American culture and its desire to conquer the stars.

That these facts might reflect badly on my own country saddens me beyond words. I was born into a nation of free-spirited individuals, where all Americans believed they were pioneers, able to forge new paths and build new communities wherever they went. Or, as stated in 1978 by one much-maligned but principled politician, born of a Jewish father and a Christian mother,

> We are the "can-do" people. We crossed the oceans; we climbed the mountains, forded the rivers, traveled the prairies to build on this continent a monument to human freedom. We came from many lands with different tongues united in our belief in God and our thirst for freedom. We said governments derive their just powers from the consent of the governed. We said the people are sovereign.[1]

Whether this describes the American nation today I do not know. If one were to use as a guide our accomplishments in space since Barry Goldwater said these words, one would not feel encouraged.

Yet, the true test of a free and great people is whether they have the stomach to face difficult truths, and do something about it. It is what the American public did in the 1860s, when it freed the slaves. It is what that same society did in the 1950s, when it ended racial discrimination. And it is what the Russian people did in 1991, when they rejected a communist dictatorship and became free. I sincerely hope that future Americans will be as courageous, performing acts as noble.

Above us, the stars still gleam, beckoning us. "A man's reach should exceed his grasp, or what's a heaven for?" said the poet Robert Browning.

Who shall grab for that heaven? Who will have the courage, boldness, and audacity to reach for the stars, and bring them down to us all?

For the last 40 years far-sighted dreamers in both the United States and Russia struggled to assemble the first interplanetary spaceships. For many political reasons, they called them space stations, and pretended that their sole function was to orbit the earth and perform scientific research in space.

Their builders, however, knew better. Someday humans will put engines on these space stations, and instead of keeping station around the earth, humans will launch them out into interplanetary space, leaving Earth behind to voyage to other worlds and make possible the colonization of the planets.

When that great leap into the unknown finally occurs, what kind of human society will those explorers build, out there amid the stars? Will it be a free and happy place, "a monument to human freedom"? Or will it be something else, something of which few would be proud? The nation that reaches for the stars will be the one to make that determination.

"What's past is prologue," wrote Shakespeare. The events in space in the past 40 years have sent the human race down a certain path. It is my hope that by telling that story, I help future generations travel that road more wisely.

As far as the eye could reach, spread vast expanses of Russia, brown and flat and with hardly a sign of human habitation. Here and there sharp rectilineal patches of ploughed land revealed an occasional state farm. For a long way the mighty Volga gleamed in curves and stretches as it flowed between its wide, dark margins of marsh. Sometimes a road, straight as a ruler, ran from one wide horizon to the other.[2]
—Winston Churchill, as he flew into the Soviet Union for the first time during World War II.

Peter [the Great] probably also experienced what many succeeding generations of his countrymen experienced when returning home from abroad: a feeling of disappointment, irritation, even resentment, at one's own nation, whose backwardness smacks one in the face.[3]
—Russian historian Aleksandr B. Kamenskii, describing Peter the Great's first trip to England.

In Russia, like nowhere else, [they] are masters at discerning weaknesses—the ridiculous—and shortcomings in a foreigner. One may rest assured that they will miss nothing, because, naturally, no Russian deep in his heart likes any foreigner.[4]
—Catherine the Great

I am not unduly disturbed about our respective responses or lack of responses from Moscow. I have decided they do not use speech for the same purposes as we do.[5]
—Franklin Roosevelt, October 28, 1942, in a letter to Winston Churchill.

We have to provide the crew with virtually everything for the entire duration of their absence from the earth—air to breathe, food and drinking water, repair tools, spare parts, heatable and pressurized quarters for the stay on the cold Martian plains, surface vehicles and fuel for them, down to such prosaic items as a washing machine and a pencil sharpener.[6]
—Willy Ley and Wernher von Braun, 1956

I've been waiting all my life for this day![7]
—Sergei Korolev, the day that *Sputnik* was launched.

1

Skyscrapers in the Sky

The East

The year was 1958, the very dawn of space exploration. The Soviet Union had already launched Sputnik, the first artificial satellite, while the United States was gearing up for its own manned space missions. The race to the moon had not yet started, no human had yet been in orbit, and no one really knew how that journey to the stars was going to unfold. There were many guesses, and wild surmises, but the future remained unknown, even to those in the center of the action. Everyone knew it was going to happen, however. The world waited with bated breath, anxious and eager to see the exploration of the heavens begin.

In this wild, unpredictable moment, at the dawn of the supersonic age, amid a Cold War that threatened to annihilate the planet and with the white-hot blast of the first nuclear explosion still burning in people's minds, a number of visionaries across the globe stepped forward to lay out the first real, concrete blueprints for colonizing the Solar System. These men wanted to go to the stars, and actually believed they could do it in their lifetime.

In the Soviet Union, the visionaries were engineers attempting to consolidate their country's lead in space. They had already built the first rockets able to place a satellite in orbit, and less than a month after Sputnik they had also proved, by launching a dog into orbit, that they could place life in space. If they moved quickly,

they could use their technological lead to dominate the colonization of the stars.

Without question, the most important Soviet visionary was
Sergei Pavlovich Korolev. Under his leadership in the early 1950s
the Soviets had designed and built the R7 rocket, able to put a
payload weighing about 5 tons into earth orbit. He then got the
okay from Khrushchev and the communist leadership to use that
rocket to launch Sputnik. Though trained as a rocket engineer,
Korolev was more a manager and a political lobbyist. As Nikita
Khrushchev himself noted, "When he expounded or defended ideas,
you could see passion burning in his eyes. . . . He had unlimited
energy and determination, and he was a brilliant organizer."[1]

A hard-driving, square-faced man who demanded the utmost
from everyone, Korolev evoked fear, respect, adoration, hatred, and
love from the engineers working under him at Experimental Design Bureau #1. He once screamed at an army general, "If you don't
fix this in ten minutes I will make you a soldier!" "He was very
strict, sometimes crude," said Mark Gallai, the test pilot who
trained the first Soviet cosmonauts. Andrey Sakharov, Nobel Prize
winner and the inventor of the Soviet hydrogen bomb, even called
him "cunning, ruthless, and cynical."[2]

At the same time, Korolev took care of his people and their
families, making sure they had food, housing, medicine—goods not
often easy to obtain in post-World War II Soviet Russia. Once a
week he made himself available for anyone to see him. "People
would come to him with all kinds of requests, and he would see
everyone," remembered Antonina Zlotnikova, his technical secretary. "It was a difficult, post-war time. . . . He got them medicine
and interceded about their housing."[3]

Above all, Korolev wanted the work done right. "If things went
badly he could not live peacefully," noted one of his biographers.
Or as Korolev himself said, "I can never forget, going home, that
something is wrong with the technique."[4]

And he wanted to send humans into space, to fly like eagles
between the planets. Since childhood he had been fascinated by
flight, designing and flying gliders before he had even graduated
high school. On the day he successfully flew his first homemade
glider, he wrote, "I feel a colossal sense of satisfaction and want to
shout something into the wind that kisses my face, and makes my
red bird tremble. . . . It's hard to believe that such a heavy piece of

metal and wood can fly. But it's enough to leave the ground to feel how the machine comes alive and flies whistling, answering to the least movement of the controls."[5]

After his triumph at building and launching Sputnik, Korolev immediately proposed a grand plan for the Soviet exploration of space. First he would use the R7 rocket to do some basic, preliminary, orbital research while simultaneously building a new, more powerful launch rocket capable of putting four to five times more mass into orbit. Then he would build "artificial settlements" in space, assembled from the larger rockets' unused upper stages. These near-Earth orbital stations—which he intended to launch by the early 1960s—would make possible the study of weightlessness and radiation on humans, plants, and animals. More importantly, Korolev and his engineers would use these stations as prototypes for learning how to build interplanetary spacecraft. These "artificial settlements" would then be assembled in orbit as spaceships able to send humans to Mars, Venus, and the moon.[6] In 1960, he proposed this grand plan to the leadership of the Communist Party, and got it approved, at least superficially. As far as Korolev could tell, under his leadership the Soviet Union was going to carry the human race to the stars.

Korolev was not the only Soviet designer with grand dreams. Two other men in particular would later become as important as Korolev, if not more so. Valentin Glushko, like Korolev, was in charge of his own design bureau in the 1950s. Reading Jules Verne as a child, Glushko fantasized about sending men into space, of going to the moon and the planets and colonizing the stars. When he was 15, he wrote to Konstantin Tsiolkovsky, the first Russian to dream seriously of space travel and thus considered the father of the Soviet space movement, who wrote back asking the boy if he was really serious about space flight. Glushko's response was enthusiastic and idealistic. "I want to devote my life to this great cause."[7] Training himself as an engineer, Glushko became the expert who built all the rocket engines that Korolev used to launch Sputnik, Gagarin, and all the early Soviet groundbreaking space firsts.

Like Korolev, Glushko was a hard-driving perfectionist who could tolerate no errors. Unlike Korolev, Glushko was more an engineer, focusing his entire energies on designing better rocket engines.[8] Tall and big-shouldered like a basketball player, Glushko had started out before World War II as Korolev's superior. After both men were arrested by the secret police during the purges under Stalin, they somehow switched places. After the war, Glushko found himself forever in Korolev's shadow, the mere engine-maker for the genius who was sending man to the stars. Though he dreamed of building gigantic rockets and space stations that would be used to colonize the moon and the planets, decades would pass before Glushko was finally in a position to implement any of these plans.

In fact, over the years the rivalry between these two men drove a wedge down the center of the entire Soviet space industry, preventing much of Korolev's grand plan from ever reaching fruition. They could not agree on the kind of propellants their rockets should use, and by the early 1960s rarely worked together. Glushko preferred engines that used storable fuels, such as hydrazine and nitric acid, because they allowed a rocket to stand fueled for long periods, an advantage for a missile that must be launched quickly and at a moment's notice. Korolev preferred cryogenic propellants like liquid oxygen—which evaporated quickly and could therefore not be left in a rocket for more than a few hours—because they were less toxic and produced a greater thrust, an advantage when the objective is to lift as much mass into Earth orbit as possible.[9]

The third engineering visionary who shaped the future of Soviet space exploration was a man who in many ways developed its most important hardware, and who even today is probably its least known and most underrated space architect. Throughout the 1950s Vladimir Chelomey had been designing cruise missiles for the Soviet navy. Born in 1914 in the Ukraine to parents who were teachers, Chelomey loved math and science from childhood. He wrote a book on vector calculus at 22, and at 26 completed his doctoral thesis on rocket engines. In between he published more than a dozen articles on mathematics for the official journal of the Kiev Aviation Institute.

Growing up in an educated family in a society where literacy was still somewhat rare, Chelomey was fiercely proud of his sophisticated roots. A stylish dresser who once spent two months designing the desk in his office, he liked to puff himself up, putting himself above men like Korolev and Glushko—both were more than a decade older—by calling them mere "constructors" while referring to himself as a "scientist." "He was very cultured," remembered Sergei Khrushchev, the son of the former Soviet leader and an engineer who worked for Chelomey during the late 1950s and early 1960s.

In 1944 Chelomey convinced Georgi Malenkov, head of the Politburo's committee on rockets, that he could build a Russian version of the V-2 rocket. Malenkov in turn convinced Stalin, who signed the orders putting the 30-year-old boy genius in charge of his own design bureau.[10] For the next 14 years Chelomey built a variety of cruise missiles. Then in 1958, shortly after Korolev launched Sputnik, Krushchev's son Sergei got a job at Chelomey's design bureau. For the next six years, Chelomey took full advantage of this direct link to the head of the Soviet Union to milk as much power and money as he could for his own space projects, which in turn helped sap support from Korolev's own initiatives.

Chelomey, even more than Korolev, wanted to build interplanetary spacecraft. In the late 1950s, at the same time Korolev was proposing space stations and new launch rockets, Chelomey proposed a winged spaceship dubbed *Kosmoplan* ("Space Glider" in Russian) to take men to other planets. It would use a nuclear-powered engine to produce a plasma or electrical pulse that would slowly accelerate the spacecraft on a trajectory toward Mars. After entering Mars orbit and completing several months of reconnaissance and research, *Kosmoplan* would refire its engine and slowly return to Earth, where a giant umbrella would unfold to protect and brake the return vehicle as it plunged into the earth's atmosphere. Once slowed sufficiently, a capsule would open and release the space plane itself, unfolding its delta-shaped wings to land normally on any airport runway. Chelomey had other grand plans, including a two-stage, reusable, winged launch vehicle somewhat similar to the space shuttle, systems for snatching satellites in orbit and returning them safely to Earth, and a whole new family of launch rockets.

When he finally got a face-to-face meeting with Khrushchev in April 1960, however, he found the Soviet leader uninterested in most of these ideas. Though Khrushchev was fiercely proud of his country's space achievements and was quite willing to approve daring space exploits to prove the superiority of communism and the Soviet Union, he knew that the Soviet Union couldn't afford to build most of what Chelomey, or Korolev for that matter, envisioned. To Khrushchev, only Chelomey's offer to build a family of

new rocket launchers seemed practical. Chelomey proposed team-
ing up with Glushko, using the storable-fueled engines that Korolev
had rejected. With these propellants, Chelomey's rockets could
serve both as space launchers and as intercontinental ballistic missiles, a flexibility that pleased Khrushchev enormously.[11]

After some negotiations, Khrushchev approved construction of
the new rockets, while at the same time giving Chelomey control
of a larger, more capable, design bureau. Later generations of the
rocket Chelomey would produce would be dubbed Proton, eventually becoming the primary launch vehicle for placing Russian space
station modules in orbit.[12]

The West

While these Soviet visionaries were competing to consolidate
the Soviet lead in space, in the West an host of dreamers were struggling to get the free world out of the space-travel starting gate. Unlike the dreamers in the Soviet Union, the Western visionaries were
not simply engineers located on military bases building missiles.
Many were scattered throughout society: writers of science fiction
and science fact, imagining the possibility of colonizing the alien
stars visible in the night sky. They filled books and pulp magazines
(like *Fantasy & Science Fiction*, *Galaxy*, and *Amazing Stories*) with
hundreds of fantastic tales about alien invasions and epic space
journeys to imagined places on Mars's desert terrain or Venus's
rainy jungles. Most of them believed that the first steps into space
would require the construction of grand orbiting skyscrapers—what
they called "space stations"—put together by spacesuited construction workers bolting girders and panels into place, creating what
looked like giant World Trade Centers circling the earth. Sleek
spaceships would flit from station to station and, after refueling,
carry colonists to settle new worlds on the moon, Mars, and
Venus.[13]

Of the many 1950s science fiction writers who popularized this
bold future—dreamers such as Isaac Asimov, Robert Heinlein, Ray
Bradbury, and Clifford Simak—maybe the most influential was
Arthur C. Clarke. As well as writing popular science fiction novels
describing the first missions into space, Clarke was an accomplished engineer who practically invented the idea of artificial communication satellites. In 1945, he wrote an article for *Wireless*

World in which he proposed a three-satellite cluster, placed in geosynchronous orbit, to provide instantaneous global communications.[14] In his fiction writing Clarke described wheel-shaped space stations where hundreds of people lived and worked. He described ungainly interplanetary spaceships, some shaped like donuts, others complex assemblages of girders and spheres, traveling from planet to planet in easy, exuberant leaps. He described colonies on the moon, on Mars, on the asteroids.[15]

Clarke's non-fiction writing was no less inspirational. In a 1951 book called *The Exploration of Space*, he tried to predict how, in the coming decades, humanity would go to the moon, build colonies on Mars, and even travel to the stars. His "deep space" ship, designed to carry humans to other worlds, ". . .would have no vestige of streamlining and could be of whatever shape engineering considerations indicated as best." Clarke figured that because such a space-based ship would be assembled in zero gravity, it would not have to be strong enough to stand up under its own weight. It could "have about as much structural strength as a Chinese lantern, and perhaps the analogy is not a bad one as the tanks could, at least for some fuels, be little more than stiffened paper bags!"[16]

Clarke also imagined a whole plethora of manned space stations, all shaped differently and designed to circle the earth in a variety of different orbits depending on the station's purpose and research goals. Some would be used to photograph the earth. Some would study the stars. Some would be used to do biological research. Some would be used as radio-relay stations, providing the equipment to make his three-satellite communications cluster a reality. And some would be used as refueling stations, not unlike a modern airport terminal hub, where manned ships coming from Earth would unload their passengers, and interplanetary ships coming from the moon, Mars, and Venus would fuel up while picking up these passengers to take them on the remainder of their voyage.[17]

As credible and influential as Clarke's writing was, he couldn't hold a candle to German-born Willy Ley, a man who could easily be given credit for creating the entire field of science writing. Ley, a passionate advocate for space exploration from his youth, had been one of the founders of the German Society for Space Travel, formed in 1927 when he was only 21 years old. Only one year earlier Ley had written his first book, *Journey to the Cosmos*, in which he outlined the future of man in space. "On the day the first manned rocket leaves the earth's atmosphere," he wrote, "mankind . . . will have taken the first step into a new age—the age of dominion over space." Throughout the late 1920s this private club launched a number of small experimental solid-fuel rockets, some rising as high as 3,000 feet, others exploding on the ground.[18]

By 1932, however, the Society was going bankrupt. The Great Depression was at its worst and the economy of Germany was collapsing. Moreover, membership in the Society, never very large, was dropping; the idea of space travel was simply too strange for most people. For example, when the Society tried to incorporate, a bureaucrat in Breslau initially rejected their paperwork, claiming that the phrase "space travel" did not exist in the German language.[19]

Ley, whose best talent was writing, not engineering, soon discovered that he could no longer even write about space exploration and rocketry. When the Nazis took power in 1933 they ordered him to cease writing for foreign publications. Fearful of the Nazis (who at this time had imprisoned several other rocket enthusiasts, accusing them of high treason), Ley made the moral and practical decision to leave Germany and emigrate to the United States. He did this despite being an author who did not speak or write English very well, thereby risking forever his career as a promoter of space exploration and rocketry.[20]

In the end, Ley succeeded in becoming an incredibly prolific and successful American writer, producing over the next 35 years dozens of books on space, science, and rocketry. For most of the 1950s and 1960s you couldn't read a science publication without coming across Ley's name. He wrote monthly science columns for several different science fiction pulp magazines. He wrote essays for encyclopedias and reference books. He wrote books. He wrote articles for some of America's most prominent magazines. "Willy Ley rallies the nation for space" was how one historian described his American writings during these years. Sadly, though he had dedicated his life's work to its achievement, Ley did not live to see humans walk on the moon; he died from a heart attack on June 24, 1969, less than a month before the Apollo 11 landing.[21]

Ley believed that the exploration of space would take place in a series of logical steps. First would come the short, manned missions, proving that humans could survive in space while demonstrating the basic technologies for doing so. Next would come the building of large manned space stations in low Earth orbit. "In all probability," he said in 1949, "the unmanned orbital rocket will be succeeded by a manned 'station in space'."

> The construction of this station would begin with a large manned rocket which would be [placed in low Earth orbit]. Additional material could then be brought up to enlarge the ship which is there, and the station would grow out of the first rocket.[22]

More than any other man, Willy Ley can be credited with establishing the wheel as the expected shape of all future space stations. In numerous books and magazine articles, always accompanied by glorious and grand illustrations, he repeatedly laid out its design and construction. "When man first takes up residence in space," Ley wrote in 1952, "it will be within a spinning hull of a wheel-shaped structure, rotating around the earth much as the moon does."[23] Elsewhere he wrote

> The space station [will be] a gigantic wheel, about 250 feet in diameter with a rim at least 22 feet thick. Three main spokes connect the rim with the hub, but there are also a number of separate pipes running from the hub to the rim. The space station will need a crew of at least 30 men to run smoothly and efficiently. There will be another 20 to 30 men aboard who are not crew members, but observers and scientists who are on temporary duty.[24]

For Ley, the space station was a required preliminary outpost for all future human space exploration. "Because of the special conditions prevailing on such a station (infinite vacuum, permanent apparent weightlessness, the possibility of creating any extreme of temperature either by concentrating the sun's rays or shielding something from the sun's rays), it could well be a most valuable laboratory. And it would also be a watchdog for the whole planet. Finally, it could be a refueling place for rocket ships."[25]

Like Clarke and many other writers and engineers of the time, Ley saw the space station as a separate entity from the interplanetary spaceships that would follow. When the station was finished, it would become the base of operations from which to study the earth and the stars, to provide military security, to do weather forecasting, and to stage the shipbuilding and refueling facilities for the construction of the more advanced interplanetary ships.[26]

Not all the Western promoters of space exploration in the 1950s were writers like Clarke and Ley. One man, Wernher von Braun, was an engineer, and had begun his rocket-building career in Germany at the same Society for Space Travel that Willy Ley had helped found. The son of a former German Minister of Agriculture, von Braun as a teenager wanted to learn everything he could about rockets, and through Ley was introduced to the Society.[27]

In many ways, Wernher von Braun was possibly the most grandiose, and the most practical, of the 1950s visionaries. The man who built the V2 rocket for the Nazis was remarkably similar to Sergei Korolev. Though an engineer, von Braun was more of a manager and lobbyist than a builder. Like Korolev, he had a charismatic personality. He was a crisp speaker whose friendly enthusiasm for space travel quickly made his audiences as enthusiastic. Unlike Willy Ley, who fled Germany when the Nazis were gaining power and the German Society for Space Travel was running out of money, von Braun decided, in his passionate and obsessive desire to build rockets and travel into space, to take a job for the German Army. "It became obvious," von Braun wrote years later, "that the funds and facilities of the Army would be the only practical approach to space travel." Von Braun did not think much about the moral dimension of his actions. "I was still a youngster in my early 20's and frankly didn't realize the significance of the changes in political leadership," he wrote. "I was too wrapped up in rockets."[28]

For 10 years he worked in the German missile program, helping to devise test rocket after test rocket, trying to figure out why some blew up and others flew wildly off course. During this time he found the site for and helped design the Peenemünde launch facility on the north coast of Germany, the first rocket spaceport ever built. Finally, in the waning years of World War II, all that work resulted in the V2 rocket, the first ballistic missile used in battle. With Hitler's firm support ("What I want is annihilation," said Hitler. "Annihilating effect!") and the use of slave labor, the Peenemünde team built and launched more than 2,500 rockets, aiming them at England and Antwerp in a futile effort to stop the Allied invasion.[29]

As Nazi Germany collapsed and the Allies closed in, von Braun was again faced with a choice: Surrender to the Soviet Union or surrender to the United States. Going to the Soviets would be easier. They were closer, and would certainly provide the German engineers with anything they needed to build spaceships.

This time, von Braun took the harder choice, and brought his team to America. After years of working for cruel overlords who were willing to starve slaves to death to get their projects completed, von Braun had had enough. He no longer could cooperate with dictators merely so that he could build rockets. As von Braun noted in 1955, "As time goes by, I can see even more clearly that it was a moral decision we made [when we chose to come to America.]"[30]

Three years later, while isolated in New Mexico teaching the U.S. Army how to build and launch the V2 rocket he had designed and built for the German army, von Braun sat down and for pleasure wrote a short science fiction book he called *The Mars Project*. In it, he described in numbing technical detail (with formulas!) the first interplanetary flight to Mars. The mission would require an armada of 10 ships, assembled in Earth orbit, carrying a total of 70 men. Each ship would weigh approximately 4,000 tons and carry the fuel, water, food, and supplies needed for a two-and-a-half-year journey, along with small tugs or ferries for transferring crew from ship to ship.

After a 260-day voyage, the fleet would swing into Mars orbit. There, a crew of about a dozen men would assemble in the nosecone section of one of the ships. This nosecone, resembling an airplane, would then detach and descend to the surface, landing on skis in what von Braun imagined as the smooth ice-covered polar regions of Mars. Once on the surface, the crew would abandon their landing craft and travel to the Martian equator, where they would build a runway for the arrival of two more nosecone ships, which would land like airplanes on this homemade runway and then launch like rockets back to the mother ships in orbit. All told, an expedition of about 50 men would stay on the Martian surface for about 15 months.[31]

For von Braun, the technical problems of building and launching such an expedition, while difficult and challenging, were always solvable. "Even now [1954] science can detail the technical requirements for a Mars expedition down to the last ton of fuel."[32] He was an engineer and a rocket scientist. If he was given the money and resources, he knew he could build the equipment to get humans to Mars.

What concerned him more were the human problems, the physical and emotional stresses space flight would put on the human body. "What we do not know is whether any man is capable of remaining bodily distant from this earth for nearly three years and return in spiritual and bodily health." Over the next decade he increasingly wondered whether the human body could withstand prolonged weightlessness. In a series of articles he wrote for *Colliers* magazine in the mid-1950s, he wrote that, ". . . over a period of months in outer space, muscles accustomed to fighting the pull of gravity could shrink from disuse—just as do the muscles of people who are bedridden or encased in plaster casts for a long time. The members of a Mars expedition might be seriously handicapped by such a disability. Faced with a rigorous work schedule on the unexplored planet, they will have to be strong and fit upon arrival."[33]

Von Braun also considered the emotional and psychological strains caused by confinement in a small space.

> Can a man retain his sanity while cooped up with many other men in a crowded area, perhaps twice the length of your living room, for more than thirty months? Share a small room with a dozen people completely cut off from the outside world. In a few weeks the irritations begin to pile up. At the end of a few months, particularly if the occupants of the room are chosen haphazardly, someone is likely to go berserk. Little mannerisms—the way a man cracks his knuckles, blows his nose, the way he grins, talks, or gestures—create tension and hatred which could lead to murder.[34]

Recognizing the problems these issues posed for space travel, in the mid-1950s von Braun predicted that the first mission to Mars could not happen as quickly as many scientists and writers like Ley, Korolev, Clarke, and Chelomey imagined. As he wrote in his last *Colliers* article, published in 1954,

> Will man ever go to Mars? I am sure he will—but it will be a century or more before he's ready. In that time scientists and engineers will learn more about the physical and mental rigors of interplanetary flight—and about the unknown dangers of life on another planet. Some of that information may become available within the next 25 years or so, through the erection of a space station above the earth.[35]

Like both Willy Ley and Sergei Korolev, von Braun had come to believe that to get to the other planets, humans would have to build space stations first and use them to learn how to live and work routinely in Earth orbit. In a book he co-wrote with Willy Ley in 1956, the two men wrote that "No expedition [to the moon or planets] can be made until after at least a temporary manned space station has been put together in an orbit around the earth, for the space station is, in a manner of speaking, the springboard for longer trips."[36] "Within the next 10 or 15 years," wrote von Braun in a 1952 issue of *Colliers*, "the earth will have a new companion in the skies, a man-made satellite. . . . Inhabited by humans, and visible from the ground as a fast-moving star, it will sweep around the earth at an incredible rate of speed in that dark void beyond the atmosphere which is known as 'space'." He added, "Development of the space station is as inevitable as the rising of the sun."

For von Braun, humanity would first build space stations to prove that they could live and work in space for long periods. Then, just as Ley and Clarke had suggested, these giant orbiting skyscrapers, manned by dozens, would be used as either a refueling stop or a shipbuilding yard where engineers and construction workers assembled the new interplanetary ships for voyages to the moon and beyond.

The Problem

Though scattered across the globe, these men, along with thousands of others, all imagined a kind of grand adventure in space, and longed to make it happen. In turn, their visions motivated a whole generation, and soon every technological culture throughout the world was caught up by the idea of traveling in space and visiting other worlds.

Soon money was allocated. Soon the first man-carrying rockets were launched, both in the United States and in the Soviet Union. Soon, men were heading to the moon.

Not surprisingly, the actual events of the 1960s only vaguely matched the predictions of the 1950s visionaries. No space stations were built, and the first manned moon ships went there directly, bypassing the so-called springboard that Korolev, von Braun, and Ley had thought essential.

These brilliant engineers simply could not control the wild bronco of history that they were riding. Building giant rockets required the involvement of politicians, and trying to steer the large political forces wielded by leaders like Kennedy and Khrushchev in the beginning, and Brezhnev and Nixon in later years, proved impossible. Thus, the first human flights in space involved, not a space station, but a race between powerful nations to fly directly to the moon.

And yet, von Braun, Korolev, Ley, Clarke, and the innumerable dreamers of the 1950s were not wrong. Their basic assumption, that the first voyages to the planets could not occur until people learned to live and work in zero gravity for long periods, has proved essentially correct. If anything, all these visionaries, except von Braun, vastly underestimated the work and time needed to make an interplanetary voyage possible. Korolev, for example, estimated that he would be ready to send his first missions to Mars by the late 1960s.[37] Ley believed that the first of his giant wheel-shaped space stations could be completed by 1970, followed soon after by the first missions to Mars and Venus.[38] And Chelomey thought it possible to skip space station development entirely and leave Earth directly in his *Kosmoplan*.

Korolev was the only one who realized the difficulty of building a grand and gigantic space station in orbit around the earth. While Westerners like Clarke, Ley, and von Braun imagined construction workers riveting the station into shape as it orbited the earth, Korolev saw his first space stations as nothing more than large, prebuilt vessels that would be ready for occupation when launched. He also realized that the first interplanetary spaceships would not be constructed like skyscrapers in Earth orbit. Instead, he planned to revise his space-station designs and link several together to quickly create a larger and more capable interplanetary ship.

Korolev understood something that few either then or since have recognized: There is little difference between an Earth-orbiting space station and an interplanetary spaceship. Once you build a habitable, manned station in orbit, capable of keeping humans alive for periods exceeding a year, there really wasn't any reason to use it as a refueling stop or a base of operations, as imagined by Ley, von Braun, or Clarke. Instead, it makes much more sense—especially considering the cost and difficulty of building it in the first place—to turn the station itself into a ship for taking people to other planets.

For example, the technical problems of creating a self-sustainable life-support system are the same in either a space station or an interplanetary spaceship. A person needs, at a minimum, about two liters (about a half-gallon) of water per day to survive. To carry enough water to stock a multi-year mission on either a space station or a spaceship makes no sense; the cost of lifting that weight is prohibitive. Instead, a small initial amount of water can be recycled, captured from both the ship's humidity and the crew's urine and turned into potable water. Similarly, a person breathes about three pounds of oxygen each day. Rather than hauling thousands of pounds of oxygen into space, the carbon dioxide that humans exhale, which in turn has to be scrubbed from the spaceship's atmosphere, must somehow be recycled back into breathable oxygen. Supplying food is more difficult. Both space stations and interplanetary ships would probably carry enough food for their journeys, just as sea-going ships do on Earth. However, getting the food into space requires new methods of food storage that are both lightweight and can keep provisions fresh and edible for months, even years.

Korolev intended to carefully study these life-support problems in his orbiting "artificial settlements." From his point of view, many of these technical problems had already been partially solved in submarines and ships. Space merely required their solution at a much higher order of efficiency, and in a manner alien to any Earth situation.

And the circumstances in space are certainly alien. Consider again the problem of food supply. Interplanetary voyages will take years, even decades. Unlike the Spanish and Portuguese sailors of the fifteenth century, who could trade or hunt for food as they went, deep space explorers will have no place in the outer reaches beyond low Earth orbit to restock their food supplies. While it might be possible to carry enough food for a journey to Mars, doing so on a journey to Jupiter or Pluto is probably impractical. Not only would growing some of the required food on the interplanetary ship reduce the weight of supplies but the plant life would also help recycle the ship's atmosphere. However, growing food is not simple. On Earth, food production involves agriculture, animal husbandry, and the extensive use of vast areas of land for planting crops. An interplanetary spaceship would not have such resources. Furthermore, it was unknown whether plant and animal life could survive, grow, and reproduce in the weightless environment of space. To find this out was going to involve many experiments in orbit over a number of years.

Then there was the problem of supplying the spaceship's electricity. Burning coal or oil made no sense. Von Braun and Korolev instead suggested using solar power. Large solar panels could be built and attached to the outside of the station, drawing the light energy of the sun and turning it into electricity. Learning how to build effective solar panels would also require years of in-orbit experimentation.

The perfect spaceship, whether floating in Earth orbit or in the endless emptiness between planets, also needs an almost limitless supply of engine fuel for propulsion. Outer space does not have gas stations for refueling. While the Russian engineers proposed using some form of nuclear or ion engines to generate the required thrust for long periods without need for refueling, it was Clarke who proposed that light from the sun could also be harnessed for this purpose. Giant sails would use the radiation pressure of this light to push the spacecraft between planets.

Other problems: How would the vessel's orientation be maintained? Letting it drift freely would not work, because radio antennas, engine nozzles, and various other sensors, for example, telescopes, must be aimed. Using chemical thrusters might work, but would be very wasteful of fuel, especially on larger spaceships. Gyroscopes were a possible solution. Such systems are completely self-contained and require little fuel. Von Braun had already proved them successful in the V2 rocket. To gain some reliable control over their rocket, he and the engineers at Peenemünde installed a gyroscope in the V2's nosecone, just below the warhead. Like a spinning top, which can remain upright on the most tenuous of foundations, the angular momentum of the spinning gyro held the rocket vertical and steady. In fact, the gyro worked so well that a ground engineer tilting the gyro's disk in flight forced the rocket to tilt as well, giving missile launchers a method for steering the rocket.[39] Both Korolev and von Braun knew that building such a system in a space vessel would require new engineering and several years of tests.

There were other challenges, too. Consider, for example, the range of temperatures the hull of a spaceship or station is exposed to, from approximately –300°F in shadow to +300°F in direct sunlight. Somehow the hull must be able, not only to withstand these extremes, but also to radiate the heat away as well as retain its thermal balance so that the interior temperature remains livable.

For the engineers in the 1950s, there was also the question of meteoroids and radiation. How strong or thick did the hull have to be to reasonably protect its occupants from small impacts and lethal cosmic rays? In 1958, no one knew. It would take years of experimentation and test flights to find out.

The challenge of solving these technical problems was further compounded in that any system, either in Earth orbit or on its way to another planet, had to function reliably for years at a time. Even if a crew merely flew past either Mars or Venus, the shortest possible flight time would still be a year. More likely, voyages would last anywhere from two to four years, as von Braun had proposed.[40] During this time, the ship's systems had to function without major breakdowns. And if there was a failure, the spacefaring crews, in Earth orbit or far from home, must be able to use the tools and supplies at hand to fix them.

Then there was the question of the human body itself, a question beside which all other technical issues paled. Could humans survive in space long enough to travel to other planets? Was gravity necessary for life to prosper? Not knowing, men like von Braun, Ley, and Clarke imagined their space stations and interplanetary spacecraft having complex systems for reproducing gravity. Ley's wheel-shaped space station would spin, creating centrifugal force that simulated Earth gravity by pushing its inhabitants outward, away from the wheel's center. Such systems were incredibly complicated. For Korolev, the idea of building them seemed impractical. Instead, he proposed using his artificial settlements to test whether humans could tolerate weightlessness for the years necessary to get to and from the nearest planets.

There also remained von Braun's concerns about whether a crew of human beings, living in cramped, alien quarters farther from home than any humans had ever been, could manage to live together for years on end without going insane or killing each other or themselves. What would people do for those endless months in empty space? How would they fill their time? As one of the first to ask these questions, von Braun was also one of the first to realize how much time it would take to learn the answers. As he wrote almost 50 years ago, "Will man ever go to Mars? I am sure he will—but it will be a century or more before he's ready." You simply couldn't learn the consequences of spending a year in space—*until you spent a year in space.* Before the first pioneers could board a spacebound fleet heading out across the vast black ocean between the planets, decades of orbital research would have to be done, the blue-white, glittering Earth never more than a few hundred miles away.

And finally, what about the enormous costs? Funding long term orbital missions like this would require the same kind of political compromises accepted by both Korolev and von Braun in order to get their first rockets launched. The exploration of space would unfold, not as the grand pioneering in-space construction project imagined by these first visionary engineers, but reshaped to fit the dreams of politicians. A generation would have to sacrifice its hunger to travel to the stars so that later generations could do so with skill and increasing ease.

This is the story of the men and women who made that sacrifice, and of the battles they had to fight, both on Earth and in space, to make it possible, and of what they learned as they did so about the limits of human endurance.

2

Salyut: "I Wanted Him to Come Home."

Propaganda

The flax plant was a little thing, not quite an inch tall. Nestled in a small square greenhouse attached to one wall of the *Salyut* space station, its soil was artificial, its light came from fluorescent lamps, and it grew in an alien universe that had no gravity. Along with some cabbage and hawksbeard, this flax plant was a pioneer, the first Earth plant to grow in outer space.[1]

Each evening, after doing his daily exercises, Viktor Patsayev glided over to the facility to water the plants. A sad-faced man with a careful and precise manner, Patsayev pushed a handle to pump water from a reservoir into the layers of artificial soil that held the seeds. After only a week or two, several flax shoots had begun to poke up through the fake soil. The weed, normally found in empty lots and garbage dumps, had been chosen for its pioneering role because it grew quickly and was small. Its space partners, cabbage and hawksbeard, had been chosen for similar reasons. However, none of the plants was prospering. Though they sprouted, their leaves seemed smaller than normal, and were growing far slower than on Earth. Something was hindering their growth.

As an engineer, Patsayev considered the engineering aspects of the problem. Was the failure of some plants to grow due to the greenhouse itself? Perhaps it needed re-engineering. Sometimes its pump sprayed water at only one part of the artificial soil.

19

Sometimes the water was distributed more evenly. Without the pull of gravity, the water had no natural guide for getting to the roots. Or was it space and weightlessness itself? Maybe some aspect of zero gravity or the radiation of space was affecting the seeds. Maybe plants themselves could not thrive without the pull of gravity.

He did not know. Yet, the answer was immensely important. If humans were going to live and work in space permanently, it was essential that plants go and live there with them.

Viktor Patsayev, along with his crewmates Georgi Dobrovolsky and Vladislav Volkov, were the first occupants of the first manned orbital space station, named *Salyut*. For more than three weeks these three men worked in their little home in space, beaming to Earth daily images of their life in the weightless environment of space. Those three weeks had been difficult and challenging. The gear on board, from clothing to food, had been measured and designed for entirely different crews. The equipment was new and untested, the circumstances unknown and untried.

Even communications with Earth were distant and detached. Because uncontrolled direct conversations with their families were not allowed in the authoritarian Soviet society, the men listened each day to short audiotapes that had been recorded earlier. In turn, the men could talk back only during television videotaping sessions, using one of the two black-and-white television cameras on the station. Their monologues were recorded so that their families could see them a day or two later.

For the wives and children of these men, the weeks in space had been exciting but wrenching. As was typical for the secretive and overbearing Soviet Union, the men had been forbidden to tell anyone, even their wives, about the flight. No one in their families knew that the men were scheduled to go into space until *after* the launch, giving them no chance to prepare for what was to come.

In that rugged, isolated, and unnatural environment, the men endured, finding ways to survive. And though Patsayev was in charge of caring for the plants, all three men found themselves drawn to them. In their artificial home—about the size of a typical city bus—covered with metal panels and instrument controls and surrounded by the endless black vacuum of space, these tiny green plants seemed their only direct link with Earth. As Volkov said during one radio communication, "They are our love."[2]

Now, after 24 days in space, the men were finally heading home. They began to pack up their gear, carefully putting samples of some of their experiments in their *Soyuz* descent capsule. Of the plants, Patsayev could take very little. However, the greenhouse had been designed so that sections could be detached and returned to Earth for more careful study. Patsayev removed the film camera, which had been filming the plants day-by-day one frame every ten minutes to show their growth, as well as single samples of the flax, hawksbeard, and cabbage plants, and transferred everything to their *Soyuz* spacecraft.

On June 29, 1971, the crew packed up and climbed into the descent module of *Soyuz*. Then, at 11:28 P.M. Moscow time, they undocked from the *Salyut* station, and quickly eased away from what had been their home in space for the last 24 days. Dobrovolsky radioed the ground, "We have checked the systems. Everything is normal. The horizon has come up for me. The station is above me."

Ground control answered, "Goodbye, Yantari,* till the next contact."[3]

That next contact was, sadly, not to be.

When it became obvious that the Soviets were going to lose the race to the moon, the Soviet rulers, led by Leonid Brezhnev, scrambled to find some way to save face. In October 1969, three months after Armstrong and Aldrin had landed at Tranquility Base, the Soviets launched the only triple spaceship mission ever flown, in which two *Soyuz* spacecraft attempted to dock while a third took photographs. The docking failed (because of a technical glitch in the rendezvous equipment on the *Soyuz* spacecraft), and the mission itself seemed somewhat pointless. Despite placing three manned spacecraft in orbit on successive days, a feat that no one has been able to match even to this day, as well as putting seven men in orbit at once, a new record, the triple mission seemed nothing more than an empty stunt that accomplished little.

---

*Every Soviet/Russian space mission commander chose a code name for his mission. For *Soyuz 11*'s crew, the code name was Yantar, or amber in English. Dobrovolsky as commander was called Yantar-1, Volkov Yantar-2, and Patsayev Yantar-3. In Russian, the plural of yantar is yantari.

Ten days later, in ceremonies at the Kremlin to honor the seven cosmonauts, Brezhnev attempted to put a new and positive spin on this and other Soviet failures in space. In his speech, which would have consequences as significant as John F. Kennedy's 1961 speech committing the U.S. to a lunar landing, Brezhnev proclaimed that, "Soviet science regards the setting up of orbital stations, with changeable crews, as man's main road into outer space." Brezhnev further claimed, quite falsely we know today, that the construction "of long-term orbital stations and laboratories" had been the Soviet goal from the very beginning, part of "an extensive space program drawn up for many years."[4]

The man who made this declaration, Leonid Ilyich Brezhnev, was the consummate propagandist. The first-born son of a steelworker, he had risen slowly and carefully from humble beginnings to become the General Secretary of the Communist Party, ruler of the Soviet Union. Born on December 16, 1906, in the Ukrainian industrial city of Dneprodzerzhinsk (then called Kamenskoye), Brezhnev's life had been beset by war, starvation, violence, and death.

He was only 11 years old when the Bolshevik Revolution took place in 1917 at the end of World War I. Over the next three years, Dneprodzerzhinsk, located on the Dneper River about halfway between the Black Sea and Kiev, became a war zone. First the Austrians took over in April 1918. Then the Red Army rolled through in December 1918. For the next 18 months the civil war between the communist Reds and the more capitalistic Whites brought violence and looting as the two sides traded control of Dneprodzerzhinsk almost weekly. Then, after the Reds took over for good in early 1920, the city saw famine and disease, with dead bodies left to rot on the streets and reports of cannibalism. Brezhnev, 11 to 13 years old during these years, wore rags for shoes, caught black typhus and barely survived, and witnessed the death of almost half his schoolmates. Both his priest and schoolteacher were murdered in the fighting. As Brezhnev later wrote, "[I] grew to manhood, so to speak, not in days but in hours."[5]

In 1923, in a desperate effort to find work, Brezhnev's family moved north, leaving the Ukraine for the city of Kursk, 300 miles south of Moscow and just north of the Ukrainian border. There, 17-year-old Brezhnev linked himself to the Communist Party, joining its youth organization, Komsomol, which gave him a four-year education as a "consolidator," someone whose job was to force peasants and small private farmers to join the collective farms. After graduation, Brezhnev executed this policy in two Soviet provinces, sometimes evicting peasants from their land, sometimes arresting them if they protested, and often exiling them or worse if they resisted. In the late 1930s, he somehow avoided arrest during Stalin's terror campaign that killed millions, and put future Soviet heroes like Sergei Korolev and Valentin Glushko in prison. By 1939, Brezhnev had become Khrushchev's propagandist, helping him execute a violent purge of the Ukrainian ruling elites.

Brezhnev's skill at propaganda was unsurpassed. In his first years under Khrushchev he organized an "army" of 8,000 agitators and propagandists to run the 200 newspapers and magazines in the Dnepropetrovsk province. He took over the campaign to Russify the Ukraine, forcing all publications to switch languages from Ukrainian to Russian. He had history books rewritten to Russify the Ukrainian past.

Then came World War II. Brezhnev, in charge of the defense mobilization, had to organize a desperate retreat for himself, his family, and the factory he had been given charge of from Dnepropetrovsk. He later participated in the campaign that recaptured the Ukraine and defeated the Germans, acting as the political officer for the 18th Army.[6]

For the next two decades Brezhnev followed Khrushchev in his rise to the top of the Communist Party and the leadership of the Soviet Union. By 1964 Khrushchev's brash and erratic style of leadership was no longer acceptable to his Communist Party cohorts. Elites like Brezhnev, Alexei Kosygin, and Mikhail Suslov could not tolerate Khrushchev's tendency to play brinkmanship with the West, risking nuclear war in places like Cuba and Berlin. For these men, the decades of violence, war, chaos, and terror had to end. Security was the most important consideration, and stable, strong, and rigid rule was best way to achieve it. Dubbing themselves a "collective leadership," they teamed up to depose Khrushchev in October 1964, and then spent the rest of the 1960s invigorating the Soviet Union's military, so that by 1971 its navy was the largest in the world, it had more nuclear weapons than the U.S., and its army had been reformed and restructured.[7]

During this time Brezhnev marshalled his considerable public relations and negotiating skills to maneuver his way into the top spot of the "collective leadership." Step by step he took power, easing his allies into positions of authority while slowly pushing aside his opponents. Much of Brezhnev's support resulted from his strong desire to maintain the status quo. Under no condition would his rule lead Soviet Russia back to the purges, the violence, the civil wars, the invasions that Brezhnev and everyone else had lived through during the past 40 years. Instead, his leadership guaranteed stability. People who followed orders and supported the goals of their superiors could slowly move up the ranks and thereby obtain greater privileges and wealth.[8]

As Brezhnev increased his power, he also began to change how the Soviet Union interacted with the noncommunist world. The invasion of Czechoslovakia in 1968 had cast a pall over Soviet international relations, and Brezhnev sought a way to distract others from thinking about the injustice of that attack. His strategy was to use diplomacy and detente (that is, his skills at propaganda) to defuse the hostility of his nation's opponents. Publicly, he would declare the Soviet Union as dedicated to world peace, disarmament, and the end of the Cold War. Privately, he and his fellow Soviet leaders would use these public relations gains, along with the might of the Soviet Union's rebuilt military, to push their communist and national agenda wherever they could.[9]

Brezhnev did not establish these policies by himself. The "collective leadership" was no lie. In the early years Politburo members like Aleksei Kosygin and Mikhail Suslov wielded great power. In later years, men like Andrei Gromyko, Marshal Andrei Grechko, and Dimitri Ustinov influenced every decision Brezhnev made. Nonetheless, Brezhnev set the tone, a tone that strongly appealed to the men who had deposed Khrushchev. These apparatchiks wanted to maintain their control, but to do so with as little risk as possible. Detente, combined with a strong defense, fit their needs very well.

His October 1969 space-station speech was one of the first glimmers of this new approach. Politician that he was, Brezhnev's primary interest in space exploration was how he could use it as a political tool. For him, space exploration served as a convenient and exciting cloak to hide from the world the aggressive nature of his foreign and domestic policy. He also conceived orbital facilities very differently than did his space engineers, who saw space stations as the first attempt to build a vessel that could sustain humans in space long enough to travel to other planets. To Brezhnev, *Salyut's* goals were entirely different. While his speech mentioned interplanetary exploration, he shifted the focus to scientific research. "Major scientific laboratories [in space] can be created for the study of space technology, biology, medicine, geophysics, astronomy, and astrophysics."[10] Then, in a rhetorical flourish, Brezhnev emphasized this focus.

> Space for the good of people, space for the good of science, space for the good of the national economy. Such in brief, is the substance of the Soviet space program—its philosophical credo.[11]

Just as he desired his rule to be stable and safe, he wanted his space program to follow the safest and most secure path. Launching orbital facilities was far less risky than planning visionary missions to Mars. Interplanetary missions were also unnecessary and irrelevant to the Soviet leadership's political goals. Thus, *Salyut* would be advertised as a laboratory in space, above all else.

However, to become a useful political tool for Brezhnev, the Soviet space program had to become efficient and organized—something it had not been in the 1960s. In this sense, Brezhnev's 1969 speech most resembled Kennedy's "man on the moon" speech. If the Soviets' objective had always been to build space stations, they had better beat the United States in launching the first space station into orbit. And with the National Aeronautics and Space Administration (NASA) gearing up to fly its own space station, called *Skylab*, in 1973, the pressure was on.

Brezhnev's public pronouncement—forcing his country to finally "put up or shut up"—thus brought focus to the previously confused and chaotic Soviet space program. In the Soviet Union no single organization like NASA ran the manned space program. Instead, the program was divided between several different "design bureaus," each competing for government support and rubles. Most manned missions were run by the Experimental Design Bureau #1, headed by Sergei Korolev. Korolev's bureau, which had built *Sputnik* and the *Vostok* and *Voskhod* manned capsules that had stunned the world with their achievements in the early 1960s, was, by 1969, trying to compete with the American Apollo program by building lunar landers, orbiters, and a giant rocket, the N1, comparable to the Saturn 5. Korolev's grand plan to colonize the Solar System had been abandoned, and his design bureau had no program, at that time, for building permanent space stations.

A second design bureau, under the leadership of Vladimir
Chelomey, *was* building space stations, but for the Soviet military.
Since the overthrow of Khrushchev, Chelomey had scrambled to
find funding for his myriad space projects. His interplanetary space
plane, *Kosmoplan*, had languished. So had his reusable shuttle system. Only his Proton rocket had reached fruition, completing its
first successful launch in 1965.

At the same time, both before and after Khrushchev's overthrow, Chelomey successfully sold the Defense Ministry on building a competitor to the American military's never-to-be-completed
Manned Orbital Laboratory (*MOL*). Originally named *Almaz*
("diamond" in Russian), Chelomey's station was, like *MOL*, conceived and built as a military facility for orbital surveillance. Once
in orbit, two cosmonauts would use the *Almaz* station as living
quarters while they aimed its telescopes and radar facilities to track
the movements of foreign military operations.[12]

Almaz, as designed, posed two problems to Brezhnev. Its work,
secretive and obviously military in scope, would not lend any cre-
dence to his claims about a peaceful Soviet space-station program.
Furthermore, in 1970, *Almaz* was not yet ready to fly. The engineering of its guidance and life-support systems was still not completed. Moreover, the space capsule that the cosmonauts would
use to go back and forth from *Almaz*, loosely copied from the
American *Gemini* capsule, was far from finished.

To validate Brezhnev's claims, and to get a civilian space station
into orbit before *Skylab*, a deal was struck within the bureaucracy of
the Soviet space program. An *Almaz* hull, built by Chelomey's bureau, would be turned over to the Korolev bureau (run, since
Korolev's death in 1966, by his former deputy Vasily Mishin), which
would build the internal systems and configure it so that a *Soyuz*
spacecraft could dock with it. And the rocket to launch the station
would be the Proton rocket, built by Chelomey's bureau. This patch-
work thus became the rough blueprint for the next two decades of
Soviet-era space stations. Weighing more than 20 tons, *Salyut* was to
be the most massive payload yet launched by the Soviet Union.[13]

For the next 18 months the two design bureaus scrambled to
complete the station. For Chelomey's bureau, things were essentially straightforward because they had been working on the station for several years. To fulfill their end of the bargain they merely

had to deliver an already designed *Almaz* hull to Korolev's bureau and prepare an already tested Proton rocket.

The Korolev bureau, however, which for years had focused on a lunar program with no space station design of its own, had to start practically from scratch. In less than two years they had to conceive, design, and build every component of the station's interior. To make their job easier, they cannibalized whole sections of their *Soyuz* spacecraft and attached them wholesale to the *Almaz* hull. Then, to save more time, they simplified and trimmed as much as they could. Computers were either rudimentary or left out entirely. Preflight testing of on-board equipment was reduced, if not eliminated. And to save weight, time, and money, much of the planned scientific equipment was omitted. For example, the scientists designing the Oasis greenhouse had originally planned a much larger facility, including incubators and a larger variety of plants. For lack of time and money, this facility was dropped, leaving just the small single Oasis greenhouse, about the size of one of today's desktop computer towers.[14]

Not surprisingly, the resulting space station was not the wheel-shaped station that engineers and science fiction writers had visualized for decades. Instead, to make feasible its launch on a rocket, *Salyut* was a series of four cylinders of increasing diameter resembling a blunt wedding-cake tower 52 feet long. At its bow was the docking port, attached to the station's narrowest cylinder and dubbed the transfer-docking compartment. Ten feet long and six-and-a-half-feet wide, this section acted as *Salyut*'s front door or foyer. Cosmonauts would dock their *Soyuz* spacecraft here, using for guidance the *Soyuz* radar antenna and a single television camera in a periscope, fixed to the compartment's outside. A probe on the *Soyuz* would insert itself into a cone-shaped receptacle, latches would close, and the capsule would be pulled in and hard-docked, creating an airtight seal with electrical link-up. The docking probe and cone could then be removed to open a docking tunnel through which the cosmonauts entered the station.

If they needed to make repairs on the station's exterior, the occupants could also use the docking-transfer compartment as an airlock. They would close the docking cone, seal a second hatch that separated the compartment from the main body of the station, evacuate the air, and then open a side hatch to go outside.

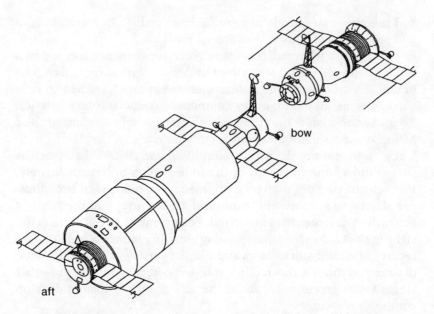

bow

aft

Salyut, with a *Soyuz* spacecraft about to dock. Note how much the *Soyuz's* aft service module resembles the aft service section of *Salyut. NASA*

Floating through the transfer-docking compartment's inner hatch, a cosmonaut next entered the station's main body. Slightly less than 30 feet long and open along its entire length, this section was made up of two cylinders of different widths stacked on top of each other and having about the same interior size and shape as the first class cabin of a small commercial jet with the seats and overhead compartments removed and the walls squared off by racks of equipment. The narrow first section, twelve-and-a-half-feet long and nine-and-a-half-feet wide, was the station's command post. Theorizing that the cosmonauts would be more comfortable if the station had a consistent "up" and "down" orientation, this cylinder had a designated "floor" and "ceiling." The control and communication equipment, along with three chairs (that also served as bunks) and a table were all oriented correspondingly. Eating facilities were also oriented this way, as were a library and a tape recorder. On the ceiling and pointing outward into space were two cameras, one still and one movie.

Just past the table and bunks *Salyut's* largest cylinder began, a little less than 9 feet long and almost 14 feet wide. Almost filling this space was a large conical housing for a solar telescope, with its base on the "floor" and its top reaching nearly to the "ceiling."* An exercise treadmill was mounted next to the telescope housing on the "floor." Along the surrounding walls were mounted storage compartments, the Oasis greenhouse, two refrigerators for food, a toilet, several control panels for operating a gamma-ray telescope, multi-spectral cameras for studying the earth, and a number of other scientific control panels.[15]

Salyut's aft section, a small 7-foot long by 7-foot diameter cylinder attached to the aft of the station's widest section, was essentially nothing more than the cannibalized service module from a *Soyuz* spacecraft. Containing no habitable space, it housed *Salyut*'s main engines and fuel tanks as well as its attitude control system.

Power for the station came from four solar panels, two attached to the transfer-docking compartment at the bow, and two attached to the aft service module. Just less than 12 feet long, these panels were taken directly from the *Soyuz* spacecraft, and produced about 3.6 kilowatts of power.[16]

The station's attitude control system was somewhat simple, and could only be operated manually. Since it was not necessary for this civilian station to maintain a specific orientation during most of its flight, it was allowed to drift into what was called a gravity-gradient position, flying perpendicular to the earth's surface with its heaviest end pointing down.

Imagine a punching-bag doll with a round, weighted bottom. No matter how hard you punch it, the weights on the bottom force it to roll back into an upright position. Gravity forced *Salyut* to behave somewhat similarly. With a heavy *Soyuz* spacecraft attached, gravity always turned it so that the station pointed Earthward, bow end down. Even if *Salyut*'s engines were fired to change this attitude or orientation, gravity would eventually pull it back. If a specific attitude was required, ion sensors, which measured the direction the station was pointing by trapping ions from the surrounding atmosphere as they flew past, told the crew the station's orientation. They then used the station's small engine thrusters to adjust the station's attitude.[17]

---

*Because the hull of *Salyut* was originally designed as a spy station with this housing holding reconnaissance spy cameras, for years the published diagrams of the station's interior did not show this housing. Soviet censors instead showed the working section with a big empty interior space.

Overall, the interior space of *Salyut* was over 3,200 cubic feet, about the same as that of a large loft apartment.[18]

Launched on April 19, 1971, the station was placed in an orbit ranging from 124 to 138 miles high, tilted 51.2 degrees to the equator. Plans called for two three-man crews to occupy it, the first crew staying for at least three weeks and the second for at least a month. However, the rush to launch the station had prevented final testing of some of its components. For example, even before launch a number of engineers believed that the shroud protecting the station's main telescope would not open after launch. Very quickly, ground controllers confirmed that this was so. Not only did this prevent the telescope's use, it also rendered useless several other research instruments. Notwithstanding this problem, *Salyut* seemed to be functioning quite well. The only other malfunction in the weeks after launch was the failure of two lithium perchlorate canisters, designed to remove carbon-dioxide from the air and replace it with oxygen. Because these were designed to be replaceable, their failure posed no problem.[19]

Confident that the program could go forward, the Soviet space planners now did something unprecedented: Immediately following the successful launch of *Salyut* they announced the launch date for the station's first manned crew, April 22, 1971, and named the crew to the public—space veterans Vladimir Shatalov, Alexei Yeliseyev, and Nikolai Rukavishnikov. They did not, however, announce that the mission was planned to last at least 22 days and, if all went well, could be extended to 30 days.

In the early morning of April 22, this crew rode the launchpad elevator to the top of the rocket and climbed into their *Soyuz* capsule. The launch itself was scheduled for dawn. The rocket was fueled and the countdown proceeded normally, despite heavy rains all through the night.

One minute before blastoff, however, part of the launch tower did not retract properly. Mishin and others in the control room delayed the launch, and then scrubbed it completely when they couldn't immediately pinpoint the problem. To make possible a second attempt the next day, they left the rocket fueled on the launchpad.[20] For the next 24 hours the engineers in mission control brainstormed the problem, eventually deciding that the high humidity from the heavy rains, combined with the super-cold temperatures of the rocket's propellants, had caused ice to form in the tower's joints, jamming it into position.

Convinced that they understood the problem, mission control decided to try again. On April 23, the countdown once again proceeded normally, and once again the tower failed to retract completely just before liftoff. At that moment Mishin decided to take the risk and launch regardless, figuring that the rocket's liftoff would shake the ice free and let the tower pull back naturally as the rocket flew past. To everyone's relief, at 2:54 A.M. Moscow time, *Soyuz 10* successfully rocketed into space. As back-up cosmonaut Alexei Leonov remembered ruefully, "it was a very dangerous moment."

Once in orbit there were more problems. *Soyuz's* automatic attitude-control system was not working properly, and commander Shatalov was forced to orient the spacecraft manually. Furthermore, on *Salyut* another four lithium perchlorate canisters had failed, meaning that unless they could replace at least two, carbon dioxide would quickly accumulate when the cosmonauts began breathing the station's atmosphere.

After two days of orbital maneuvers, *Soyuz 10* moved within 10 miles of *Salyut*, where the *Soyuz's* Igla automatic rendezvous system took over and guided the manned capsule to within 600 feet of the station. At that point Shatalov took command, successfully bringing the two vessels together manually, completing a soft docking with the docking probe on *Soyuz* inserted and linked to the receptacle cone on *Salyut*. Fifteen minutes later, however, he radioed that he was unable to confirm a hard dock. A closer look at ground telemetry in mission control showed the two spacecraft separated by a three-and-a-half-inch gap. At first, Shatalov tried to push the two spacecraft together by firing his *Soyuz* rockets. This did nothing. Then, with permission from the ground, he attempted to undock so that he could try again. Now, however, he could not get the docking latches to unlatch.

The *Soyuz* spacecraft is made up of three sections: an orbital module, a descent module, and a service module. At the bow is the orbital module, an almost spherical section with the docking port at the front end. Attached to this is the descent module, a bell-shaped capsule with a heat shield at its base. The cylindrical service module, attached to the rear of the descent module, carries the batteries, fuel, and oxygen supplies for the entire spacecraft, and is the only part of the *Soyuz* spacecraft that has no habitable space. When a *Soyuz* crew returns to Earth, they abandon both service and orbital modules, and use the descent module as their return capsule.

If the *Soyuz 10* crew could not get the orbital module undocked from the station, they could still get home by leaving it in place and returning to Earth in their descent module. If they did this, however, *Salyut* would become useless, its only docking port blocked.

Luckily, a second undocking attempt worked, and *Soyuz 10* backed off. Though none of the crew could see anything out of the ordinary in the docking port, they no longer had sufficient fuel or oxygen to risk another docking. After only two days in space, they returned to Earth, landing safely at night on the empty steppes of Russia, 75 miles northwest of the city of Karaganda.

Soyuz 10's failure meant that the following mission, *Soyuz 11*, had to scale back its mission goal from a minimum of four weeks to three. The launch date was scheduled for June 6, 1971. If all went well, a newly proposed third mission would follow, scheduled to last as long as 30 days, depending on what supplies remained on the station.

On the ground, the program's difficulties continued. For example, confusion and conflict surrounded the picking of the crew for *Soyuz 11*. The prime crew was supposed to be commanded by Alexei Leonov, who had flown on *Voskhod 2* and was the first human to walk in space. His crewmates were flight engineer Valeri Kubasov and test engineer Petr Kolodin. Only days before launch, however, Kubasov had been sent to the hospital with an infection in his right lung. Doctors were worried that he might have tuberculosis.

According to Soviet policy, when one member of a crew had to be replaced the crew was replaced wholesale, even though Leonov and Kolodin were in fine shape and completely prepared to fly. When Leonov and Kolodin learned of this decision they were furious. At the cosmonaut hotel where astronauts were housed before missions, Leonov threw furniture in anger, and Kolodin got drunk. The next day Leonov appealed to his bosses, begging for the chance to fly the mission. Though some hesitated, Major-General Kuznetsov, director of the Gagarin Cosmonaut Training Center, eventually convinced everyone that the crew-replacement decision should stand. Kuznetsov argued that the risks of mixing crews who had not trained together far outweighed the risks of flying a slightly less-trained crew.[21] Two days before launch the crew was officially replaced.

Later that evening Kolodin got into a shouting match with Mishin, who, ironically, had originally been against replacing the entire crew. "History will punish you for this decision," Kolodin shouted harshly. Leonov had by this time resigned himself to the crew change, and tried to ease Kolodin's mind, pointing out that they would instead get to fly the longer, second, mission to *Salyut*. "Don't worry, you'll be in space in just a few months," he told Kolodin.[22]

The situation was even more awkward because the back-up crew of Georgi Dobrovolsky, Viktor Patsayev, and Vladislav Volkov had been training together for only four months. Because another cosmonaut had been grounded because of a drinking problem, crews had been shuffled extensively in February. Moreover, all the personal items on board the already orbiting station, from clothing to food supplies, had been measured and planned for the first two crews, neither of which was ever to enter the station.[23]

Of the new crew, only Volkov, who at 36 was also the youngest, had been in space before, having flown during the triple *Soyuz* mission in October 1969 and spending just less than five days in orbit. Trained as an engineer at the Moscow Aviation Institute, Volkov had worked for years at the Korolev design bureau. A handsome, emotional, and exuberant man, Volkov had a touch of Cary Grant in his manner and look. He liked to play the guitar and sing and, according to Svetlana Patsayeva, Patsayev's daughter, "He was the life of every party." In childhood, airplanes had filled his life. His father was an aeronautical engineer, and his mother worked in aircraft plants. Both before and after the war the family lived close to one of Moscow's airports. His father liked to joke how the boy's first word was not "mama" but "airplane." In 1966 Volkov was one of eight engineers chosen as part of the first official civilian class of Soviet cosmonauts. Despite resistance from the air force, which did not want its monopoly on cosmonaut selection weakened, Mishin was able to force the selection of five of his engineers, including Volkov.[24]

Volkov's commander, Georgi Dobrovolsky, was seven years his senior, a military test pilot, but untried yet in space. Born in the port city of Odessa on the Black Sea, the boy's father had abandoned the family when Dobrovolsky was only two years old. Poor and with few resources, his mother had first struggled to support herself and her two children by working as a store clerk. Later, her brother took over, providing enough money to feed and house the tiny family. Then World War II arrived and the Nazis invaded Russia, sweeping through Odessa in a wave of violence. From 1941 to 1944, the family struggled to survive amid the chaos of an occupied city. In February 1944, 14-year-old Dobrovolsky was arrested for possessing an illegal pistol. The Nazis tried the teenager and sentenced him to 25 years of hard labor.

By then, however, the German army was retreating from Stalingrad and it was clear that they could only hold the Crimea for a few more months. Dobrovolsky's uncle offered a bribe to the prison officials, pointing out that if they accepted the money and released the boy immediately they would at least get something out of it. If they waited, they would get nothing when Soviet troops arrived and freed the prisoners. On March 19, 1944, the boy was smuggled out of prison, and went into hiding until the city's liberation a few months later.

With the war over, Dobrovolsky returned to his prewar dreams, and in 1947 he enrolled in the air force school in Chuguyev. For the next 15 years he dedicated himself to flying, becoming a fighter pilot with the highest rank in the Soviet air force. During these years he supported his mother and sister, while also getting married and raising a family of his own. Then, in 1962, he was suddenly told to report to his commander's office. Baffled, he wondered if the meeting had something to do with his academy work. To his astonishment, he was instead ordered to take the medical exams to see if he was physically qualified to become a cosmonaut. Though Dobrovolsky had never dreamed of flying in space, the honor of being chosen made him proud and pleased.[25]

The third crew member, Viktor Patsayev, was also older than Volkov and a space rookie. Thirty-seven years old at launch, Patsayev had worked for years with Volkov as an engineer in the Korolev design bureau. Known as a calm, soft-spoken man with firm and strong opinions, Patsayev never hesitated to express himself if he thought he was right. For example, during Volkov's first space flight, one experiment had failed to operate. At his debriefing Volkov claimed the failure was due to weightlessness. Patsayev disagreed, calmly stating to Volkov's face that he believed the failure was caused by Volkov's sloppiness.

Patsayev was born in a tiny village in Kazakhstan, where his upbringing was in many ways similar to Dobrovolsky's. He was eight years old when the war started and his father, an infantry-man, was killed while defending Moscow. Left a widow in a war-torn country, his mother was barely able to feed her family. None-theless, the boy flourished, developing a passion for astronomy and physics at school. Because the career of spaceman did not yet exist, Patsayev chose to become a geologist. Unable to get into the Mos-cow Geographic Institute, however, he enrolled in the Penza Indus-trial College, studying math and early computer technology.* After graduation in 1955 he went to work at the Central Aerological Ob-servatory near Moscow. There, he met his wife, got married, and started a family. Together, they divided a small, communal, cold-water flat with three other families, sharing a kitchen and bath-room with more than a dozen people.

There he also began working as a subcontractor, designing in-struments for the satellites being built by Korolev's design bureau, intended to study the upper layers of the atmosphere. Ironically, the engineer who had built a telescope as a child but had never dreamed of space flight was suddenly working for the man who would soon put the first satellite in orbit. When Korolev himself came to the observatory and gave a lecture describing how he intended to send men to the moon, Patsayev decided that Korolev's design bureau was where he wanted to work. By 1958, soon after *Sputnik*, he had changed jobs, joining Korolev as an engineer. Over the next decade he worked on numerous space projects, becoming friends with fel-low engineer Volkov. Together they applied to become cosmonauts when in the mid-1960s the opportunity was first offered them. While Volkov was immediately accepted, Patsayev was rejected at first and was not accepted until several years later, in 1968.

Truth

On June 6, 1971, these three men finally lifted off in *Soyuz 11*. Like the crew before them, they planned to complete a minimum of 22 days in space. On the ground, the secretiveness of Soviet soci-ety was all-pervasive. Unlike the failed first launch to *Salyut*, the

---

*Even today, 50 years later and a dozen years after the fall of the Soviet Union, no one, not even Patsayev's wife and children, has a precise idea what he studied in college. In the Soviet Union during and just after Stalin's rule, all such scientific work was considered top secret.

launch time was not announced. Even their families knew nothing
of the launch until the men were in orbit.

This time the docking proceeded without difficulty, though the
crew spent almost 30 minutes making sure that the hard dock was
solid and that the docking tunnel between *Soyuz 11* and *Salyut*
was hermetically sealed. Engineers had concluded that the *Soyuz
10* docking had failed because the force of impact during the first
docking had damaged *Soyuz 10*'s docking latches. To prevent a
reccurrence, *Soyuz 11*'s docking probe and latches had been reinforced.[26] Dobrovolsky stayed at the *Soyuz* controls to maintain
communications with the ground while Volkov opened the hatch.
To the consternation of all three men, the smell of burning insulation wafted in at them.

Salyut had been orbiting unmanned for almost two months.
The odor could be from a fire, from toxins building in the atmosphere because of the failed lithium perchlorate canisters, or from
almost anything. For a few minutes the men argued about whether
they should go in. Volkov was supposed to be first, but he hesitated, unsure. Patsayev decided not to wait and drifted into *Salyut*,
immediately floating to the communications system and switching it on, where he reported to mission control that *Salyut* was
now occupied. Volkov and Dobrovolsky then followed him in, and
very quickly the three men activated the station's systems, ignor-
ing the faint stench of burnt insulation. They turned on the air
regeneration system and replaced two of the six failed lithium perchlorate cartridges. Then, for safety, they retreated to the cramped
Soyuz capsule, where they spent a tense first night trying to sleep.

By morning the odor was gone, its cause a mystery that even
today remains unsolved. The three men floated back into *Salyut* to
continue turning on and checking out its systems.

In 1971 the record for the longest space flight was the 18-day
mission of *Soyuz 9* in 1970. Unlike most of the American astronauts who had flown from ten days to two weeks and returned in
reasonably good health, the two *Soyuz 9* cosmonauts had serious
physical problems after landing on Earth. Neither man had the
strength to stand, and both had to be carried from their capsule on
stretchers. It then took their bodies 10 days to readapt to Earth
gravity.

The experience of weightlessness seemed to have serious and
unknown consequences for the human body. For example, there

were questions about the prevalence of space sickness. While very few American astronauts had gotten sick in space, a number of Russians had, including Gherman Titov (their second cosmonaut after Gagarin) and the crew of *Soyuz 9.* No one knew what caused the sickness. No one knew who was susceptible. And no one knew if the sickness was curable. Would it prevent space flights longer than the two weeks it took Americans to get to the moon?

For Dobrovolsky, Volkov, and Patsayev, space sickness was never an issue. The only obvious physical symptom they noticed was the sudden and significant shift of blood from their feet to their heads, making them feel heavy-headed, as if they had colds. In fact, the change was so pronounced that the men noticed it at once. As Dobrovolsky reported the first night in space, "Vadim [Volkov] and I looked in the mirror and then at each other and laughed: 'Mugs like bulldogs!' "[27] The more significant health concerns centered on the long-term effects of weightlessness on the human skeleton, heart, muscles, and balance system.

The skeleton: Almost every human who had been in space up to this moment had experienced some form of bone loss, with the flushing of calcium from the bones reaching as high as 15 percent in a few cases.[28] No one knew whether this weakening of the bones could be countered and stopped, or whether the body would eventually adapt and the bone loss would cease on its own.

The heart: Without gravity the cardiovascular system shifted drastically, with more blood gathering in the head and upper body, causing the face to puff up, as noticed by Dobrovolsky and Volkov. For some, such as Gherman Titov, the change was so pronounced that he felt as if he were hanging upside down and blood was rushing to his head. Upon return to Earth this change put serious stress on the heart, and caused dizziness and nausea for the first few hours.

Muscle strength: No one knew at all whether a human being would have the strength to walk after several months of weightlessness. What *Soyuz 9* as well as some of the longer American missions had indicated was that Wernher von Braun's concerns were apparently justified: Zero gravity seemed to cause the muscles to atrophy, similar to what happened to patients after several months of bed rest. Whether this degradation could be countered was unknown.

To try to avoid these difficulties, as well as to learn what caused them, the *Soyuz 11* cosmonauts had been given a detailed exercise program. They were supposed to use *Salyut*'s treadmill a minimum of two-and-a-half hours a day. Samples of their blood and exhaled air were to be taken before, during, and after exercise.[29] And all three men were to periodically wear special suits that attempted to simulate Earth gravity. These inflatable "Penguin" suits forced the wearer to use his muscles because the inflated tubes that lined the suit tried to bend the body into a fetal position.

To counter the unnatural shift of blood from the legs to the upper body, the men were supposed to periodically put their legs inside an apparatus designed to force blood to circulate into the lower limbs. Dubbed "Chibis" (the Russian name for the lapwing bird), this device was a variation of an American design used during several *Gemini* missions, in which the astronauts had worn cuffs that attempted to force blood into the legs.

For their first few days on *Salyut* the men struggled to establish a routine. Their schedules were staggered, each man working a different eight-hour shift, so that at least two men were on duty at all times. This system, however, meant that each man had to try to sleep while the other two worked and moved about the station. Because *Salyut* had no private cabins, the planned rest areas, sleeping bags strapped to the walls around the large telescope housing in the largest section, put a sleeping man in the midst of everything. To get some quiet, Patsayev moved his sleeping bag first into the docking-transfer compartment, and later into the *Soyuz* spacecraft itself.[30] The others soon copied him.

The men also discovered that weightlessness was far stranger and more difficult to adapt to than anyone had imagined. Everything took far longer to do than on Earth. Tools wouldn't stay where they left them. Equipment wouldn't operate as planned. And weightlessness made it harder to do anything, because the slightest force caused a person to drift away from his work. For example, on June 12, their fourth day on board, the men attempted to measure their blood pressure and take blood samples. They found the task laborious and time-consuming. "The difficulty is that the man is not tied to the chair," Volkov tried to explain. "Everything flies away: you get one thing, and another flies away."[31] Ground controllers found it hard to believe that weightlessness made even this most ordinary task so time-consuming. Repeatedly they asked the men to explain why things were taking so long and why they were having so much difficulty keeping to their schedule.

After almost a week in orbit, the crew taped their first television press conference, showing the world their home and how they lived on it. Volkov, cameraman and narrator, described how the three men had adapted to space, as he panned his camera across the control section and hatchway leading to the docking tunnel and *Soyuz 11*, carefully avoiding any view of the large telescope housing behind him. As he filmed, Dobrovolsky "sat" at the main control panel, while Patsayev floated out from *Soyuz 11*. Noting that Dobrovolsky appeared unshaven, while Patsayev was clean-shaven, the capcom (or "capsule communicator," the person in charge of radio communications with the crew) asked why Dobrovolsky hadn't used the station's special electric razor, attached to a vacuum cleaner to suck in the shavings. All three men laughed, with the always more-buoyant Volkov explaining that he and Dobrovolsky had decided to grow beards, while Patsayev insisted on shaving every morning.

Two days later, the men were on television again. Volkov demonstrated how he used the station's treadmill to jog, despite the lack of gravity. As he "ran," he kept himself in place by holding onto two vertical cords in front of him. These televised broadcasts were so popular in the Soviet Union that the crew began doing them regularly. During different shows they demonstrated how they ate, the experiments they did, the clothes they wore, and the nature of weightlessness.

Unbeknownst to their public audience and despite their seeming good humor and enthusiasm, the crew of the first space station were having a wide range of engineering problems. For example, it was impossible to do any astronomical research because of the stuck shroud covering the main telescope. The best the cosmonauts could do was test its pointing system, something they did several times during the flight, proving that they could maneuver the station and hold it in position accurately enough for useful astronomy. While one cosmonaut pointed the telescope another reoriented the station using *Salyut's* engines.

The Chibis system had technical problems and was rarely used. The Penguin suits also turned out to be too fragile, quickly tearing with use. The inflatable tubes that were supposed to create resistance and exercise the muscles could not hold air. And the buckles for securing the suits to the body were difficult to use.

The treadmill also remained unused most of the time, not because the men didn't want to exercise but because the vibrations

caused by its use interfered with other experiments. More importantly, the treadmill's design was impractical and awkward. Simply holding onto the two hand cords was insufficient to keep the
exerciser's feet pressed to the treadmill as he ran. And the hand
cords kept breaking. The men had to change sets several times, a
time-consuming task that prevented them from doing their other
work. As a result, Volkov stubbornly refused to do his exercises,
despite his performance on the treadmill on television. He also
spurned the station's meat dishes, a decision that reduced the variety in his diet.[32]

Then, in the early morning hours of June 17, on the mission's
10th day, [33] Volkov radioed to the ground that the men once again
smelled smoke and suggested that a fire was smoldering behind
one of the station's panels. The crew was immediately ordered
to retreat to their *Soyuz* spacecraft and begin preparations for
undocking.

On board things were a bit more complicated. Despite a generally positive attitude during most of the mission, there had been
personality clashes between Volkov, the only crewman to have previously flown in space, and Dobrovolsky, the commander. During
moments of crisis, such as when they had first opened the hatch
and smelled smoke, Volkov seemed reluctant to follow orders, preferring to act on his own. In that crisis, this willfulness resulted in
Volkov's refusal to enter the station first, which was his job.

Faced again with the threat of a serious fire, Volkov once again
challenged his commander's authority. While all three men had
smelled smoke, only Volkov wanted to call mission control and
ask for permission to return to Earth. After some arguing Dobrovolsky told him to go ahead and call the earth from the relative
safety of the *Soyuz* descent capsule. He and Patsayev were going to
continue searching for the problem. Thus, when Volkov's early
morning message was received, Dobrovolsky and Patsayev were
not with him in *Soyuz*, but were instead opening panels and trying
to find the smoke's exact source. When ordered by mission control
to withdraw to *Soyuz*, they had no choice but to abandon their
effort and prepare for evacuation.

In the capsule the three men once again argued about what to
do. Volkov wanted to evacuate immediately, while Dobrovolsky
and Patsayev were more willing to stick it out a bit longer and try
to fix things. They switched *Salyut* to a back-up electrical system,

assuming the fire was electrical in nature. In addition, they cut off the station's fans; without air circulation the fire would quickly burn up the oxygen around it and then strangle on the soot and carbon dioxide that in weightlessness gathered in its stead. As they did this work, Volkov kept arguing that they should evacuate. Finally, Mishin himself got on the radio and told Volkov to back off and let Dobrovolsky make the decisions. Dobrovolsky and Patsayev returned to *Salyut* and continued the search while Volkov remained behind in *Soyuz*. After a bit more searching they found the smoldering cable, disconnected it from the electrical system, and watched as the smoke dissipated with no serious damage to the station.[34]

The next day the men were given the day off. To the world the Soviets simply announced that it was a "rest day," even though it came on a Thursday.[35] Though the mission was able to continue, crew morale after this was never quite the same. For the remainder of the mission there lurked an inescapable undercurrent of nervousness, especially from Volkov.

Despite these problems, the overall design of the *Salyut* station seemed vindicated. Its internal atmosphere, 25 percent oxygen and 75 percent nitrogen, was comfortable and helped reduce the risk of fire. The use of lithium perchlorate canisters to scrub carbon dioxide from the air and replace it with oxygen worked without difficulties. Even the failure of the six lithium perchlorate units turned out to be a good thing, because they were quickly replaced with working units, proving that the system was repairable.

For drinking water, the station had several water tanks, carrying enough water to supply each man two liters per day. *Salyut*'s temperature control system, made of two coolant loops filled with antifreeze, were also used to pull humidity from the air, making it a preliminary test of a prototype water-recycling system. The water condensed on the loops and was then drawn off and expelled from the station. Since the human body releases a great deal of water, both in sweat and in exhalation, this system was essential to keep the station from becoming overrun with moisture. Its success proved that more sophisticated recycling systems were possible. The toilet, though simple and noisy, also functioned well. A loud fan worked to suck odor and material down into a bag, which was then closed, sealed, and removed once the cosmonaut was done. This waste, plus other small bits of garbage, were disposed of in small 1-liter garbage cans which, when full, were inserted in a small airlock and ejected from the station.[36]

By its third week, the voyage of *Soyuz 11* was becoming a worldwide triumph, especially because the various crew conflicts and engineering failures were not made public in the government-censored Soviet press. The frequent televised press conferences were personal and entertaining, showing the crew in a human light. Each night the Soviet press glowingly reported the day's successes, with lengthy commentary on the mission's importance for the human exploration of space. After years of American victories in space, it appeared that the Soviet space program was finally moving forward.

On June 19—their 13th day in orbit and two days after the fire—the three men celebrated Patsayev's thirty-eighth birthday on television. Dobrovolsky and Volkov had secretly brought a lemon and an onion from Earth as birthday gifts. With mock seriousness the men laid out a "feast" of food tubes and cans, including curd-cheese paste, juice, veal, sugared fruit, and nuts. Then, with their television audience watching, they presented their gifts to Patsayev, holding up tubes filled with prune paste as a toast. Volkov, despite the disagreements two days earlier, enthusiastically sang a birthday song for his longtime engineering friend. Even as they ate Patsayev looked at the television camera and with his usual deadpan seriousness said, "I especially enjoyed the onion."[37]

Three days later, in another broadcast, Patsayev lovingly showed off his Oasis greenhouse. "These are our pets," he explained, describing how all three men repeatedly checked the greenhouse throughout each day, watching to see if the plants were growing. Though the different plants had produced leaves, all remained puny compared to Earth plants. Their roots, also much tinier than normal, seemed to develop wildly, as if without gravity they did not know which way to grow.

On Earth, Galina Nechitailo, one of the experiment's chief designers and in charge of supervising the cosmonauts as they worked with the greenhouse, watched and wondered. Maybe the plants weren't getting enough water. When, later that day, she was given a few minutes to talk to Patsayev about it, she asked him to change the watering routine: "Water the plants twice a day, morning and evening."

"The instructions say once a day," Patsayev noted.

"I know, I know," said Nechitailo. "However, they must be watered twice a day."[38]

For Nechitailo and her husband, Alexander Mashinsky, both young space botanists only a few years out of college, this small experiment began a 20-year quest to grow and harvest plants in space. Nechitailo, with her fiery black eyes and long black hair, had worked in Korolev's design bureau since graduating college in the mid-1960s. Her husband, Mashinsky, worked in the Academy of Sciences of the U.S.S.R. Both were part of a team of biologists that worked in the Korolev design bureau, laboring long and hard to get a variety of plant greenhouses into space. Though their larger and more ambitious original design had been rejected because of cost and weight during the hurried effort to build *Salyut*, the little Oasis greenhouse was for now an ample replacement.

Listening to and watching Patsayev's reports these young scientists were baffled, asking some very fundamental questions: Why had some of the seeds prospered better than others? Why were they all doing poorly? Why was the lack of gravity making the roots grow so wildly? To get answers to these questions, Nechitailo and her colleagues in the biology team would simply have to wait until the men brought back their live plant samples.

By June 25, the crew's 19th day in space, the mission was beginning to wind down. All three men were eager to get home. During that day's television broadcast, the last scheduled public show, Dobrovolsky described how the crew was beginning their preparations to return to Earth, packing equipment and experiments into the *Soyuz 11* spacecraft. Then he added, "We, as humans, miss the earth, and we are of course impatiently awaiting our return."[39]

For Dobrovolsky, the desire to return was made more poignant in that it had been his decision to stick it out during and after the fire. When he listened to tapes of his 12-year-old daughter, Marina, all she could say was that she missed him. Even today, more than 30 years later, the most important thing that Marina Dobrovolsky can remember about her father's space flight was how much she missed him. "I wanted him to come home," she recalls sadly.

For Patsayev, the desire to get home and see his wife and children was just as strong. Listening to the audiotapes from home, he heard his nine-year-old daughter, Svetlana, play a simple piece of piano music, then describe her schoolwork and how well she was doing. It made him feel so proud.

For Volkov, the fire problems and crew discord made him long to get home as soon as possible. And he had a strong foreboding about being on the mission. In fact, there were stories about how he had seen a fortuneteller before launch, who told him that the flight of *Soyuz 11* would be his last space mission. Whether true or not, the sooner he was back on Earth with his wife and kids, the better.[40]

The circumstances surrounding the fire had left all three men touchy and nervous. For the last week they had temporarily stopped their workouts, and were now trying to catch up. Then, beginning on June 27, two days before their return, ground control began to shift the men's sleep schedules so that they went to sleep and got up at the same time. Such a schedule, however, meant that for long periods each day no one was on watch. Dobrovolsky radioed the ground to question this policy. "We do not want to do this," he said.

"Do it, do it," the capcom insisted. "Everything is going well. Everything is in order onboard. Don't sigh, it must be done."[41]

Finally, on June 29, 1971, the crew climbed into the descent module of the *Soyuz* spacecraft and closed the hatch that separated it from the orbital module. To everyone's chagrin, the light indicating an open hatch failed to turn off, suggesting that the hatch had not sealed properly.

Volkov nervously shouted into the radio, "The hatch is not pressurized, what should we do, what should we do?!"[42]

After calming Volkov down, cosmonaut Alexei Yeliseyev, the capcom, had Dobrovolsky and Patsayev repeat the entire hatch closing procedure. Still the "hatch-open" light did not go off. Now all three men were getting anxious. In the *Soyuz* descent module they wore no spacesuits. If the hatch had really not closed properly, when the men undocked and then discarded the orbital module in preparation for re-entry, their atmosphere would leak out and they would suffocate.

For the next 20 minutes they and ground controllers debated what should be done. Eventually mission control decided, based on ground telemetry from the spacecraft, that the problem might be some dirt on the gasket that sealed the hatch. They instructed the men to open the hatch a third time, wipe the gasket clean, and then seal the hatch again. As back-up cosmonaut Alexei Leonov notes even today, "Even one human hair could prevent the hatch from sealing."

This time, when they closed the hatch the light finally shut off. Further pressure tests indicated that the hatch was properly sealed.[43]

With his crew still edgy from the hatch problem, Dobrovolsky undocked from the *Salyut* station and eased away. Two hours later they fired their retro-rockets. Twelve minutes thereafter both the *Soyuz* orbital and service modules separated from the descent module, allowing it to return to Earth on its own.

Death in space can come suddenly, and without warning. It is not required to foreshadow its arrival, nor give its victims time to prepare. When it arrives, it is sudden, surprising, and terrifying.

As the descent module they occupied separated from *Soyuz 11*'s orbital module, a vent designed to equalize the descent module's atmosphere with the earth's after the release of the main parachutes accidentally opened. The hole was small, about the size of a quarter. All three men could hear a hissing sound, and immediately knew that something was wrong. Dobrovolsky immediately released his seatbelt so that he could get to the hatch, thinking now that the "hatch-open" light had been correct, and that the leak was coming from there.

It was not. Now they scrambled desperately to find the leak. The whistle was being fed into their microphones and back through their earphones, making it more difficult to find the hissing's source. Less than 20 seconds had passed since the orbital module's separation.

Volkov and Patsayev also unfastened their belts and together turned off the communications system to stop the feedback. They could hear where the hiss was coming from: a valve under Dobrovolsky's couch. Dobrovolsky and Patsayev tried to close it manually, but there wasn't time.

By now the men were no longer rational. Dobrovolsky floated back to his seat, and struggled to refasten his seatbelt, leaving it tangled. Seconds later all three were unconscious. Less than two minutes later they were dead.

The *Soyuz* spacecraft meanwhile continued its automatic descent, and 30 minutes later was plummeting over Kazakhstan 125

miles east of the city of Dzhezkazgan. As planned, its chutes
opened at the right time. As planned, it fired its secondary rockets
to slow the spacecraft's impact when it hit the ground.

And as planned, the military rescue crew arrived very shortly
thereafter. When they pried the hatch open, however, they found
the three men dead in their couches.

After this disaster, more than two years passed before the Soviets
attempted another manned mission. According to the official accident investigation, the valve opened because the powder in one of
Soyuz's explosive bolts had become unstable over time. When the
bolt blew as planned, separating the descent capsule from the orbital module, the force was enough to jerk the valve open.

Not everyone agreed with this conclusion. A few people believed instead that it had been Volkov's job to check the valve, and
that he had failed to do it. Which explanation is right will probably
never be known.[44]

Regardless, over the next two years the Soviets did a major
redesign of the Soyuz spacecraft, changing both the valve and the
explosive bolts. In addition, the program also set new rules for
manned flights, limiting all Soyuz missions for the next decade to
only two crewmen, thereby leaving enough room in the descent
capsule for the men to wear spacesuits during ascent and descent.
A short unmanned test flight in 1972 verified these changes. In
addition, the spacecraft's solar panels were replaced with internal
batteries, simplifying its operation but reducing the ship's orbital
life to only two days, as demonstrated by a test in 1973.

The three men were buried in Red Square, with full honors. For
the Soviet Union, the tragedy of their deaths was made even more
painful because of its public nature. This secretive, authoritarian
society insisted on hiding all its failures, making believe they didn't
happen. The deaths of Dobrovolsky, Volkov, and Patsayev, however, could not have been more visible. For Russians, the accident
seemed almost shameful, as if they had hung their dirty laundry
out for all to see.

For the wives and children of the three men, their deaths became the delineating moment of their lives. "Our whole life
turned," remembers Marina Dobrovolsky. "Before was a bright
happy childhood. After, darkness and tragedy." The shame that
many in the space program felt because of the mission's failure

caused the families to be abandoned by friends, and ignored by the tight-knit, tiny space community in which they lived. Some, such as Patsayev's wife, were able to adapt. She wrote a book about the space mission, making herself one of the world's experts on what it had tried to accomplish. For Dobrovolsky's wife, however, the loss and abandonment was too much. She died at the young age of 47. Even today, many years later, her daughter believes that death was due to a broken heart.[45]

For Galina Nechitailo and the biology team at the Korolev design bureau, the depressurization killed their specimens of space-grown flax, cabbage, and hawksbeard, preventing them from finding out exactly how they had been changed by weightlessness. The only knowledge they could salvage from the mission was the belief, based on what Patsayev had told Nechitailo during the flight, that the watering system of the Oasis greenhouse needed redesign. Getting water to plant roots in zero gravity was apparently more complicated than they had expected.

The disaster also prevented any further manned missions to *Salyut* before the space laboratory was de-orbited on October 11, 1971. As it burned up over the Pacific Ocean, so did what was left of Patsayev's tiny garden. What happened to its remaining plants in the three-and-a-half months after he abandoned them will remain forever unknown.

Finally, the deaths of Dobrovolsky, Volkov, and Patsayev meant that the first successful space station was not a Soviet one. Despite Brezhnev's claims, the Soviet Union was about to lose another race into space.

3

Skylab: A Glorious Forgotten Triumph

More of Everything

Like a glorious torch, the largest and grandest ever created by the hand of man, the Saturn 5* rocket inched its way skyward. Generating more than seven million pounds of thrust, enough power to lift 3,200 tons off the ground and accelerate more than 125 of those tons to faster than 17,500 miles per hour, the rocket hardly moved at all to begin with. Creeping upward from its launchpad, it seemed to hang gently in the air. Then it rose faster, and faster, and faster, its five F1 engines consuming fuel at a rate almost 30,000 pounds per second.[1]

The roar from the rocket's engines rolled across the Florida swamps like thunder. Even now, three decades later, the impact of that roar is legendary. During the launch of the first Saturn 5 in November 1967, the sound wave almost blew over the television booth from which news anchor Walter Cronkite was broadcasting. Reporters crouched to the ground, their hands clamped over their

---

*The rocket's official name was Saturn V, using a Roman numeral. However, considering how rarely Roman numerals are used today as well as to avoid confusion, I use the number "5," since that is how the name was pronounced.

ears to escape the roar. Anne Morrow Lindbergh described it as "a trip-hammer over one's head, through one's body. The earth shakes; cars rattle; vibrations beat in the chest. A roll of thunder prolonged, prolonged, prolonged."[2]

This launch, however, on May 14, 1973, was to be its last. Though designed and built by Wernher von Braun as a tool for colonizing the Solar System, once the Saturn 5 had gotten men to the moon, the American public and politicians decided its cost was too great for further use. For this last mission the Saturn 5 was lifting into orbit the United States' answer to *Salyut*, the first American space station, *Skylab*. If all went as planned, the station's first crew would be sent into space one day later, on May 15.

If all went as planned.

From 1964 until 1969 the Apollo Applications Program, NASA's follow-up program to the lunar landing, had struggled to find a focus. No one in Congress, NASA, or the general public had really put much thought into what to do in space after the first men landed on the moon. Initially, many in NASA wanted to use the 1950s space-station concepts as a guide for future space exploration. When NASA did several technical studies in the early 1960s to see if a rotating, wheel-shaped station could be built, however, it found the cost of lifting the necessary mass into Earth orbit simply prohibitive.

By the mid-1960s NASA had lowered its sights, deciding to make its first operating space stations smaller and simpler, and to use them to study the long-term effects of weightlessness.[3] The disturbing loss of bone tissue during the early *Mercury* and *Gemini* space missions had to be studied, and ways found to either control or counteract it if humans were going to survive long journeys in space.

The problem was that NASA didn't have the money to build even a simplified station. Though the space agency had the Saturn 1B rocket—able to lift about 20 tons into orbit, about the same payload as that of the Soviet Proton rocket that had launched *Salyut*—it didn't have the funds to design and build a U.S. version of *Salyut* to put on top of the Saturn 1B.

Instead, the agency had to improvise from existing parts. Von Braun's engineers in Huntsville, Alabama, proposed what they called the "wet tank" concept, using the hydrogen tank of the Saturn 1B rocket's second stage as the habitable area of the workshop. After the stage had used its fuel to get the stage in orbit, astronauts would enter the empty tank and install various pieces of equipment to make it habitable and usable as a space station. To put it mildly, this solution was technically complex. Much of the workshop's equipment could not be launched inside the tank while it was filled with hydrogen fuel. Upon arrival the astronauts would have to unpack the equipment, move it into the workshop, and then set it up. Furthermore, the Saturn 1B rocket, while still quite powerful, placed such severe weight limits on this complicated plan that engineers constantly found themselves exceeding its capabilities.

Many in NASA, including von Braun, knew that all these problems could be solved if they could simply use a Saturn 5 rocket to launch the station. With its gigantic lift capacity, at least 100 tons or five times that of either the Saturn 1B or Proton, the Saturn 5 could easily place all the needed equipment in orbit, with room to spare.[4] Until the lunar program was assured of success, however, Jim Webb, NASA's administrator, was not willing to commit any Saturn 5s to a follow-up program.

In fact, the real obstacle for this first American space-station program was John F. Kennedy's looming, unalterable deadline for landing a man on the moon and returning him safely to Earth before the end of the decade. Until that lunar landing, no one at NASA was willing to focus effort on a follow-up program, especially if it would distract from the more immediate and urgent task at hand. This desire not to be distracted became even more intense after the January 27, 1967, launchpad fire that killed astronauts Gus Grissom, Ed White, and Roger Chaffee. NASA's engineers had enough problems fixing the *Apollo* program. The agency's priority was landing men on the moon, and if there were any money shortages it was programs like the space station that had to suffer.

After the early and very successful flights of *Apollo 7, 8, 9,* and *10* in late 1968 and early 1969, however, this bleak picture began to change. Jim Webb had stepped down as NASA's administrator in late 1968, and the new administrator, Thomas Paine, was looking eagerly to jumpstart a follow-up program to the lunar landings. Under his leadership, NASA had reconfigured its space-station program, replacing the wet tank concept with a dry, ground-built facility launched on a Saturn 5. Two days after Neil Armstrong stepped on the moon, Paine made the public announcement. An upper stage of a Saturn 5 rocket would be refitted as the orbital workshop and, if all went as planned, be launched sometime in 1972.

At that moment, the goals of the space-station program subtly shifted as well. Just as Brezhnev moved the public focus of the Soviet space program away from the riskier human exploration of the planets toward much safer scientific research, President Nixon and most Congressmen had been doing the same. Paine wanted the first space station, assembled by some form of reusable space shuttle, to be the jumping-off point for later lunar and planetary exploration. Neither Congress nor Nixon had any interest in such an ambitious program. The only station concept that Paine could get approved was that proposed by von Braun's team using the Saturn 5, dedicated, not to exploration, but to science—studying the earth, the sun, weightlessness, and the human reaction to it.[5]

This station concept eventually became *Skylab* which, until 1995, was the largest and heaviest single space vehicle ever placed in orbit. Weighing more than 82 tons and having about 12,700 cubic feet of habitable volume, *Skylab* was a monster compared to every other space station flown in the twentieth century. It was four times heavier with almost four times the habitable area of every *Salyut* station, and had an interior space so large that the *Salyuts* could almost have been garaged inside it. Even *Mir*, as large as it eventually became, did not exceed *Skylab* in interior space and mass until it had been in orbit for almost a decade. *Skylab*'s interior space was so large that astronauts used it to test jetpacks for flying around in space, even doing complex maneuvers and aerobatics.[6]

Yet, like *Salyut*, *Skylab* bore no resemblance to the wheel-shaped space station imagined by Wernher von Braun and Willy Ley in the 1950s. Like the *Salyuts*, *Skylab*'s design was shaped by the requirements of the rocket that launched it, making it an 86-foot-long, wedding-cake tower of decreasing-diameter cylinders mounted on top of one another. In fact, much of *Skylab*'s basic shape was remarkably similar to that of *Salyut*, not because NASA and Soviet engineers were copying each other but because certain design choices made simple common sense. *Skylab*'s size, however, allowed it to have more of everything.

Skylab's bow, for example, also started with a docking-transfer module, called the multiple-docking adapter, but this module, at

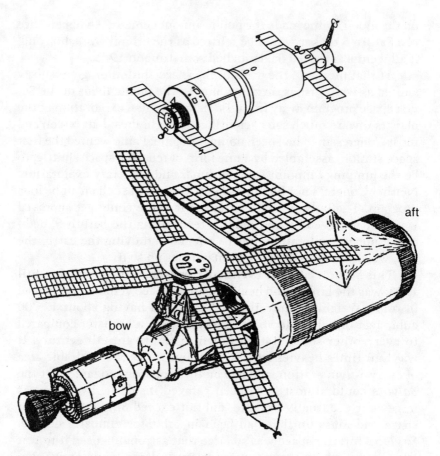

Skylab, with an Apollo spacecraft docked in the bow. Salyut is also shown at approximately the same scale to illustrate the size difference. Also shown are the repairs performed by Skylab's first crew, including the umbrella sunshade and the opened solar panel. NASA

17 feet long by 10 feet wide, was almost twice as large as Salyut's and it had *two* docking ports for parking two Apollo capsules. Attached to the multiple-docking adapter was the airlock module, about the same size but appearing much wider because it was protected by a 22-foot-wide cylindrical shroud held in place by a gridwork of metal beams and struts. Inside the airlock the astronauts would don their spacesuits, seal the hatches on either end,

evacuate the air, and exit through an exterior hatch on the side to climb out along the shroud's struts to do space walks.

Attached to the docking adapter and airlock modules with an additional gridwork of beams was the large *Apollo* telescope mount and its four solar panels, giving the station its windmill look. Upon reaching orbit the mount cantilevered sideways so that it extended at right angles thirteen-and-a-half feet from the rest of the station. This mount held the 12-ton solar telescope complex, which included an ultraviolet spectrometer and spectrograph, an X-ray telescope, and a visible-light chronograph, all designed to look at the sun in ways not possible previously.

The mount also contained three large gyroscopes for controlling *Skylab*'s orientation, the first time such a system, long proposed by engineers and science fiction writers alike, was used for a manned spaceship. Each gyro was almost 2 feet across, weighed 155 pounds, and spun at a rate of approximately 150 revolutions per second. Positioned so that each spun around one of the x, y, and z axes of three-dimensional space, the gyros could thus control *Skylab*'s roll, pitch, and yaw to within a tenth of a degree. Moreover, by tilting any one of the gyros slightly, an astronaut could literally force the entire station to turn in the desired direction. Then, for more precise control, nine smaller gyros (three for each dimension) acted as sensors, providing positional information so that the station's six small attitude thrusters, firing nitrogen gas, could make fine adjustments.[7]

Beyond the airlock module was *Skylab*'s main section, the gigantic orbital workshop built from the unused third stage of the Saturn 5 rocket. A cylinder 46 feet long and 22 feet wide, it alone had more than 9,500 cubic feet of interior space, carved from the space once occupied by the stage's hydrogen tank. Two more large solar panels were attached to its outside to augment the power provided by the four panels on the telescope mount, all six producing a total of ten kilowatts of power.[8]

Because the workshop's interior space was so large, it was divided into two sections, or decks, separated by a gridwork, instead of a solid, floor (thereby giving astronauts something to hold onto in zero gravity while also saving weight and allowing air to flow freely). Entering the station from the airlock module, the astronauts floated into the 21-foot-wide by 20-foot-high upper deck, the largest open space on the station and the largest interior room ever placed in orbit. Here, walls were ringed with storage cabinets, water tanks, food freezers, and two small research airlocks for safely placing experiments outside in the airlessness of space.[9]

The lower deck was *Skylab*'s living quarters, divided into four rooms: wardroom, galley, toilet, and sleeping quarters. The wardroom held the exercise bicycle, shower, and medical equipment. It had at its base a central hatch connected to a small airlock that opened into the upper stage's empty oxygen tank, which the crews used as a garbage dump. The shower was a collapsible tube fastened to the "floor" of the lower deck. An astronaut climbed in, pulled the tube over his head, attaching it to the lower deck's "ceiling." Then he filled a pressurized bottle with heated water and attached a hose so that he could spray himself.[10]

The astronauts ate in the galley. Here they also spent much of their time gazing at the earth from *Skylab*'s one large window. At the galley's center was a customized small hexagonal table. Instead of chairs, three stations unfolded from the table with bars to hold the astronauts' thighs, and floor straps to anchor their feet. Each food station had recessed compartments where the crew placed prepackaged freeze-dried meals to be heated, the food designed to stick together and not float away in zero gravity.

The toilet facility was a particular challenge to the station's designers. Not only did the system have to work comfortably, it had be able to collect both urine and feces so that medical research could track bone loss and monitor blood chemistry. Extensive samples were to be collected before, during, and after each *Skylab* mission. The toilet was a form-fitting opening, mounted on one "wall" of the station, with handles on either side, footholds on the "floor" below the toilet, and a seatbelt for securing the user to the seat.* A fan on the inside of the unit sucked the smell from the station, and also pulled the solid waste into a bag for drying and storage.

Tests of the toilet system on the ground seemed successful, though its designers at the Fairchild Hiller Corporation had no way to know for sure if it would work adequately in zero gravity. To try to find out, they performed two days of tests using several specially picked volunteers and a KC-135 cargo jet making parabolic arcs that produced about 30 seconds of weightlessness. Though the vol-

---

*Because it was in zero gravity, the "wall" also functioned as a "floor" when, as in this case, necessary.

unteers said they could do what had to be done in this brief time, and supposedly produced nine successful attempts, the engineers were still unsure how well the toilet would work once *Skylab* reached orbit.

The urine system also caused NASA's engineers endless problems during design and construction. To do medical research, samples had to be returned to Earth in an undamaged and uncontaminated state. For months, medical researchers and NASA engineers argued about how to do this. The standard method on Earth was to freeze the sample. NASA wanted instead to dry the samples in order save the weight and cost of a freezer. After much discussion, NASA decided to add a freezer and go with the standard method.[11]

Unlike *Salyut*, *Skylab* actually had a separate sleeping compartment for each astronaut. The three cabins were lined up side by side, each about the size of a small closet, with a stiff, pleated curtain to provide privacy, a set of six small cabinets for storing personal items, and a mummy-type sleeping bag attached to one wall.

The ship's atmosphere was a mixture of 74 percent oxygen and 26 percent nitrogen, the reverse of the ratios on Earth, at 5 pounds pressure per square inch (compared to sea-level air pressure of 14.7 pounds per square inch). As the men breathed, the station's ventilation system circulated the air across a set of molecular sieves, which used activated charcoal filters to remove carbon dioxide and other trace impurities. This atmospheric purification system also contained water condensers, not dissimilar to ordinary home dehumidifiers but smaller, lighter, and more energy efficient. These pulled moisture from the air and collected it in waste tanks. Rather than recycling this water, *Skylab* was instead launched with more than 800 gallons of drinking water, more than enough to supply three men for more than six months. Building a system to completely recycle water, and air for that matter, would have to wait until later.

Because its atmosphere was so thin and the station so large, *Skylab* had an intercom system with 12 communications stations scattered throughout, allowing astronauts to talk to each other as well as to ground control. In such a thin atmosphere, a voice simply does not carry very far, no matter how loudly one yells.

Skylab's exterior was covered with a thermal insulation shield to keep the inside at a comfortable temperature as well as protect it from meteoroid impacts. It also had two separate coolant loops filled with a kind of antifreeze. This coolant, by also flowing through a radiator on the station's exterior, regulated its interior temperature.[12]

Because *Skylab* was planned to be a science laboratory, the three crews occupying it had almost 90 different experiments to perform. The station's exterior had eight telescopes for studying the sun and three cameras for observing the earth. Inside, it had a material-processing chamber able to do a variety of metallurgy experiments. Unlike *Salyut*, *Skylab* carried no significant plant experiments.* For the same political and funding reasons that caused NASA officials to shift *Skylab*'s focus from exploration to science research, the agency decided to emphasize only research with practical Earth applications. Studying how to grow plants in space, while certainly worthwhile and important if one wishes to colonize the planets, has no real use on Earth.

Despite *Skylab*'s ambitious design, immediately after liftoff the station very quickly appeared to be a complete failure. During launch part of its meteoroid or thermal heat shield, wrapped around the main workshop to protect it from meteoroids and shade it from the sun, was ripped away. Furthermore, though the four windmill solar panels on the telescope mount had unfolded properly, the two large solar panels attached to the orbital workshop were not functioning. One panel seemed completely inoperative while the other worked but could not generate electricity. Engineers guessed that the loss of the heat shield had somehow prevented the panels from unfolding as planned.

Without the heat shield, temperatures in the workshop immediately rose to as high as 130°F.[13] The failure of the two solar panels meant that the laboratory had only about 40 percent of its power supply. With the station's internal temperature rising, ground controllers needed to orient the laboratory so that as little direct sun-

---

*The one plant experiment on *Skylab* was a student experiment, testing whether rice seeds could grow in space. It focused less on the actual science learned and more on training the student. Moreover, few of the seeds developed, preventing any significant conclusions. See Summerlin, 67–73.

light as possible hit its exposed surface. With the loss of two solar panels, they also had to maximize the sunlight hitting its four remaining solar panels to generate enough electricity to keep the station functioning.

While controllers juggled with these contradictory requirements, shifting *Skylab*'s orientation again and again, NASA mission control delayed by 10 days the launch of the first manned flight to the station in order to figure out how to repair the damage, with thousands of engineers throughout NASA and industry working 20 hour days trying to save the project.

To shade the station from the hot sunlight, NASA engineer Jack Kinzer at the Johnson Space Center came up with a simple and fast solution: an umbrella. From inside the station, astronauts would extend and open this sunshade of mylar and nylon, inserting it through one of the small research airlock tubes in the main workshop. To repair the solar panels required a visual inspection by the astronauts to find out what had happened. Then they would attempt a repair, performing several space walks if necessary.[14]

"We Fix Anything!"

On May 25, 1973, after 10 days of brainstorming, the first manned mission to *Skylab* lifted off. Crewed by *Apollo 12* veteran Pete Conrad and rookies Joe Kerwin and Paul Weitz, the flight's mission of space research had shifted to repair and reconstruction because of the station's technical problems. Only if they could fix *Skylab* would they try to complete their full tour of four weeks.

For commander Pete Conrad, this mission, dubbed *Skylab 2* by NASA, was his fourth and last space flight. A flamboyant and outgoing man, Conrad had been flying since he was a teenager. Later he became a pilot for the Navy, specializing in that most dangerous type of flying, nighttime landings on aircraft carriers.

Conrad was quick-witted, enthusiastic, and exuberant. Astronaut Mike Collins described him as, "Funny, noisy, colorful, cool, competent; snazzy dresser, race-car driver. One of the few who lives up to the image [of hotshot astronaut]. Should play Pete Conrad in a Pete Conrad movie." When he first applied to become an astronaut in 1959, Conrad was rejected. He later suspected this rejection occurred because of his attitude during the arduous and almost absurd psychological tests. During one test, the psychologists gave the astronauts a blank sheet of paper and asked them to describe what they saw there. Conrad looked at it and quipped, "But it's upside down." Another time, he obtained a spiral notebook and pen and began mimicking the psychologists, taking notes of their every move, just as they did to him.[15]

Years later, when he, a somewhat short man, stepped onto the lunar surface, the third human ever to do so, he loudly announced, "Whoopie! Man, that may have been a small one for Neil, but it's a long one for me," thereby winning a bet with a French journalist who did not believe that American astronauts could choose their own words in space.

Conrad's pizzazz did not interfere with his skills, however. On *Apollo 12*, he proved that human-guided precision landings were possible, landing his lunar module on the moon less than 600 feet from its target, an old, unmanned *Surveyor* probe.

Conrad always considered his *Skylab* mission his greatest achievement, far more challenging than the *Apollo* landing. The lunar mission had been thoroughly tested and planned, and went like clockwork. *Skylab*, on the other hand, required the crew to use every ounce of creative energy to save the mission.[16] At liftoff Conrad expressed the thrill and joy he felt at being given this task, yelling into his microphone "We fix anything!" as the Saturn 1B rocket slowly rose from the launchpad.[17]

As Conrad brought his *Apollo* capsule close to *Skylab*, the patter of enthusiastic commentary continued. "Tally-ho the *Skylab*!" Conrad chortled when he first spotted the station. Then as he got closer, he described her condition in detail. "[Solar wing] 2 is completely gone off the bird. Solar wing 1 is, in fact, partially deployed . . . there's a bulge of meteor[oid] shield underneath it in the middle, and it looks to be holding it down. . . . It looks, at first inspection, like we ought to be able to get it out."[18]

Based on what the men could see, one main solar panel was completely gone—torn off somehow when the meteoroid shield had come free during launch. The second panel was intact, but a single metal strap from the lost meteoroid shield had tangled with it. Instead of unfolding a full 90 degrees, it had opened only 15 degrees.

The crew soft-docked with the station so that they could eat before starting repairs. They then undocked, donned spacesuits, and Conrad maneuvered the capsule close to the jammed solar panel. Paul Weitz opened the *Apollo* capsule's hatch and, holding a telephone repairman's 10-foot-long tool for pulling and prying at wires, attempted to free the stuck panel. For more than a half-hour Weitz worked on the strap without success, with Joe Kerwin holding his legs to give him some leverage. His effort rocked the entire 82-ton *Skylab* laboratory back and forth, requiring Conrad to ease the *Apollo* capsule away for fear of a collision. Finally, the crew gave up, closing the hatch so that Conrad could bring the capsule around to re-dock with the laboratory.[19]

Now, however, their spacecraft refused to re-dock. Six times Conrad tried, using five different techniques. All failed. In a last-ditch attempt, the astronauts dismantled the equipment for achieving a soft docking and tried a direct hard dock. This technique worked. The astronauts now began their first sleep period in space, still confined to the small *Apollo* capsule.[20]

The next day, their second in space, the astronauts finally entered *Skylab*. In both the multiple-docking adapter and airlock module the temperature was a comfortable 50°F. In the workshop, however, the air temperature was a blistering 130°. Weitz, the first to go in, noted calmly how the laboratory climate reminded him of being "in the desert."[21] During the next few hours, with Kerwin standing by in the *Apollo* command module, Weitz and Conrad deployed the umbrella sunshade through the research airlock, periodically taking breaks from the heat. With the parasol in place the laboratory's temperature immediately began to drop, and for the next 12 hours the temperature dropped by about 2° per hour. By the end of the next day the laboratory was a balmy 90°, with the temperature still dropping.[22]

For the next two weeks the three astronauts unpacked equipment, set up experiments, and discussed plans for releasing the stuck solar panel. Without that panel, the laboratory had only about 40 percent of its electrical power, curtailing research. With it, the power output would almost double, from four kilowatts to seven.

The medical experiments, however, could go on. Each day the men weighed themselves, using the first zero-gravity "scale," or body-mass measuring device (as it was officially called). The astronaut sat on its seat and latched himself in. The seat then began oscillating, at a speed that varied depending on the mass placed upon it. Measuring the pulses determined his weight.

Other medical tests included use of a lower-body negative-pressure device, similar in concept to the "Chibis" suits used on *Salyut*.

Not only did it force blood to circulate in the astronaut's lower limbs, thereby duplicating the body's circulation on Earth, it also measured the state of each man's cardiovascular system. Also tested was an exercise bicycle, which the astronauts found difficult to use because the harness for holding them on the bike was inadequate. Instead, the men discovered that running along the curve of the upper deck's inside wall was a more effective exercise routine. As their speed increased, centrifugal force drove their feet against its surface.[23]

As they settled in to their daily routine, the astronauts were pleasantly surprised that they experienced no motion sickness at all. Though American researchers had assumed that the very large interior space of *Skylab* would cause vertigo and space sickness, for reasons no one understood, the three men, like their counterparts on *Salyut*, never felt the slightest nausea. They did notice, as had the *Salyut* cosmonauts, that zero gravity caused the fluids in their body to shift, with blood moving up to their heads and out of their legs. "One feels a strange fullness in the head," Kerwin noted. "[It's like the] sensation of having a cold. . . . One sees the puffy look on the faces of his fellow crewmen and hears their nasal voices." As the head became bloated, the legs became sticklike. "One can almost see the fluid draining out of the legs of his fellow crewmen, making them look little and skinny like crow's feet."[24]

Finally, on June 7, after 14 days in space, Pete Conrad and Joe Kerwin donned their spacesuits for an audacious space walk to free the jammed solar panel. At 6 A.M. all three men climbed into the airlock module, with Weitz moving past into the docking module. There he closed the hatch behind him so that he stood by just outside the *Apollo* capsule. (If access to the airlock module became impossible while Conrad and Kerwin were outside, he could don his own spacesuit and open the Apollo capsule's hatch to provide all three men with an escape route back to Earth.) Meanwhile, Conrad and Kerwin closed the second airlock module hatch and evacuated its atmosphere. By 10:30 A.M. they had opened the outside hatch and floated out into space.

By the time they exited, the station had entered the earth's night side. Everything was dark, the huge planet below them pitch black and blocking out the stars. Conrad looked about and burst out, "Where the hell's the world, anyway?"

Astronaut Rusty Schweickart, the capcom in Houston, answered, "*Skylab*, Houston, we're right here. We're listening in loud and clear."

"Oh," Conrad said, realizing that Schweickart had thought his question had been directed at mission control. "I didn't mean the *world* world, I meant the clouds and the earth and sea world underneath."[25]

Because the stuck solar panel was located on a part of *Skylab* where no space walk had been planned, no handholds had been installed there. To help the men safely reach the offending strap, ground engineers had designed a special 25-foot-long extension tool made of 5-foot lengths with an ordinary heavy-duty wire cutter at its far end. A long cord ran from the wire cutter along the length of the pole, allowing an operator to snap the cutter's jaws closed by pulling on it.

Inside the airlock Kerwin passed the 5-foot lengths to Conrad, who assembled them into a single pole. Then the two men, attached to the station by 55-foot-long umbilical cords, climbed onto the side of *Skylab*'s gigantic exterior surface. With Conrad holding him in place, Kerwin struggled to slip the wire cutter's jaws around the strap that pinned the solar panel. Without good handholds, and with the unwieldy pole flailing about in space, their effort was difficult at best.[26]

After an hour they finally got the jaws secured around the strap. Before Kerwin tried cutting the strap, Conrad crawled out along the partly open solar panel—using the pole as a handrail—to hook a cord to the panel. Pulling the cord would swing the panel open toward them once the strap was removed. He also made sure that the wire cutter's jaws were properly placed around the strap. They were. Now all Kerwin had to do was pull.

At that moment *Skylab* moved out of radio contact with Earth. In fact, for the next 63 minutes the two men were on their own, out of touch with ground control. Holding himself in place with his umbilical cord, Kerwin pulled on the wire cutter rope, trying but failing to get its jaws to snip the strap. "Man, am I pulling," he said in frustration. As Kerwin pulled, Conrad climbed back along the pole to the wire cutter to see if he could help increase Kerwin's leverage. At that moment, Kerwin finally got the jaws to close, snapping the strap.

The panel immediately popped open, flinging both Conrad and Kerwin off the station and into space. For a few seconds they were both independent satellites, *Sputniks* all by themselves. Then they each grasped at their tethers and slowly pulled themselves, hand over hand, back to the station.[27]

The solar panel, though free, had opened only about 20 degrees. The two astronauts grabbed the second cord that Conrad had hooked to the panel and pulled, trying to force the panel to unfold and fully deploy. However, the length of the cord and its angle—almost flat against the side of the station—reduced their leverage. They couldn't get the panel to open.

Once again Conrad climbed out along the side of *Skylab*. By placing the cord over his shoulder and standing upright against the station's wall where the panel was hinged to the station, Conrad figured he could increase the leverage enough for Kerwin to swing the panel open.

With Kerwin heaving near the airlock, and Conrad leaning his shoulder against the cord and pushing, the panel suddenly snapped free, releasing the tension on the cord and once again sending both men tumbling out into space. By the time they were back in control and hauling themselves to the station, the solar panel was slowly unfolding, as intended. By the next day it would be in its planned 90-degree position.

"Would you look at that!" Conrad enthused.

Kerwin replied, "I expected it to come, but I expected to lose you too. By gosh, we got you and the solar panel!"[28]

The final task on this space walk required Kerwin to climb the struts leading onto the *Apollo* telescope mount. There, he would replace one camera's film magazine, and pin open the balky aperture door of a second camera. As Kerwin climbed up the mount's superstructure, he was overwhelmed by a feeling of exposure, as if he were climbing on the girders of an unfinished skyscraper. All previous space walks had taken place on the outside of a small space capsule, so that the men who went outside had nothing to judge distances by. All they could see was the glowing blue-white Earth below them, far away, and the glittering stars.

Kerwin, however, stood 25 feet above a structure about 5 stories tall, with the earth 270 miles below him. For the first time, an astronaut had a sense of how high above the earth he actually was. As one of the astronauts from a later *Skylab* mission noted, "It's like being on the front end of a locomotive as it's going down the track! But there's no noise, no vibration; everything's silent and motionless."[29] Keeping his eyes focused on his work and not on the vastness around him, Kerwin quickly changed the film magazine, pinned the aperture door open, and headed back down, relieved to get back close to *Skylab*'s main body.[30]

With the solar panel deployed, the orbiting laboratory now had sufficient power to complete most of its experiments. The men got down to work. They started a variety of metallurgy and welding tests. They took extensive multi-spectral photographs of the earth. They made repeated solar observations using the station's solar telescope, capturing the first good pictures of a solar flare from space.[31]

But it was the *experience* of weightlessness that proved to be their most compelling scientific discovery. Like the crew of *Salyut* before them, they discovered that space was simply a much stranger and more alien place than anyone had imagined.

For example, Kerwin discovered that up and down was determined solely by how he was oriented. He found that a bulkhead could function as a wall, a floor, or a ceiling, depending solely on how he oriented his head. While one person could be looking out the window, thereby perceiving it as a wall, another person could see the window as the floor, if his feet pointed to it. "All one has to do is rotate one's body to a [new] orientation and whammo! What one thinks is up *is* up!" said Kerwin, who thought about this sensation more than the other men. "It turns out that you carry with you your own body-oriented world, independent of anything else," Kerwin described in post-flight debriefing. "*Up* is over your head, *down* is below your feet, *right* is this way and *left* is that way, and you take this world around with you wherever you go."

Because perspective was so important in weightlessness, the men actually found that they could feel lost if they entered a room upside down. One time Kerwin even got lost without going anywhere. It was the middle of his sleep period when he was awakened in his cabin by a radio call from Earth. "It was pitch black," he said. "When I scrambled out of bed, I had no way of determining up from down; I had no visual reference in the dark." Having no idea where the light switch was, all he could do was to grope along the walls, trying to identify something so that he could get himself oriented. "It took me a whole minute to get the lights on."

In the big, open, upper deck, the strangeness of weightlessness left the men feeling both exposed and exhilarated. That gigantic empty space allowed them to do acrobatics undreamed of on Earth, soaring from wall to wall while doing endless flips and twists. In their spare time, if they weren't running along the circumference of the deck, they had contests to see who could do the most in-air somersaults. "We were tickled to death," Conrad said afterward. "We never went anywhere straight; we always did a somersault or a flip on the way, just for the hell of it."[32]

At the same time, if a man let himself drift away from a wall, floating gently in the middle of that empty space, there was little he could do to get back to where he wanted to be until he drifted across and touched another wall—or a crewmate gave him a shove as he flew past. Thus, the men generally preferred the lower deck, where the room heights were more normal and the space was divided up into smaller units.

Meanwhile, hard as Earth engineers had worked to make the station a livable environment, the strangeness of space made it impossible for them to guess the exact right way to build things. Just as the men on *Salyut* had discovered, without gravity, nothing behaved as you expected.

Things floated away. Whenever a cabinet was opened, the objects stored inside took on a life of their own and drifted all over the place. Though the gridded floor gave the crew attachment points for their own feet, the station did not have enough attachment points to keep everything else in place. Often, objects simply disappeared, blown away by the station's air currents. Finding them was also made harder because objects didn't move or look the way one expected. Conrad even found that he could look right at a tool and not recognize it if it was floating at an angle that was impossible on Earth.[33] To find lost items the men invariably flew to the upper deck, where miscellaneous objects gathered, pulled against the inlet screen of the air vents by the ventilation fans.

The shower, though usable, generally took too much time to use. To spray one's body required keeping the nozzle close to the skin, because the water spritzed only about 6 to 8 inches from the nozzle before gathering into a floating blob. If the spray did hit the body, it scattered into many little blobs of water floating everywhere, requiring a considerable time to vacuum up before opening the shower and getting dressed.[34]

The difficulty of designing things for living in space became most evident during meals. The food table was set at waist height, as if the crew would eat sitting down—a natural expectation on Earth. In space, however, it was a strain on the stomach muscles to bend at the waist. Instead, the natural position was half bent, somewhat akin to a relaxed fetal position. A dining table set at chest height, tilted like a drafting board, would have worked much better. When the men spooned out some food, they had to move it directly and quickly from the package to their mouth, or else bits of food drifted all over the place. The designers of the dining area in turn had forgotten to provide a method for securing utensils when not in use. If a man let go his fork or spoon, it took on a life of its own, soaring away. As one of the astronauts noted, "I've got a spoon stuck on the [ventilating screen] upstairs right now."[35]

Weightlessness also changed their bodies. Not only did their faces become bloated and their legs skinny, their spines changed shape as well. No longer pressed down by gravity, the spine lengthened by one to two inches as it straightened out and the disks relaxed.[36]

Though they all slept well, the men seemed to need less of it, averaging between five and seven hours a night.[37] And though Conrad found the cabins and sleeping bags far better than anything provided on any previous spacecraft, all three men thought they were too close together: if one man got up at night, he invariably woke the other two. The men also disliked having the sleeping bags attached to the walls, preferring the freedom to set them as they wished in their cabins. Conrad, for example, didn't like how the ventilation fans blew air up his nose as he tried to sleep. (The fans were necessary because, without gravity, a man could suffocate on his expelled carbon dioxide if it was allowed to build up around his face.) To avoid it, he flipped his bag over so that he slept in the other direction. In this arrangement, however, the fans blew air into his bag, puffing it up like a balloon so that it no longer could hug his body and keep him warm.[38]

After 28 days in space, the men came home on June 22, 1973, splashing down safely in the Pacific Ocean. As they walked unsteadily across the deck of the aircraft carrier that had plucked them from the ocean, serenaded by "Anchors Aweigh" and the cheers of the crew, they could proudly say they had succeeded in fulfilling Conrad's prediction at launch.

"We can fix anything!" Conrad had sung out enthusiastically, expressing in that short phrase the bold, old-fashioned, American exuberance that had put him and 11 other men on the moon. Given the right tools and the courage to use them, Conrad, Kerwin, and Weitz had proved that humans could enter a damaged space station and fix it, even while it orbited Earth in the hostile vacuum of space. *Skylab* was now a functioning space station, ready to receive its next two crews.

Nonetheless, the unknowns of space remained. Though the men insisted on walking the entire distance from the helicopter to the carrier's medical facility, it took them almost three weeks to completely recover from their month in space. Each experienced lower-back pain and muscle soreness below the waist. Moreover, they all suffered badly during the period they were at sea following their recovery. Adjusting to gravity was hard enough without the rolling and rocking on board an aircraft carrier. Future recoveries were timed to reduce the period on board ship to less than a day.[39]

Of the medical results, the bone-loss measurements were the most worrisome. Though the amount of calcium loss was mixed, and one man even gained mineral content in some bones, the overall data indicated that weightlessness weakened at least some of the body's weight-bearing bones. "In general, [this loss] followed the loss patterns observed in . . . bed-rested subjects," one researcher explained.[40] Whether this constant bone loss would stabilize with time, or continue unabated as long as the human body was exposed to weightlessness, remained an unknown of significant importance.

Like a Machine

One month later, on July 28, the second *Skylab* crew, commanded by *Apollo 12* veteran Alan Bean, with rookies Jack Lousma and Owen Garriott, took off from Cape Canaveral. Scheduled to last anytime from four to six weeks, depending on the week-to-week medical status of each man, the mission eventually ran more than eight weeks, becoming one of the most productive space missions in history.

Right from the beginning, however, things did not go well. During the very first orbit, less than an hour after launch and well before they had docked with the station, Jack Lousma started to feel nauseous. He immediately took an anti-nausea pill, which seemed to help, allowing him to eat lunch.

At the same time, commander Alan Bean noticed something worrisome out one window. "We've got some sort of sparklers going by the right window, over by Jack, but we don't have any going by the left," he told mission control. "Maybe we've got something spraying out that side."[41]

Ground controllers immediately pinpointed the problem. The *Apollo* spacecraft was made of two modules, the cone-shaped capsule that carried astronauts to and from space, and the cylindrical service module, which carried the spacecraft's fuel, supplies, electrical systems, and engines. Attached to the outside of the service module were four sets of four small rocket thrusters, used to orient the spacecraft properly both for docking and for main-engine firings. For safety, each set was designed to work independently of the others, and only two sets were necessary to steer the spacecraft. By the time Bean noticed the fuel "sparklers," ground engineers had already spotted a fuel leak in one thruster set. They told Bean to shut it down, leaving the spacecraft with three working sets.

Docking and reactivation of the station seemed to go well, but by the afternoon of their first day on board *Skylab* all three men were sick, with Lousma having vomiting fits and Bean and Garriott dizzy and nauseous. The pills no longer seemed to help.

Nor did sleep. By the next morning the men struggled to eat breakfast, and by noon the nausea and dizziness had returned. For the next three days these symptoms forced them to reduce their workload, including postponing a planned space walk to replace the deteriorating sunshade parasol installed by Conrad and Weitz. Doctors on the ground suggested that the astronauts restrict their head motions to help their bodies' system of balance adapt to weightlessness. In addition, because the worst symptoms seemed to occur soon after eating a big meal, ground doctors suggested the men eat many small meals instead.

By the sixth day in orbit, the symptoms had subsided and the three men were beginning to eat better and work more effectively. On this same day, however, a second set of attitude thrusters on the *Apollo* service module sprang a leak. Suddenly, the entire mission was threatened. If the second set of thrusters failed, the *Apollo* spacecraft would have only two left, with no backups remaining in case a third failed. Immediately mission control began to gear up for a possible rescue mission, preparing the next available Saturn 1B rocket with a specially outfitted *Apollo* capsule. Two astronauts would fly it to *Skylab*, where it would dock at the second port and be used to carry all five astronauts back to Earth.

As it turned out, the two leaks were caused by independent problems, and mission controllers decided to wait things out, figuring that the chances of a third failure were slim. In the meantime, there was plenty of time to prepare and launch a rescue mission if necessary. The astronauts weren't going anywhere, and *Skylab* had more than enough supplies to keep them alive—for months if necessary.

With all these difficulties, by the end of the first week the crew was significantly behind schedule, and ground controllers were pressing them to catch up. Furthermore, the men had not yet done the planned space walk to replace the temporary parasol, and had instead spent a significant amount of time doing unplanned basic maintenance, trying to fix the station's dehumidifier and the urine system centrifuge. On top of all this, the men, just like previous long-term spacefarers, found that living and working in weightlessness was more challenging than expected. Even the simplest task was made complicated, especially because objects kept wandering away unexpectedly. "Every time you go to do something like get your kit out and shave," Bean explained, "you find there are no shaver heads there, and you have to go hunt . . . somewhere."[42]

Irritated by all these annoyances, Bean finally got testy with ground controllers. "We've been working from sunrise right now, and we're still not finished. Tell [the flight planners] to give us a little more [slack] if they possibly can, because I've looked out that window five minutes in five days. The rest of the time we've been hustling."[43]

To make matters worse, on August 5, the seventh day of the mission, both of *Skylab*'s coolant loops began leaking. Without a fix, engineers figured that the primary loop would be dry within three weeks, with the secondary loop lasting at least until the next station occupancy, meaning that the third crew would have to arrive prepared to fix the problem.[44]

Despite these problems, the crew began preparations for their first space walk. On August 6, rookies Garriott and Lousma donned their spacesuits and climbed into the airlock module, with Bean standing by in the docking module. After almost three months exposed to the harsh vacuum of space and direct sunlight, the umbrella-like parasol that Conrad and Weitz had installed had deteriorated significantly. Because the parasol had to be thin enough to fit through the research airlock, its protective canopy, made of nylon, mylar, and aluminum, had to be very thin, less than four-thousandths of an inch thick. To make it that thin, the nylon, which quickly disintegrates when exposed to ultraviolet light, had not been given a protective coat of thermal paint.

A second heat shield had been devised using sturdier materials, but installing it required a space walk. Mission control had decided to go with the parasol for the first mission (because Conrad and crew had enough work to do to fix the stuck solar panel), and let Bean's crew build the replacement shade. The new shield was designed somewhat like two masts with a sail between. The men were to rig two 55-foot-long poles on the side of *Skylab*, and then pull the shade up along the poles with cords, covering both the deteriorating parasol and the exposed part of the station.

While Garriott began assembling 22 5-foot sections into the two long poles, Lousma attached to the station several extra foot restraints as well as a special base plate for securing the two poles. Once assembled, the two poles were fitted into the base plate to form a "V," then placed against the side of the station. Though there were some minor difficulties at the beginning of the space walk (first, the men had to untangle their tethers from those that kept the equipment from floating away; second, the equipment didn't behave exactly as predicted by the simulations on Earth, and required improvisation), the whole operation took only four hours.[45] Almost immediately ground controllers could see an improvement in temperature inside the station.

On the same space walk Lousma climbed to the top of the telescope mount to reload several of the film magazines and deploy and retrieve a number of experiments. Hanging high above *Skylab*, he felt the same sense of height and distance that Kerwin had. The experience became most intense when the station flew into night. "[You are] hanging by your feet as you plunge into darkness," Lousma remembered. "You can't see your hands in front of your face—you see nothing but flashing thunderstorms and stars."[46]

Before climbing down, he paused to take a good look at the leaking attitude jets on the *Apollo* capsule, in plain sight below him about 30 feet away. As far as he could tell, there were no signs of leakage.

The success of this space walk seemed to somehow change the whole tone of the mission. From this moment on, the problems seemed to evaporate and the three men began to operate with an efficiency that was downright startling. Each morning, they gathered in the galley and had breakfast together. Then, like a machine that was steadily building momentum as it rolled downhill, they went about their business, doing one task after another with great enthusiasm and efficiency, foregoing their later meals and grabbing food on the run as they worked into the night.

As a crew, the personalities of the three men fit together perfectly. Bean, who had been Pete Conrad's crewmate on the moon, was as enthusiastic and hard-working, if not quite as entertaining a talker, as Conrad. "We're working less hard at the moment than we were prior to [launch and could] do a little bit more," Bean told mission control right after the space walk, trying to get them to give the crew more work. "We've got the ability, and time, and energy, and I know y'all do down there."[47]

Garriott, a bemused-looking, thin-faced man with a distinctive mustache that made him look like a western cowpoke, was even more eager to do more. Not only did he urge his crewmates on, he continually requested more work from scientists on the ground.[48]

Lousma, meanwhile, was a hard-working, fun-loving guy who took to space like a duck to water. "God! I'm glad that Saturn worked," he enthused during one space walk. "It's great to be here!" In charge of the television broadcasts, he opened each with the words, "Hello space fans!", then described the activities of his crewmates as he filmed them, performing almost like a television sports announcer.[49]

Over the last six weeks of their mission these three men completed 150 percent of their scheduled flight program. They completed 48 separate runs of Earth observations, three times more than the first *Skylab* crew, filming large areas of the United States and 28 other countries on all seven continents. They did many man-hours of observations of the sun, using the telescope mount, taking some of the best pictures of the first coronal mass ejections ever seen by man. One investigator called one such ejection "the most significant [solar] event since the launch." (By far the largest eruptions on the sun's surface, each mass ejection barrels more than 10 billion tons of hot, electrically charged gas traveling at more than a million miles per hour out into the Solar System.)[50]

Then, on August 13, Bean used the vast interior space of *Skylab* to test several jetpack designs that a space walking astronaut could use for maneuvering in space. One was a modified backpack unit which had been intended for a 1966 *Gemini* mission but not used.

Not only did the unit work, but Garriott, who had not trained to fly the unit before launch, was later able to pick up its operation after an hour of review, flying it easily. Other jetpack designs were more radical. In one, Bean flew about using a pistol that sprayed nitrogen gas. In another, he used jets attached to his feet. Both methods were less than successful. The foot-jets worked well only when pointed down. If the astronaut tried to fly forward or backward, the jets caused him to rotate in place, his feet moving while the rest of his body stayed stationary.

As Bean took these units for a spin, Lousma filmed his acrobatics for the public on Earth, giving his "space fans" a blow-by-blow description. "No apparent difficulties whatsoever," he described Bean's actions enthusiastically. "Rotating slowly but surely. . . . Ooooh, he just blasted me."[51]

Of the many experiments that these workhorses completed, one in particular attracted a great deal of attention on Earth, partly because it had been devised by 17-year-old high-school student, Judith Miles from Lexington, Massachusetts. To see if weightlessness affected spiders' ability to weave webs, she had sent two, Arabella and Anita, to the station with its second crew. Both spiders adapted well, building ordinary webs and demonstrating that they did not use gravity to determine direction. However, their web strands were far thinner than those seen on Earth, indicating that the lack of gravity caused the spiders to believe their webs did not need to be as strong.[52]

Arabella and Anita were not the only experimental animals on board. Garriott had brought a plastic bag filled with water and two minnows. Attaching it to one wall of the station, he watched the fish struggle to orient themselves. Just like human astronauts, the fish were unused to relying on just vision for balance. At first they swam in circles, totally confused. Then they settled down and oriented themselves by sight, with their bellies toward the station's wall as if it were a river's bottom.[53]

In addition to research, the crew did a significant amount of routine maintenance. When six small gyroscopes overheated and failed, they did a space walk to modify and install a set that they had brought with them. They fixed several tape recorders. They replaced four circuit boards from a video recorder. They even repaired one of the pedals on the exercise bicycle.[54]

Amid all their hard work, the men had time for practical jokes. One day, astronaut Bob Crippen was working as capcom in Houston. Suddenly, to his astonishment, a distinctly female voice came over the loudspeakers from *Skylab*. "Hello, Houston this is *Skylab*. Are you reading me down there?. . . Hello Houston, are you reading *Skylab*?"

Crippen, unsure what was happening, cautiously answered, "*Skylab*, this is Houston. I hear you all right, but I had a little difficulty recognizing your voice. Who have we got on the line here?"

The woman responded, "Houston, roger. I haven't talked with you for a while. Is that you down there Bob? This is Helen here in *Skylab*. The boys hadn't had a home cooked meal in so long, I thought I'd just bring one up. Over."

Crippen recognized the voice as that of Owen Garriott's wife, Helen, though he had no idea how she could be broadcasting from *Skylab*. "Roger, *Skylab*," he answered doubtfully. "I think somebody has got to be pulling my leg. Helen, is that really you? Where are you?" By now a baffled crowd of mission controllers had gathered around Crippen.

After a pause Helen replied. "Just a few orbits ago we were looking down on the forest fires in California. You know the smoke sure does cover a lot of territory. And, oh Bob, the sunrises are sure beautiful. . . Oh-oh, I have to cut off now, I see the boys floating up towards the command module, and I'm not supposed to be talking to you. See you later, Bob."

Speechless and confused, Crippen could only say, "Bye-bye."

On *Skylab*, Bean, Garriott, and Lousma were hysterical with laughter. In an earlier private conversation with his wife, Garriott had given her the lines and recorded her saying them. He then waited for the appropriate moment to play the tape, starting and stopping it to create the illusion of a conversation.[55]

After 59 record-breaking days in space—more than twice as long as any previous single mission—the astronauts returned to Earth on September 25, 1973.

Based on the condition of the first *Skylab* crew at recovery, the in-space daily workout for Bean, Lousma, and Garriott been increased from 30 minutes on the exercise bicycle to 60. As a result, their heart and cardiovascular systems showed no ill effects from weightlessness, and were more or less back to normal in about a week. However, the one hour of bicycle exercise was apparently insufficient to keep their muscles in shape in zero gravity. All three men were weaker than hoped for upon return, their leg muscles having lost 25 percent of their strength. The bicycle didn't seem to provide the best kind of workout—some of the men found it easier and more fun to rotate the unit's pedals with their hands rather than their feet.[56]

More importantly, the men once again showed a steady loss of bone mass during their 59-day flight. Despite the increased exercise, the production of bone material seemed somehow tied to the presence of gravity. Scientists began to wonder if there was an outside limit of time in space beyond which humans could not go before their bones became too brittle.[57]

Finally, the motion sickness at the beginning of the mission posed a different problem that also had to be solved. As NASA Deputy Administrator George Low noted, "Were we to lose three or four days out of each seven-day space-shuttle flight because of motion sickness, the entire shuttle effort would be in jeopardy."[58]

Getting an answer to these questions, however, was going to be difficult. On August 13, 1973—the same day that Bean was flying around inside *Skylab* in a jetpack and showing the world what Americans could do if they put their mind to it—NASA canceled all work on a second *Skylab* orbital workshop. Designed to duplicate the first station, this follow-up station had languished in studies and bureaucratic indecision for years. Now, citing lack of funding from Congress, the space agency announced that, effective August 15, all work on a follow-up station mission would cease. The hardware for the second *Skylab* was eventually cut up, reconfigured, and installed at the National Air & Space Museum as a museum display illustrating what the first *Skylab* had been like.[59]

Somehow, if Americans were going to answer every one of these remaining questions—vital to the success of future interplanetary space travel—the last crew of the first *Skylab* would have to do it on their own.

Reality Strikes

As the last crew closed in on *Skylab* on November 16, 1973, three men, the first all-rookie American crew in five years, pressed their faces to the windows to stare at both the shining Earth below them and the gleaming giant space station creeping up before them. "Oh, there's the mother!" laughed Bill Pogue. "We is dere. Yeah, yeah, yeah, yeah." Commander Jerry Carr then brought his *Apollo* capsule into its docking port, only to find that he couldn't get a hard dock. Like Conrad, after two failed attempts, he had to back up and gun his engines to force the two spacecraft to lock together.[60]

In order to avoid space sickness, this third crew had been told not to enter the orbital laboratory on their first day in space and sleep instead in the *Apollo* capsule. Doctors theorized that maybe the Pete Conrad crew's delay in entering the vast interior space of the orbital workshop—caused by the heat shield problems—was why they experienced no space sickness. In comparison, Bean's crew had entered the station immediately and had gotten sick. Doctors thought that confining Carr's crew to the small *Apollo* capsule for their first 22 hours in space would help them adapt.

Unfortunately, the theory was wrong. Soon after reaching orbit, pilot Bill Pogue experienced nausea and took anti-nausea pills. Then, once docked, both Carr and Ed Gibson, the science-pilot, felt uneasy and took pills as well, while Pogue felt so unsteady he had to stop working and rest.[61] As the three men prepared to bed down for the night, Pogue suddenly felt much worse. "Just hold on," he said suddenly, "No kidding, I am going [to be] sick." He then unceremoniously vomited his dinner into a vomit bag while his crewmates cringed. For the next few minutes he felt quite sick, though he didn't vomit again.

The situation was very ironic. Pogue, a pilot his entire life, was known as "Iron Belly," the man who never got sick. He had flown 43 combat missions during the Korean War, then spent three years with the air force's aerobatic flying Thunderbirds in the mid-1950s. He had also graduated from Empire Test Pilot School in Farnborough, England, and had been a flight instructor at Edwards Air Force Base in California. During all these activities, he had never felt nausea, and his stomach had never bothered him. Now, however, the iron had turned to mush. Green-faced, he could do nothing but hold out and hope the nausea would fade.[62]

At that moment this last *Skylab* crew did something that illustrates better than anything how the goals and perspectives of the humans in space and the people on the ground can never be identical, and are often quite at odds. Out of radio range when Pogue vomited, Carr and Gibson wondered whether they should say anything about it. Both felt inclined to keep the incident a secret, realizing how much of a big deal the doctors would make of it. After much discussion, they decided to mention the nausea but leave out the vomiting. They would keep the incident "between you, me, and the couch," as Carr said. To hide the evidence, Carr would take the vomit bag and throw it in the trash airlock once they entered the space station.[63]

The trouble with this plan was that they forgot that an automatic tape recorder was running during their entire conversation. The next day, ground controllers routinely downloaded the tapes and the day after that, mission controllers could read the tape transcripts, neatly typed and mimeographed for their perusal.

The reaction was swift and firm. Astronaut chief Alan Shepard, the first American in space and the fifth man to walk on the moon, got on the open radio with Carr to give him a firm and very public reprimand. "I just want to tell you that on the matter of your status reports, we think you made a fairly serious error in judgment."

An embarrassed Carr could only agree. "Okay, Al. I agree with you. It was a dumb decision."[64]

For the rest of the mission there was a never-ending undercurrent of tension between the ground and the men in space. Because the men knew that everything they reported would eventually get into the press, they were generally very careful about what they said on the public radio. At the same time, ground controllers felt unsure whether the crew was telling them everything.[65]

To make matters worse, the crew was faced with meeting the very high expectations caused by the phenomenal efficiency of the previous *Skylab* crew. Their work schedule and 12-hour workday had been determined based on what the second crew had accomplished, a pace unrealistic for most crews. In fact, both ground controllers and the third crew began the mission expecting the astronauts to come off the starting line moving as fast as the second crew. Not only had more experiments been added to the schedule, the flight duration itself had been increased. By bringing with them a supplementary supply of about 400 energy-food bars, Carr, Pogue, and Gibson hoped to extend their flight to as long as 12 weeks.[66]

Making things even worse, ground controllers once again completely forgot how long it takes to adapt to weightlessness. One flight controller called the task of getting the station activated "only a little more complicated than when you come back from vacation."[67] They nonchalantly packed the schedule, and didn't let up, even when the men fell behind.

The strangeness of weightlessness, however, could not be ignored. Even the most ordinary tasks took longer than expected.
Moreover, after two long missions, *Skylab* was beginning to get a
bit disorganized. Sometimes the men had no idea where an item
was, and had to ask ground controllers, who could only tell them
where it was *supposed* to be. Then they searched, seemingly forever, for the simplest things, time that ground controllers refused
to consider when they were organizing the astronauts' schedule.
"There's not enough consideration given for moving from one point
in the spacecraft to another," Pogue noted in frustration at one
point.[68]

This heavy workload and their unfamiliarity with space caused
them to make errors. For example, during a set of medical tests, the
men had to photograph each other with infrared film. When Pogue
was photographer, he found it difficult to keep from drifting, and
accidentally kicked a valve on a water tank and almost flooded the
workshop.[69]

Working frantically, the men still managed to get a myriad of
complicated tasks finished in the first week. Gibson and Pogue
did one space walk, reloading cameras in the telescope mount.
Pogue attached an improvised valve to the leaking primary coolant loop, allowing him to recharge it when necessary and thus
bring the system back into operation.[70] And all three completed
an extensive set of medical tests, creating a baseline of data for
later comparison.

Nonetheless, by the end of the first week they were far behind
schedule, though unwilling to challenge what ground controllers
were asking them to accomplish. After their *faux pas* with Pogue's
vomiting incident, they had become defensive. The most they were
willing to do was hint at how overwhelmed they felt, and suggest
gently that changes be made. "If we're ever going to get caught up,"
Carr noted at one point, "we're going to have to whack something
out [of the flight plan]." Later he said, "The best word I can think of
to describe [the situation] is frantic. . . . No matter how hard we
tried, and how tired we got, we just couldn't catch up with the
flight plan. And it was a very, very demoralizing thing to have happen to us."[71]

On the ground, these hints fell on deaf ears. Assuming that the
crew was steadily getting more adapted to weightlessness, ground
controllers started to tighten the schedule, squeezing in more tasks

in less time. In fact, in an effort to get the crew to catch up, controllers even began scheduling some tasks *during meal times*, while grudging the crew's desire to have at least one day off a week. "We've got to get this crew organized," griped flight director Neil Hutchinson.[72] This tension between ground control and men in space illustrated how a permanent space facility could not function like a short-term mission to the moon. As the weeks piled onto weeks, Carr, Pogue, and Gibson found themselves becoming increasingly overwhelmed by the demands being made on them. Flight engineers at NASA expected them to work continuously, during every waking hour, six-and-a-half days a week, much like the previous crew had.

But Carr, Pogue, and Gibson were not Bean, Garriott, and Lousma. As much as the men themselves wanted to get their full program completed, they were a different crew, with a different style of working. By expecting them to perform like their predecessors, ground controllers only helped feed the distrust and tension created by the mistaken effort to hide Pogue's nausea on the first day.

Finally, about a month into the mission, the men lost patience. On December 15, Jerry Carr described at length how the procedures he was being ordered to follow for managing the food were wasting his time. "I get snowed under and there doesn't seem to be any way I can catch up." Then on December 20, both Bill Pogue and Ed Gibson recorded a long, carefully prepared tirade, describing how the scheduling was unrealistic and causing them to make mistakes and lose morale. An exasperated Pogue said, "The time since we've been up here [has been] nothing but about a 33 day fire drill. I found we've been chasing quantity rather than quality." Pogue then added, "So I plan from here on to take the time that it takes, regardless of what happens."[73]

For the next 10 days the men worked more at their own pace, less willing to cooperate with the demands of mission controllers. Then, on December 28, the halfway point of their flight, Carr taped and downloaded to the ground a six-minute message, asking that the crew and ground controllers have a frank and detailed discussion the next day about scheduling and the status of the overall mission. "Where do we stand?" he asked. "What can we do if we're running behind and we need to get caught up? . . . We'd like to have some straight words on just what the situation is right now."[74]

The next day, ground controllers used the teleprinter to send up a message outlining their view of the situation and scheduled a two-hour communications session the next night. Then, in full view of the public, the crew and flight controllers hashed out their problems.[75]

To the crew, the most important issue was their desire to have more control of their time, rather than having every second of their day scheduled precisely, in a manner not of their choosing. Communications between the ground and space also needed improvement. Ironically, the crew's concerns about being behind schedule were unfounded. An analysis by ground controllers found that, despite their frustration and sense that they were not getting things done fast enough, the third crew had actually accomplished about the same as the second crew. Yet, because ground controllers had previously neglected to check these facts, everyone had been working under a false impression, damaging morale. After an hour of conversation, both sides felt the air had been cleared. And in fact, for the rest of the flight the crew worked with increasing efficiency, their morale significantly improved.[76]

The astronauts returned to Earth on February 8, 1974, after completing 84 days in space, setting a human longevity record that was to last for 4 years, and an American record that was to last more than 20 years.

Whether Carr, Pogue, and Gibson could have survived a trip to Mars in zero gravity remained an open question. Most of the data seemed hopeful. Despite the disagreements between the third crew and mission control, none of *Skylab*'s crews had experienced any of the internal crew conflicts that had concerned Wernher von Braun in the 1950s. In fact, all three *Skylab* crews got along just fine, with only very minor and inconsequential disputes. The largeness of *Skylab*, with separate cabins for each man and a vast interior, probably helped keep human relations normal and easygoing.[77]

For the last crew, it did take several weeks for them to mentally readjust to life under gravity. Walking was unsteady. (Pogue, for example, had a tendency to drift to the right, even when trying to go straight ahead.) Side rails were added to their beds to keep them from trying to "float" off when they woke up. And they all had to break the habit of simply letting go of objects when they were done with them. As Pogue noted, "[We] had become used to releasing objects and having them float nearby until needed again."[78]

Physically the men came through with flying colors. Despite their long flight, Carr, Pogue, and Gibson recovered far faster than the previous two *Skylab* crews. Within 24 hours they were walking without help, something the men on the previous flights had not been able to do. Their leg muscles showed, as one doctor reported later, "a surprisingly small loss in leg strength," with the muscle size far less diminished and quickly returning to normal size. "Muscle in space is no different than muscle on Earth. If it is properly nourished and exercised at reasonable load levels, it will maintain its function."[79]

Bone loss, however, continued to be a concern. Carr, Pogue, and Gibson all experienced a continuous loss of bone mass throughout their flight. With the completion of the *Skylab* program, scientists still did not know if humans could keep their bones strong enough during the long periods in weightlessness necessary to travel to the nearest planets. "Assuming that [the loss] would continue," noted the scientists in charge of bone research, "the calcium loss rate of 0.3 to 0.4 percent per month observed in *Skylab* . . . takes on clearer and more ominous significance when it is realized that flights to Mars and return, when ultimately conducted, will take one-and-a-half to three years."[80]

In total, *Skylab* had been occupied for 171 days, almost half a year. The station had functioned well, though there had been failures both in the thermal systems and in the gyros (one large gyro failed and several of the smaller gyros had repeated problems). Yet when the last crew came home, the station still had enough oxygen for another 200 days of use, and water for 90 days.[81] If its goal had been to get a crew of three to Mars instead of going nowhere as it circled the earth, *Skylab* could almost certainly have been configured to do it.

Skylab's success was in large part because Wernher von Braun and his fellow engineers at the Marshall Space Flight Center in Huntsville, Alabama, had built their rocket well. With its gigantic ability to put huge amounts of mass into orbit, von Braun's Saturn 5 gave the United States an unprecedented opportunity to forge a path to the planets and the stars. Using the Saturn 5 aggressively in the 1970s and 1980s, as von Braun vainly lobbied for in his later years, might have not made the exploration of Mars and the planets commonplace, but it would have made the construction and occupation of permanent manned space stations a fast and ubiquitous achievement, enabling planetary exploration at a pace far exceeding anything we can imagine today.

The Saturn 5's Achilles heel, however, was that it cost so much to launch—more than half a billion dollars. And by the 1970s the American public had grown timid and shortsighted, uninterested in grand exploration and noble acts of heroism. Disco was in, space was out. The money and will were simply not there.

Though NASA had intended to incorporate use of *Skylab* with the space shuttle program, delays and lack of funds prevented any shuttle mission to the station before atmospheric drag caused *Skylab*'s orbit to deteriorate. On July 11, 1979, five years after its last crew departed, *Skylab*, still weighing 77 tons, burned up in the atmosphere over the Indian Ocean, with pieces crashing into the outback of Australia—the largest being a one ton piece recovered near the town of Kalgoorlie.

Meanwhile, the United States was committed to building a reusable space shuttle, under the false pretense that this first-generation experimental launch vehicle could operate like a truck—fast and cheap. According to this naive theory, the ease of launch would eventually make the construction of space stations cheap and easy as well, leading to the colonization of the planets.

While the United States dithered, the Soviet Union moved forward. Ironically, the winner and loser in the race to the moon were about to trade places.

4

The Early *Salyuts*:
"The Prize of All People"

Summit

On May 29, 1972, Leonid Brezhnev and Richard Nixon stood together at the main entrance into the 198-foot-long St. George's Hall, the largest room in the Kremlin complex. Filling the hall before them were the powerful, the curious, the artistic, and the influential, milling in a tense anticipation of the two leaders' arrival. As both men stepped into the room, the Kremlin Orchestra began playing "The Star-Spangled Banner," followed immediately by the Soviet "Internationale." Both anthems echoed through the cavernous white-marble chamber as the two leaders worked their way through the long line of dignitaries.

After 10 days of intense talks, preceded by several years of increasingly complex negotiations, this march down the lengthy floor of St. George's Hall marked the successful completion of the 1972 Moscow summit, an event that, according to both Nixon and Brezhnev, heralded the official beginning of "detente." Nixon would come home to the United States and declare that his Moscow summit was the harbinger of "a generation of peace," an event that would "lead the world up out of the lowlands of constant war, and onto the high plateau of lasting peace." Brezhnev in turn described how this summit demonstrated his ". . . conviction that no foundation other than peaceful co-existence is possible . . . between [our] two countries in the nuclear age."[1]

This Moscow summit, the first time since the beginning of the Cold War that an American president had visited Soviet Russia, produced a number of unprecedented agreements. Amid the glitter of daily receptions, trips to Brezhnev's private dacha, and a ride up the Moscow River on a hydrofoil, the summit announced cooperative agreements on environmental protection, on medical research and public health, on avoiding incidents at sea between the two navies, and on establishing better trade relations. The most significant accord was the Strategic Arms Limitation Talks, or SALT agreement. This arms treaty was the first to place detailed limits on the numbers of weapons, from missiles to submarines, that each nation could manufacture and deploy.

To demonstrate as visibly as possible their commitment to peaceful co-existence, Nixon and Brezhnev also used the Moscow summit to agree to a joint American-Soviet space mission. After 15 years of intense competition in space, an *Apollo* capsule would dock with a *Soyuz* capsule sometime in 1975, thereby illustrating Brezhnev's and Nixon's commitment to detente.[2]

At the concluding reception in St. George's Hall, Brezhnev and Nixon said good-bye shaking each other's hands in a four-handed clasp. Brezhnev good-humoredly practiced the use of the newly learned English word "okay," while Nixon responded by practicing how to say "khorosho," the Russian word for "good."[3]

Despite the high-sounding words and numerous agreements, the true consequences of this summit had more to do with public relations than fundamental change. Nixon, forever the politician campaigning for votes, saw the Moscow summit as a way to cement his lead in the 1972 elections. Though he recognized that there could never be true peace while Russia was ruled by communist dictators, Nixon was also aware of the 1970s American exhaustion with conflict and war. After seven years of bitter fighting in Vietnam as well as more than twenty years of Cold-War tension, the citizens of the United States desperately wanted relief. To get it, both they and Nixon were willing to ignore the violent and totalitarian nature of the Soviet Union and make believe that peace was possible.[4]

Brezhnev, for his part, had learned how to take advantage of this American desire for harmony. By 1972, after almost eight years in power, Khrushchev's old propagandist realized that he could score enormous public-relations points with Europe, the Third World, and his own rivals in the Kremlin by positioning himself as the world's peacemaker. Brezhnev understood that his claims of peace and goodwill between nations would resonate in the West. The claims would also disarm his opponents by making it appear as if they wanted to provoke a nuclear war if they opposed him. In the past two years, he had called for a ban on chemical and bacteriological weapons, for the creation of nuclear-free zones in various parts of the world, for an end to all nuclear-weapons testing, and even for total nuclear disarmament. His expanding campaign of detente had resulted in a treaty with West Germany, getting that country to accept the existence of East Germany and the Soviet-German border established by Stalin after World War II.

The 1972 Moscow summit with Nixon was more of the same. Besides the SALT and space treaties, Nixon and Brezhnev signed a joint communiqué declaring that their nation's primary goal was "peaceful co-existence," that

> ... they will do their utmost to avoid military confrontations and to prevent the outbreak of nuclear war. They will always exercise restraint in their mutual relations, and will be prepared to negotiate and settle differences by peaceful means.[5]

Similarly, the agreement to fly the first U.S.-Soviet joint manned mission allowed Brezhnev to painlessly and without risk illustrate his so-called peaceful intentions. No longer just a symbol of his country's scientific achievement, as stated in his 1969 space-station speech, space was now made to serve foreign policy objectives.

Under negotiation since the fall of 1969, the space agreement called for the United States to build a small docking module to connect the American *Apollo* capsule with the Soviet *Soyuz* capsule. The Soviets in turn would design and build a universal androgynous docking mechanism. Instead of having an active male probe that docked with a passive female receptacle cone, the androgynous port had flaps on its periphery that overlapped and interlocked with each other, allowing either craft to be the active port. According to the agreement, all future spacecraft were supposed to use this mechanism. At some magical moment in 1975, an American astronaut would open a hatch and reach out to shake the hand of a Soviet cosmonaut while they both orbited the earth in weightless bliss.

Ironically, at the same time Brezhnev was signing this space treaty and mouthing high-sounding words about "peaceful cooperation," his country was forging ahead with a nonstop effort to outdo the United States in outer space. Over the next year the Soviets made three desperate bids to beat *Skylab* into orbit.

Their first try, a civilian station launched July 29, 1972, didn't even get into orbit, breaking apart when the second stage of its Proton rocket failed. The second, *Salyut 2*, a military *Almaz* station built entirely by Chelomey's design bureau, was placed in orbit on April 3, 1973, one month before *Skylab*. It failed after 13 days when some disaster, caused by either a fire or the failure of one of the attitude engines, caused the station to overheat, which led to a catastrophic rupture of its hull. The third, a civilian station dubbed *Cosmos 557* and launched May 11, 1973—only three days before *Skylab*—failed when its attitude engines began firing spuriously immediately upon reaching orbit. Ground control was slow to respond, and the station's tanks were soon empty, making the station useless.[6] Russian space designers could only watch helplessly as the three American crews on *Skylab* came and went, setting record after record.

These failures finally forced Brezhnev and the Politburo to overhaul the Soviet space program, replacing many of its top people in May 1974. In mission control, the flight director who had let the third station's tanks go empty was fired, while others were demoted or encouraged to leave the program. Vasily Mishin, who had been Sergei Korolev's first deputy for decades and had taken over the design bureau when Korolev died in 1966, was dismissed. The entire program was revamped, with the Korolev design bureau being merged with a number of other bureaus to form the Energia (pronounced e-NER-gee-ya, with the "g" pronounced like the "g" in "get") Scientific-Production Association. This new, restructured, design bureau was then placed under the control of Korolev's chief rival, Valenti Glushko.[7]

Finally, after decades of being in Korolev's shadow, Glushko had the chance to shape the Soviet space program. He immediately canceled Korolev's N1 lunar rocket, ending a program that had cost billions, had employed the efforts of thousands of engineers, and had never been able to launch the N1 without it blowing up. In its place he lobbied to build a Soviet space shuttle and a new superpowerful rocket to launch it. Later named Energia after the design bureau that built it, this rocket was intended to match the Saturn

5, putting as much as 100 tons into orbit. According to Glushko's plans, the rocket would provide the lift capability for building a lunar base.[8]

Brezhnev's overhaul of the Soviet space program, however, didn't immediately eliminate the competing military and civilian space station programs. Unlike capitalism, where private companies compete for the cash of customers who can take their money wherever they wish, and who usually have limited resources and are, therefore, more cautious about how they spend their earnings, the Soviet government that financed the design bureaus (like national governments everywhere) could print however much cash it needed, had no real understanding of "the value of a dollar," and was therefore much more easygoing about financing a number of different projects, sometimes for incredibly trivial reasons. For example, Glushko got funding for his shuttle and rocket programs, but was refused funding for the lunar base, even though the base was Glushko's reason for building the shuttle and rocket. Though Brezhnev and the communist leadership were completely uninterested in going to the moon, they were quite willing to spend billions of rubles to prove that the Soviet Union could do anything the U.S. could do, even if it accomplished nothing.

In another example, Brezhnev hated firing anyone or canceling any program. His political strength rested on making sure everyone's job was safe, which was why, even after numerous disasters, it had taken him so long to dismiss Mishin. It was also why, in 1970, he chose to build two parallel and competing strategic missile systems designed by two different bureaus, thereby throwing away billions. He couldn't bring himself to cut anyone's job.[9] Similarly, canceling Chelomey's *Almaz* program outright would create enemies. Instead, the safe and secure route was to let *Almaz* die slowly (even though that decision was enormously wasteful), which was why the three failed missions had alternated military with civilian launches and why, over the next three years, the next three stations continued that rotation. Brezhnev and the men around him couldn't decide which way to finally go.

First came the military station. Launched June 25, 1974, *Salyut 3* was similar in many ways to the first *Salyut* station. It weighed approximately 20 tons, was cylindrical, had one docking port, and had about 3,200 cubic feet of interior space. Unlike *Salyut 1*, whose guts had come from the Korolev bureau's *Soyuz* spacecraft, *Salyut 3* was an

Almaz space station through and through. Built by Chelomey's design bureau and funded by the Soviet military as a military project, the details of its design were significantly different.*

Unlike *Salyut 1, Salyut 3* had its docking port at its aft end, using a small spherical docking module which also contained the station's airlock. This module also carried a drum-shaped, 3-foot-wide, recoverable capsule capable of returning about 260 pounds of reconnaissance film to Earth. Because *Salyut 3* did not have an aft *Soyuz* service module for maintaining its orbit, it used twin engines built directly into its main body, the two nozzles of which pointed out the aft end on either side of the docking module. Also attached like wings to this rear docking module were *Salyut 3*'s two solar panels, significantly larger than *Salyut 1*'s, and able to rotate to face the sun. Two more curved solar panels were permanently affixed to the station's "top." All told, the four panels could produce about 5 kilowatts of power (compared to *Salyut 1*'s 3.6 kilowatts and *Skylab*'s planned 10).[10]

Because of its military nature, *Salyut 3* (as well as the later *Almaz* station *Salyut 5*) were the only manned space vehicles ever launched carrying actual military weapons. A 23-mm caliber rapid-fire cannon was built into the hull of the station. The cosmonauts could aim this cannon by looking through a gun sight and then rotate the station as required.[11]

An *Almaz* station was originally designed to be launched with a manned capsule attached to its bow—called *Merkur* by later historians because it was remarkably similar to the American *Mercury* and *Gemini* capsules. Once in orbit, cosmonauts would open a hatch in their heat shield and enter the station. When the time came to return to Earth, the heat of re-entry would seal the seams in the hatch door, thereby protecting its passengers. (This exact design had been successfully tested by the U.S. Air Force's Manned Orbital Laboratory program in November 1966, using a *Gemini* capsule.) However, *Salyut 3* did not use the *Merkur* capsule. Instead the cosmonauts were launched separately on *Soyuz* spacecraft.

---

*Because of the military nature of Almaz stations like *Salyut 3*, the Soviets were reticent about revealing its design, contents, or the manned missions that were launched to them. For example, it was decades before historians knew for certain whether *Salyut 3*'s docking port was at its aft or bow end.

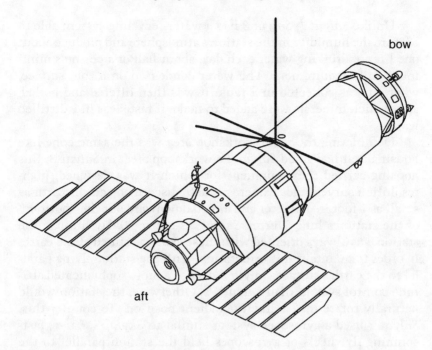

Salyut 3 with its aft docking port. *NASA*

The interior of the station was also somewhat different from *Salyut 1*, partly because its purpose was different, and partly because it had been built by a different design bureau. Unlike *Salyut 1*, the forward narrow-diameter section did not have the look of a command center with chairs and control panels, nor was there a docking-transfer compartment in the bow. Instead, this area, isolated from the large working section by a bank of twelve oxygen tanks, was the station's living quarters, with a shower, toilet, four portholes, two bunks (one of which folded against a wall to save space), a table, a small collection of books, chess set, tape recorder, and water tank.[12]

The shower, an addition from *Salyut 1*, was similar to *Skylab*'s. A cylindrical curtain, attached to the "ceiling" of the living quarters, was pulled down and fastened to the "floor." A jet sprayed water from one end of this waterproof compartment, while a pump sucked it out the other end.[13]

Unlike *Salyut 1*, *Salyut 3* had a water-recycling system able to capture the humidity in the station's atmosphere and produce about one liter of drinking water each day, about half of a person's minimum daily requirement. The water condensed on a cold surface, where it was collected in a tank. It was then filtered and boiled, after which minerals were added to make it taste less like distilled water.[14]

Dominating the main workshop area was the same cone-like housing that had filled *Salyut 1*'s workshop area. In *Salyut 3*, this housing carried the equipment for which it was designed, high-resolution surveillance cameras able to discern objects on Earth as small as 1 foot in diameter and aimed downward out the "bottom" of the station's hull. During *Salyut 3*'s entire 90-day lifespan the station was always oriented with these cameras aimed at the earth. In order to maintain this orientation, with the station flying parallel to the earth's surface, *Salyut 3* needed a more sophisticated attitude control system than *Salyut 1*'s. Otherwise, the station would naturally rotate into a gravity-gradient position. To counter this, *Salyut 3* used a gyroscope system similar to *Skylab*'s. Large, fast-spinning flywheels or gyroscopes held the station parallel to the earth, and thereafter resisted any drift from that position. If the crew needed to point a camera, they could use this flywheel to adjust the station's position by as much as 1 degree per second.[15]

Despite these improvements and the knowledge gained from both *Salyut 1* and *Skylab*, Chelomey's engineers still included many features that were pointless in the weightless environment of space. For example, the two bunks were unnecessary. With no gravity, there was no need to provide a platform for a person to rest on. A small compartment like those used on *Skylab* with a simple, lightweight sleeping bag worked better and saved weight. Furthermore, psychologists had the station's interior decorated so that it had a distinct floor and ceiling. The floor was made of a Velcro carpet (allowing the men to walk if they wished) and the walls were painted a different color than the ceiling.[16] Yet previous space explorers could have told the psychologists that this was all unnecessary: Your orientation was determined solely by how you positioned your head.

Of the two missions launched to *Salyut 3*, only the first was able to successfully dock with the station. Launched eight days after *Salyut 3*, *Soyuz 14*'s crew was two military men. The commander was veteran cosmonaut Pavel Popovich, the eighth man to fly in space, now returning to space after a lapse of almost 12 years. His crewmate was flight engineer Yuri Artyukhin, on his first and only space mission. Because of the mission's military nature, the two men spent most of their time taking photographs of the earth. To the public it was announced that they were studying the resources of the Soviet Union, including ore deposits, water pollution, crop resources, and changes in ocean ice sheets. Though they certainly did this research, the focus of their observations was military surveillance. Working an 8-hour day, they followed a staggered schedule so that the cameras were manned for 16 hours out of every 24.

Still, some medical research was done. The men took electrocardiograms as well as blood and respiratory samples of each other. They also tested their lung and heart capacity as well as the blood circulation to their brains. And both men used an improved version of the Penguin suits first tested during *Salyut 1*.

Though their 16-day mission was significantly shorter than any of the *Skylab* flights, both Popovich and Artyukhin experienced some of the same discomforts felt by the *Skylab* astronauts immediately after their return to Earth. Their blood pressure and pulse were low, and though immediately after landing they were able to walk without assistance, it took them four to six hours before they could do this in a normal manner, and several days for their bodies to return entirely to normal.

The second mission to *Salyut 3*, *Soyuz 15*, had been planned as a three-week-plus mission. Launched on August 26, 1974, the cosmonauts, rookies Gennady Sarafanov and Lev Demin, were unable to dock with the space laboratory when their Igla radar-docking system failed. For their first two attempts using this automatic docking system, their speed was too fast, and they had to swerve aside to avoid a violent collision. On the third attempt they tried to dock manually, but used too much fuel, forcing them to abort once again. Because the *Soyuz* ferry depended on the station for power— it had no solar panels and carried batteries with only two days' power reserve—they no longer had the energy or fuel for further docking tries. For safety they had to scrub their mission, returning to Earth after only two days in orbit.

Because *Salyut 3* had a life expectancy of only about six months, there was no time for a third flight. On September 23,

1974, the station's recoverable module returned to Earth. The laboratory was de-orbited four months later, burning up in the atmosphere.[17]

The weak showing from *Salyut 3*, only 16 days in space compared to almost a half-year by American astronauts on *Skylab*, illustrated once again how much the Soviet space program was struggling to find its way. As a military surveillance satellite, *Salyut 3* produced little for its cost. For the many millions of rubles spent to launch one station and two manned capsules, *Salyut 3* was able to give the military only 16 days of reconnaissance. Launching a cheap unmanned *Cosmos* reconnaissance satellite when needed seemed much more practical.

More significantly, the secretive military nature of the mission contrasted starkly with Brezhnev's public stance of "peaceful co-existence" and detente that he had so aggressively touted at the Moscow summit and at numerous other political events throughout the world for the last four years. Newspapers gave the station little coverage, and when they did, it was accompanied by confusion and contradictory comments from Soviet officials, who at first hinted and then denied that the missions involved training for the joint *Apollo-Soyuz* mission scheduled in two years.[18] Thus, *Salyut 3* was in general a public-relations disaster.

To get Brezhnev what he wanted required a successful civilian space station. And after four years of redesign and two failures, Energia, the former Korolev design bureau, was ready to try again.

Summer of Triumph

Georgi Grechko looked like a teddy bear of a man. Round-faced, with an easy grin, he had spent his entire life working in the Soviet space program. In 1956, he had done the first preliminary calculations for launching an artificial satellite, calculations eventually used by *Sputnik*. Later, in 1966, he belonged to the first class of civilian Soviet cosmonauts. For the next nine years he was assigned to a variety of projects, none of which flew. In 1967, he trained for a lunar orbital mission that never flew. In 1969, he trained for a lunar-landing mission that never flew. In 1971, 1972, and 1973, he trained for various *Salyut* missions that never flew.[19] Finally, on January 11, 1975, after almost 10 years of waiting, Georgi Grechko finally flew into space, one of two men to first occupy *Salyut 4*, the Soviet Union's first truly successful space station.

On launch day Grechko was 44 years old. Born and raised in Leningrad, he, like so many other cosmonauts, had direct experience with the violence of war during the Nazi invasion of Russia. In early June 1941, when Grechko was 10, his parents sent him to spend the summer with his grandmother in the Ukrainian city of Chernigov, several hundred miles north of Kiev. Two weeks later, the German invasion of the Soviet Union began. His mother managed a bread factory, and, therefore, lived under military rules. His father immediately volunteered for the army, and fought at the front. Neither could come and get the boy. By September, the Germans had driven through Chernigov in their march to Kiev. By December the Ukraine was completely conquered.

For the next two years the boy lived in occupied territory. He and his friends played amid bombed wreckage, picking up abandoned weapons and firing them at German planes as they flew overhead. Several boys were killed by these "toys." Once, an artillery shell went off within a few feet of Grechko, throwing the boy back against a water barrel, a 6-inch piece of shrapnel puncturing the wood only inches from his head. Another time, the boys snuck up to the German barracks and fired at them. To their abject terror several soldiers came out, mounted motorcycles, and began searching for the snipers. "We were so frightened, I think we ran faster than the motorcycles," Grechko remembers. [20]

After the war Grechko went back to school, dreaming of becoming an engineer, a calling that the boy had chosen years earlier when he was only eight and a man who taught students how to draw blueprints came to visit his parents. During dinner he showed the child how an engineer drew a nut and bolt. "I was surprised there were such strict rules," Grechko recalled. "Each line had to have a precise angle, a precise thickness. Everything had to be to a precise scale." The boy was so impressed that there and then he decided to become an engineer.

After the war he read science fiction books about traveling to other worlds, and engineering books about the German V2 rockets. From what he read, it seemed that it might be a hundred years before the first person went into space. "I didn't think I could live so long," Grechko thought. "I decided instead to make rockets, large rockets. I would spin those nuts on rockets." Working hard at school, he enrolled in what was then called the Leningrad Institute of Mechanics. Graduating with high marks, Grechko was hired by Sergei Korolev, who was building the R7 rocket that would place *Sputnik* in orbit. Grechko's first job was to calculate the rocket's trajectory in sub-orbital flights across the Soviet Union. When Korolev lobbied in 1966 to add engineers to the cosmonaut ranks, Grechko was first in line.

His partner and commander on *Salyut 4* was air force pilot Alexei Gubarev. Unlike Grechko, Gubarev had become a cosmonaut somewhat by accident, as did many of the early military cosmonauts. A good-natured man who, like Grechko, was quick to smile, Gubarev loved aviation, and learned to fly at a young age. Joining the military immediately after high school, he later switched military academies in order to get the pilot training he craved. During the Korean War he flew combat-support missions for the Chinese and North Koreans. Later he was stationed on the Black Sea as a squadron leader. When, like Dobrovolsky, he was selected in 1963 to take the cosmonaut medical tests, Gubarev did so eagerly. He hadn't planned on flying in space, but when the opportunity was offered, he was quite happy to grab it.

By 1974, the Soviet space program had settled on a compromise about who would fly each space mission. For military flights like *Salyut 3*, both the pilot and the flight engineer had to be military men. For civilian flights such as *Salyut 4*, the pilot would come from the military, and command the mission. The flight engineer would be a civilian, and be in charge of experiments and station maintenance.

For Grechko and Gubarev, their military-civilian partnership was a pleasure and a joy. Both men worked hard and with dedication. Both had a good sense of humor. And both were completely committed to setting a record in space and getting back to Earth safely.

The space station that they were to occupy, *Salyut 4*, was actually a civilian follow-up to *Salyut 1*, not the about-to-be-deorbited military *Salyut 3*. Launched the day after Christmas 1974, *Salyut 4* was also the Soviet Union's first completely successful space station, giving the Russians their first hint of how they might finally triumph in space.

Like the previous Soviet stations, *Salyut 4*'s main hull was based on the *Almaz* design, weighing more than 20 tons and having

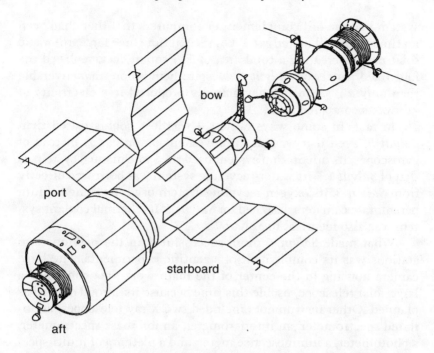

Salyut 4. Note how the approaching *Soyuz* spacecraft no longer has solar panels, a change initiated after *Salyut 1*. Note also how the aft service section is still nothing more than a *Soyuz* service module. *NASA*

about 3,200 cubic feet of habitable space and one docking port. However, as a civilian project built by Energia and dedicated entirely to scientific research, it had more in common with *Salyut 1* than *Salyut 3*. Like *Salyut 1*, its docking port was in its bow, at the end of a similar docking-transfer compartment with airlock. Like *Salyut 1*, the narrow-diameter section of the main workshop contained a command center with computers, chairs, table, and galley. Like *Salyut 1*, a small *Soyuz* service module was attached at the aft end, containing fuel tanks and main engines. Like *Salyut 1*, the large-diameter section of the workshop was dedicated to scientific experiments. For this mission, the large camera housing contained a 10-inch solar telescope.

    Salyut 4's innovations included a teletype machine like *Skylab*'s, allowing ground controllers to relay information and instructions without the need for oral communication. Its guidance system had the full complement of computers that there had been no time to install on *Salyut 1*. On its exterior three large, rotatable solar panels, creating a total area of 215 square feet, replaced the four fixed panels on *Salyut 1*. Despite being more maneuverable than *Salyut 3*'s panels, they actually produced less electricity, 4 kilowatts compared to 5.[21]

In fact, in some ways *Salyut 4* was less sophisticated than *Salyut 3*, even if it was an improvement on *Salyut 1*. Instead of gyroscopes, its orientation system used the same manual system as that of *Salyut 1*.[22] Its water-recycling system was borrowed directly from *Salyut 3*. Its oxygen-recycling system used the same lithium perchlorate canisters designed for *Salyut 1*. Its thermal coolant system was also identical to *Salyut 1*'s.

What made *Salyut 4* more successful than the previous two stations was its complement of scientific experiments. The large camera housing in the center of the main work area contained a large solar telescope, usable this time because its shroud opened as planned. Other instruments included two X-ray telescopes, an infrared spectrometer, an interferometer, an ion-mass spectrometer, a photometer, a luminescence meter, and a plethora of multi-spectral cameras. Twice, the cosmonauts used *Salyut 4*'s solar telescope to observe the sun, obtaining spectrographs and optical photographs of sunspots and faculae.

Salyut 4 also had a wide range of medical instruments, included a blood analyzer, a bone-tissue density monitor, a pulmonary ventilation recorder, brain-blood-level monitor, and an electrical muscle stimulator. As part of their routine, the men dedicated two-and-a-half hours each day to physical exercise, split between a bicycle and a treadmill. Both machines were more thoughtfully designed than the treadmill on *Salyut 1*, and worked well, though boringly. Instead of holding onto cords, they put on a harness that clipped them down to the gear, and held them in place.

The men also used a new version of the "Chibis" leggings, first tried on *Salyut 1*, to force blood into the body's lower half in order to simulate the earth's environment. Two new Penguin suit designs for forcing the men to use their muscles were also tested. Samples of blood and exhaled air were repeatedly taken, and the cosmonauts' bone density was measured regularly.

Like the *Skylab* astronauts, both Gubarev and Grechko took about a week to completely adapt to the weightless environment, though Gubarev seemed to suffer more than his partner. Both men complained of colds, though the stuffy noses might have instead been caused by the redistribution of blood from the legs to the head and upper body, as experienced by the men on both *Salyut 1* and *Skylab*.

Their adjustment problems might also have been worsened by their refusal to follow the recommended sleep schedule. Both men worked too hard, averaging between 15 and 20 hours of work per day. At one point Grechko took even more time out of his sleep schedule to repair the solar telescope, fixing its orientation system and recoating its mirror with aluminum.[23]

For Grechko, overwork became the routine. As flight engineer he was in charge of all the experiments, which made his workload somewhat larger than his partner's. And since he liked what he did, he did it passionately and without pause. Many years later he described how, to save time, he would stuff a chocolate bar in his pocket so he wouldn't have to stop working to eat lunch. Then, at bedtime he would discover the candy bar still in his pocket, uneaten.[24]

Grechko's work program not only involved extensive medical and astronomical research, but also responsibility for a variety of biological specimens. For the first time since *Salyut 1*, plant life had returned to space. *Salyut 4* contained an updated version of the same Oasis greenhouse experiment that Viktor Patsayev had so lovingly cared for three years earlier. Energia's biology team had redesigned the watering system so that the water doses could be administered automatically and precisely. Moreover, the seeds were preplanted on Earth in removable cartridges. For this mission, Grechko would try to grow green peas. The next mission to *Salyut 4* would try to grow both peas and onions. Grechko would also attempt to grow single-celled amoebas, which, like the plants, could not only provide food but also recycle human waste products. In addition, the station carried the larvae of 95 *drosophila* flies and the eggs of both guppies and tadpoles.[25]

On their first day aboard *Salyut 4*, Grechko activated the Oasis greenhouse. He quickly discovered that the new automatic watering system was inadequate. "First the water didn't go in it, then it went the wrong way," Grechko wrote after the flight. Then the plants got too much water, and mold developed. "So I disconnected the automatic system, installed more seeds, and watered them by hand."[26]

Grechko needed answers to a lot of questions if he was to get the system to work. Asking the capcom, who was usually either another cosmonaut or an engineer, was generally useless. They would have to find Galina Nechitailo—whose job it was to train and supervise the cosmonauts in their use of the experiment—ask her, and then come back with the answer. Grechko would then have more questions, requiring more searches and more delays. Finally Grechko began to beg them to let Nechitailo speak to him directly. "Give me Galya," he said, over and over. "Let me speak to her." After getting approval from the psychologists, who decided the contact would be a good thing, Galina Nechitailo was given air time with Grechko. In fact, from this point on, she had permission to talk directly to every cosmonaut on every subsequent space station, through *Mir*.[27]

One week after Grechko had rebuilt the greenhouse, 10 pea shoots had sprouted. By the mission's third week, however, these shoots had withered, the leaves had brown stains, and the roots were failing to grow. Nor were the plants the only biological specimens to suffer. Later generations of amoebas were smaller and less active, as if something in weightlessness was stunting their growth. To the two men the guppies and tadpoles seemed normal, swimming about in their tiny aquarium. Once back on Earth, however, the biologists discovered that the fish had not developed the air sac that fish use for buoyancy. "They just crawled like insects," Nechitailo recalled. And while the tadpoles had fared better, studies of their balancing organs showed deformities. Both problems were linked by the scientists to some disturbance in calcium production while in space. The *drosophila* flies fared far worse. While most seemed to pass normally from the larval stage to the cocoon stage, only 17 of the 95 flies emerged from their cocoons, and 10 of these died almost immediately. All showed some mutations and abnormalities.[28]

Finally it came time to go home. On February 9, 1975, after 30 days in orbit—almost a week longer than the record set by the ill-fated *Salyut 1* crew—Gubarev and Grechko mothballed the station, climbed into their spacesuits, and strapped themselves into their *Soyuz* spacecraft. Gubarev undocked it, steered it clear of *Salyut 4*, and fired its engines, sending their descent capsule on a perfect trajectory toward Earth.

At six miles elevation, the system was designed to automatically release its parachutes. Grechko watched the clock anxiously for this event. To his horror, nothing happened.

For the Soviets, the return from space was the most nerve-racking moment of any space mission. While Americans remain paranoid about fire since the deaths of Grissom, White, and Chaffee on the *Apollo 1* launchpad in 1967, the only Russian space deaths have occurred on the return to Earth. First there had been Vladimir Komarov, killed when the parachutes of his *Soyuz 1* craft did not open properly and he hit the ground at more than 400 miles per hour. Then there were the deaths of Dobrovolsky, Volkov, and Patsayev on *Salyut 1*. By 1975, Soviets engineers had done everything they could think of to make sure nothing went wrong during descent. Even so, Gubarev and Grechko had donned their spacesuits, just in case.

For Grechko, the moment when he thought the parachutes had failed remains seared in his memory. "It was absolutely terrifying," he remembered with a smile decades later. "Sweating, panting like a dog. The fear was so strong you could go crazy. Fear blocked everything, thoughts, movement. It was paralyzing. I realized I had no more than five minutes of life left."

Rather than panic, Grechko steeled his mind. He was a spaceman, it was his duty to face these situations squarely and coolly and figure out what was wrong. "I decided not to shriek 'Mama'!" He began to check the systems, one by one. "Time passes very slowly in these kinds of situations. Maybe it was a minute, maybe a few seconds."[29]

And then, the drogue chute popped out, followed by the main chute. Everything had operated as designed, and, to Grechko's immeasurable relief, the descent module floated to Earth exactly as planned. The Soviets had finally succeeded in completing a record-setting space-station mission, albeit only a Soviet record.

Once under the care of doctors, both men fared much better than any previous Soviet cosmonauts after a long space flight, and more in line with the experience of the nine American *Skylab* astronauts. Each man had lost a few pounds, but recovered quickly, and had no difficulty walking immediately after landing. However, the medical data showed, as with the *Skylab* astronauts, that their bone density had declined; the calf bones of each had shrunk as much as an inch in circumference.[30] Just as had been found on

Skylab, lack of gravity seemed to cause a thinning of the human skeleton.

Because Grechko's and Gubarev's mission goal was scientific rather than the military surveillance operation of *Salyut 3*, the flight received significant worldwide attention. Articles about its scientific findings were published in respected Western publications like *Nature*, and Western press coverage was far greater than for *Salyut 3*, including front-page stories and positive editorials.[31] Despite the flight's success, however, no Soviet space station had yet been occupied by more than one crew. To catch up to the United States a second mission *had* to dock with *Salyut 4* and occupy it.

Easier said than done. The next mission to *Salyut 4* was probably one of the most frightful spaceflights ever. Intended as a 60-day mission timed to end before the launch of the joint *Apollo-Soyuz* mission, *Soyuz 18* instead became the world's longest suborbital flight, and the first manned mission to make an emergency landing during launch.

Liftoff was on April 5, 1975, less than two months after the return of Gubarev and Grechko. The crew, Oleg Makarov and Vasili Lazarev, had sailed together in space once before, flying for two days on *Soyuz 12* in September 1973, testing the redesigned *Soyuz* spacecraft following the deaths of Dobrovolsky, Volkov, and Patsayev. Four-and-a-half minutes after liftoff, just before their rocket's first stage was to stop firing, engine vibrations caused three of the six explosive bolts holding the first and second stages together to fire prematurely. With the first-stage hanging halfway free, the second-stage rockets kicked in, throwing the entire rocket off balance and out of control.

By this time in the launch sequence the rocket was already more than 110 miles high moving at more than three miles per second. The emergency escape tower on top of the rocket had long since been jettisoned. To save the crew, ground controllers activated the next planned abort sequence, first separating the *Soyuz* capsule from the rocket, then separating the descent module that carried the men so that it could make a ballistic re-entry.

For eight short minutes—with their descent capsule still arching upward into the sky at more than 5,500 miles per hour—the two men were weightless. Then, at about 100 miles altitude, they began to fall, faster and faster. "It was all very sudden," Makarov remembered years later. "Everything was a big surprise." Plummeting Earthward, they were pressed down in their couches, experiencing 10 g's for one minute and 20 g's for a few seconds. "Above 10 g's you can't breathe," Makarov described. All they could do was hang on and wait.

Their parachutes released, as scheduled, at six miles, but they were dropping far faster than planned, amidst high mountains and forests. As they came in, the wind blew them sideways against a slope, their capsule hit the snow hard, then rolled and bounced downhill for several seconds before the parachutes became tangled in trees and stopped their fall. Gingerly the two men climbed from the capsule, falling into snow about chest high on a 45-degree slope. To their alarm they discovered they had come to a stop only a short distance from the edge of a cliff.

Their landing site was in the western Siberia, in the Altai Mountains close to the Mongolian border, territory controlled by the hostile Chinese. The weather was below freezing. Their spacecraft had no heating. As rescue crews, seriously hampered because of the landing site's remote mountainside location, tried to reach them, the two men built a fire and waited. As they gathered wood, they were further startled to find that every tree around them was rotten and brittle. Even trunks a foot across could be snapped apart with their bare hands. How these fragile trees had kept their capsule from rolling off the cliff was a mystery.

About an hour later local villagers arrived with torches. By the next day the two men were on their way home. Lazarev never flew in space again, and both men had to appeal directly to Leonid Brezhnev when the Soviet bureaucracy decided to deny them their 3,000 ruble bonus for flying in space. The bureaucracy said their mission hadn't entered space. Brezhnev decided that it had.[32]

The Soviets' manned program marched on. Almost immediately the program began preparing another *Soyuz* mission to *Salyut* 4. They had still not succeeded in flying a second crew to a space station. If they hurried, they could complete a second long-term mission within *Salyut 4*'s expected operational lifespan.

Moreover, another manned launch was necessary to ease American concerns caused by the aborted mission. During the preparations for the joint *Apollo-Soyuz* mission scheduled for July, the Soviets had been forced to reveal more about the failure than they were previously wont to do, including a full briefing to NASA engineers in which they described what had caused the three latches to blow prematurely, and how they had been rewired to

prevent a repeat of the problem.[33] Challenged by American doubts
about the reliability of their equipment, the Soviets had to send up
another crew, if only to prove that the repairs to their rocket had
been fixed before the *Apollo-Soyuz* mission began.*

But most important of all, the Soviet space program was under
intense political pressure from Brezhnev to get *Salyut 4* manned
during the summer months of 1975. For Brezhnev, 1975 was a year
of triumph and success. More than any other time in the short, sad
history of the Soviet Union, these few months resonated with victory and unbridled possibilities for the communist movement.
Only the year before, the United States had been badly rocked by the
scandal of Watergate and the resignation of President Nixon. Then,
on April 30, 1975, mere weeks after the *Soyuz 18* launch abort,
Saigon fell to the North Vietnamese army and became Ho Chi Minh
City. After more than a generation of resistance—aided by the full
force of the United States military—the small southeastern part of
Vietnam had finally succumbed to communist rule. Meanwhile, in
Helsinki, Finland, after three years of difficult negotiations, Brezhnev
was about to score what he considered his biggest foreign policy victory with the signing of the Helsinki Accords.

Since the mid-1950s the Soviet leadership had repeatedly tried
to get the West to accept borders established in eastern Europe after the war, thereby giving official acceptance of Soviet control over
Poland, Hungary, Bulgaria, Romania, Czechoslovakia, and most
importantly, East Germany and East Berlin. Before detente, Soviet
"diplomacy" had been harsh and brutal, using either intimidation
or military force to try and get the United States and the western
European nations to agree to a peace settlement. In 1948, Stalin
had blockaded Berlin. In 1956, Khrushchev sent troops to both Poland and Hungary. In 1961, he had built the Berlin Wall. In 1968,
Brezhnev invaded Czechoslovakia. None of these actions had
worked.

After 1968, Brezhnev tried detente instead of intimidation. By
the end of 1972, this new policy had finally brought the United

---

*Unfortunately, the Soviet habit of hiding failure did not go away so easily,
and in the end helped feed many American doubts. Rather than give the new
mission to *Salyut 4* its own number, the Soviets instead designated it *Soyuz
18*, renaming the aborted mission of Makarov and Lazarev "the April 5th
anomaly." For clarity, historians have since renamed the two missions *Soyuz
18-1* and *Soyuz 18-2*.

States and Europe to the table to negotiate a comprehensive treaty on European security. Even as Makarov and Lazarev gathered wood in the freezing cold of the Altai Mountains, the treaty negotiations in Helsinki were racing toward completion. The signing ceremonies, which Brezhnev, President Gerald Ford, and all the major leaders of Europe would attend, were tentatively scheduled for the end of July. If a second long mission could arrive at *Salyut 4* in May, it would be in space at the same time as the *Apollo-Soyuz* mission. For Brezhnev, the propaganda value of having *Salyut 4* occupied while the joint *Apollo-Soyuz* mission was going on was immeasurable. The U.S.-Soviet mission would allow him to highlight his policy of detente, while the *Salyut* mission would be apparent proof of his country's dedication to fulfilling the promise of his October 1969 space-station speech.

Eager to have the signing ceremonies coincide with both space missions, Brezhnev put personal pressure on his negotiators in Helsinki. In early March 1975, only weeks before the aborted *Soyuz* flight, he sent a letter to every political leader involved and proposed that the negotiations be completed by the end of June.[34] At the same time, he made sure the Soviet space program had whatever resources it needed to fly both missions, including two fully functional and separate mission control centers.

Manned by two veterans, Pyotr Klimuk and Vitali Sevastyanov, *Soyuz 18-2* was launched on May 24, 1975, only seven weeks after the aborted mission, and only seven weeks before *Apollo-Soyuz*. Scheduled to last more than 60 days, *Soyuz 18-2* would set a new Soviet space endurance record, second only to the last American *Skylab* mission. One day later, Klimuk and Sevastyanov docked their spacecraft with *Salyut 4*. As they opened the hatch and glided into the station, they were pleasantly greeted by a homemade sign left behind by Gubarev and Grechko: "Welcome to our common home!"[35] After years of trying, the Soviets had finally docked a second crew to a single space station. Now the question was: could the second crew remain in space for a significant time period?

Soyuz 18-2's commander, Pyotr Klimuk, 32, was one of the youngest cosmonauts and, in fact, had always been an early starter. He was accepted into the Communist Party at 20, unusually young. He was selected as a cosmonaut at 23, also very young. His first space flight, *Soyuz 13*, a 7-day mission to test the redesigned *Soyuz* spacecraft, occurred when he was only 31. This lightning-quick ability to get a lot done in a short time and ahead of everyone else earned him the nickname among cosmonauts of "Klimuki," or "many Klimuks."[36]

Sevastyanov, older at 39, was an engineer and designer. When the Soviet space program decided to train its spacecraft designers as cosmonauts, Sevastyanov, like Grechko, was one of the first in line, spending several years in the late 1960s training in the Soviet lunar program that never got off the ground. He flew his first mission, on *Soyuz 9*, in 1970, the first long term Soviet space flight, lasting 18 days.[37]

Picking up where the crew of *Soyuz 17* had left off, Klimuk and Sevastyanov began an extensive program of scientific research. For their first two days on board, however, they worked to get *Salyut 4* reactivated and shipshape after being in orbit for six months, five of which had been unmanned. As Sevastyanov reported, "It was necessary to replace and repair some equipment."[38]

The next day was supposed to be a day off. Though required only to do two hours of exercise, Sevastyanov was eager instead to start his scientific work. Ground controllers told him to relax; he kept insisting that they get started. Finally, the capcom suggested that, as a reminder to calm down, Sevastyanov should put up a sign telling him to "Beware of Weightlessness!"

Once they did begin their research, they did it differently than the previous mission. The work schedule for Grechko and Gubarev had included a variety of experiments each day, often requiring them to repeatedly re-orient *Salyut 4*, depending on what they were doing. This arrangement was wasteful of both time and fuel. For *Soyuz 18-2*, the schedule was more sequential. Klimuk and Sevastyanov spent several days working on similar experiments, then moved on to a different set of similar research. For example, at one point during the flight the cosmonauts performed 10 straight days of Earth observations, taking 2,000 photographs with the laboratory's optical and multi-spectral cameras. Their instruments covered 3.3 million square miles of the Soviet Union, measuring pollution levels in the atmosphere and detecting ore deposits and crop levels. Later, they performed almost two weeks of astronomical research, taking more than 600 spectrographs and photographs of the sun and using the station's X-ray telescope to study several suspected black holes in the constellations Scorpio, Virgo, Cygnus, and Lyra.[39]

Not all their research followed this scheduling. Their plant experiments required continuous attention as they attempted to grow both peas and onions in their Oasis greenhouse. Once again, they had problems with the watering system, causing the first set of seeds to quickly wither and die. Of the second set, watered by hand, none of the peas grew normally, and most died after a few weeks. The onions, meanwhile, managed to sprout to a height of 6 to 10 inches, but they did so more slowly than Earth plants, and were far smaller. Something, whether lack of gravity or the equipment in which they were housed, was preventing the plants from prospering.[40]

Yet even these stunted plants provided the men with some joy. July 8 was Sevastyanov's 40th birthday. Two days later Klimuk was to be 33. To celebrate, Nechitailo gave the men permission to pull several onion sprouts from Oasis. They then spiced their birthday meal with these space-grown scallions, the first space-grown vegetables ever eaten by humans.[41]

Several other experiments, started by Grechko and Gubarev, were reactivated, including the fruit-fly studies. This time more than 700 larvae, descendants of the few living flies brought back by Grechko and Gubarev, were brought into space. Unfortunately, most showed abnormalities as they changed to pupae in the cocoons. Only 64 came out of their cocoons, one of which Sevastyanov nicknamed Nyurka. Shortly thereafter all but a handful of the flies were dead. The few that remained showed significant mutations when brought back to Earth. No one knew whether these aberrations were caused by zero gravity.[42]

And both men continued the same medical studies performed by other spacemen on earlier space flights. They did regular blood and respiratory tests, as well as electrocardiograms. They exercised a minimum of two hours a day on the treadmill and bicycle machine. They used the "Penguin" suit to exercise their muscles and the "Chibis" suits to draw blood into the lower parts of their bodies. Then, in the last 10 days of the flight, their daily workout was increased significantly.

As their mission wound down, however, the condition of *Salyut 4* began to deteriorate. The portholes had become permanently fogged over (damage caused somehow by a long-term interaction with the upper atmosphere), condensation was everywhere, and a green mold coated much of the station's interior. The water-recycling and the atmosphere filtering systems needed redesign: As built, they simply were not efficient enough.[43] The situation had become unpleasant enough that both men requested terminating their flight just before the launch of *Apollo-Soyuz*. They were politely told to hang in there for just a little longer. Very obviously the Soviet political leadership wanted them in orbit when the joint American-Soviet mission was flown.

On July 15 Alexei Leonov and Valery Kubasov climbed into the *Soyuz* capsule and lifted off from the Baikonur Cosmodrome.* Unlike every previous Soviet space mission, this launch was televised live around the world; NASA officials had insisted that the mission be given full coverage. Seven-and-a-half hours later, Tom Stafford, Vance Brand, and Deke Slayton, in the last *Apollo* capsule ever launched by the United States, followed them into orbit. The *Apollo-Soyuz* Test Project had finally begun.

Because of orbital mechanics, the actual docking could not occur until July 17. While waiting for the arrival of the Americans, the cosmonauts did an orbital correction, adjusted the air pressure in their capsule to match that of the docking module being brought into space by *Apollo*, turned on some miscellaneous experiments, and then, on *Soyuz*'s 20th orbit, the two Soviet control centers set up a radio link between it and Klimuk and Sevastyanov on *Salyut 4*. After hearing Leonov's voice, Klimuk responded, "It is pleasant, it is very pleasant for us to hear a few words."

For about five minutes the two crews chatted and joked. At one point Klimuk and Sevastyanov described how they had listened to the launch of *Soyuz 19*, and were quite willing to do repairs for them if needed. At another, Sevastyanov noted that ". . . there are seven people in space right now," calling them ". . . the magnificent seven."

Leonov asked, "How are things out there?"

Klimuk responded, "Everything is fine. Everything is normal." Almost as if he was trying to convince himself he added, "Our

---

*Though the actual location of the cosmodrome was near the village of Tyuratum, not Baikonur (located almost two hundred miles away), for decades the Soviets tried to hide this fact by falsely placing it near Baikonur on maps and in press releases. Even though the deception was soon uncovered, the usage became so accepted that the cosmodrome's name today is officially the Baikonur Cosmodrome, the name by which it became famous.

station is working perfectly, and we have become so accustomed [to it] that it's just like home. First we flew a whole month, and now our second month is ending, so it's just like at home."

Twelve hours later, the gold-and-white cone and cylinder of the *Apollo* spacecraft approached. On its nose was attached the ungainly boxlike docking module. Tom Stafford eased the two spacecraft together, the cabin pressures were equalized, and the hatches opened for the first international handshake in space. Almost immediately, Brezhnev was sending his personal greetings to the crews, calling it "a new page in the history of space exploration."

For two days *Apollo* and *Soyuz* stayed docked together, performing a variety of scientific experiments in between the almost innumerable public-relations telecasts. Then, on July 20, they separated. After another very short conversation with the *Salyut 4* crew—consisting almost entirely of garbled "Do you read us?" and "How do you read?" comments—*Soyuz 19* returned to Earth. Knowing that there would not be another American space mission until the space shuttle began operations at least three years hence, the *Apollo* spacecraft remained in orbit for another five days, performing a variety of scientific experiments. It returned on July 24.

Last to come home were Pyotr Klimuk and Vitali Sevastyanov, landing two days later. Despite 63 days in space, both men walked from their capsule, refusing to be carried. They had lost only a few pounds, and within two days had regained this weight and were apparently back to normal, playing tennis, running, and swimming.

Their biological samples were immediately handed over to the scientists. Nechitailo took the peas and onion plants back to Moscow so that the biology team could probe, measure, and dissect them, trying to figure out what caused their decline. At the same time, she and her partners did another redesign of Oasis, while also coming up with several completely new greenhouse systems.

Soyuz 18-2 was the last manned mission to *Salyut 4* before it was finally deorbited a year and a half later. However, another *Soyuz* spacecraft, dubbed *Soyuz 20*, arrived in November 1975 and remained docked to *Salyut 4* for 90 days, testing the *Soyuz* spacecraft's ability to complete an automatic rendezvous and docking as well as function in space for long periods. *Soyuz 20*'s "crew" included flies, tortoises, vegetable seeds, maize, cacti, and other plants, and their successful recovery in February proved that longer missions with this craft were now possible.

Not surprisingly, the 63-day mission of Klimuk and Sevastyanov was significantly overshadowed in the West by the joint *Apollo-Soyuz* flight. Americans were not interested in reading about a Soviet success in space. Moreover, detente and peaceful cooperation, as demonstrated by *Apollo-Soyuz*, was considered much more agreeable to read about than missions like *Salyut 4* and *Soyuz 18-2* that illustrated the competitive nature of space exploration.

The Soviet Union, on the other hand, gave the mission wide coverage. Forced to open their doors for the joint mission, the Soviets decided to widen coverage of their space-station mission as well. The Soviet press gave copious daily reports, as it had during *Salyut 1*, of the cosmonauts' experience (conveniently leaving out details about any technical problems, such as the failing water-recycling system).

For Brezhnev, the triumph of this double space mission was then topped by the signing of the Helsinki Accords. Five days after the return of Klimuk and Sevastyanov, he stood at the podium in Helsinki, with the leaders of the North Atlantic Treaty Organization watching, and proudly declared the Accords as "the prize of all people who cherish peace and security on our planet," adding, "Uppermost in our mind is the task of ending the arms race and achieving tangible results in disarmament.[44]"

For Leonid Brezhnev, this moment was clearly the apex of his rule. For one shining moment, the Soviet Union and the communist movement seemed to dominate the future of human history.

Unintended Consequences

Eleven months later, the Soviet military got its second try at orbiting a manned space station, launching *Salyut 5*, the last military *Salyut* mission. Practically identical to *Salyut 3*, its military nature kept its design and the work performed within it secret for decades. For 411 days, *Salyut 5* orbited the earth. Three different crews were sent to occupy it, with the first and third staying 50 and 18 days respectively. The second crew had docking troubles, and had to return to Earth after only 2 days in space. Each mission, however, had its share of adventure and surprises.

The first mission was manned by space veteran Boris Volynov and rookie Vitaly Zholobov. Planned as a 9-week mission, the two men devoted almost all their time to military reconnaissance. They also observed a massive Soviet military maneuver staged in the eastern part of the Soviet Union. As with *Salyut 3*, press coverage of the mission was minimal. *Pravda* often did not mention the mission for days, and when it did, the reports offered relatively little detail.

The cosmonauts did perform some scientific work, though much less than was done on *Salyut 4*, and much of what they did appeared to be nothing more than window-dressing to hide the military priorities of their mission. The most interesting experiment involved an aquarium holding two guppies, one pregnant. Each day the men watched their weightless swimming antics. At first the fish were badly disoriented like the minnows on *Skylab*. After several days, however, they seemed to adapt. Instead of using their internal balancing organs, the fish used their eyesight, like humans, orienting to an air bubble on one wall of their tank as if it was the surface of the water.[45]

Though there was no greenhouse on this flight, they attempted one plant experiment with hawksbeard seeds to see if weightlessness would cause mutations. Though the seeds showed slightly fewer mutations than the control seeds on Earth, scientists were unsure what this meant.[46] The most significant engineering experiment involved tests of several fuel pump designs; if a working pump could be built, future space stations could be refueled in space, thereby extending their orbital life. And in addition to the standard exercises and monitoring of blood, heart, brain, and lung, a new mass-meter or "scale," similar to that used on *Skylab*, allowed the cosmonauts to weigh themselves.

Originally planned to last 66 days,[47] the flight was cut short after 50 days when the two men began having serious psychological problems, both with each other and with the ground. Volynov, the commander, was a hard man who drove both himself and Zholobov, his flight engineer, harder than necessary. In order to maximize their reconnaissance of the ground, the men kept to a staggered schedule, working 16 hours a day. Overworked and under stress because they were doing extra reconnaissance work in an effort to compensate for the loss of two *Cosmos* spy satellites, the men became ill and exhausted. They began to argue with each other and the ground. And they both became increasingly unstable.

Volynov reported that there were problems in the system that recycled the station's atmosphere. In the last few weeks he repeatedly described how an "acrid odor" was becoming worse. By mid-

August he thought it so bad that it made him sick. Zholobov, meanwhile, experienced a growing distress at being cramped within a
single metal room surrounded by the endless blackness of space. In
later years he described his fears. "When I saw [the distant stars] I
realized that space is a bottomless abyss. . . . And that's not the end
of our world. One can travel further and further and there is no
limit to that journey. I was so shocked that I felt something crawling up my spine."[48]

Because the interior of a *Salyut* station was so small, without
private cabins or a series of large isolated sections like *Skylab*, there
was no place either man could go to avoid the other. By the third
week of August and the seventh week in space, the situation finally became untenable. With less than a day's notice, the men
were ordered to pack up what they could and come home. On August 25, 1976, they made an emergency night landing in the middle
of a farm field in northern Kazakhstan. Despite being underweight
and weak (the early return, their unstable mental condition, and
the focus on reconnaissance work had all interfered with their exercise schedule), both men recovered quickly. Within a few days
they were exercising and taking extended walks.[49]

The second mission to *Salyut 5* was another Soviet docking
failure, the third in five years. Launched seven weeks after Volynov
and Zholobov came home, *Soyuz 23* was manned by two rookies,
Vyacheslav Zudov and Valery Rozdestvensky. Because of either a
failure of the Igla automatic rendezvous system, designed to bring
the *Soyuz* spacecraft to within a few hundred yards of *Salyut*, or
the crew's decision to ignore it and instead guide the capsule in
manually (even today it is unclear which), rendezvous maneuvers
wasted so much fuel that by the time *Soyuz 23* was close enough to
dock, it had only enough for an uncontrolled ballistic re-entry to
Earth.

The mission was aborted, and the two cosmonauts made an
unscheduled landing the next day—at night, in a blizzard on ice-
covered Lake Tengiz in the middle of Kazakhstan, unintentionally
achieving the Soviet's first and only manned splashdown. Dragged
upside down by its parachutes, the capsule's interior was soon
soaked with freezing cold water. The men were saved from drown-
ing only by the quick arrival of search and rescue helicopters with
divers who were able to attach a flotation collar. Still, the capsule
floated almost 5 miles from shore in shallow, cold water clogged

with ice floes. Evacuation had to wait until dawn, when the divers were finally able to attach a line so that it could be dragged to shore.[50]

Soyuz 23, despite its failure, set one significant precedent. For the first time, the failure of a Soviet manned mission was announced *prior* to its return to Earth.[51] The public nature of space exploration, as well as Brezhnev's desire to use it to demonstrate detente and the peaceful intentions of the Soviet Union, was forcing the insular Soviet society to come out of its shell.

The third and final flight to *Salyut 5* lifted off four months later, on February 7, 1977. *Soyuz 24* lasted 18 days and was crewed by veteran Viktor Gorbatko and rookie Yuri Glazkov. Though the crew entered *Salyut 5* wearing breathing masks, and did several tests to analyze its atmosphere, they found no evidence of the "acrid odor" that had supposedly made Volynov sick. Once, during a television broadcast, they even partially purged the station's atmosphere, replacing it with air from stored tanks.[52] For many of the psychologists in the Soviet space program, these tests proved that Volynov's "acrid odor" had been nothing more than a delusion.

For the rest of their flight, Gorbatko and Glazkov concentrated on military photography. Once again, very few details of their mission were reported to the public—despite two television broadcasts. To make it seem that the mission was scientific in nature, some research was performed, including Earth resource photography of the Soviet Union as well as infrared spectrographs of the amounts of water vapor, ozone, nitrogen oxide, and pollution in the upper atmosphere. Because of the shortness of the flight, 18 days, only a few biological experiments could be done. More hawksbeard seeds were exposed to space, only to show no significant mutations. Growing mushrooms and other fungi in test tubes revealed that in weightlessness they developed into shapeless blobs.[53]

The focus remained on military surveillance, which was why, according to some Western experts, the mission lasted only 18 days. By bringing the crew home sooner, military experts could quickly analyze the flight's reconnaissance data. On February 24, 1977, the cosmonauts packed the small *Almaz* cargo capsule with film and experiments. One day later they undocked their *Soyuz* craft and returned to Earth. The day after that, the cargo capsule undocked and came home as well.

The six years following the *Salyut 1* disaster had been, in general, a failure for the Soviet space program. Of the six stations launched in that time, one was destroyed during launch, and two failed immediately after reaching orbit, making it impossible for any cosmonauts to ever occupy them. Eight *Soyuz* missions attempted to dock and occupy the three stations that cosmonauts could visit. Three failed: two because they could not dock with the station, and a third because its launch rocket failed at launch. Of the five successful missions, only three lasted longer than *Salyut 1*, and none was able to top the record of the last *Skylab* crew. Furthermore, much of the work was dedicated to a wasteful military program. To get funding to fly in space, Chelomey and his engineers sold their manned space program to the military. The result was neither good surveillance nor good research into manned space-station construction.

The most useful engineering knowledge that came from the two military stations was proof that the aft of the *Salyut* hull could accommodate both a docking port and engines. Furthermore, the fuel pump tests on *Salyut 5* showed that these engines could be refueled in space. Using this knowledge, Soviet engineers were able to redesign the next station, giving it two ports, the civilian bow port from *Salyut 4* and the military aft port from Salyuts 3 and 5. The redesign also attached to that aft port fuel lines for refueling the station's engine tanks.

Nonetheless, it was with the two missions to *Salyut 4* that the future of the Soviet space program manifested itself. After years of political infighting between their different design bureaus combined with confused political management, the successful occupancy of *Salyut 4*—with its experiments, heroic cosmonauts, and high-sounding ideals about conquering the stars—gave the Soviet Union its first real hint of what its future in space should be: the careful, scientific, peaceful, and permanent long-term occupancy of space.

Salyut 4, combined with the joint *Apollo-Soyuz* mission, also demonstrated to Brezhnev and his collective leadership the propaganda value of civilian space missions—a use of space that fit so well with Brezhnev's policy of detente. After *Salyut 5*, all Soviet space stations were civilian-run, and had non-military goals. And they were all built by Energia, following the basic overall plan for colonizing the planets first set out by Korolev back in 1960. After almost two decades of confusion, the Soviet space program finally had the same kind of clarity and single-minded purpose that the United States program had obtained in 1961 when Kennedy gave NASA the assignment of putting a man on the moon in less than a decade.

Ironically, at this moment of greatest triumph, with two simultaneous civilian space missions followed immediately by the signing of the Helsinki Accords, Brezhnev sowed the first seeds of the fall of the Soviet Union.

It began in Helsinki. For one thing, the negotiation process there had placed the Soviet negotiators at a disadvantage. Though the Soviets managed to keep some control of their Eastern bloc allies, Romania and Yugoslavia took very independent public positions, while the Polish and Hungarian negotiators showed a willingness to act on their own in private. Meanwhile, Western diplomats showed surprising unity in demanding the inclusion of many human-rights clauses in any agreement. The Soviet negotiators were further handicapped by Brezhnev's fervent desire to get an agreement signed at the completion of both *Soyuz 18-2* and the joint *Apollo-Soyuz* missions. Brezhnev wanted to crown both space achievements with a triumphant agreement in Helsinki.

On May 28, only three days after Pyotr Klimuk and Vitali Sevastyanov successfully docked with *Salyut 4* and had begun to activate it, Western diplomats were startled by a sudden change in tactics by Soviet negotiators. For months, the Soviet negotiating team had stalled negotiations over the many human-rights clauses that the West wanted included in the agreement. No matter what the Western negotiators proposed, the Soviets stonewalled. On that May 28, however, the Soviet negotiators had an abrupt and dramatic change of heart, almost certainly triggered by the successful occupancy of *Salyut 4* two days earlier. During a break in a negotiation session in Geneva, the French ambassador had invited the negotiators to a working lunch in a private room in one of Geneva's most elegant restaurants. For most of the meal the conversation was difficult and unproductive. The chief Soviet negotiator, Anatoly Kovalev, said little, spending most of his time smoking. Then, in the middle of the main course, Kovalev was called away to take a phone call. Upon returning he was clearly upset, and he and the other Soviet negotiators excused themselves to huddle together on the other side of the room to argue vehemently about something. After a few minutes, they returned to the table. Kovalev, ". . . in tones that were a mixture of triumph and spiteful anger," suddenly listed a series of concessions that made the final agreement possible.[54]

In fact, the Soviet shift that followed was downright startling. Over the next few days they quickly accepted language saying they would ". . . respect human rights and fundamental freedoms, including freedom of thought, conscience, religion or belief for all, without distinction as to race, sex, language, or religion." They also agreed to ". . . promote and encourage the effective exercise of civil, political, economic, social, cultural, and other rights and freedoms, all of which derive from the inherent dignity of the human person and are essential to his free and full development." Other, more detailed, clauses promised the reunification of families, the easing of restrictions on immigration, and an increased freedom for the press, both inside and outside their borders. Another clause even rejected the Brezhnev doctrine—used in 1968 by Brezhnev himself to justify the invasion of Czechoslovakia—which stated that fellow communist nations had the right to invade other communist nations if a threat to communism was perceived. Instead, signatories of Helsinki were to ". . . refrain from any acts constituting a threat of force for the purpose of inducing another participating State to renounce the full exercise of its sovereign rights." Such acts of violence or interference were prohibited ". . . regardless of mutual relations."[55]

No one had imagined the Soviet Union ever agreeing to such clauses.[56] In fact, Brezhnev and the Soviet leadership probably never intended to abide by them anyway. The Helsinki Accords included other, vague, language that implied the interpretation of these clauses was entirely up to him and the Soviet leadership. Moreover, to Brezhnev such clauses were merely public-relations ploys that he used to disarm Western suspicions about his intentions. Just he and his fellow communist rulers routinely ignored many similar clauses in their own constitutions, the new clauses in the Helsinki Accords were mere window-dressing that they could ignore if they wished. The only thing that mattered to Brezhnev was to get the West to accept the borders established after World War II, and to do it publicly at the same time the Soviet Union was flying two triumphant missions in space.

Nonetheless, the accords required that every signing nation publish them in full within their borders. Proud that the agreement seemed to formalize permanent communist rule over eastern Europe, Brezhnev made sure they were heralded loudly and prominently in the Soviet and eastern European press.

Brezhnev, however, seriously underestimated the effect of the Helsinki Accords. Only weeks after their signing, the first cracks in his program of detente began to appear. On October 10, 1975, Andrei Sakharov, the father of the Soviet hydrogen bomb, was awarded the Nobel Peace Prize. Since the 1950s Sakharov had fallen out of favor, eventually being placed in internal exile because of his willingness to take a strong public stand demanding freedom within the Soviet Union. As noted in a *New York Times* front-page article, the decision of the Nobel committee "was seen as a test of the Soviet Union's sincerity in fulfilling the spirit of the Helsinki agreement."[57] The accords had served to focus Western interest on how the Soviet Union routinely harassed and arrested its citizens.

Behind the Iron Curtain, the Helsinki Accords helped generate protests. To dissidents like Sakharov, "the prize of all people" guaranteed by the accords was not peace and security as Brezhnev declared, but *freedom*. Almost immediately, they began organizing, demanding that the Soviet Union live up to its signed promises. In Moscow, a group led by physicist Yuri Orlov declared that they intended to ". . . promote observance of the humanitarian provisions" of the Helsinki Accords. They resolved to make regular reports to the public of any violations by Soviet leaders of their signed promise to give Soviet citizens, as stated in Helsinki, ". . . human rights and fundamental freedoms, including freedom of thought, conscience, religion, or belief."[58] Similar groups quickly formed in Czechoslovakia, in Poland, in Lithuania, in Soviet Georgia, and in Armenia.

Suddenly, Leonid Brezhnev's high-sounding words of peace and goodwill, heralded both from space and from the treaty table, were being put to the test. To convince the world that his detente was something more than a superficial propaganda campaign, Brezhnev would have to do more than speak high-sounding words.

And once again, he was to use space to try to prove the legitimacy of detente.

5

Salyut 6: The End of Isolation

"I Could Work in Space 24 Hours a Day."

The hatch wouldn't open. No matter how hard they pulled, Vladimir Dzhanibekov and Oleg Makarov couldn't get the docking hatch of their *Soyuz* capsule to release. On the other side, inside *Salyut 6*, the newest Soviet space station, Georgi Grechko and his crewmate Yuri Romanenko anxiously waited. Both men had already been in space for 30 days, and were eager to greet their guests. On Earth almost 200 miles below, ground controllers, historians, engineers and, most importantly, Leonid Brezhnev, also waited impatiently, watching through *Salyut 6*'s television camera aimed at the bow hatch where Grechko and Romanenko floated.

Suddenly, almost unexpectedly, the *Soyuz* hatch popped free, throwing both Dzhanibekov and Makarov back a few feet. Almost instantly Grechko dived in and grabbed Makarov, joyously pulling him into *Salyut 6*. Dzhanibekov followed, and the four men gathered in the station's main section, hugging and laughing, toasting each other happily with tubes of cherry juice.

To the rookie Dzhanibekov, the station had that "new-car" smell. Launched just three months earlier, *Salyut 6* still hadn't lost that aroma of freshness. To the more experienced and cynical Makarov (whose last flight had been the terrifying aborted launch of *Soyuz 18-1*), the station exuded a more complex ambience. Mixed in with the new-car smell of fresh metal and plastic and equipment

114

were the smells and sounds of a working space vessel: the whir of fans, the smell of chemicals and fuel combined with sweat and flatulence.

Nonetheless, at that moment the men felt mainly joy and happiness. For the first time, three separately launched spacecraft, carrying two separate crews, were linked together in space, producing an orbital complex almost 100 feet long. After a decade of failure and struggle, the Soviet space program finally had clear direction, and was going somewhere.

Launched on September 29, 1977, *Salyut 6* was the direct descendent of the last civilian station, *Salyut 4*. Using the same 20-ton *Almaz* hull, it had three solar panels producing the same 4 kilowatts of power. It had the same temperature control, atmosphere-recycling, and attitude-control systems. Its interior was laid out in much the same manner, with the main compartment dominated by the cone-like telescope housing, the narrow-diameter section holding the station's control center, and the bow of the station func-

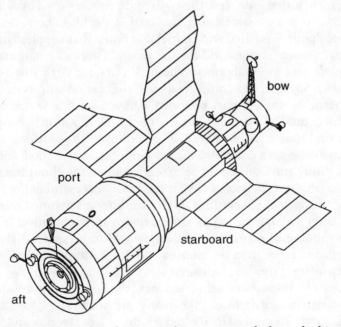

Salyut 6. Note how the aft service donut now includes a docking port and engines. *NASA*

tioning as the docking-transfer module with airlock and docking port.

Salyut 6 did incorporate some significant improvements on *Salyut 4*. The docking-transfer module was larger, as was the hatch for exiting the station during space walks. It used an improved water-recycling system, based on what had been learned on the previous three *Salyut*s. Engineers hoped that this Rodnik system ("spring" in Russian) would be more successful than *Salyut 4*'s in preventing humidity from accumulating inside the station. To make its inhabitants' lives more comfortable, the soundproofing on operating equipment had been improved. A shower, similar to the one that had been on the military *Salyut*s, had been added, and the variety of the food increased. Many food items were also made as bite-sized chunks, to make them easier to eat in weightlessness.

However, the station's most significant improvement was that it had two docking ports, one on each end. To add a second port, designers had extended *Salyut 6*'s aft end several feet with the addition of a donut-shaped service section. The new docking port, with an entrance tunnel leading into the station, was in the new section's center. On either side were the twin main engines used by *Salyut 3* and *Salyut 5* for maintaining the station's orbit. The rest of the service donut was filled with refuelable tanks that supplied fuel for the twin main engines and for the four clusters of small attitude jets placed at intervals on the donut's outside. With this second port, two *Soyuz* crews could dock with the station simultaneously. Furthermore, the aft port provided a haven for a new unmanned freighter, dubbed *Progress*, which could dock automatically and bring fuel that could be pumped directly into the station's tanks. Because *Progress*'s cargo also included oxygen, water, food, and supplies, future missions could be extended almost without limit.

Salyut 6 was to become by far the most successful of the Soviet Union's *Salyut* space stations. Eighteen different manned missions were launched to it during its almost five years of operation, achieving a number of significant milestones. Human space flight was extended to more than six months and the station periodically carried two crews totaling four men.

Despite these later achievements, the first *Soyuz* mission to the station was a failure. Like many previous failed *Soyuz* missions, *Soyuz 25*, crewed by rookies Vladimir Kovalyonok and Valeri Ryumin, was unable to dock with the station. Once the Igla automatic system had brought the spacecraft close to *Salyut 6*, Kovalyonok took manual control and piloted it into a soft docking in *Salyut 6*'s bow port. At that point, the docking latches refused to engage, preventing a hard docking. Under orders from the ground, Kovalyonok backed *Soyuz 25* up a few feet and tried again. No luck. He backed up again, and this time tried to jam *Soyuz 25* forward with much greater speed. The two spacecraft banged together, but still the latches refused to engage.

Some obstruction or damaged equipment, either in the bow docking port of the *Salyut* station or in the docking gear of *Soyuz 25*'s orbital module, was keeping the two spacecraft from completing a hard dock. Their fuel reserves low, Ryumin and Kovalyonok had no choice but to retreat, returning to Earth after only two days in space.

Because their *Soyuz* orbital module was discarded as planned, burning up during re-entry, there was no way of knowing if the problem was in its docking port or worse, in the bow port of *Salyut 6*. In order to use the station's most important new capability, its two ports, the next crew was going to have to make repairs, including the first Soviet space walk in nine years.

Before this rescue mission could occur, however, Leonid Brezhnev intervened. Though the docking failure of *Soyuz 25* had had nothing to do with its rookie crew, Brezhnev stepped in personally to forbid any future all-rookie crews.[1] To accommodate this command, and to bring someone to *Salyut 6* who had the specific abilities to fix the problem, required the immediate shuffling of all crew assignments.

Thus, Georgi Grechko returned to space as part of the first crew to occupy *Salyut 6*. His experience as an engineer and his previous success on *Salyut 4* made him the best choice to try to fix the problem on *Salyut 6*. Within days of *Soyuz 25*'s landing, 46-year-old Grechko was paired with 33-year-old rookie Yuri Romanenko.[2] With Romanenko providing support, Grechko would do a space walk to inspect and, if necessary, repair the bow docking port. If all went well, the two men would then attempt to break Skylab's record for the longest mission in space.

This crew shuffle brought Romanenko and Grechko together only two months before launch, an unusually brief training period for Soviet crews. The circumstances were further complicated in that, as pilot and military officer, rookie Romanenko was officially the mission commander, even though Grechko, the hero from *Salyut 4*, was more experienced. Moreover, their personalities clashed. Grechko, an engineer, was a friendly, good-natured man who was quick to smile and laugh. Romanenko, a military man, was far more serious and focused. "He is highly strung and temperamental," Grechko later noted. "I am not fond of people who give orders."

Days before launch, the two men worked out a compromise. As Grechko remembered, "We would act as if Yuri was not the commander and as if this was my first time in space." They even tried to get mission control to not give them separate orders and instead assign duties to the whole crew, and let them decide who would do what.[3]

For the rescue work, however, such egalitarianism could not work. Romanenko was still a pilot, and Grechko was still an engineer. On December 12, after one day of orbital maneuvers, *Soyuz 26* was finally close enough to *Salyut 6* for Romanenko to take manual control. With the condition of *Salyut 6*'s bow docking port still unknown, Romanenko piloted the *Soyuz 26* capsule into *Salyut 6*'s aft docking port.

Then, after spending the first eight days activating the station's systems and experiments, the crew prepared for the first Soviet space walk in almost nine years, and the first ever from a Soviet space station.

Grechko was to take the lead. The plan called for the men to don their spacesuits and seal and depressurize the docking-transfer compartment. Grechko was to open the front docking port, inspect it for damage, and then climb out to inspect the Igla radar antenna on the outside to see if it was in good working order. If all was right, Grechko was to attach a cassette filled with biological samples to the outside of the station and come back inside. Romanenko was to wait at the docking port in case something went wrong and hand Grechko tools and hold him in place if necessary.

On December 20, the two men climbed into their spacesuits. Unlike the American spacesuits in use then and now, which are custom fitted to each person, the Soviet Orlan-D spacesuit was adjustable. Within the strict limits on body sizes allowed for Soviet cosmonauts, an Orlan suit could fit anyone. A cylinder of metal covered the chest and torso like a suit of armor. At the holes for the arms and legs flexible fabric was attached. To don this semi-rigid suit, a cosmonaut unhinged a permanently mounted oxygen pack attached to the back of the suit's rigid section, inserted his feet and body through this hatch and climbed inside. His partner then closed the hatch, sealing him in. With this suit, a cosmonaut could get dressed for up to three hours in space (later increased to five) in about five minutes.[4]

Once dressed, the men closed the inner hatch that separated the docking-transfer compartment from *Salyut 6*'s main chamber and depressurized the docking compartment. Rather than use the airlock hatch, Grechko was going to open the port itself, swinging its receptacle cone assembly into the docking compartment so that he could inspect its parts. At first, the cone did not open. Because it hinged inward, the tiny pressure from the small amount of residual atmosphere in the docking compartment held it in place. It took several tries before the two men, pulling hard together, could get it to pop open.

As soon as the cone opened, however, the last bits of air inside the compartment rushed out. To Grechko's surprise, he found himself being sucked out as well. Instinctively he grabbed at the edges of the opening, holding himself inside and waiting for things to settle down. Then he swung the docking cone into the station to study it, quickly reporting that it seemed in perfect working order. "The butt end is brand new—just as when it was machine-tooled. There are no scratches or dents or traces. The cone is clear: not a scratch."[5]

Then he floated through the port, carefully inspecting its latches, plugs, and various mechanical parts. Periodically Romanenko handed him a tool so that he could take something apart and reassemble it. Once again, everything seemed to work fine.

Next Grechko climbed out the port onto the station's exterior where he could inspect the Igla automatic docking system. Romanenko followed him into the port, holding his ankles so that he could work without drifting. Grechko again reported that everything seemed fine. He climbed farther out so that he could attach a small cassette of experimental compounds that engineers on Earth wanted to expose to space and see how they degraded.

As ebullient as Grechko normally was, he took this space walk somewhat nonchalantly. "It was nothing exciting," he remembered. "The same starry sky, the same Earth below. And some structures around me."

Romanenko, meanwhile, waited in the port, finding himself fascinated with the blue-and-white view of Earth below him. A serious and hard-working man, Romanenko committed himself totally to a project once he decided to do it. As a child, the son of a navy destroyer captain, he had been fascinated with ships, engineering, and planes, so in high school he built both model ships and airplanes. Then, after graduation, in order to learn construction techniques, he worked for a year, first as a concrete mixer on construction sites and then as a locksmith. Only then did he enroll in the Chernigov Higher Air Force College, becoming a pilot-engineer and flight instructor for the college. Then, a visit to the school by cosmonaut Gherman Titov convinced him to become a cosmonaut. He applied, went through the stringent medical exams, and was selected. By 1970 he was a member of the cosmonaut corps where he worked for the next six years as backup on a variety of missions and as capcom in Soviet mission control during the *Apollo-Soyuz* mission.[6]

Now he floated in zero gravity, watching the bright blue Earth drift past him 200 miles below. As Grechko worked a few feet away, Romanenko got the urge to get a better view. For safety he was attached to the station in two ways, first by a rigid back-up cable that kept him safely within the docking port, and second by a 60-foot umbilical cord that provided their spacesuits both communications and electrical power.[7] Unable to restrain his curiosity, Romanenko disengaged the back-up cable so that he could drift further out into the port. With the outer rim of the hatch still about 3 feet away, he could see the vast blue Pacific ocean gliding by, streaked with white clouds.

Outside, Grechko had finished attaching the cassette, and was using his handheld camera to beam back pictures of the earth to mission control. Then, after about five minutes, he pulled himself into the hatch as Romanenko backed up out of his way. Together again within the docking-transfer compartment, they had some extra time because Grechko had finished his work in only 20 minutes, well ahead of schedule.

At that moment the station moved out of range of Earth communications. Free from ground control, Romanenko looked at Grechko. "I'm only a meter from space," he said. "Maybe I'll never get a chance to go outside. Let me take a look."

Grechko agreed, though he said, "Do it fast. We don't want to fall behind schedule."

Eagerly, Romanenko shot past his partner, aiming for the open port. At that instant Grechko's heart skipped a beat. Seeing the detached rigid back-up cable, he somehow imagined that Romanenko would fly out the opening and drift away. Aghast at the possibility, Grechko grabbed at Romanenko's tether, yelling "What's your hurry?!" Grechko had forgotten that Romanenko was still tethered by his umbilical cord, which *had* to be attached to the station. Without the electricity provided by that cord, their spacesuits couldn't function.

Abashed, he spent the next few minutes playing out Romanenko's tether so that his crewmate could poke his head and body out the port and look out into space. Then Romanenko eased back inside the docking-transfer compartment and closed the docking cone. Neither said anything to ground controllers about Romanenko's little excursion.[8]

At the post-flight press conference, the jovial and imaginative Grechko ("He is an amusing chap," noted Romanenko.) jokingly told interviewers that Romanenko's safety line had become detached, and that he would have been lost in space if Grechko had not grabbed the end of his line in the nick of time. To Grechko's surprise, and Romanenko's chagrin, the journalists took Grechko seriously, a circumstance that has since forced them to repeatedly explain what really happened.[9]

With the bow port apparently operational and the receptacle cone shut, the two men patiently awaited orders from the ground to start repressurizing the docking-transfer compartment so that they could climb out of their spacesuits. But the order did not come.

Unbeknownst to the cosmonauts, engineers in mission control were getting telemetry from the station indicating that the docking cone had not shut properly. The situation, with a warning hatch light and the threatened suffocation of the crew, seemed frighteningly reminiscent of *Salyut 1*. In this case, the men's lifeboat, *Soyuz 26*, was docked to the other end of *Salyut 6*, with the station's main body between them and it. If the transfer compartment did not repressurize they could not get back inside the station, because the compartment hatch opened *into* the main compartment, and the difference in air pressure between the station interior and the docking compartment made it impossible to push the hatch open. Furthermore, a space walk on the exterior of *Salyut 6* back to *Soyuz 26* was useless. Even though they could open the hatch on the *Soyuz* spacecraft to get inside, they couldn't close it on their spacesuit umbilical cords. The only option was for mission control to remotely depressurize the entire station so that the men could open the interior hatch and then repressurize everything. Though theoretically feasible, no one on the ground wanted to try it.

For about 10 minutes, ground controllers scrambled to try and figure out what was wrong with the closed docking cone. Then, in the hope that the problem was just a failed sensor, ground controllers told the cosmonauts to begin filling the transfer compartment with air. If there was no leak, the pressure would hold and then increase steadily. The men opened the valves on several oxygen tanks. To the relief of mission control, the pressure quickly began rising. A second test confirmed that the docking cone had shut, and the compartment was quickly repressurized. Grechko and Romanenko, entirely unaware of the crisis, climbed out of their suits and re-entered the main body of *Salyut 6*.[10]

With the port fixed, the two men settled into their planned daily routine. Their sleep schedule, unlike that of crews on earlier civilian stations, was not adjusted to keep them awake when the laboratory was over Earth-based ground stations. Neither was their schedule staggered so that someone was on duty for most of the day, as had been the case on the military *Salyut*s. Both of these schedules had been unnatural and tiring, and had caused problems. Instead, the men lived on Moscow time, with a normal five-day work week and two days off on the weekend.[11]

Each day they ate four meals, made up of freeze-dried meals and bite-sized chunks of food, from meat to candy bars. Each day they did about two hours of exercises, half on the treadmill, half on the bicycle. Each day they had a list of scientific chores to perform, from snapping detailed photographs of the earth's surface to using the station's various telescopes to study the stars. For example, their biological experiments included an aquarium holding tadpoles and an incubator holding *drosophila* flies—the same kind of flies that Sevastyanov had given pet names on *Salyut 4*. There were two sets of tadpoles, one hatched on Earth, the other in space. While the space-born tadpoles tended to swim in spirals, the Earth-born tadpoles swam about randomly, showing a greater inability to orient themselves to weightlessness.[12]

During these early weeks the two men mostly focused on checking out the station's systems in preparation for the impending arrival of the second *Soyuz* spacecraft in mid-January. Dzhanibekov and Makarov's flight was actually a quickly improvised six-day mission with only two real goals. First, the mission was to accomplish the first docking of two crews to a single space station. Second, and more important, the mission was to free the aft port, blocked by Grechko's and Romanenko's *Soyuz 26* spacecraft, so that *Progress* freighters could use it to refuel the station's tanks. Mission control was unwilling to simply undock *Soyuz 26* and have Romanenko fly it around to the bow port. Doing so would have required the complex task of shutting down the station as if it were being left unmanned, just in case the crew had trouble redocking and had to return to Earth.

Instead, controllers planned to use *Soyuz 27* to get the aft port clear, albeit in a roundabout fashion: After docking *Soyuz 27* to the bow port, Dzhanibekov and Makarov were to return home in *Soyuz 26*, thereby clearing the aft port. The switch would not only give the Soviets a new space first—the first time two vehicles had docked to a third in space—it would also give Grechko and Romanenko a fresh return vehicle.

As Dzhanibekov steered *Soyuz 27* towards *Salyut 6*, he did things differently than Kovalyonok and Romanenko, both of whom had taken manual control during their dockings. He instead decided that it was essential to let the automatic docking system complete the docking. Many unmanned *Progress* freighters were to come after his flight, and mission control needed to know if *Salyut 6*'s radar docking systems could be trusted to work.

Making this decision, however, took some nerve. During docking operations cosmonauts were required to stay inside *Soyuz*'s descent module, even though it was impossible to see anything useful through the descent module's three tiny windows. To give the pilot some visual guidance, a periscope was attached to his window. He also had a 3-inch-square black-and-white television screen, which showed a split-screen image from the two cameras pointing out the front of *Soyuz*.

Dzhanibekov's docking was scheduled to occur at night. To make *Salyut 6* visible, the station had four lights on the perimeter of its docking port, two steady and two flashing. There were also two steady lights, one each on the outside ends of *Salyut 6*'s port and starboard solar panels. If the automatic system was on course, each screen image should show Dzhanibekov six lights, for a total of twelve lights on his 3-inch-square screen.

As the *Soyuz* spacecraft edged closer, it began to twist slightly off course, its lights drifting sideways. Dzhanibekov braced himself to take control. Then, in a leap of faith, he did nothing. "Instinctively I knew that everything was right," he said later. On his viewscreen he could see 12 lights, some blinking, some not, "like a Christmas tree." If the ship was going to miss the port, the lights would have been more askew.

Dzhanibekov's instincts were right. At 20-feet separation, the automatic system made some adjustments, righted itself, and slid *Soyuz 27* precisely into the port.[13]

Grechko and Romanenko were thrilled to have company after a month in space. The men hugged and laughed for the camera. Then, after Makarov and Dzhanibekov had presented their hosts with newspapers and letters from home, Grechko and Romanenko offered their guests a simple space meal, small crackers and salt tablets, the closest they could get to emulating the Russian tradition of feeding guests bread and salt. They washed these down with toasts of cherry juice squeezed from tubes.

Unlike American society, which has few hospitality rituals other than to shake hands and to bring the host a small gift (usually a bottle of wine), Russian society places great importance on its welcoming rites. Bread and salt are eaten as a symbol of fellowship and good luck. Toasts of vodka are required before every meal. On special occasions, the toasts must be frequent, copious, and deeply savored. For Grechko and Romanenko, therefore, the toast of cherry juice was not merely for public relations. They were truly happy to have guests, and felt a strong obligation to show their happiness in a manner that everyone in Russia would understand. The isolated, lonely, artificial, and dangerous existence on the station had already begun to wear on the two men, especially Grechko. "Grechko is an extremely emotional man," noted the crusty, hard-edged Makarov. "With only a few days to go in a mission, he works fine. With a month or more left, he can be nearly sick with enthusiasm and anxiety."[14] The arrival of visitors gave Grechko and Romanenko a welcome break from their daily routine.

Salyut 6 now comprised three separately launched spacecraft, with a habitable volume of almost 3,900 cubic feet. In order to test this complex against any unexpected vibrational resonances, the four cosmonauts held onto the *Salyut 6* treadmill and "bounced" up and down together, seeing if the complex amplified this vibration (just as soldiers marching in unison across a bridge can cause resonances that can shake the bridge apart). Such a test could not be simulated in Earth gravity. While they noticed that the solar panels outside the station fluttered up and down about a foot, nothing significant happened, implying that large structures made of multiple modules could be stable in space. The Soviets would repeat this experiment whenever a space station achieved a new configuration.

As they bounced up and down, the newest space rookie, Vladimir Dzhanibekov, watched everything around him with an almost wide-eyed, innocent joy. "Everything was new and interesting," he remembered. A softspoken, gravelly-voiced man, Dzhanibekov's last name had been shortened to "Johnny" by the American astronauts when he acted as the commander of the Russian back-up crew during the *Apollo-Soyuz* project, a nickname his Russian friends use even today. Afterward, he was paired with Petr Kolodin, who had been bumped from the ill-fated mission to *Salyut 1*. In the ensuing years Kolodin had struggled to get into space. The flight with Dzhanibekov was to be his chance. For two years the two men trained together, planning to fly one of the first missions to *Salyut 6*. Then Brezhnev gave his order forbidding all-rookie crews. Since both Dzhanibekov and Kolodin were rookies, one had to be rescheduled. Kolodin was once again bumped, and Dzhanibekov was partnered with Makarov, who had already flown twice, including the aborted *Soyuz 18-1* launch in 1975.

Kolodin, meanwhile, faded from the active cosmonaut program and never flew in space. Though he was apparently a competent flight engineer like Grechko and Makarov, he was part of the military's cosmonaut corps. The cancellation of the military *Almaz* program, combined with the crew-selection compromise between the civilian and military halves of the Soviet space program (that is, military pilots and civilian flight engineers), meant there was no place for military flight engineers. Kolodin was a cosmonaut without a program to support him. As Alexei Leonov once noted, "He was just low on luck."[15]

"Johnny" Dzhanibekov in turn went on to be one of the Soviet space program's most successful cosmonauts, although his first flight was short and simple. The two crews stayed together on the station for only five days, working hard every minute and getting very little sleep. Grechko and Romanenko found themselves restless, talking incessantly with their visitors about home, friends, and the earth. Even after Dzhanibekov and Makarov went to sleep, they stayed up, reading the letters and newspapers from home, over and over.

What neither the personal letters nor any of the other men told Georgi Grechko was that his father had passed away 10 days earlier. Makarov and Dzhanibekov knew about it, and at one point pulled Romanenko aside to tell him as well. However, the three men had been ordered by psychologists to withhold the information from Grechko. Though everyone agreed with this decision, Romanenko insisted that he be the one to break the news to Grechko after their return to Earth. "It was hard, but I think it was the right decision," Grechko remembered years later. "I had two months of difficult work still ahead of me."[16]

The single experiment of note during these five days was a French experiment that tested how weightlessness affected the ability of protozoa cells to divide. Single-celled paramecia were carefully unfrozen and allowed to divide; the results showed that the in-space specimens were nearly identical to the control specimens on the ground, with only subtle changes in cell metabolism.

Finally, on January 16—having swapped the custom-fitted couches between the two *Soyuz* descent modules, thereby allowing them to also swap home-bound spacecraft—Makarov and Dzhanibekov undocked *Soyuz 26* from the aft port and returned to Earth. Grechko and Romanenko watched forlornly from *Salyut 6*, describing to mission control how they could see the spacecraft's re-entry engines firing.[17]

The two men had decidedly mixed feelings about the departure of their guests. On the one hand, they were glad to have the station to themselves again. "Anyone who has to put up guests knows how we felt. You are always as happy on the first day as on the last," noted Grechko with humor. "You are glad to see them. You are also glad to see them go. It's human nature, whether in space or on Earth." On other hand, the visitors had brought Grechko and Romanenko a taste of Earth. With only two men on board again, the station seemed a lonelier and more artificial place. Three weeks after Dzhanibekov and Makarov left, Grechko and Romanenko held a televised press conference, where they talked about how much they missed the earth. They had been in space almost two months, and were feeling quite homesick. They talked about how they dreamed of woods and rivers, and of going skiing. They talked about how beautiful the earth looked from space, describing the Volga River, the Ural Mountains, and the vast empty plains of Siberia.[18]

Their work went on, nonetheless. With the aft port now open, the first *Progress* freighter could be launched. Essentially a modified *Soyuz* spacecraft, this first-generation cargo ship had no solar panels and carried only two days of power in its batteries. Once docked to *Salyut 6* it used the station's solar panels to recharge its batteries, usually remaining attached until the aft docking port was needed by another craft.

To turn *Soyuz* into a freighter, the spherical orbital module at its bow had been modified into a cargo hold that cosmonauts could enter to get supplies and deposit trash. The middle descent module was replaced, its unnecessary re-entry and life-support systems eliminated and its necessary control systems moved to the service module in the rear. In its place, an equipment compartment contained the pumps, fuel, and tanks for refueling *Salyut 6*. Propellant flowed through pipes on the outside of *Progress*, through intake valves on the periphery of the aft docking port, and into the engine tanks in *Salyut 6*'s aft service donut.[19]

Launched four days after Dzhanibekov and Makarov had departed and docking with the station two days later, *Progress 1* remained attached to *Salyut 6* for 18 days. In that time, Grechko and Romanenko unloaded supplies, including some fresh fruits and bread, books, a cassette recorder and music tapes, and more letters from Earth. They also carefully inspected the fuel lines that connected the cargo ship with the tanks on *Salyut 6*. Then, beginning on February 2, they and mission control performed the first refueling of a spacecraft in space. For two days, ground controllers fed both fuel and oxidizer from *Progress 1* into *Salyut 6*'s tanks. To do the pumping, nitrogen was pumped into the freighter's fuel tanks at high pressure, forcing a flexible bladder holding the fuel to contract and push the fuel out and along the pipes to *Salyut 6*'s tanks.[20]

The day after the refueling, February 6, *Progress 1* was undocked, its retro-rockets fired, and it and the garbage inside were destroyed as planned during re-entry. Five days later Grechko and Romanenko broke the 63-day record set by Klimuk and Sevastyanov on *Salyut 4*. If all went as planned, their mission would break the American *Skylab* record as well.

Their final month in the relatively cramped interior of *Salyut 6* became an endurance test for both men, as the flight lurched from the absurd, to the lethal, and then to the banal.

First the absurd. At one point the men discovered that about 15 of their tiny drosophila flies had escaped from their incubator. Ground engineers, terrified that they would somehow get caught in the station's electrical connections and cause a short circuit and a fire, immediately convened an emergency commission, bringing together engineers, scientists, and even high government ministers to discuss what to do. They even tracked Nechitailo down at her country dacha and dragged her back to mission control to talk to Grechko about it.

"Zhora," she said to Grechko, calling him by the diminutive of his first name, "we are in big trouble. Try to think of something."

Grechko told her, don't worry, no problem.

Then an important government minister got on the radio and demanded to know if the situation was dangerous. Grechko joked, "Even if we accidentally swallow the flies, we can't be hurt."

"How do you know?" the minister demanded.

"They are fruitflies," Grechko explained. "They feed on fruit and honey."

Then to the engineers, Grechko pointed out why he and Romanenko were reluctant to capture the flies. "Yuri's doing aviation research," Grechko explained. "He's studying how flies fly in zero gravity, doing ballistic dive-bombs."

Grechko's light-hearted approach to the problem defused the panic felt by mission controllers. Nonetheless, after several orbits of discussion and consultation, the emergency commission decided that, for safety, the flies had to go. Grechko and Romanenko were ordered to use their vacuum cleaner to suck them up.[21]

Then came the lethal: There *was* a fire—far more serious than anything that had occurred on *Salyut 1*. Grechko and Romanenko were intently working together near *Salyut 6*'s bow, where its docking port linked to the *Soyuz* spacecraft. Suddenly they smelled something burning, and looking behind them, were astonished to see the station's main body filling with billowing smoke.

The men sprang into action. While Romanenko scrambled to gather their notes and scientific results and get them into the *Soyuz* spacecraft, Grechko became a fireman, attempting to put out what was now one of the worst in-space fires in history. Before he could do anything he had to find the source of the blaze. To do that, however, he had to be able to swim through the thick smoke without suffocating. Because there was no gravity, the smoke hung thick and dense, like a brown, noxious fog that was slowly expanding toward them. And he couldn't get rid of it by simply opening the windows.

So, Grechko dove away from the smoke, through the docking tunnel into the orbital module of the *Soyuz* capsule where the air was still clear. There he took deep breaths in order to hyper-ventilate. Then he filled his expanded lungs, held his breath, and dove back into *Salyut* and into the cloud of smoke.

Bouncing from wall to wall, he desperately searched for the fire. It wasn't behind one wall panel. Nor was it near the main telescope controls. His lungs bursting, he finally spotted it—burning insula-tion from a cable connected to a sensor for monitoring electrical use. He flicked the instrument off, and pushed himself back toward the *Soyuz* and into the clear air, passing Romanenko, who had been fly-ing back and forth between the station and *Soyuz*, trying to get the two spacecraft ready for separation and evacuation.

Grechko took several more breaths and flew back into the smoke. As far as he could tell, the fire had ceased, with no new smoke billowing from it. He swam to the ventilation system and turned it up as high as he could. Then he dove back to *Soyuz* to help Romanenko.

Within a few minutes the air began to clear. Shortly thereafter, *Salyut 6*'s orbit brought it over Russia so that communications with mission control could resume. The two men described the fire and what they had done, and were told, quite firmly, to not use the suspect piece of equipment again.[22]

Finally came the banal, as Grechko and Romanenko struggled with an increasing sense of isolation, loneliness, and claustropho-bia. Cramped in a metal room about the size of the coach section of a small commercial passenger jet, both men found that the only way they could keep their sanity was to squash their emotions and take each day as it came. The necessity for this austere attitude was made very clear one day when a simple disagreement about the skills that future spacefarers would need turned into a raucous argument that almost came to blows.

Grechko, an engineer and designer, firmly believed that future space explorers would generally be engineers and designers. "Pilot-ing a spacecraft basically amounts to docking and undocking, whereas the flight engineers have much more to do on a long flight."

Romanenko, an ace pilot and passionate about flying, disagreed. The argument, between two men with strong but very different

personalities, turned into a shouting match, with neither willing
to give an inch. Finally, in anger and disgust, Grechko pushed himself away, shouting "I've had enough of this argument! I'm leaving!" He floated to the aft of *Salyut*, near the cone-shaped telescope
housing where they usually slept. There, he stuck his face to the
window, his back to Romanenko.

After a few minutes of silence Romanenko floated to Grechko
and put his hand on his shoulder. "Zhora," he said to Grechko.
"What have we got ourselves so wound up about? The flight is
going well. If we argue, we will mess everything up."

Grechko agreed. Better to avoid such conflicts, keep a cool
head, and hold back your emotions until you returned to Earth.
From this moment on, each man kept his emotions in careful
check, while also keeping close tabs on his partner's feelings.[23]

And it wasn't easy to stay cool-headed. As the mission wound
down, Grechko had trouble sleeping, getting only about three hours
of sleep a night.[24] Then, two weeks before they broke the American
record of 84 days, Grechko began to notice Romanenko taking repeated doses of painkillers. At first he was reluctant to ask Romanenko what was wrong. "I thought he would tell me himself,"
Grechko remembered later. He became further alarmed, however,
when he realized Romanenko was taking more than the recommended dosages. Something was putting his crewmate in excruciating pain. Finally Grechko confronted him, "Yuri, tell me what is
going on."

Romanenko reluctantly explained that he had a toothache, and
that the pain had become almost unbearable. Grechko suggested
they radio the doctors on Earth to find out what they could do.
Romanenko, determined to be the serious, proper, *Russian* spaceship commander, refused, saying it would be a disgrace for the captain of a ship to complain about a toothache. After trying vainly to
change Romanenko's mind, Grechko finally struck him a deal:
Grechko would tell the doctors that *he* had the toothache, and
Romanenko would agree to do whatever they said.

The call really didn't do much good. All the doctors could suggest was that he wash his mouth with warm water and keep warm.
Romanenko wrapped his head in a scarf and wore a hat, neither of
which did much to ease the pain.[25]

At the same time, the end of the mission was approaching, and
both men knew it was even more important to complete their re-

quired two hours a day of exercise. They were also being asked to use the Chibis suit at least 10 to 12 hours a day in an effort to strengthen their cardiovascular systems.[26]

Romanenko's situation was made even more stressful in that during these last few weeks in space they were to receive their second visitors. *Soyuz 28* was to bring Alexei Gubarev, Grechko's partner during his *Salyut 4* 30-day mission, and Czechoslovakian Vladimir Remek, the first person from a country other than the Soviet Union or United States to fly in space. For propaganda reasons, this international flight had been timed so that Grechko and Romanenko would break the 84-day record of the last American *Skylab* crew while Remek was aboard.

By 1976 the Soviet leadership presented two faces of their country to the world, each in direct contradiction to the other. On one hand, Brezhnev and the unelected leaders of the U.S.S.R. had spent the 1970s trying to convince the world that they wanted to live in peaceful co-existence with their noncommunist opponents, signing treaties and declaring themselves in favor of nuclear disarmament. On the other hand, the barbed-wire fences that cut Europe in half remained strong and firm, despite the generous words of the Helsinki Accords. Travel from one side of the Iron Curtain to the other continued to be severely restricted. The independence of the Eastern bloc nations was tightly controlled.

In the Soviet Union, despite increasing pressure to abide by the clauses demanding freedom of speech and travel, dissenters like Andrei Sakharov and Anatoly Sharansky had been imprisoned or exiled, while others were sent to psychiatric wards, where they were heavily medicated or worse. For example, Yuri Orlov, the man who founded Helsinki Watch, was arrested and subjected to months of brutal interrogation. He was then sentenced to seven years imprisonment, followed by five years of exile to Siberia.[27]

Brezhnev, marshalling his public relations skills, had been looking since the signing of the Helsinki Accord for some way to make the free world forget these events. And once again, space presented itself to him as the ideal propaganda tool for distracting the West. The engineering work that the *Salyut* cosmonauts did, from resupply to long-term missions, gave him a perfect opportunity to use *Salyut 6* for foreign policy.

The *Soyuz* spacecraft in use in the mid-1970s had an in-space life expectancy of only about three months. For a space-station mis-

sion to last longer, its *Soyuz* ferry had to be replaced periodically
with a fresh vehicle. Ground engineers planned to solve this problem by flying short visiting missions such as Makarov's and
Dzhanibekov's to bring the long-term crew a fresh spacecraft and
take the visitors home in the old one.

Brezhnev immediately saw a public-relations opportunity. He
reasoned, why not use these short shuttle missions for foreign
policy goals? Why not offer one seat on each short mission to one
of his communist allies?[28] The idea would not only distract the
world from Soviet human-rights abuses, it would also scoop the
American space program once again. By the mid-1970s NASA was
beginning to plan the space shuttle's first missions, scheduled to
begin before the end of the decade. These plans included flying astronauts from European countries. If the Soviets sent foreigners to
Salyut 6, they'd grab another first in the exploration of space.

So, in 1976, during two meetings of the world's communist
parties, one held in East Berlin and the other in Moscow, Brezhnev
offered space and the *Salyut* stations to his communist allies. In
this Intercosmos program, each member of the communist alliance
was to fly one cosmonaut to a Soviet space station over the next
five years, thus proving how open and beneficial communism and
the Soviet Union were to the world's people.[29]

Brezhnev's offer, once accepted, gave him a carrot that he could
withdraw if an ally began doing things he did not like. For example,
Romania had been a troublesome ally since the early 1960s. In 1966
its leaders decided that their foreign policy had to be determined by
them based on their own particular national interests, rather than
by the Soviet Union and the general communist movement. That
same year Romania became the first communist nation to establish relations with West Germany. In 1967, Romania maintained
relations with Israel after the Six-Day War, despite the Soviet policy
to break all official contact. After the Czechoslovakian invasion in
1968, Romania's leaders passed rules that forbade the troops of
other Warsaw Pact countries to enter Romania, in an attempt to
prevent the kind of east-European military maneuvers that the Soviets had used to move their troops into Czechoslovakia. Then,
during the Helsinki negotiations, Romania had been especially
troublesome, playing a very independent role that made negotiations difficult for Soviet diplomats. Thus, while Brezhnev offered
Romania a seat on a *Soyuz* spacecraft, the sequence of flights was

carefully arranged so that they got the last scheduled seat, delaying their cosmonaut's flight until 1981.[30]

Beginning with Czechoslovakian Vladimir Remek's visit to *Salyut 6*—with Romanenko trying to hide his excruciating toothache—eight international flights flew to *Salyut 6*, each acting as support for the various long-term missions already in orbit. Remek's flight was probably the most historically significant, because it was the first manned space flight involving a crew from two different countries, and the first to involve someone from outside the Soviet Union or the United States.

Over the next three years cosmonauts from Poland, East Germany, Hungary, Vietnam, Cuba, Mongolia, and Romania visited *Salyut 6* during other long-term missions. Each international flight lasted eight days. Each completed a variety of cardiovascular and medical experiments. Each took multi-spectral photographs, usually of the visiting cosmonaut's native country. Each included at least one or two experiments, from metallurgy to microbiology, developed by scientists from that native country. However, the scientific results from these international flights were mostly insignificant. Often, the hardware from each flight merely added more clutter to the station, interfering with life in that small, claustrophobic place.[31]

What *was* considered important about these missions at the time was the political and propaganda value of the televised pictures of Soviet cosmonauts hugging cosmonauts from other communist bloc nations. Again and again, Brezhnev and the Soviet press trumpeted the love and unity that existed between them and their allies. Again and again they claimed these missions as proof that communism stood for international brotherhood and peace.

Because its propaganda value in both the Soviet Union and Czechoslovakia was so high, there was no chance that mission control would cancel the international mission and bring Romanenko and Grechko home early because of a toothache. Romanenko would have to perform on television repeatedly, acting happy and thrilled to have this first visitor from another country with him in space.

Grechko offered one other solution to the suffering Romanenko. Because his former crewmate Gubarev was commanding the international mission, Romanenko could trade places with him and go back a week earlier. "The tooth was giving him horrible pain. The nerve had been open for close to two weeks."

Romanenko refused. "If I go back early, it will somehow leave our mission incomplete. I can't do that."[32]

Thus, for seven days the two crews capered together for television cameras. Once again, the four men drank toasts of cherry juice. Once again, Grechko and Romanenko offered their visitors crackers and salt tablets as a substitute for bread and salt. On the wall of the station Grechko and Romanenko posted pictures of Brezhnev and Gustav Husak (the unpopular communist leader of Czechoslovakia) and they and the visitors listened to congratulatory telegrams from both men. Romanenko, speaking for all four cosmonauts, returned the favor with a short thank you speech addressed "personally" to "Comrades Brezhnev and Husak." "We shall apply all our strength and knowledge to defend the great honor of this international crew, which has started to carry out a joint program of socialist countries' research and utilization of outer space for peaceful purposes."[33]

Remek subsequently described his experience as a visiting cosmonaut on a Russian spacecraft at a scientific conference. Language differences at times made communications slow and difficult, compounded by a general Soviet lack of confidence in his knowledge or skills. For example, soon after landing, the joke among Russians was how Remek's hands had inexplicably turned bright red while he was in space. Doctors were baffled, until they supposedly asked Remek himself, who explained that every time he reached to touch a dial or switch, the Russian cosmonauts yelled, "Don't touch that!" and slapped his hand.[34] This joke, though untrue, nicely illustrated the disdain felt by many Russians toward outsiders like Remek, a disdain that reappeared later during many other international missions, both with their communist allies and later with astronauts from Europe and America.

These doubts were compounded by the difficulties faced by foreign cosmonauts trying to learn both a new language and a new technology in a very short time. When the Intercosmos program was first announced, Vladimir Shatalov, former *Soyuz* commander who had become director of the Soviet cosmonaut-training program, explained that it would take "about a year and a half to train the [foreign] crews, provided, of course, that there is no language barrier."[35] For many foreign cosmonauts, the training period was often less than two years, far too short for them to become fully qualified. Ironically, in the case of Remek this lack of confidence was unfounded. Even before the Intercosmos program had been announced, he had spent four years training in Moscow, and had graduated from the Gagarin Air Force Academy.[36]

On March 10, Remek and Gubarev returned to Earth. Six days later, to Romanenko's relief, Grechko and Romanenko entered *Soyuz 27* and also came home, landing in the empty plains of eastern Kazakhstan after completing a record-setting 96 days in space. The men were carried from their capsule and gently placed in lawn chairs, where they reveled in the sunlight and the warm spring air. Even though they needed help to stand and found walking very difficult, and even though Romanenko's tooth still throbbed horribly, both men could not suppress grins. They had done it! They had finally put the Soviet Union ahead of the United States! From that moment until the present, Russians have held almost every space endurance record.

Despite their joy, the weight of Earth pressed down upon them. At one point Grechko needed to urinate. With two men helping him to stand, he pissed into a bottle. To his chagrin, his urine was the color of coffee.

From then on, Georgi Grechko believed that flying in space had shortened his life. Not that he minded, however. As he said years later, "I could work in space 24 hours a day, [despite the loss of] a few years of life. . . . Which years? What years did I lose? What years have I gained? During the flight [maybe] one day counts as three. Perhaps it should be 100 for one day in space, because you'll never have such possibilities on Earth as in space."[37]

Nonetheless, upon their return the two men experienced physical fatigue, muscular pain, and mental difficulties in re-adapting to Earth gravity. Because of the bad tooth and the late visit by Gubarev and Remek, they had not been able to complete their full exercise routines during the mission's last few weeks. Once on Earth they kept trying to "swim" out of their beds in the morning, and even simple actions required thought. Even after Romanenko's tooth was pulled, it took two weeks before the cosmonauts' weight, heart rate, and other physical conditions returned to their preflight numbers. Each had lost between two and five pounds. And again, both lost bone density, as much as 7 percent in some of their weight-bearing bones. And for Grechko, the re-adjustment this time was clearly more difficult. His first flight on *Salyut 4* had only been 30 days long, and so he had recovered faster. After the *Salyut 6* flight, the changes to his body were more pronounced, and took longer to go away.[38]

Nonetheless, both men did recover, confirming what had been learned on *Skylab*: Humans could survive in weightlessness for at least a few months.

Interplanetary voyages, however, could not be completed in three months. Before humanity could initiate the real exploration of the Solar System, much longer flights, lasting a year or more, were absolutely required.

"Prisoners of Space"

On April 10, 1979, *Soyuz 33* lifted off from Baikonur, carrying Nikolai Rukavishnikov and Bulgarian Georgi Ivanov into orbit. On this, the fourth international mission to *Salyut 6*, they planned to spend about a week visiting the station's third long-term crew. At first, everything seemed to be going well. Four miles from *Salyut 6*, the spaceship's Igla radar system locked on, and *Soyuz 33* began creeping toward its docking port.

On *Salyut 6*, Vladimir Lyakhov and Valeri Ryumin watched and waited. They had already been in orbit for almost six weeks.

Less than a mile from *Salyut*, *Soyuz 33*'s main engine began a 6-second burn to help align the spacecraft with the station's port. Suddenly, after only a few seconds, the engine cut off. After some hurried analysis on the ground, ground controllers decided to try firing the rocket again in order to complete the burn. The second time, the engine cut off immediately, followed by a violent shudder that caused Rukavishnikov to reach out and grab his instrument panel in an effort stop it from shaking. In their portholes the men could see a lateral glow, possibly a right-angle plume, coming from a direction from which nothing was supposed to come.[39]

Rukavishnikov and Ivanov looked at each other. Something had gone seriously wrong.

For Rukavishnikov, this moment was like deja-vu. Eight years earlier he had been the junior member of the crew of *Soyuz 10*, the first ship to try docking with *Salyut 1*. That time, he had been unable to enter the station because the spaceship and station had failed to achieve a hard docking.[40] Now, floating a mere 3,000 feet from *Salyut 6*, Rukavishnikov asked ground control if instead of using the untrustworthy main engine, he could complete the docking with the *Soyuz*'s small attitude thrusters.

No way, said mission control. Docking with *Salyut 6* was out of the question. In fact, ground engineers were very alarmed by the situation. They considered the main engine too dangerous to use and feared that during its last firing it had also damaged the back-up descent engine. This back-up engine, not very powerful, had to burn for more than three minutes in order to slow the spacecraft enough to return to Earth. Its ignition system, however, could be fired only once. If it didn't fire, or shut down prematurely, it could not be restarted. The two men would be stranded in space—until their oxygen ran out and they suffocated five days hence. Or as the flight engineer on *Salyut 6*, Valeri Ryumin, wrote in his diary, "They might even become prisoners of space."[41]

The next window for re-entry was the following day, so Rukavishnikov and Ivanov settled down for a long, long wait. Ivanov went to sleep. Rukavishnikov waited sleeplessly. Later, he described how this 24-hour period seemed to last a month. "I was scared as hell. . . . I thought I was no longer a living being." The next day, as scheduled, they fired the back-up engine for a planned 3-minute, 8-second burn. Though the engine did not cut off early, it also failed to stop firing when scheduled. Rukavishnikov had to scramble to shut it down manually. Though they were returning to Earth, the extended burn, 33 seconds too long, made their descent far steeper and faster than intended, subjecting them to forces in excess of 10 g's, and exterior temperatures far exceeding 5,000° F. To Rukavishnikov it was like being inside a blowtorch.

They landed in the empty grassy plains of Kazakhstan, about 350 miles east of the Baikonur Space Center. When rescue crews arrived, Rukavishnikov was already out of the ship, hugging the ground. Though he tried to get onto another mission in 1984, a lingering flu caused him to be grounded one month before launch. After this, he decided that, at age 55, his spacefaring days were over.[42]

Ironically, this failure, along with a number of others during the same period, did more than any previous success to signal the maturation of the Soviet space program. How the Soviets faced the technical problems, and solved them, demonstrated that they could improvise as well as the Americans in space, and succeed, no matter the difficulties.

In the year and a half since the record-setting three-month
flight of Grechko and Romanenko, the Soviets had smashed that
record by six weeks, keeping two men alive and working on *Salyut
6* for almost five months, during which two international crews
visited them. Near the end of this five-month flight, however, a
fuel leak on *Salyut 6* threatened to disable the station. Ground telemetry indicated a leak in one of the station's three fuel tanks that
contained unsymmetrical dimethylhydrazine, or "hydrazine" for
short. Each tank was built with a solid outside body and an internal membrane bladder. Normally, to push the liquid hydrazine fuel
from the tank, gaseous nitrogen was pumped into the main body,
putting pressure on the bladder and forcing the hydrazine into the
fuel lines. Telemetry indicated that this membrane had broken,
allowing hydrazine to leak into the main body of the tank and mix
with the nitrogen gas. Hydrazine is highly corrosive, and could
damage the valves that controlled the nitrogen gas and even spread
and further disable *Salyut 6*'s entire engine system. Even though
the system had a second set of completely independent fuel tanks,
ground controllers did not want to use *Salyut 6*'s main engines before getting control of the leak.

However, without the main engines, the station's orbit would
eventually decay. Moreover, the station's smaller attitude engines
used the same tanks, and were therefore also unusable, making it
impossible to orient the station precisely. If the suspect tank inside
the donut-shaped service section could not be drained and isolated,
the space station would eventually have to be abandoned. To attempt any repairs, a *Progress* freighter had to be docked to the station so that its pumps could drain the tank. Furthermore, a crew
had to be present to monitor operations and manually adjust the
station's orientation.

In the past, the Soviets would have almost certainly abandoned the station rather than risk the lives of a new crew. This
time, they moved forward aggressively. Despite *Salyut 6*'s age,
more than a year, mission controllers decided to see if the next
crew could do repairs and keep it running. More and more, the
engineers, cosmonauts, scientists, and administrators in the Soviet space program accepted crises like this with a certain aplomb
and cold-hearted self-confidence. So what if there was a fire on
board? So what if a tank leaked? If the station still functioned,
there was no need to panic. You made what repairs you could and
moved on.

On February 25, 1979 they launched Vladimir Lyakhov and Valeri Ryumin to the station with plans to try and fix the leak. Moreover, the original plans called for Lyakhov and Ryumin to stay in space for half a year, occupying a space station that had already orbited the earth for almost a year and a half, far longer than any previous operating space station. In addition, during that time two international missions would visit them, the first bringing a Bulgarian, the second a Hungarian.

Valeri Ryumin, the flight engineer, had been in space once before, flying as part of the last all-rookie crew that had failed to dock with *Salyut 6* back in late 1977. He very much wanted to atone for that failure. Born in 1939 only a few miles from the Pacific coast several hundred miles north of Vladivostok, Ryumin had not followed the normal route for becoming a cosmonaut. Instead of attending a military or academic university upon graduating high school, Ryumin was drafted into the military, serving three years as a tank commander in Azerbaijan. A blunt, hard-nosed, but intelligent man with a shock of black hair that stuck up almost like Eraserhead's, Ryumin never backed down from any challenge, real or imagined, large or small. For him, tank command was a perfect metaphor for his personality.

In fact, Ryumin's sometimes passionate, sometimes belligerent, personality in many ways epitomized the Russian character as seen by foreigners. To his friends and equals, he could sometimes seem a saint, doing whatever was necessary to solve whatever problem. In later years, when Ryumin had become an important manager in the Russian space program, he worked long hours, late into the night, searching, cajoling, persuading people to stay on the job until the problem was fixed. "He'd use his own car, driving from home to home to try and get the help of retired cosmonauts and engineers," remembered one cosmonaut.[43] At the same time he was tough, brutal, and unforgiving to those who worked under him, the epitome of Russian no-compromise, bull-headed determination.

After completing his military service in 1961, Ryumin went to college and upon graduation in 1966 he got a job at the Korolev design bureau. At first he worked in the failed Soviet lunar program, helping design, build, and test the electrical system for the *Zond* spacecraft that flew around the moon in the late 1960s. When, in 1969, the program shifted to space-station work, he became deputy designer on the project to refit the *Almaz* station to use *Soyuz* equipment, thereby creating *Salyut 1*. The death of Dobrovolsky, Volkov, and Patsayev affected Ryumin deeply. How could he continue to build spacecraft that could kill his friends if he wasn't willing to fly in them himself? More importantly, how could he design them right if he didn't experience how they actually functioned?[44] He applied to become a cosmonaut, and went through several years of tests and training before he was finally accepted.

Now he was in space, on board *Salyut 6*, and preparing to spend six months in orbit. For the first two weeks, he and his commander Lyakhov spent their time reactivating *Salyut 6*, while also struggling to adapt to the strangeness of weightlessness. As Ryumin commented ruefully in his diary, "All the 'charms' of weightlessness are apparent. Our faces swelled—you wouldn't recognize yourself if you looked in the mirror. . . . My movements lack coordination, and I keep bumping into things, mostly with my head. Everything swims out of our hands, and wires get all tangled up."

Both he and Lyakhov discovered, as had previous spacefarers, that nothing was as easy in weightlessness as it looked. Things kept flying away from them. For example, the first time they did a stress test on the bicycle for the doctors on the ground, they found it ridiculously complicated to get everything organized. With humor Ryumin described their difficulties: "Everything gets all snarled and tangled up, and floats around, like a ball of snakes; and you yourself are floating. And with all those wires all around, you have to turn the pedals of the bicycle."

Ryumin also discovered, as Grechko and Romanenko had, that working together in the cramped isolation of *Salyut 6* was not the same as working together on Earth. "Here, every word had a meaning," Ryumin wrote later. "Even the tone was important. [You have] to be really sensitive to your friend, even more than to yourself."[45]

His commander, Vladimir Lyakhov, was a friendly, quiet man who liked fishing and was uncomfortable speaking in public. He was born during World War II in the heart of the Ukraine's coalmining region. When he was a year old, the Germans rolled through his hometown on their drive toward Stalingrad; when he was three his father, a miner, was killed in the fighting. Over the next few years his mother was forced to work in the mining industry to earn a living. Then, when he was seven, she married another miner and the boy was raised by his stepfather. At 18, rather than getting drafted into the military service as Ryumin was, he enrolled at the Kharkov Higher Air Force School, becoming a fighter pilot.

Like many cosmonauts, he had a passion for flying. When the chance came, in 1965, to apply to become a cosmonaut, he did so without hesitation, getting accepted in 1967 when he was 26.[46]

Two-and-a-half weeks after Lyakhov and Ryumin's arrival on *Salyut 6*, the next *Progress* freighter docked. After two days unloading cargo, the crew could finally begin repairing the leaking tank. First, they activated the cargo ship's engines and sent the entire *Soyuz-Salyut-Progress* complex tumbling slowly end over end, thereby using centrifugal force to separate the nitrogen gas from the hydrazine fuel. At the same time ground controllers pumped fuel from *Progress* into *Salyut 6*'s other tanks in order to empty one of the freighter's tanks. They then pumped any remaining hydrazine from the leaking tank into this empty tank. With the damaged tank empty, they sealed its fuel lines to isolate it, then opened the nitrogen gas valves to vent any remaining fuel into space. Finally, the two men used the freighter's engines to stop the complex's rotation.

During the operations there were some heated conversations between the cosmonauts and ground controllers. Ground control wanted to maintain complete control over the operation, while the cosmonauts wanted to be more than mere spectators.

Over the next few days ground controllers "rinsed" the isolated tank dry, purging it with blasts of nitrogen and leaving its valves open so that any remaining fuel could leak out into space. Then the valves were closed, the tank filled with nitrogen, and the system declared fully operational again, albeit short one tank.[47] Even with this repair, *Salyut 6*'s main engines were rarely used after this to maintain its orbit. *Progress* freighters kept the station's fuel tanks topped off in case of need, and the station's small attitude jets were used to orient it if necessary, but generally *Salyut 6*'s orbit was raised by firing the engines of one of the docked spacecraft, either *Progress* or *Soyuz*.

With the tank drained and isolated, ground controllers felt confident enough to proceed with the next Intercosmos international mission, bringing Bulgarian Georgi Ivanov to *Salyut 6*.

'Twas not to be. The failure of *Soyuz 33*'s main engines not only aborted Ivanov's mission—almost killing him and Rukavishnikov— it made the next international flight impossible. *Soyuz 34*, carrying a Hungarian, could not get off the ground before Lyakhov's and Ryumin's *Soyuz 32* spacecraft had exceeded its accepted in-orbit shelf

life of three months. To launch the next mission with a crew meant
that two men would have to come home in the outdated *Soyuz 32*
spacecraft, a risk no one in mission control wanted to take.

There were three options. Cut the Ryumin-Lyakhov flight short
and bring both men home early. Launch *Soyuz 34* manned, but
late, and risk one crew on the aging *Soyuz 32*. Or cancel the second
international mission and launch *Soyuz 34* unmanned, so that
Soyuz 32 could return to Earth uncrewed, filled instead with twice
as much experimental data and photographs as normal. Very
quickly, mission control chose the third option. Just as they had
refused to abandon the station after the fuel leak, no one in the
Russian program wanted to accept failure and cut short Ryumin
and Lyakhov's six-month mission.

To Ryumin and Lyakhov, however, this decision meant that
they must spend their entire six-month tour in space completely
isolated from other human contact. To use Ryumin's own words, it
made them prisoners of space.

The emotional atmosphere on board *Salyut 6* now became silent and moody. Both men were badly let down by the cancellation
of the visiting missions. Both also found that though they didn't
dislike each other they didn't have much to talk about either. "We
rarely talked about things other than work," Ryumin wrote.[48] The
days rolled by silently, the earth rolling under them as each man
worked wordlessly and alone.

Ryumin struggled to keep from getting depressed, noting that
his mood seemed reflected by the plants in his greenhouses, of
which he had many. In addition to a new Oasis greenhouse packed
with pea and wheat seeds, its artificial soil refined so that the roots
were forced to grow downward, he also had a small greenhouse
dubbed Fiton, which grew *arabidopsis*, a wild weed often found in
junkyards and waste dumps and chosen because it goes from seed
to full bloom and then seed again in only 40 days. Fiton was about
the size of a coffee-table book with a clear plastic panel for a front
cover. Its "soil" was five glass containers filled with a nutrient
material. When it came time to plant seeds, Ryumin released a
spring, which caused a plunger to push the seed into the nutrient
material. By having the nutrient pre-measured in this way, the watering of the plant was automatic and more precise.[49]

Ryumin also had cucumber and lettuce seeds installed in a
1-foot-wide rotating contraption called Biogravistat. Resembling a

metallic starfish, Biogravistat could be rotated to create centrifugal force, thereby allowing Ryumin and Lyakhov to grow their plants in a kind of artificial gravity. Finally, he maintained several Vazon ("vase" in Russian) containers, designed to grow bulbs like onions and tulips. Vase-like in shape, a bulb was placed in its base and then a cone-shaped cover was screwed on top to keep the bulbs inside. The cover controlled the direction that sprouts grew, while the base provided water by a hose that could be attached directly either to the station's Rodnik water-recycling system or to the tanks of water shipped from Earth.[50]

During the first three weeks of the mission, when things were going well, Ryumin had been thrilled to see a small leaf appear on the cucumber plant. "We are very hopeful that the cucumber plants will mature and color our life in this machine-filled hall," he wrote in his diary. However, immediately after the failure of *Soyuz 33* and the cancellation of all visiting missions, Ryumin wrote that "the plants that got off to such a good start growing have begun to wither." He told television viewers that same week that ". . . it is an established fact that plants won't grow in weightlessness." In fact, despite his depression, Ryumin was not entirely incorrect. During the previous five-month mission something—perhaps the violent vibrations during launch and re-entry, the harsh radiation of space, the weightlessness in orbit, or a failure of the artificial greenhouses to compensate for these conditions—had prevented the plants from reproducing.[51] As far as Ryumin could see now, the same thing was happening again.

The crew's depression was further accentuated because much of their other research was no longer possible because the damaged fuel tank limited the use of the station's attitude-control system. To avoid using the attitude engines, *Salyut 6* was allowed to naturally drift into a gravity-gradient orientation, the station's bow with *Soyuz 32* attached pointed downward to the earth. The station's three solar panels made this upright stance even more stable. As the station flew sideways through the earth's thin upper atmosphere, the two outside panels acted as wings, while the middle dorsal panel functioned as a rudder.

For most of the two weeks after the cancellation of all visiting missions, Ryumin spent his time staring out the windows. While the station's upright orientation prevented much astronomical research, it was excellent for studying the earth and its atmosphere.

Various Soviet bureaucrats and scientists, from fishermen to geologists, demanded answers to a wide range of questions, all of which they thought the men could spot easily from space. Ryumin, however, found it difficult to comprehend what he saw. The earth below him looked so alien. Even with maps, he discovered that many features were easily lost if he looked away to check a map, even for only few seconds. The earth was simply too big, and moving too fast below him. Searching to find something to do with his time, he focused on studying the beautiful but mysterious layers of the atmosphere. Two high thin layers seemed to come and go, and he obsessively searched to figure out why.

The crew's moodiness was not helped by the arrival of the next *Progress* cargo ship in May. Eager to read letters and newspapers from home, they were upset when ground controllers forbade them from opening the hatch immediately after docking. The controllers wanted to wait 24 hours to make sure the docking seal was secure. In a huff, the crew told the ground they were going to bed, and cut off communications.

Then, the next morning, they were told to wait several more orbits before opening the hatch. Still in a foul mood, Ryumin began a test of the *Soyuz* spacecraft's main engine, and erred badly. When not in use, *Soyuz*'s engine nozzle was covered with a shield to protect it from the harsh vacuum and radiation of space. During the test sequence, Ryumin forgot to open the shield, and ground controllers failed to spot his error. When he fired the engine, it burned a hole in the shield, destroying it.[52]

Though the plan was to come home in the yet-to-arrive unmanned *Soyuz 34* spacecraft, the older *Soyuz 32* had to act as their lifeboat until then. With its protective shield gone, *Soyuz 32*'s reliability became even more questionable. After several hours of discussion, ground engineers decided that as long as the station was kept in the gravity-gradient position, with the main *Soyuz* engines pointing down to the earth, the engine was protected reasonably well.

Battling depression amidst these failures and problems, Ryumin decided that he had two choices: give in to "despondency," or "forge ahead," as had the rest of the Soviet space program. He decided to work ". . . from morning till night, so that no kinds of extraneous, unnecessary thoughts [could] get into your head. So that you're not tortured with doubts."[53]

Ryumin spent hours installing a new Oasis greenhouse. He also demanded and had sent to him on the cargo ships several soft artificial soil packs. He could attach these to any wall of the station, inject water into them periodically, and let the seeds inside grow outward, the roots and sprouts dangling about the station like so many tentacles. At one point he proudly reported in a conversation to Nechitailo that ". . . we've set up quite a farm here. It looks nice and grows peas, wheat, radishes, carrots, coriander."[54]

Nonetheless, the arrival of a replacement *Soyuz* had become even more imperative. For two months engineers had been trying to pinpoint the cause of *Soyuz 33*'s failure. Eventually they decided that the *Soyuz* main-engine burn cut-off had occurred because a sensor had detected that the engine had not reached the correct operating pressure. Though new *Soyuz* engines were redesigned to eliminate the problem, the engine of *Soyuz 32*, still in space, could not be overhauled.[55]

On June 6, with a refitted engine, *Soyuz 34* was launched unmanned, docking with the station two days later. Eagerly the two cosmonauts opened the hatch and floated in. To their delight, beside the packages of fresh food and letters, they found flowers. As an experiment, Nechitailo and her cohorts had shipped them a Vazon container with blossoming tulip bulbs. The hope was that these 8-inch-tall blossoms would flourish and bud in space. Also on board was an adult Kalanchoe tree, growing from a Vazon and sent up merely to decorate the station. Very quickly, the two men adopted it, naming it the "life-tree" and making sure it was prominently displayed during every television broadcast.[56]

However, when Ryumin tried to turn on *Soyuz 34*'s instrument panel, it refused to light up. Without these controls, the spacecraft could be steered only from the ground, a condition that, while fine for an unmanned capsule, was too risky when men were aboard. After an evening's brainstorming, ground controllers suggested the men swap the switches from *Soyuz 32* to *Soyuz 34*. When they did this, the console lit up, solving the problem.

The mission rolled on. Ryumin, on his own initiative, had taken with him several quail eggs that he wanted to try hatching. He, along with everyone else, knew how important it was to find out if animal life could give birth and reproduce in weightlessness. Moreover, if he could breed birds in space, he could provide spacefarers like himself with an inexhaustible supply of food. He improvised an incubator from other equipment, and placed the eggs within it. Then, when a small insulation fire forced him to cut off the power to the incubator, he ran cables from a surplus ventilation fan to keep it working.

Despite his efforts, the embryos developed far more slowly than normal, and when several eggs finally hatched, the chicks appeared deformed and headless and quickly died. Then, to the men's further distress, the tulip petals soon fell off. The plants did not die and, in fact, their stems and leaves eventually doubled in height to more than a foot and a half. Yet, the flowers shriveled and the buds withered.[57]

As the days passed, the men wrestled with lethargy and tedium. Both became less interested in doing their daily exercises. The workouts, an hour before breakfast and an hour before dinner, were long, strenuous, and exceedingly boring. "You had to force yourself every time," Ryumin wrote. Ryumin also despised his weekly shower. "Ye-ech, that shower bath," he wrote. The set-up was even more complex and time-consuming than the shower on *Skylab*. And the cleansing soap was so caustic that if it got under his goggles it caused his eyes to burn and fill with blood.[58]

The loneliness and boredom even caused Ryumin to develop irrational fears about getting appendicitis or an abscessed tooth like Romanenko had, giving him periodic nightmares. In one case, the dream was so intense that when he jerked awake, his tooth actually ached. He found that tending the station's garden helped him overcome these obsessions. So did watching videotapes of natural scenery sent up by doctors.[59]

Despite these struggles, the men had no intention of giving up and asking to come home. In fact, they were adamant about staying and finishing their six month flight. "Can we hold out?" Ryumin wrote. "Yes, we can hold out!"[60]

To boost the crew's morale, a two-way television had been sent up on a cargo ship soon after they arrived. Once installed, the team of psychologists used it to keep the men linked with people on the ground. On the weekends a variety of famous personalities, from female singers to hockey players, were brought to mission control to chat with the cosmonauts. These weekend social calls also included visits by other cosmonauts, by friends, and by long visits with the men's families.

As the mission's end approached, Ryumin discovered that his Earth-observing skills had improved significantly. "We are beginning to [really] see the earth," he wrote in his diary. "Earlier when we looked through the window and then down at a map, we would 'lose' whatever object we were looking at. But now we are able to 're-catch' an object within seconds, as well as identify the places on Earth we are passing."[61]

After months of quiet routine and boredom, the final weeks of the mission suddenly became tense and exciting. The last *Progress* freighter had brought with it a 33-foot-diameter radio antenna, shaped something like a giant umbrella. Ryumin and Lyakhov unloaded it from the cargo ship, then assembled and attached it to the *Salyut 6*'s aft docking port so that, while still furled, it filled the inside of the docking tunnel in the center of the aft service donut. When the *Progress* freighter undocked, the antenna unfolded like a flower attached to one end of *Salyut 6*. Observing its deployment with cameras on *Progress*, ground controllers saw the wire-mesh dish spread out and entirely obscure their view of the station. For three weeks the cosmonauts and radio astronomers on the ground tested the antenna in combination with a 230-foot-diameter radio antenna in the Crimea, attempting to create with interferometry a radio antenna array hundreds of miles across.

However, when it came time to jettison the antenna on August 9, it did not drift free. Though the dish separated from the aft docking port, something held it to the station. Looking out the nearest window, the men spotted the problem. Positioned at 6 o'clock on the service donut was a 2-foot-long post, topped with a short crosspiece and used as a visual target during dockings. Somehow the wire mesh of the antenna had become entangled in this T-shaped post, either when the antenna unfolded or when it was jettisoned.

The cosmonauts could still return safely to Earth, but with the aft port blocked, *Salyut 6*'s future was limited. Ground controllers first tried pushing the station forward in hope of shaking the antenna free. Then they tried rocking the station. Neither attempt worked. The only remaining option was to have the cosmonauts do a space walk and free the antenna, a rescue attempt somewhat similar to the work done by Conrad and Kerwin on *Skylab*. The flight plans had originally called for Ryumin to do a short space walk to recover several of the experiments attached by Grechko to the docking-transfer compartment, close to the airlock hatch. The an-

tenna, however, was at the station's opposite end, 50 feet away. No
Soviet space walk had ever been planned for that part of *Salyut 6*.

Both men were willing to make the attempt, even though it
would extend their flight by about a week.[62] Ground controllers
were more reluctant. The station's two spacesuits were old, having
been in space for more than two years. The cosmonauts were tired,
having been in space longer than any previous human. And though
they could see how the antenna was caught in one place, other,
unseen, protrusions might also need release. The men would need
to improvise and bring with them a variety of tools.

After much discussion, mission control finally agreed to try it.
Simulations by other cosmonauts in a water tank at the Gagarin
Training Center indicated that the antenna could be freed. Moreover, the pressure from Brezhnev and other top communist officials to save the station so that the remaining international flights
could be flown was unrelenting.

And finally, the Russians no longer feared failure. With confidence and assurance, both the crew and the ground controllers were
determined to finish the mission as planned.

On August 15, Ryumin and Lyakhov donned their spacesuits,
evacuated the air from the airlock, and tried to open its hatch. It
was jammed, once again held in place by the slight amount of residual air still in the airlock. Both men had to pull hard to get it to
pop open, and by the time Ryumin was outside and ready to begin
the climb to the station's aft end, the sun was setting and he was
quickly plunged into darkness. Mission rules said that he could
only proceed during the 45 minutes of daylight during each 90
minute orbit of the earth. As they waited for dawn, he and Lyakhov
simply hung there, surrounded by the utter blackness of space,
sprayed by the light of billions of stars. "They look like huge diamond pins on black velvet," Ryumin thought. "We saw stars so
close it seemed we could reach out and touch them."[63]

For 30 minutes the two men waited. Then, just before sunrise,
Ryumin, impatient to get started, climbed up onto the main hull of
Salyut 6, carefully scrambling from handhold to handhold as he
worked his way aft. Behind him Lyakhov climbed out to watch
from the airlock hatch, paying out Ryumin's 60-foot-long tether. In
less than 15 minutes Ryumin reached the edge of the service donut. Above him the 32-foot-wide dish hung, rising 10 feet above his
head only a few feet away. Peering over the edge, Ryumin was re-

lieved to see that the only thing holding the antenna to the station was the post, though the antenna's metal frame had torn some insulation on the rear of the service donut.

Ryumin figured that to free the antenna he had to cut the four closest steel wires in the antenna's mesh. Each wire was "about a millimeter thick and as taut as a string." He slipped wire-cutters around the first cable to begin cutting and was startled when it snapped almost instantly, swinging the antenna violently toward him. Lyakhov screamed, "Look out! Move to the right!"[64]

Ryumin flinched away, watching the huge antenna bounce about above him. When it had settled into position he reached out again and cut the second wire. Once again the antenna swung wildly, and Ryumin dodged its huge metal frame and whipping wires.

Twice more he snipped a wire and ducked. The antenna now hung loosely on the post. To get it to float away, Ryumin took a barbed, 5-foot long pole and pushed, and the antenna slid free easily, drifting away quite fast.

Both men cheered. Then reality struck. "Everything had just gone too well," Ryumin thought. Hurriedly he began his retreat, scrambling back along the station's hull. Then he regained his composure, forcing himself to slow down, to inspect the station he had helped design. He noticed how much of the hull had become discolored or damaged from exposure. At the hatch he stopped to wipe clean one porthole with a cloth. "I thought, some specialist will want to examine this space dust." Maybe it would help explain why the portholes clouded over with time.

The next day was Ryumin's 40th birthday. As congratulations poured in from the ground, from friends, family, and co-workers, he found that he didn't have the time to take the day off and celebrate. They were heading home in only three days, and he needed to finish packing. Only that evening did he and Lyakhov stop and have a holiday supper, eating what Ryumin called the last remaining "delicacies" left on the station.[65] As they put *Salyut 6* in mothballs, the two men wrote a letter for the next crew, giving them advice on how things worked and what problems to watch out for. Then they tacked it to a wall, plainly visible for the next crew to find.

Three days later Lyakhov and Ryumin came home. During their six months in orbit they had religiously exercised two hours a day, a regimen that had paid off. They recovered as fast as the previous crew (140 days in space), and far faster than the first (96 days in space). In fact, both men had actually gained weight during the flight, a first. Nonetheless, both showed muscle and bone-mass loss, and both had a slight difficulty speaking immediately after landing as the fluid distribution in their bodies re-adapted to gravity. The bone loss was the most troubling result. Both men had lost between 3 and 8 percent in bone calcium (rates of 0.5 and 1.2 percent per month), most of which seemed to come from weight-bearing bones such as the legs and hips. Soviet scientists announced their belief that, with proper exercise, the bone loss was manageable. (Ryumin for example had exercised more than Lyakhov, and had less bone loss.) American scientists had more doubts.[66]

Ryumin and Lyakhov also brought back with them samples of their space-grown wheat, peas, onions, and cucumbers, as well as their flowerless orchids. To the delight of the biologists, the tulips almost immediately began to bloom again, almost as if the flowers needed gravity to appear. To their disappointment, the other plants were sterile, and had produced no seeds. Many had withered, or were far smaller than their Earth-grown counterparts. Even the plants on the rotating Biogravistat had not prospered. Both parsley and cucumbers had grown quickly, but then died. Despite improvements to the watering system, the plants still struggled to get water. It seemed that without gravity to guide the water down through the layers of artificial soil, it often failed to reach the roots, thereby starving the plants. More design changes were necessary.[67]

The most important knowledge gained from the mission was probably psychological. By the end of Ryumin and Lyakhov's flight—the third consecutive mission lasting more than three months—Soviet scientists were beginning to realize that the isolated and alien environment of their small *Salyut* space station was more isolated and alien than they had imagined. Generally, behavior seemed to follow a three-stage pattern. During the first two weeks in space, men found themselves disoriented and overwhelmed with an overload of sensations. They were launched in a rocket, and then bounced around in zero gravity. Then they went from a small capsule to a big station. Their blood circulation was jerked around, shifting from their legs to their faces and heads. Many got sick from the experience.

After about four weeks they became acclimated to the space environment, and the second stage, one of endurance, began. The men were isolated and alone. They had little contact with Earth, and lived in an artificial environment about the size of a single large living room, a "metal hall" as Ryumin described it, with a partner they did not choose, and from whom they could not escape. For some, such as Volynov and Zhobolov on *Salyut 5*, the stress was simply too much. They had to come home. For others, like Grechko and Romanenko as well as Lyakhov and Ryumin, the emptiness and isolation at first caused tension and sometimes conflict, followed by a mature effort to get past these problems and endure. In the case of Grechko and Romanenko, the effort made them much closer friends. "In the past we shook hands when we met," said Grechko. "Now we embrace."[68] For Lyakhov and Ryumin, the solution was to simply not talk to each other, making much of their time on the station boring and interminable.

The final behavioral stage occurred during a mission's last few weeks, when crews began to eagerly look forward to their return to Earth. Most overworked themselves, like Grechko, doing more exercises than scheduled on the treadmill and bicycle and sleeping little. For Ryumin and Lyakhov, the stress of this final stage was mitigated because the problems with the radio antenna distracted them and kept their minds focused.

For the next eight months *Salyut 6* remained unmanned. During that time an unmanned test *Soyuz-T* spacecraft was docked with the station for 95 days, testing a next-generation, modified *Soyuz* capsule that could allow three-man missions for the first time since the deaths of Dobrovolsky, Volkov, and Patsayev.

During those same eight months, while Soviet engineers were putting *Soyuz-T* through its paces, Soviet troops rolled into Afghanistan.

"Gross Interference"

On Christmas Eve, 1979, the 105th and 103rd Guards Airborne Divisions of the Soviet air force began an intense airlift, bringing, in just over two days, thousands of troops into the airfields surrounding Kabul, the capital of Afghanistan. At the same time, tanks and armored vehicles poured south from the Soviet border, moving on the cities of Kushka, Herat, and Kandahar. By December 27, with Soviet troop concentrations at more than 5,000 men, the operation rolled into Kabul. At 7:00 P.M., ground forces blew up the city's main telephone exchange. Fifteen minutes later the Ministry of the Interior was captured. By dark, Soviet personnel controlled key intersections, post offices, ammunition depots, the radio station, and all government buildings.[69]

Several hundred Soviet commandos—wearing Afghan army uniforms and driving vehicles with Afghan markings—attacked Darulaman Palace, where the communist ruler of Afghanistan, Hafizullah Amin, lived. He had moved there only a few weeks earlier for safety, acting on the "advice" of his Soviet advisors. In a battle that lasted four hours, Darulaman Palace was occupied and Amin was killed. Soviet troops found him drinking at one of the palace bars, and shot him instantly. That same evening, using a Soviet transmitter on Soviet soil but broadcasting on Kabul Radio frequency, Barak Karmal, one of the founders of the Afghan Communist Party but exiled in the Soviet Union for the past two years, announced that he was now in charge.

In a move that put the final nail in the coffin of his policy of detente, Leonid Brezhnev and the fellow members of the Soviet Politburo had decided to invade and occupy Afghanistan in order to prop up its failing communist dictatorship. In power for only 21 months, the rule of the communist People's Democratic Party of Afghanistan, led by Nur Mohammad Taraki, had been brutal, violent, bankrupt, and unsuccessful, a failure that even today, a quarter of a century later, causes problems worldwide. Its leaders had imprisoned 30,000 people and executed 2,000. They had tried to force through a land-reform program that bankrupted the small farmers. They had threatened the practice of religion, in a country 100 percent Muslim and devoutly so.[70] By September 1979, the populace of Afghanistan was in an uproar, ready to rebel.

In the Soviet Union, Brezhnev and the Politburo were also losing their patience. As oppressive as the Soviets usually were, Taraki's government had exceeded even their tolerance. In September they had even suggested that Taraki fire his prime minister, Hafizullah Amin, whom they blamed for most of the problems. Instead, Amin found out about the plot and had Taraki killed so that he could take over the government himself. He then accelerated the purges and executions. If the Soviets did not take action, Afghanistan could become the first communist government in history to fall. Worse, Afghanistan was a Soviet neighbor, and its downfall carried dreadful consequences for Soviet security.

In the past, Brezhnev could have accepted a neutral but agreeable neighbor in Afghanistan. However, after communist rule was established he could not tolerate a retreat to neutrality. As he declared defiantly in Moscow on the occasion of Barak's first trip outside Kabul after the invasion, "The revolutionary process in Afghanistan is irreversible. Time works for new revolutionary Afghanistan." Less than a month after the start of the Soviet invasion, more than 50,000 Soviet troops, including 1,750 tanks and 2,100 infantry combat vehicles, were in Afghanistan, controlling all the major airports, military installations, and cities.[71]

Brezhnev badly misjudged the international consequences of the Soviet invasion of Afghanistan, however. By 1979, detente between the United States and the Soviet Union had already showed signs of failing. In the four years since the signing of the Helsinki Accords, the Western nations had become increasingly impatient with the Soviets' truculent refusal to institute human-rights reforms. Refuseniks and dissidents continued to be arrested and imprisoned, and the borders to the communist bloc remained closed. The Afghan invasion brought about the complete and final end to detente. U.S. President Carter withdrew the SALT II Treaty from Congress. He imposed a grain embargo. He blocked the sale of computers and high technology to the Soviet Union. He postponed the renegotiation of a cultural exchange agreement. And he pulled the United States out of the 1980 Moscow Olympics. As Carter declared in a nationally televised broadcast within days of the invasion, "The world cannot stand by and permit the Soviet Union to commit this act with impunity." He further added that ". . . neither the United States nor any other nation which is committed to world peace and stability can continue to do business as usual with the Soviet Union." Or, as one senior administration official put it, "The probability is that U.S.-Soviet relations will be at a very low level for years to come."[72]

The "generation of peace" and "peaceful co-operation" proclaimed by Nixon and Brezhnev seven years earlier was officially over.

Meanwhile, Soviet engineers spent the spring of 1980 finishing their tests of the first *Soyuz-T*. The new spacecraft was going to significantly increase the capabilities of the Soviet manned program. It carried additional fuel and electrical capacity, including the reinstatement of two solar panels to the outside of the service module. Its propulsion and computer systems were also redesigned.

These changes meant that *Soyuz-T* could remain in orbit undocked for four days, twice as long as the older *Soyuz*.

Two days after *Soyuz-T* returned to Earth, a *Progress* freighter lifted off, docking with *Salyut 6* on March 29, thus heralding the beginning of another manned mission. *Soyuz 35*, planned to last six full months, lifted off two weeks later, carrying rookie Leonid Popov and, surprisingly, Valeri Ryumin!

After his return the previous August, Ryumin had intended to spend several years working on the ground in mission control. Since graduating from school, he had always changed jobs every few years. "Otherwise," Ryumin explained in his memoirs, "a person begins to 'fall asleep' at the job." After finishing his tour of military duty he trained to be an electrical engineer. Then he spent three years building the *Zond* spacecraft. Next he spent three years help- ing to design and test *Salyut 1*. Then, for the last eight years or so, he had been a cosmonaut.[73]

Now he wanted to try running the missions from the ground. The idea of spending several years as a backup in order to get an- other flight assignment didn't appeal to him. Instead, he figured management would be more interesting, since it would put him directly in charge of each new flight. The next launch crew was supposed to be Leonid Popov and Valentin Lebedev. Ryumin was eager and ready to work from mission control with both men.

Six weeks before launch, however, Lebedev injured his knee exercising on a trampoline. At first everyone thought it was merely a sprained knee. Then the pain got worse, and doctors told him that an operation was necessary. Two days after this accident Ryumin got a call from Aleksei Yeliseyev, mission flight director at the time, under whom Ryumin was working as he prepared to join mission control. The two men were supposed to attend a conference to- gether, and had to work out a meeting place beforehand. After chat- ting about this for several minutes, Yeliseyev mentioned offhand- edly that Lebedev had torn his knee ligaments. The cosmonaut had been grounded, and that Yeliseyev wanted Ryumin to fly in his stead.

Ryumin was astonished. He didn't know what to say.

"You have tonight to think about it, and tomorrow we'll talk," Yeliseyev said as he hung up. In other words, Ryumin was being ordered to go. In Soviet society, one could rarely refuse a direct "request" like this.

Despite the objections of his wife and children and despite having never trained with Popov, Ryumin decided it was in his best interest to do it. He knew the station and was familiar with its problems. He was still fascinated with studying the sporadic and unpredictable glows in the atmosphere's upper layers, and thought his previous experience might help him discover their cause.[74]

Unlike his previous flight, which had no visitors, Ryumin's second mission to *Salyut 6* saw the station transformed into a traveling motel. Four different crews arrived at regular intervals, including three international missions and the first manned flight of the redesigned *Soyuz-T* spacecraft. The three international missions brought a Hungarian, a Vietnamese, and a Cuban to space. Each was accompanied by the typical pronouncements of peace and international cooperation. Just before the Vietnamese visitor's arrival, which had been timed to coincide with the 1980 Olympics in Moscow, Ryumin and Popov participated in the opening ceremonies, reading an uninspired message to the crowds on Earth. "Let the Olympic fire of friendship burn on Earth always. Let people vie with one another only in sports arenas."[75]

Most of the world ignored the event. When Brezhnev first announced these international missions three years earlier, their propaganda value seemed priceless. Now, with numerous Russian dissidents in prison, with Soviet troops in Afghanistan, and a U.S. boycott of the Moscow Olympics, the statements seemed nothing more than empty symbolic gestures.

Before, between, and after these visits Popov and Ryumin spent their time maintaining the station, doing minor repairs and equipment checks. They replaced control panels, wiring, ventilation fans, anything that had to be fixed. They discovered that with a little innovation they could keep even the most complicated piece of equipment working. For example, rather than struggle with numerous and clumsy 10-pound tanks, they figured out a way to pump water directly from a *Progress* freighter into the Rodnik water tanks on *Salyut 6*. They attached a hose to the tanks and ran it from *Progress*, through the docking port that separated the two ships and to the Rodnik system.[76]

Unlike his first flight to *Salyut 6*, Ryumin's second was a much livelier and happier mission. Leonid Popov was a warm, easygoing man with a sense of humor that fit well with Ryumin's. For example, when the two finished their last breakfast on Earth before climbing into their spacesuits for launch, they found they still had two uneaten cucumbers. Ryumin, remembering his failure to grow cucumbers on the previous flight, suggested they pocket both and take them into space. He wanted to have some fun with Galina Nechitailo, who so passionately wanted to harvest plants in space and had so far had so little success. He also knew that at that moment she was in the hospital. A joke might cheer her up.

Three days later, during their first space telecast, Popov aimed the camera at one of the station's greenhouses while Ryumin showed off his new garden. Among the dead stalks and seeds left over from eight months earlier lay one full-size cucumber. As Popov filmed, Ryumin innocently explained that they had been shocked to discover this cucumber when they first came on board. He thought the cucumber must have grown by itself during the last eight months. Everyone in mission control was speechless. Then they began peppering the men with questions. Eventually, ground controllers decided that it must be a plastic cucumber, which was what was reported on television that night.

Later, when Ryumin heard about this report, he kicked himself. "We should have taken a bite while we were on television."[77]

Before launch Ryumin had asked a radio reporter if he would help him create a fake news report of the cucumber discovery and take it to Nechitailo's hospital bed. The man, who knew her, heartily agreed. That night the reporter brought a tape recorder to her bedside and played the report. "So, here you are, wasting time, while Ryumin grows huge cucumbers in space. What are you going to do about it?"

Almost a quarter of a century later, Nechitailo remembers this moment with affection. "There weren't a lot of people in our space program," she remembered. "We all had close and warm relations."

This was not the only joke Ryumin and Popov played on Nechitailo. Later, during the Vietnamese visit, Ryumin joked that one of the Vietnamese plants had blossomed in honor of the visit. To his astonishment, Galina and other ground controllers believed him. Nechitailo, who, as a scientist, was actually Ryumin's superior in the Soviet hierarchy and could get him grounded if she wanted to, got very excited about the possibility of the first blooms in space. She gave him careful instructions on how to preserve and pack the flower so that it could be sent back when the visiting crew returned to Earth a few days hence. Then she arranged to fly to the landing in the remote emptiness of Kazakhstan so that she could be on hand to get the flower as soon as possible.

Rather than reveal the truth, Ryumin decided to fashion a fake flower out of some pink paper he happened to have on board. When Nechitailo got on the radio after the return of the Vietnamese, Ryumin nonchalantly asked her, "Galina, how's the flower?" All she could do was laugh, and demand that he grow her some real flowers in space.[78]

Nor did Ryumin and Popov limit their jokes to Nechitailo. Late in the mission, they told the director of the medical group ". . . to have the same girls at our landing who put the medical belts on us during launch." They explained, deadpan, that neither man had been able to figure out how to take the belts off, and that they needed these women to help.

The doctor was credulous. "You mean, you've been in [the belts] for the entire five months?" Ryumin and Popov saw no reason to explain, insisting that they needed those girls to help. The next day the director began to question them about whether they were having nightmares and hallucinations.

Soon after, Popov decided to play a joke during a television broadcast. He and Ryumin rigged an empty spacesuit with cables and a tape recording. Halfway through the broadcast, a knock was heard coming from the closed docking hatch behind them. The two men turned in surprise, saying "Who's there?"

A prerecorded voice answered, asking for permission to enter. Then Ryumin pulled a cord, opening the hatch and pulling the spacesuit into the station, directly at the camera. Once again ground controllers were speechless. Then everyone broke out into laughter.[79]

Despite these on-board shenanigans and visits from four different crews, Ryumin and Popov still had to work hard to get through the long months of isolation and loneliness. The time cramped together in such a small and artificial space once again began to wear on them, draining both men of enthusiasm. Two months into the mission their treadmill broke. Rather than fix it, Popov and Ryumin simply stopped exercising. Ryumin, who had hated the workouts on his first mission, wrote in his diary how the repair ". . . meant unscrewing a lot of bolts and would take a lot of time to repair." When their doctors found out, the cosmonauts were ordered to make the repair anyway. Five months into the mission, they decided to cancel their weekly shower. After two missions, Ryumin had had enough. The work required to set up and dismantle the shower, almost a whole day, made the effort seem pointless.[80]

Their loneliness became most obvious with the arrival of each new visiting crew. The two men stayed up late talking with their visitors, giving toasts, and eating with relish the newly arrived fresh food. Then, after their tired visitors had finally gone to sleep, Ryumin and Popov spent many additional hours reading and re-reading their mail. "How wonderful it was to get letters in orbit," Ryumin mused happily in his diary.[81]

Research helped keep them focused. Once again, they tended a variety of greenhouses. Once again, Ryumin turned *Salyut 6* into a veritable garden, with plants hanging everywhere, cultivating on-ions, peas, radishes, garlic, cucumbers, parsley, and dill. And once again, they found that space seemed a difficult environment for plant life. The flowers on their orchids fell off. Though onions and garlic shoots seemed to prosper long enough to produce seeds, no seeds appeared. Moreover, some *arabidopsis* and hawksbeard seeds that were brought to space at the beginning of Grechko's mission showed significant chromosomal damage when Ryumin brought them home nearly three years later.[82]

Nonetheless, in their fifth month in space, the two men suc-ceeded in getting an *arabidopsis* plant to bud, the first time buds had ever grown in space. Ryumin wondered whether the redesign of the greenhouse, which kept the plant's atmosphere separate from the station's, could have helped the plant's growth. However, when Energia's biologists studied the plants on the ground, they were sadly disappointed. Though the plants had grown from seed and developed seeds, the new seeds were sterile. Whether it was weight-lessness, the design of the greenhouse, or some other factor that caused the sterility, no one yet knew.[83]

Popov and Ryumin returned to Earth on October 11, 1980. The first few minutes after landing were difficult. They were carried from the capsule to lounge chairs, where Soviet reporters photo-graphed them and asked them some questions. Once again Ryumin was overwhelmed with joy at smelling the clean air and seeing the grass. Then they were carried to a temporary medical tent set up nearby. To Ryumin, everything felt twice as heavy, as if he were in a 2-g environment. "I didn't feel very good," he remembered.

When it came time to walk the 1,000 or so feet from the tent to the helicopter, however, Ryumin insisted on walking. Not only did this short stroll feel like weights were pressing down on his body, his coordination and balance were confused.[84]

Within a month, however, both men were completely back to normal. Their circulation readjusted, their muscles regained their strength, their sense of balance returned. Ryumin's return to normal was especially significant, because he had spent 12 of the last 20 months in space.

The loss of bone calcium remained the only significant medical unknown. Ryumin's exercise routine once again seemed to slow the rate of bone loss, though the workouts didn't stop it. During the second mission he lost 4 percent of his bone tissue in his heelbone, around 0.75 percent per month, a slow rate of loss compared to others.[85] Whether a human skeleton could remain strong enough over longer periods was still unclear.

The final two crews to occupy *Salyut 6* stayed in space for relatively short periods, focusing more on testing new equipment or keeping the aging station operating so that the last few Intercosmos missions could be flown. Originally designed to operate between 18 and 24 months, by late 1980 the station had been in orbit more than three years. Every additional mission was a bonus.

Soyuz-T 3, launched November 27, 1980, was very short, only 13 days long. Its main purpose was to re-initiate Soviet three-man space flights. The crew, veteran Oleg Makarov and rookies Leonid Kizim and Gennady Strekalov, devoted their time to either testing the *Soyuz-T* spacecraft or performing maintenance and repair work on *Salyut 6*. The cosmonauts installed a new hydraulic unit for the laboratory's temperature-control system, as well as replacing several electrical components.

After their return to Earth on December 10, 1980, *Salyut 6* remained unoccupied for three months until the launch of *Soyuz-T 4* on March 12, 1981 and the orbiting laboratory's sixth and last manned occupancy. Crewed by Vladimir Kovalyonok and Viktor Savinykh, this 74-day mission's goals were simply propaganda and maintenance. The last two international flights in the Intercosmos program brought cosmonauts from Mongolia and Romania to the station to complete the program.

Though they focused on repairs and maintenance, the cosmonauts also tended the station's greenhouse, attempting, as had previous crews, to grow plants and flowers from seed. While an *arabidopsis* plant flowered once more, they still could not get seeds to bud. Nechitailo suspected the cause was lack of light. Maybe if they increased the light to the plants, they might finally germinate.[86]

Kovalyonok's and Savinykh's return to Earth on May 26, 1981, signaled the end of *Salyut 6*'s manned operations. By now the station had the smell of an "old car." Its fabric-covered walls were stained with food. Its gear needed continuous maintenance. Debris had begun to accumulate in its nooks and crannies.[87]

The station's last task was to see if two several-ton modules could dock together and function as a unit, using a transport-support module conceived by Chelomey to support his *Almaz* station. On April 25, 1981, one month before the return of Kovalyonok and Savinykh, this transport-support module, dubbed *Cosmos 1267*, was launched from Baikonur. *Cosmos 1267* weighed 15 tons and was about two-thirds the size of a standard *Almaz* station. Docked to its port was a *Merkur* capsule (weighing another five tons), originally designed by Chelomey to carry humans to and from orbit.[88] After four weeks of orbital tests, the *Merkur* capsule separated and returned safely to Earth.

After two more months of orbital maneuvers, the module docked with *Salyut 6* in mid-June, several weeks after Kovalyonok and Savinykh had returned home, creating an orbiting facility weighing about 35 tons with a total volume of about 5,000 cubic feet, less than half the mass and half the volume of *Skylab*. For four months the two modules flew in orbit together, using *Cosmos 1267*'s engines to make several orbital changes. Then the complex was allowed to drift in orbit for an additional nine months while ground engineers tested its combined systems to see how they were holding up in space. Finally, on July 29, 1982, *Cosmos 1267*'s engines were fired one last time, bringing both modules out of orbit to burn up over the Pacific Ocean.

This *Salyut* complex demonstrated that a many-ton structure made of two spacecraft modules could be operated in space safely for long periods. The next Soviet station would attempt to build such a structure, and this time put men inside.

Morality sometimes carries with it a momentum unintended. Sometimes those who claim high moral positions for superficial reasons later find themselves trapped by those positions, and have no choice but to follow them, against their will.

For more than a decade Brezhnev and those who ruled the Soviet Union with him had been claiming that they were for peaceful co-existence. Yet again and again, they gave the lie to this statement. They built barbed-wire walls that divided cities and kept children from parents, husbands from wives. They invaded and occupied their neighbors—Hungary, Poland, East Germany, Czechoslovakia, Afghanistan—using brutal military force to impose their will on populations that simply wanted to live their lives in peace.

By the early 1980s, when *Salyut 6* was finally de-orbited, the contradiction between what Brezhnev said and what he did was clear. Few believed his words. Instead, it was President Reagan's words that moved people, as he called the Soviet Union an "evil empire" and demanded, standing before the Berlin Wall, that Gorbachev "tear down this wall." In western Europe, political resistance to the installation of more-sophisticated military missile systems weakened. In the Middle East, countries like Egypt, Syria, and Iran, though often hostile to the West, decided that they wanted even less to do with the Soviet Union, kicking Soviet advisors and military troops out of their countries.

The military actions of the Soviet Union also made the nine international missions to *Salyut 6* seem nothing more than empty gestures. So what if a Czechoslovakian, a Pole, and an East German had flown in space, if the citizens of these countries remained imprisoned and oppressed? The Soviet Union and the communist movement were simply not as open as these space missions tried to suggest. Instead, they were far more restrictive and oppressive than the tiny "metal hall" in which Ryumin had spent a year.

And yet, hollow as Brezhnev's high-sounding words sounded, they had an effect—one that was quite unintended. For 70 years the communists, culminating with Brezhnev, had devoted enormous energy to preaching the lie to the Soviet public that communism stood for prosperity, freedom, democracy, and justice. The Soviet public had listened to this propaganda, and learned. As Russian historian Rachel Walker has noted, "The many peoples of the Soviet Union, who had never experienced democracy and had no history of democracy, learned the language of democracy from the [Communist P]arty itself."[89]

The Soviet space program illustrates this perfectly. Brezhnev intended his international program as nothing more than a series of superficial publicity stunts. The Soviet people, however, took

him at his word, learning from these flights the lesson that "peaceful co-existence" and "world peace" were more important than establishing a global communist utopia. Moreover, the openness that
the Americans had demanded during the *Apollo-Soyuz* missions
was like a drug. Even though press coverage during the *Salyut 6*
missions was never as unrestricted as during the joint mission, it
was far more open than it had been before *Apollo-Soyuz*. The genie
was out of the bottle. Soviet citizens had seen a glimpse of freedom, and liked it.

Under Brezhnev's status quo, play-it-safe leadership, communism had failed to give the Soviet people freedom, or fulfill any of
the promises it had made since the October Revolution in 1917.
Instead, the entire centralized communist system had steadily crept
toward bankruptcy. Both agricultural and industrial production had
declined. The lack of consumer goods encouraged a black market
and extensive corruption. And the citizens of the Soviet Union remained trapped in a burdensome, overwrought bureaucracy that
worked incessantly to stifle their freedom. "Stagnation" was the
code word used by most Soviet citizens to describe the decaying
system under which they lived.

Having listened to Brezhnev's words, the Soviet public more
and more wanted the words to be more than empty propaganda.
The consequence of this desire was literally inconceivable, for
Brezhnev, his comrades in the Communist Party, the citizens of
the Soviet Union, its nation's space program, and every person
worldwide.

6

Salyut 7: Phoenix in Space

Birth

On November 10, 1982, 75-year-old Leonid Brezhnev died. After 18 years of trying to persuade the world that peace and justice could be obtained by image and good feelings rather than action and justice, a weak heart finally put an end to Brezhnev's life.

Now Yuri Andropov, the new leader of the Soviet Union, led a contingent of Politburo members, including Konstantin Chernenko, the man many thought Brezhnev had favored as his successor, to Brezhnev's funeral bier. Nearby sat Brezhnev's wife, Viktoriya, and his daughter, Galina. The day before, Andropov had been chosen as the new head of the Communist Party, and immediately declared his intention of continuing Brezhnev's policies. Despite a sagging economy, crumbling infrastructure, high debt, and growing corruption, the Soviet Union was going to continue its journey down the road of communism and centralized rule.

The arrival of Andropov and his cohorts at Brezhnev's funeral bier began three days of mourning. Brezhnev's body lay in state, surrounded by flowers, serenaded alternately by an orchestra and a string quartet, playing dirges. Then, in a ceremony of pomp and ritual exceeding anything since the death of Stalin almost 30 years before, Brezhnev was buried in the front of the Kremlin Wall, in a position second only to Lenin's.

Fittingly, the master propagandist left the real world with a parade of propaganda. In order to guarantee that his funeral was attended by a large gathering of mourners, thousands of Communist Party members, factory workers, soldiers, students, and farmers had been deputized by their bosses to come to Moscow and view the body. Silently they lined up, three abreast, to trudge past the body of a leader none of them had chosen.[1]

At that same moment, high overhead in space, the first two occupants of the *Salyut 7* space station had other concerns. Neither man could stand the other, and with only days to go before setting a new in-space flight record, one of them suddenly became very sick.

Valentin Lebedev awoke on November 10 to see Anatoli Berezovoi gripping *Salyut 7*'s treadmill, writhing in pain. "What's wrong?" he asked with sudden concern.

"I don't know," Berezovoi responded. "I don't feel well; maybe it's something I ate. My left side hurts."

Ironically, this was the first civil conversation between the two men in days. Two days earlier they had argued about how much time each spent talking to his family. Lebedev felt that Berezovoi had dominated the last short communication session, robbing him of his fair share of talk time with his wife and son.[2]

For six months, Lebedev and Berezovoi had struggled to complete the first mission to *Salyut 7*, their effort to work together becoming at times the equivalent of a fingernail scratching on a blackboard. They had fought with ground controllers, with their doctors, and with each other. Somehow, despite the conflicts, they managed to persevere, stretching their mission to six months and getting approval to extend it long enough to set a new record in space.

Now Berezovoi's illness threatened that record. When Lebedev told mission control about the problem he was immediately told to prepare for a quick return to Earth. "It was terrible!" Lebedev wrote later. "Ten years of training and just one week before setting a new record we might have to land."[3]

Launched on April 19, 1982, *Salyut 7* was very similar to its predecessor, *Salyut 6*. It weighed essentially the same, about 20 tons, provided about the same volume of habitable space (about 3,300 cubic feet), and included a similar assortment of telescopes, cameras, materials, furnaces, plant greenhouse experiments, and other equipment. Its impractical and bulky chairs had finally been replaced with collapsible stools resembling bicycle seats and designed for weightlessness. In the narrow-diameter section a refrigerator had been added, as well as a water heater. The walls were covered with a removable and washable fabric, so that if stained cleaning was easy. The ventilation fans were quieter and had dust collectors. The food was improved, depending more on freeze-dried meals organized so that each man could pick and choose any meal combination, rather than having a pre-planned day-by-day menu set before launch.[4] Besides these small changes, however, *Salyut 7*'s interior was almost identical to that of *Salyut 6*. It still had no private cabins for the crew, and its largest room was still dominated by the telescope housing originally designed for military surveillance telescopes.

The station's exterior had the most changes, though they, too, were subtle. In an effort to keep the windows from clouding, hinged covers had been fitted to each porthole to protect them when not in use. To make space walks more efficient, numerous handholds had been added to the station's exterior. During several planned space walks cosmonauts were to use the handholds to manually attach two supplementary solar-panel strips to each main panel, the panels already provided with special attachment clips on their outside edges. These supplementary strips would increase each panel's surface area and power production by 50 percent. Later space walks would also test several experimental welding techniques. The station's docking ports had been strengthened. The engineers at Energia wanted to assemble separately launched large modules to form a larger overall station. Based on the test docking of *Cosmos 1267* to *Salyut 6*, they had redesigned *Salyut 7*'s ports, making them stronger and more substantial.

The plan was to fly incrementally longer missions to *Salyut 7*, from seven months to eight to ten and finally year-long missions. After the first seven-month shakedown flight by Lebedev and Berezovoi, later eight-month and ten-month missions would each be augmented with supplies and space from transport-support modules like the module that had been docked to *Salyut 6* near the end of its life. This transport-support module, launched with a *Merkur* capsule, would be docked temporarily to the bow port. Later, a

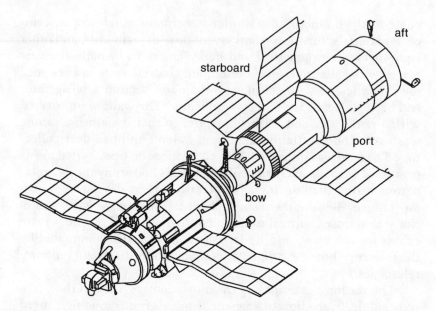

Salyut 7, with a transport-support module docked to the bow port. The side strips have not yet been added to any of *Salyut 7*'s solar panels. *NASA*

shortened version, dubbed *Kvant* (meaning "quantum" in Russian) and launched without *Merkur*, would be added permanently to the aft port.[5]

The addition of modules was expected to significantly increase the station's capabilities, almost doubling its interior space. The extra solar arrays would bolster the electrical supply, while its more than 5 tons of cargo would be triple the payload of a *Progress* freighter. And the *Merkur* capsule would allow the safe return of more than 1,000 pounds of experiments, specimens, and exposed film. *Kvant*, meanwhile, would add about 1,400 cubic feet of habitable space to the station.[6]

The problem was that all that capability would be useful only if the humans occupying the space station could successfully work together. For the first crew of *Salyut 7*, however, even talking was difficult, and avoiding a fight barely possible. The first indication of a problem appeared a year before launch. From the first day Anatoly Berezovoi and Valentin Lebedev started training together they found it difficult to avoid arguing. Their verbal dueling became so bad that flight controllers at one point demanded a promise from the two men that they would not argue during the flight.

A moody and emotional man, Valentin Lebedev had struggled for more than a decade to become a cosmonaut. Born in Moscow during World War II with a military officer for a father, he had grown up entranced by the heroic propaganda of the war and postwar period. Though his parents wanted him to attend college, he imagined himself a pilot and instead applied to pilot school. After failing his first medical exam (his pulse was too high), he asked and got permission to take it again. Purposely arriving an hour early, he found a quiet spot on the grass nearby and stretched out under a birch tree to calm his mind and heart. This time he passed. While at pilot school Lebedev applied for the cosmonaut program, only to be rejected because his blood pressure was too high. Three times he applied, and three times he was rejected for the same reason.

After graduation, he got a job at the Korolev design bureau, working on the giant N1 rocket that was supposed to compete with the Saturn 5. There, he applied to the cosmonaut program again, only to be rejected again. For seven years he tried again and again to get into the program. Over the years he took the blood pressure test eight times, failing each time. Meanwhile, he continued his pilot training, learning to fly gliders, helicopters, jets, and turboprops. Finally, in early 1972, he passed the medical exam, and was accepted as a cosmonaut. He considered himself "calmer and more mature." In addition, the year before he had joined the Communist Party, which certainly helped to grease the wheels and get his application accepted.[7]

Ironically, once he was finally accepted into the program Lebedev got his first flight into space remarkably fast. Chosen as third backup for *Soyuz* 13, he quickly moved up to second backup and then to the prime crew when the original crew was grounded because of an inability to work together or with anyone else. At one point during training the original crew wouldn't even sit at the same lunch table together.[8]

The situation was somewhat absurd. For example, the Energia biological team had built an experiment for cultivating hydrogen-breathing bacteria in space in the hope that these bacteria could help recycle a spaceship's atmosphere. The original crew called this device a "hydrogen bomb," and demanded that its back-up unit be blown up, to prove that an atomic explosion would not occur in space. Then they resisted doing the scientific homework necessary to prepare for the mission. "I gave them poor marks," Nechitailo remembered.

Thus, only 15 months after becoming a cosmonaut, Lebedev was in space. He and Pyotr Klimuk formed an all-rookie team, flying the short eight-day research mission of *Soyuz* 13 in December 1973.

Following this mission, Lebedev took a four-year break to focus on other projects, including getting his PhD. Though he wanted to fly in space again, his demanding and emotional personality made getting picked for another mission difficult. Often Lebedev would get passionate about some petty issue, and drive everyone around him crazy until those passions were satisfied. For example, as part of the civilian cosmonaut corps, Lebedev noticed one day that military cosmonauts had uniforms, and were instantly recognizable because of it. "People salute them and not us," he complained. He decided that the civilian cosmonauts needed their own uniform, and spent the next few months harassing every Energia official he could think of to get the uniforms designed and made. Only when he had a uniform was he satisfied. That no one wore it did not seem to matter to him.[9]

By the late 1970s he was finally getting new assignments, acting as backup for two *Salyut* 6 missions. Then, picked to fly with Popov in 1980, Lebedev was grounded shortly before launch when he tore ligaments in his knee exercising on a trampoline. Recycled into the crew rotation, he was picked to fly the first mission to *Salyut* 7 a year later.

His crewmate, Anatoli Berezovoi, was making his first space flight. Born only three days before Lebedev, Berezovoi was a neat, tidy, and hard-nosed military man raised by a peasant family in the mountainous regions of the North Caucasus. It was only after Gagarin's space flight that he decided to become a cosmonaut. He became a military pilot but it wasn't until 1970 that he was finally assigned to the military *Salyut* program. Though more stable than Lebedev, Berezovoi's tough approach to work could make him just as hard to work with. As he himself wrote, "I remember that some comrades [took] offense at me for allegedly great faultfinding, but I [believe] that the attitude of forgiving your partner for everything only hampers things."[10]

Once in space, these two men struggled to get along. While they had many pleasant and happy moments in the first few months, the seeds of discord lurked everywhere. For example,

Berezovoi, though a rookie, understood that everything had to be returned to its proper place, or else the station would become a chaotic mess and they would waste a lot of time looking for things. Lebedev was sloppier. One week into the mission Berezovoi complained to him about his habit of leaving his tools and equipment (sextant, cameras) floating about the station. "How long will I have to keep catching your stuff?" he complained.

Then, a few days later, Lebedev excitedly noted that the first pea shoots had become visible in the Oasis greenhouse. He tried to get Berezovoi to take a break from photographing the agriculture off the northeast coast of the Black Sea, near his home town in the Caucasus, to see Lebedev's discovery. "Come and look! The pea plant in our garden has put out shoots. While you study the fields of Krasnodar, you miss the crops in your own garden."

Berezovoi scoffed. "Don't confuse the issue. That's biology, not agriculture."

The friction, though mild in these early days, made Lebedev moody and unsatisfied. The next day he had trouble sleeping, and woke up early. "For a long time I tossed and turned. Thoughts kept racing through my head."

Then there were problems with equipment and communications with the ground. The Rodnik water system needed repair. Instruction manuals were inaccurate or out of date. Ground controllers gave too many orders, many of which were arbitrary or incorrect. And the workload was unrelenting. The day an alarm went off by mistake, Lebedev lamented, "It's a shame that instead of being masters of the equipment we are its slaves."

Yevgeny Kobzev, the crew's doctor, tried to alleviate their loneliness and frustration by playing them recorded music. Once he played a tape of Lebedev's 10-year-old son Vitali playing the piano. To Lebedev, the most heart-wrenching moment was near the end of the tape, when the boy finished playing, got up, and yelled, "Mom, I'm going out!"[11]

To relieve his boredom, Lebedev nursed the plants in the station's experimental greenhouses. Like Patsayev on *Salyut 1*, Grechko on *Salyut 4*, and Ryumin on *Salyut 6*, Lebedev found that tilling small plants provided him with an immense, inexplicable pleasure. At the start of the mission he had planted peas in Oasis. Then, with the arrival of the first *Progress* freighter, he had arabidopsis in a Fiton container, lettuce in a Biogravistat, an onion

bulb in a Vazon, and tomatoes in a new, larger, and more sophisticated greenhouse called Svetoblok.[12]

By this time the scientists in Energia's biology program were
certain that the difficulties in growing plants stemmed from the
design of their greenhouses. The problem of hydroponics in zero
gravity was surprisingly more difficult than expected—what was
taken for granted on Earth did not happen in space. Water did not
flow naturally down to the roots, and oxygen and atmospheric toxins built up around the plant, suffocating it. Thus, except for the
arabidopsis weeds grown during Ryumin's second tour on *Salyut*
6—which had produced sterile seeds—no plant had yet flowered or
generated buds or seeds. To attack the problem, each greenhouse
had been improved in some manner. For example, to keep Fiton's
atmosphere pure, its ventilation system had been sealed and used
anti-bacterial filters. In addition, from the very first day the *arabidopsis* seeds were given continuous light. On *Salyut* 6 the plants
had gotten only 14 hours of illumination per day during the first
two weeks after planting.[13]

As flight engineer, it was Lebedev's responsibility to care for
these plants, and to his joy and exhilaration, after only two weeks
many seeds had sprouted. One pea plant had even grown longer
than 1 inch and sprouted nine shoots.

Lebedev's sleeping bag was attached to *Salyut* 7's ceiling right
next to the Oasis greenhouse. Each night when he went to bed, he
stared at the growing sprouts. "The stems are in bud with leaves
like small blue bells, still weak, but fresh and green. They make
me happy; they were born in space."

Then Oasis began to leak. Many of the system's hoses and gaskets had dried while sitting in the dry desert air of Kazakhstan
before launch. Once in space they cracked. Desperate to save his
plants, Lebedev spent hours replacing parts. "I have to do it; it's our
farm. I water the plants regularly, happily. I spoil them."[14]

By early June, the tallest pea plant in Oasis was almost a foot
tall, with many leaves and six branches. Its roots, however, had
grown wildly, some down into the artificial soil, some up into the
sky. The leaves, meanwhile, had an unnatural white or brown
"bloom" or "mold" on them. The possibility that he might be the
first man to get plants to bloom and produce seeds in space was
almost intoxicating to Lebedev. "When I smell [our little space gar-
den]," he wrote, "it seems I can smell the Earth."[15]

Relations between Lebedev and Berezovoi, however, continued to be tense. Their personalities just did not mix well, so they spent much of their time working separately, saying little to each other. Unlike Ryumin and Lyakhov, however, Lebedev's demanding personality made this kind of separation difficult. Even if Berezovoi wanted to avoid his partner, he couldn't. At some point Lebedev would appear, hovering near him to complain, or worry, or discuss some matter or another. And in the small station, there was no place to go to escape.

As they struggled to get along in their tiny home in space, their effort was complicated by the arrival of the first person from outside the communist bloc to fly on a Soviet rocket. *Soyuz-T 6*, launched on June 24, 1982, carried a three-man crew of two Russians, Vladimir "Johnny" Dzhanibekov and Alexander Ivanchenkov, and one Frenchman, Jean-Loup Chretien, for an eight-day mission.

The Frenchman was there because once again, Brezhnev had wielded space as a diplomatic and propaganda tool. Though he intended using *Salyut 7* as an instrument for international diplomacy, Brezhnev saw no reason to repeat the Intercosmos program of *Salyut 6*. The cost was great and the benefits limited, especially since flying cosmonauts from other communist countries like Poland and Czechoslovakia had done little to help their puppet leaders control an unhappy populace. Outside the communist bloc the Intercosmos program produced even fewer benefits, since instead of demonstrating the success, prosperity, and openness of the Soviet system, it served instead to illustrate the restricted life within the communist bloc.

Moreover, the American space-shuttle program was finally getting underway, with a stated policy that foreign astronauts would fly on the shuttle. Already a West German and a Canadian were in training and scheduled to fly in 1983. As in the 1960s, the Soviets felt compelled to compete.

Thus, in the late 1970s Brezhnev, in an effort to show that the communist bloc was as open as the West, decided to raise the ante and make his space program available to noncommunist nations. Brezhnev made his first offer in 1978 to then Indian Prime Minister Morarji Desai, who immediately rejected it. In April 1979, Brezhnev tried again when French President Valéry Giscard-d'Estaing came to the Soviet Union for trade talks. During these meetings, Brezhnev proposed that a Frenchmen fly to a Soviet space station, just like Eastern European cosmonauts were doing currently. Giscard-d'Estaing promptly accepted.[16]

Two months later, in June 1979, Brezhnev repeated the offer to India, this time making it to Indira Gandhi, who, though not in power, was campaigning for office. As soon as she became prime minister in January 1980, she and Brezhnev began negotiations, with Gandhi repeatedly trying and failing to get Brezhnev to end the Soviet occupation of Afghanistan. In the end, Gandhi's desire to fly the first Indian cosmonaut won out over her desire to free the Afghani people, and—despite the presence of Soviet troops in Kabul—she and Brezhnev agreed in early 1980 to fly an Indian to *Salyut 7*. Brezhnev applauded the decision, noting that the joint international space mission would be "a symbol of our friendship."[17]

Once again, Brezhnev's opening of the Soviet space program to noncommunist nations carried with it unintended consequences. The extensive links forged between East and West during these missions further widened the cracks in the communist wall, cracks that had first appeared during earlier international space missions. Just as the United States had insisted on full disclosure and live coverage for the *Apollo-Soyuz* mission, the French and Indians demanded the same thing for missions that included their citizens. For both missions, the launch dates were announced in advance and the launches were broadcast live. Specific details about Soviet space operations were publicly revealed for the first time. Soviet engineers and cosmonauts were given the chance to meet and work with foreigners.

As time passed, the Russians involved in the space program became more and more used to the West's open way of life. To them, the secretive and clannish behavior of the past seemed to make less and less sense. Not only did it rob them of the credit they deserved, but by hiding their failures it also diminished the greatness of their achievements. "The honesty of the American press made it more persuasive, more influential," noted cosmonaut Alexei Leonov, commander of the Soviet half of the *Apollo-Soyuz* mission. "Every little problem was written about in great detail, so that the image of American astronauts grew, making them heroes. It was a much more clever approach."

Once these communication links with the West were established, they wouldn't go away, no matter what walls Brezhnev and the communist leadership tried to build. Men like Leonov were famous and powerful, and able to influence others. With time, the links acted as bridges, and helped shatter the Iron Curtain that kept the Soviet empire intact.

However, when the French mission was launched in 1982, the end of the Soviet Union was inconceivable, and detente had fallen apart. Giscard-d'Estaing had lost the 1981 election, the Soviets had invaded Afghanistan, and martial law had been declared in Poland. In early 1982, many influential members of the French Academy of Sciences signed a letter of protest, urging that the joint mission be called off.[18]

It was not. Though years had passed since Brezhnev was able to interest any other nation in joint space missions, the French mission already had its own momentum. The science would be worthwhile, and the French saw no reason to lose this chance of sending one of their own into space.

Even so, Chretien's flight almost ended tragically during docking. Dzhanibekov, its commander and pilot, was the man who, in 1978, had coolly allowed the automatic docking system to dock his *Soyuz 27* spacecraft with *Salyut 6* rather than take manual control. As *Soyuz-T 6* approached *Salyut 7* everything seemed normal. At about 3,000 feet the automatic Igla rendezvous system turned the spacecraft so that the aft engines faced the station in order to make the last orbital burn. Then the computers, using the spacecraft's smaller attitude thrusters, began rotating *Soyuz-T 6* back around to bring its docking port in line with *Salyut 7*'s. Halfway through this maneuver, however, with the spacecraft only about 1,300 feet from the station and moving in at about 23 feet per second (about 15 miles per hour), the spacecraft's computer system crashed, leaving *Soyuz-T 6* in an out-of-control, end-over-end tumble.

"I can't see the station," Dzhanibekov remembered. "I don't know my position." In less than a minute *Soyuz-T* was going to ram *Salyut 7*, killing everyone on both spacecraft.

This time Johnny took manual control immediately. With *Soyuz-T*'s computers dead and the spacecraft tumbling, he could not use its large engine to brake. Instead, with each rotation he fired the small attitude thrusters, trying to slow the spin as well as ease the spacecraft sideways so that it would miss *Salyut 7*.

The side thrusters, however, were not very powerful. "It was like a little child pushing a big car." With each tumble, he could see *Salyut 7* flash by in his screen, getting larger and larger.

Then they shot past, separated from the station by less than 30 feet, flying between *Salyut 7*'s starboard and dorsal solar panels.

The *Soyuz-T* spacecraft drifted another 1,000 feet before Dzhanibekov could get it under control with the small thrusters. He then waited patiently until he reacquired ground signal, explained what happened, and got permission to finish the docking manually.[19]

Dzhanibekov's position as commander of this international mission had been a late decision. Chosen only four months before launch, he had replaced the original commander, Yuri Malyshev, when Malyshev took ill.[20] Johnny's easy manner with the Americans during the *Apollo-Soyuz* mission plus his success during two *Salyut 6* missions, one of which was an international mission, made him an obvious choice to replace Malyschev for the ground-breaking first Soviet mission with a noncommunist nation.

Two hours later, Dzhanibekov, Ivanchenkov, and Chretien opened the hatch into *Salyut 7* and joined Berezovoi and Lebedev, placing five humans in an orbiting station for the first time. In what had become a tradition of political public-relations gestures dating back to the *Apollo-Soyuz* mission, the crews exchanged telegrams from their respective political leaders, then shared crackers and salt tablets—enhanced by a meal of six French dishes, including crab pate, processed cheese, and ragout of rabbit and lobster. Unfortunately, the meal lacked French wine (only vodka was permitted on board) and garlic (the station's air filters couldn't handle the odor).

Chretien's eight days in space were filled with many of the same public-relations gestures seen on the previous Intercosmos flights. Besides performing the usual potpourri of biological experiments, Chretien spent every orbit over Europe doing television broadcasts home, a routine so absolute that he never had a chance to view his own country from space.

After six days, Dzhanibekov, Ivanchenkov, and Chretien returned home, carrying with them letters from both Berezovoi and Lebedev to their families. Alone together again, the two men struggled to get back into their routine. The hectic pace of the visit, the strain of dealing with a foreigner on board a Soviet spacecraft, and the tension between them, had left them exhausted and depressed. To help them recover, they were given the next three days off.

The break helped, but not much. As the weeks stretched out, the loneliness, boredom,. and friction became more pressing. At one point in mid-July Lebedev became obsessed again with fixing the

Oasis greenhouse. For almost two days he replaced the system's numerous gaskets.

Berezovoi thought it was a waste of time, and said so. "Why are you putting up with this thing? Why don't you just tell the engineers to shove it if it was made so poorly?"

Lebedev tried to explain. If he could fix it, he could grow peas and wheat.[21] The two men argued.

With their mood ugly and impatient, they eagerly awaited the arrival of the next *Progress* supply ship. Desperate to read letters from home, they were furious when they opened the hatch and discovered that the cargo had been packed so that the letters were buried at the back, forcing them to unpack everything else first. For the next few days they vented their anger on ground controllers rather than each other.[22]

The tension finally eased during the last week in July, as they spent a week preparing for their only space walk. Equipment had to be prepared, spacesuits checked, and routines practiced. There was no time for argument. With Berezovoi backing him up, Lebedev spent two-and-half hours retrieving several experiments from the outside of *Salyut 7*. He also tested a newly designed special wrench, tightening a series of bolts made of different materials to see how each functioned in space.

Lebedev couldn't help staring at the emptiness around him. As he wrote later that night in his diary, "Space is very beautiful, with the dark velvet of the sky, the earth's blue halo, big lakes and fast rivers, the masses of clouds, and complete silence all around. There is no sensation of flight; no wind whistles in our ears, nothing to hold us down. This panoramic scene is very peaceful and majestic."[23]

And then they were back inside, confined to the metal chamber of *Salyut 7* and its ceaseless chores.

In the Oasis greenhouse the peas were dying. The lower pods had dried up, no new ones had appeared, and the leaves were withering. The wheat plant, however, was almost 8 inches long, "tall, fragile, and swaying under the fan as if it were being blown by the wind," wrote Lebedev in his diary. In the new Svetoblok greenhouse, the tomato plant struggled, barely reaching 3 inches in height after almost three weeks of growth. In Fiton, which Lebedev had watched in vain for months, the *arabidopsis* seeds had finally sprouted. The tiny, string-like stems weaved their way upward out of the layers of artificial soil, looking less like a plant than a loose floating jumble of thin twine. Each of the four plants was about an inch tall, and each had four green leaves. "Like little children, fresh and fragile," Lebedev wrote in his diary.[24]

For Lebedev, the focus since the beginning of the mission had been on the peas and wheat in Oasis. He had devoted hours working on the greenhouse, trying to keep it running. Fiton he had mostly ignored, because its watering system was automatic and because nothing seemed to grow there. Now, however, it was Fiton that produced the real excitement. Within two weeks of sprouting, the *arabidopsis* did something no plant in space had ever done before. On August 4, he asked Nechitailo, "Does *arabidopsis* develop pods?"

"Sure," she answered excitedly. "They might have seeds inside, you should look carefully."

Then on August 14, Lebedev announced to Nechitailo with glee, "Hurrah! A pod has burst: it spilt seeds!" The seedpods of the *arabidopsis* plant had ripened and burst, revealing many seeds within. To Lebedev the seeds reminded him of "small fish teeth." To Nechitailo, the seeds were worth more than gold. "Keep them safe," she told him anxiously. "We need them all alive."[25]

Two days later, the station was graced by new guests. Unlike earlier visiting missions, however, *Soyuz-T 7* did not bring a cosmonaut from another nation. Instead, Brezhnev and the Soviet leadership had decided it was time to once again put a woman in space.

Almost 20 years had passed since the first woman cosmonaut, Valentina Tereshkova, had flown on *Vostok 6*. Despite the Soviet claim that their society treated men and women with complete equality, women had never been given much respect in the Soviet Union. In fact, in Russia the woman's liberation movement had never arrived. Though most women worked outside the home, the roles of men and women remained unchanged and traditional: the man was the breadwinner while the woman took care of the home and children.[26] After Tereshkova's mission, her backups were sent back to their regular jobs, and the idea of putting women in space was consigned to history books.

By the early 1980s, however, the United States was gearing up to fly the space shuttle regularly with large crews, including a large contingent of women, all of whom would fly repeatedly, doing work as pilots, engineers, and scientists. Once again, as in the 1960s, the Soviet leadership felt compelled to compete. Glushko, head of Energia, suggested that they fly female cosmonauts again, and the Politburo, led by Brezhnev, immediately agreed. Openings were created for women, and a significant part of the *Salyut 7* program was dedicated to putting a woman into orbit. Glushko even started organizing an all-woman cosmonaut mission.[27]

Launched on August 19, *Soyuz-T 7* carried two men, Leonid Popov and Alexander Serebrov, and one woman, Svetlana Savitskaya. Eager to see a female after being locked together for almost 100 days, Lebedev and Berezovoi scrubbed the station clean and laid out the best freeze-dried meal they could. In fact, they worked so hard at getting ready that they were almost late turning on *Salyut 7*'s radar system so that the *Soyuz-T* spacecraft could dock with them. When they pulled open their hatch, however, only Popov and Serebrov floated in, laughing and hugging them.

Where was Svetlana? For Lebedev and Berezovoi, less than halfway through their planned seven-month mission, a visit by a woman seemed almost mouthwatering. They poked their heads into the docking tunnel, calling to her. She shouted back, "Just a moment, just a moment!" Lebedev could see her inside the *Soyuz-T* capsule, combing her hair.

Serebrov explained. Savitskaya had actually prepared a "dress" for this historic moment, and wanted to put it on. "We had to wait a long time for her," Lebedev wrote later. "Like any woman, [she] was primping [herself]."[28]

Once they were finally all together in the main compartment, Lebedev and Berezovoi presented Savitskaya with Fiton and its arabidopsis flowers. "It seemed appropriate to give these first space flowers to the first woman on our station as a symbol of human settlement in space."

They then offered her an apron. "Even though you are a pilot and a cosmonaut, you are still a woman first," Lebedev said. "Would you please do us the honor of being our hostess tonight?"[29] Savitskaya agreed, gritting her teeth. Though she was probably one of the world's best female pilots, her Russian background made it impossible for her to refuse.

During the next few days Lebedev and Berezovoi slept little. Like Grechko and Ryumin before them, the chance to see others, the opportunity to read letters from family and friends, the break

in the utter monotony and boredom, was too much to give up.
Every minute with their visitors was precious.

In between their socializing, the work still went on. For
Lebedev, the most significant moment came when he carefully removed the plants from Oasis and Svetoblok and packed them as
well as Fiton so that the visiting crew could take everything back
to Earth for study. All told, 200 space-grown *arabidopsis* seeds were
returned to Earth. On television later he admitted that he felt "sad
and uncomfortable" when he looked at the empty garden. "It was
such a pleasure to take care of them," he explained.[30] And then
Serebrov, Popov, and Savitskaya were gone, returning to Earth and
leaving Berezovoi and Lebedev to hunker down for their last two
months in orbit.

The shock of once again being alone on the station was shatter-
ing. Suddenly it was an effort to do anything. "We felt such languor
in our bodies that we didn't want to get up," wrote Lebedev in his
diary. They began sleeping 12 hours a day. Often Lebedev had headaches. Several times he dreamt he was back on Earth, with his
family, eating real food. Frequently he argued with ground controllers, with his doctors, with his wife.[31]

Their irritation with each other continued. They argued over
little things. They worked apart, as much as possible. And when
they did work together, they were disorganized.[32]

To fight the boredom Lebedev planted some new seeds in the
Oasis and Svetoblok greenhouses. Soon he saw small shoots,
though he couldn't identify the plant and he couldn't get the set of
plants to grow as well as previous sets. One plant grew 3 inches,
then died. Another died immediately. And somehow he just didn't
have the energy or enthusiasm to care for the plants as he had earlier in the flight.[33]

Then Berezovoi got sick, and Brezhnev died. With only days to
go before they were to set a new record in space, the whole mission
was threatened. Both men were horrified at the idea that after six
months of suffering together in space, their hardships might all be
in vain.

After much discussion doctors told Lebedev to give Berezovoi
an injection of atropine to ease the muscle spasms. "For the first
time in my life," Lebedev wrote, "I was giving someone a shot."
Filling the syringe, Lebedev decided to administer the injection in
Berezovoi's hip.

"Valia," Berezovoi said nervously. "Be very careful."

Lebedev stuck the needle in about halfway and squeezed. [34]

By the next day, Berezovoi began to feel better. By that evening both men decided that they could stick it out and finish their mission. The possibility that they might fail so close to setting a new record helped them to focus.

It also caused Lebedev's demanding character to drive ground controllers crazy. With a new space record only days away, he radioed mission control and suggested that they play the Soviet national anthem during the celebratory television broadcast. Ground crews searched for a copy of the anthem, and couldn't find one anywhere in mission control.

In space, Lebedev kept asking, "Will you be ready? We must play the national anthem the moment we break the record."

The controllers were getting desperate. They didn't want to upset an orbiting cosmonaut on a record-breaking mission. They called the nearest radio station, figuring *they* must have a recording of the anthem. The sound engineers there couldn't find a copy either.

Lebedev kept bringing up the subject. "You must play the national anthem. You must!"

Next they called a television station, which had a tape of the anthem, but couldn't provide mission control with a usable copy.

By this time, Valeri Ryumin had taken over from Yeliseyev as flight director in mission control, with responsibility to supervise everything that happened on *Salyut 7*.[35] Watching how Lebedev's demands were driving his ground controllers crazy, he gave them permission to ignore the cosmonaut. Rather than getting him a copy of the national anthem, the controllers decided to tell him to "stuff it." Because radio stations play the anthem when they go off the air late at night, they decided to play *that* for him.

So, in the wee hours of the record-breaking morning, when the two men were still asleep, mission control turned on the radio, upped the volume as loud as possible, and fed the radio station's signal up to space. Whether it woke Lebedev, they couldn't care less.[36]

Finally, on December 10, 1982, Lebedev and Berezovoi came home, landing amidst a snowstorm on the plains of Kazakhstan, with howling winds and temperatures far below freezing. Only two helicopters were able to land before the storm worsened, one of which lost a wheel and crashed in a dry riverbed. Inside the capsule the men struggled from their harnesses and huddled in the cold in their spacesuits, waiting for rescue. It took 10 hours for a troop transport to arrive and carry them back to Baikonur, which was the nearest airbase.

As with previous long missions, it took them several weeks to regain their strength and walk normally again. And as with earlier flights, they had lost bone tissue. The mineral density in the legs of one man had dropped 7 percent, about 0.9 percent per month. Yet, the Soviet doctors were encouraged. Over the length of the mission, the bone loss rate seemed to decline, and once again the low loss rate indicated that exercise helped counteract the desire of the body to weaken its own bones.[37]

To the emotional Lebedev, their seven-month mission was symbolized by a drawing he said was hung on the station's wall.

> A lonesome cowboy [is] tied to a cross, with a gun mounted above, pointing toward him. There is a string tied from the trigger of the gun to an unmentionable spot. In front of the cowboy sways a beautiful naked woman, torturing him with a teasing look. In the background stands the cowboy's stallion with sympathetic tears dripping from his eyes, because he understands his master's dilemma.[38]

For Lebedev, this is what it was like to live for seven months in space. They were trapped. They had to control their emotions. And they were continuously threatened by death. The isolated confinement of the tiny *Salyut* space station made life on board almost intolerable. How were others going to travel to Mars and beyond in such circumstances? What had to be done to make such long journeys more humane and tolerable?

In the two decades since that mission, both men have had as little to do with each other as possible.[39] If their mission had been much longer, or if it had sent them to the isolated, distant regions of the outer solar system, they might have consummated von Braun's fears of murder in the confined space of a tiny interplanetary spaceship.

As difficult as their mission was, Lebedev and Berezovoi produced one discovery of profound significance. About half the seeds from the arabidopsis plants that Lebedev had sent back with Savitskaya, Popov, and Serebrov were analyzed and found to be alive and viable. Scientists planted them and grew *arabidopsis* plants from them.[40]

For the first time in history, a plant had produced healthy seeds in space. Weightlessness might not be a barrier to reproduction after all. Life could survive in space. The stars might still be open to humanity.

Things Go Wrong

September 9, 1983, was a day like many others on *Salyut 7*. Vladimir Lyakhov and Alexander Alexandrov awoke at around 7 A.M. After eating a breakfast of bread, juice, and cottage cheese with blackcurrant puree squeezed from a tube, they floated to the aft docking port, where sat the 17th *Progress* freighter to dock with a Soviet *Salyut* station.

For the last three weeks the two men had been unloading the freighter's cargo. In space for more than two months, they had eagerly read the letters from family and friends as well as the Soviet-censored newspapers filled with local and national news. They read that Western actresses Judy Davis and Jessica Lange took Best Actress awards at the 13th Annual Moscow International Film Festival. They read stories about the athletic events at an Olympic-type sports tournament in Moscow. They even read the long and sometimes boring speeches by Andropov and other communist leaders, because these speeches gave the men a taste of life back on Earth.[41]

This morning the two men were going to do some basic science. Lyakhov would turn *Salyut 7* so that Alexandrov could beam a radio signal to a precise spot on the ground. By measuring how the signal changed as it traveled through the earth's atmosphere, scientists would have a better idea of how the atmosphere conducted radio waves. As Lyakhov fired the station's engines to rotate it into position, he noticed that the spin was not quite right, its velocity slightly off. Then, at a little before 9 A.M., he fired the main engines, and was startled to discover that the pressure of one fuel tank was close to zero.

Alexandrov floated to the nearest aft porthole and tried to see if anything looked different. To his surprise he saw a glittering spray of frozen drops. He took pictures, hoping that the motion of the spray through space could tell him something about where it came from.

At the next communication session at 10 A.M., they described what had happened. For the next 90 minutes, engineers pored over the telemetry, trying to figure out the cause of the problem. By 11:30, they knew. There was a leak somewhere in the *Salyut 7* service donut and extremely corrosive nitrogen tetroxide was gushing into the station's structure. The situation was alarming to ground controllers, and far more serious than the *Salyut 6* fuel leak five years earlier. Not only was nitrogen tetroxide toxic, it could breach the integrity of the station. Ryumin ordered the two men to retreat into their *Soyuz-T 9* spacecraft, seal the hatch behind them, and prepare for an emergency return to Earth.[42]

At that moment, the operation of *Salyut 7* changed forever. Building a spaceship that could operate for years at a time in space was not as simple as the success of *Salyut 6* had implied. For the rest of *Salyut 7*'s life, cosmonauts and ground engineers struggled mightily to keep the station afloat, battling mechanical failures, technical problems, near disasters, and challenging construction projects. Often they had to improvise at a moment's notice. Repeatedly, their repair work was so complicated that it required them to tear the station apart and put it back together. And all the work was far more dangerous than anything done by anyone previously, forcing men to labor for long hours outside, separated from the deadly vacuum of space by only their spacesuits.

For hours, the two men waited impatiently in their *Soyuz-T* capsule for mission control to tell them whether they had to abandon the station. Neither wanted to leave. For Lyakhov, this second mission in space was even more challenging than his first with Valeri Ryumin. In the years since, the quiet, softspoken military pilot had calmly waited his turn for another flight, working backup on several missions. He finally moved up to prime crew after the *Soyuz-T 8*, the follow-up mission to the one crewed by Lebedev and Berezovoi, had tried and failed to dock with the station in April. (That docking failure might have been considered the harbinger of the failures to come. During launch, *Soyuz-T 8*'s radar antenna had been ripped off when the rocket shroud was released. The crew, Vladimir Titov, Gennady Strekalov, and Alexander Serebrov, asked and got permission to attempt a manual docking, using just their eyes to judge distance and speed. As they approached *Salyut 7*, pilot Titov guessed that their speed was too great, and aborted the rendezvous, trying to brake the *Soyuz-T* spacecraft to avoid a collision. In truth, their speed was much greater than even he guessed, and the best that Titov could do was deflect their path so that *Soyuz-T 8* raced past *Salyut 7* mere feet away, barely missing a deadly crash. Serebrov, watching out one window, could see the painted strips on the outside of the station flash past, faintly illuminated by *Soyuz-T*'s dim headlights. As he described later, it was a "real brush with death."[43] With their fuel low, *Soyuz-T 8* was forced to return to Earth.)

Lyakhov's crewmate was Alexander Alexandrov, a goodnatured, fast-talking man born of the space age. Both his parents had worked for Sergei Korolev, his father building the launchpad for Korolev's first rocket, his mother designing rockets. Rocket engineering had always been part of Alexandrov's life. Even in college he did side work at Korolev's design bureau. When he graduated in 1964 he got a fulltime job there, helping to adapt the *Vostok* spacecraft to carry the first three-man crew into space. Later, he helped build the *Soyuz* and *Salyut* control systems while studying for his PhD. At the same time he applied to become a cosmonaut, only to be turned down as too young. Two years later he tried again, only to fail again. He became an unofficial cosmonaut, allowed to participate in all cosmonaut training without officially being a candidate. At the same time, in 1976, he started working in mission control. Finally, in 1978, he applied again to the cosmonaut corps, and was at last accepted.[44]

Sent into space on June 27, 1983, for a five-month mission, Lyakhov and Alexandrov's first and most important job was to unload the large, 20-ton transport-support module dubbed *Cosmos 1443* docked to *Salyut 7*'s bow port. Launched in March, *Cosmos 1443* had originally been intended as a supply ship for the failed *Soyuz-T 8* mission. The module increased the interior space of the complex by one third, from 3,300 to 5,000 cubic feet, and the total mass of the complex to more than 51 tons (about half that of *Skylab*), not including cargo. The cargo itself weighed more than 3 tons, and included the first set of solar panel extension strips.

If Titov, Strekalov, and Serebrov of the *Soyuz-T 8* had been able to dock with the station, they had planned to use the module's extra space and supplies to extend the in-space flight record to at least eight months. Titov and Strekalov had also trained to attach the solar panel extension strips during a space walk. The failed docking of *Soyuz-T 8*, however, ruined these plans. To keep the orbit of the *Salyut 7-Cosmos 1443* complex from decaying, it had been raised after the docking failure, and was now too high for a fully loaded *Soyuz-T* spacecraft carrying three men to reach. To lower that orbit would consume too much fuel. Instead, Lyakhov and Alexandrov were sent up knowing that they would have to do the work of three, working twelve-hour days and five-and-a-half-day weeks to unload *Cosmos 1443*.[45]

For almost six weeks the two men labored to unload the module. They then filled its recoverable *Merkur* capsule with film, experiment results, and a failed computer memory circuit board and an air regenerator that had been replaced by a newer unit. In the rest of the module they packed garbage and miscellaneous abandoned gear. On August 14, 1983 *Cosmos 1443* undocked, and nine days later the *Merkur* capsule returned to Earth safely.* One month later, the transport-support module was de-orbited, and burned up in the atmosphere over the Pacific Ocean.

The two men then flew their *Soyuz-T* spacecraft from the aft to the bow port, a maneuver that was now becoming routine, and *Progress* 17 arrived. Three weeks later, the propellent line burst, threatening the entire mission. After several hours reviewing their data, ground controllers decided that the station was habitable. Lyakhov and Alexandrov were given permission to go back inside, where they could smell a faint odor, similar to formaldehyde, and see small stains on the gasket rim of the aft hatch of *Salyut* 7.

Like the *Salyut* 6 fuel leak in 1978, this leak made useless *Salyut* 7's 2 main engines as well as its 32 small attitude thrusters. Without these engines, the crew could re-orient the station only by using the engines of either their *Soyuz-T* spacecraft or a *Progress* freighter, techniques that were both wasteful of fuel. The station drifted into a gravity-gradient position.

Unlike the situation with *Salyut* 6, however, *Salyut* 7 was far more deprived of power. The equipment on the new station was more demanding, requiring power from either the solar panels on a transport-support module or from at least one set of extra strips on its own three solar panels. Without either, the station's basic three solar panels could generate only about two-thirds of the power required by *Salyut* 7 when in a gravity-gradient orientation.[46]

Lacking about one-third of their power, the cosmonauts were

_____

*In 1993, this recovered capsule was sold at a Sotheby's auction in New York for $552,500.

forced to keep the station's temperature at about 55°F, which, in turn, was too cold for the Rodnik water-recycling system to work. The station's humidity was soon at 100 percent, with globs of water everywhere. "It was very cold, very hard," Alexandrov remembered. They couldn't exercise, because their sweat accumulated in the cabin. And the water and limited power made it impossible to do much research besides Earth photography.[47]

More power could be generated if the extra solar panel strip already on board could be attached to one of the station's solar panels. Because Lyakhov and Alexandrov had not trained to do the necessary space walks, mission control decided to try to get the men from *Soyuz-T 8*, who had trained to do them, into space as fast as possible. Strekalov and Titov were quickly scheduled for a September 27 launch for their second attempt in six months to dock with *Salyut 7*.

It was not to be. Ninety seconds before blast-off, with Titov and Strekalov waiting at the top of their fully fueled *Soyuz* rocket, a fuel valve at the base of the rocket malfunctioned, opening and spilling fuel uncontrollably onto the launchpad. A fire broke out and flames engulfed the rocket with its 180 tons of very flammable fuel.[48] At that moment, the automatic launch-escape system should have kicked in, executing the following steps: First, explosive bolts fire, flinging the *Soyuz-T* capsule free of the three-stage rocket. One second later, solid-fuel engines in a tower attached to the top of the capsule ignite, lifting the *Soyuz-T* orbital module and descent module away and clear. Five seconds after that, more explosive bolts fire to separate the manned descent module from everything else. Its parachutes then release and its retro-rockets fire, slowing the capsule enough for a safe landing.

The automatic launch-escape system did not kick in, however. The fire had burned the system's wiring, preventing it from being activated automatically.[49] Feeling strange vibrations and seeing black smoke and yellow flames outside their window, Titov and Strekalov tried to fire the launch-escape system manually, only to get no response. To fire the escape system manually from mission control required each of two different operators, located in two separate rooms, to press separate buttons at the same time. With flames rising from the launchpad and the entire rocket already leaning 20 degrees to the side, controllers scrambled madly to get the system to fire.

Just 10 seconds after the flames first appeared, controllers miraculously managed to somehow do this, activating the escape system and throwing Titov, Strekalov, and the *Soyuz-T* capsule more than 3,000 feet into the air. For five seconds the emergency engines fired, subjecting the two men to forces exceeding 15 g's. Then the engines cut off, the descent module separated, and its parachutes unfolded.

At that moment, the entire rocket and launchpad exploded. The blast was so intense that the capsule, three miles away, was thrown sideways, and launchpad workers in underground bunkers felt the pressure wave.

Strekalov and Titov landed safely, their capsule hitting the ground with a hard bump that shook both men up but did them no damage. Rescuers quickly pulled them from the capsule, then gave them a glass of vodka to calm their nerves as everyone watched the nearby launchpad crumble in flames and clouds of smoke. It took 20 hours to put the fires out.[50]

This failure caused more problems for the *Salyut 7* mission. Not only were Lyakhov and Alexandrov not trained to add the solar panel extensions, but their *Soyuz-T* capsule had already been in space for three months, longer than any previous *Soyuz* capsule. Lacking a replacement spacecraft, they had to either trust the reliability of their old *Soyuz-T* capsule, or abort the rest of their mission and return immediately to Earth. After much discussion, flight director Ryumin and his engineers decided to let the mission continue until its planned completion at the end of November. After more than a decade of use, they were confident that the *Soyuz-T* spacecraft would be safe, even after as long as five months in space. Ryumin and the other mission controllers even felt confident enough with the situation to improvise and let Lyakhov and Alexandrov attempt the space walks to install the two solar panel extensions.[51]

Nor was the improvisation limited to letting the men do a space walk for which they were untrained. As the men prepared their equipment, they were shocked to discover a rip in the outer layer of Alexandrov's suit, behind the knee of one leg. Rather than cancel the space walks, they fixed the suit. First they cut a section out of an aluminum duct and fit it into place around the suit leg so that it covered the rip. Then they scavenged loose-leaf-book rivets from their medical manuals and used these to clamp the duct down.

Then they wrapped everything with silk cord and duct tape. Though the repair made this suit-leg slightly longer and impossible to bend, Alexandrov didn't mind. "You use your hands in space," he noted. Legs were pretty useless in zero gravity.[52]

Finally, on November 1, the two clambered outside to attach the first extension panel. To Alexandrov, the moment was "tense, emotionally charged." He took the lead, crawling along the small 6-foot-wide docking/transfer module. Between them the two men lugged the first extension panel, packed like a strip of paper folded back and forth over and over again and pressed flat to form a pleated cube. The idea was to attach the cube to the base of the solar panel along one edge, and use a winch to unfold the pleats, raising the strip along that edge like a flag on a flagpole.

Hanging on the outside of *Salyut 7*, Alexandrov felt as if he were riding a barrel flying 200 miles above the earth. Below them the world was in darkness, a giant black shadow blocking out the stars. Close by, he was fascinated watching nuts and bolts and small items drift away, glistening against the blackness of Earth. "They looked like stars," he thought. He even released a few bolts to watch them glitter—to the chagrin of ground controllers.[53]

The two men reached the station's mid-section and stopped at the center array, sticking up from *Salyut 7* like a dolphin's dorsal fin. Alexandrov fitted his feet into a foot restraint and, with Lyakhov's help, attached the cube to the edge of the panel. Then he inserted the winch handle, and began unwinding the pleats.

"The winch was tight," he remembered, difficult to turn in zero gravity. Slowly, inch by inch, he heaved at the handle, Lyakhov holding his body down to give him leverage. After 20 minutes of struggle, the strip finally reached its full extent. At that moment, the handle of the winch broke off. "If it had happened only a second sooner," Alexandrov noted, "I wouldn't have been able to finish raising the strip."

Lyakhov then took the cables attached to the extension and plugged them into a pre-built electrical outlet. All told, the space walk lasted about three hours.

After one night's sleep both men went back outside and installed the second extension to the dorsal panel's other edge. This time things went so smoothly that they had time to stop work and watch a sunrise and sunset as Salyut 7 whipped around the earth every 90 minutes. In fact, the hardest part of the whole job was after the space walk, inside *Salyut*, when they had to connect the new strips to the station's electrical system. The job took hours, because they had to check the voltage on every pin in every plug, numbering in the dozens, and they could do this work only when the station was in darkness, because in daylight the panel began charging and the pins became live electrical wires.[54]

Three weeks later, both men returned to Earth, using the same spacecraft that had put them in orbit, the oldest *Soyuz-T* spacecraft ever used to return humans from space. Both recovered in what was now an almost routine manner. Concerning bone loss, the data was less encouraging. While one man showed no obvious sign of bone loss, the other (Soviet documents do not say which is which) had experienced an 11 percent loss, that is, a very high, monthly rate of 2.2 percent.[55]

Above, in space, orbited a crippled *Salyut* 7. The leak in its propellent system left its main engines disabled. Without a *Progress* freighter or a *Soyuz* spacecraft, there was no way to maintain its orbit. From the end of November until the arrival of the next crew and the next *Progress* tanker in February 1984, the station's orbit slowly decayed, dropping about a half mile per week. Furthermore, even with one set of solar-panel extensions installed, the station did not have enough electrical power. Unless extension strips could be put on the two other solar panels, the station's usability would be significantly limited.

Despite these problems, *Salyut* 7's next crew would not only attempt to rebuild the station, they would also try to set a new in-space endurance record of eight months. The audaciousness of the Russian space program was about to take an exponential leap forward.

Things Go Right

On February 8, 1984, commander Leonid Kizim, flight engineer Vladimir Solovyov, and Dr. Oleg Atkov lifted off from Baikonur. After the many psychological problems Lebedev and Berezovoi experienced during their 211-day flight, the Soviet space program had decided to try something different. Maybe long-term missions would operate better if the crews were made up of three instead of two men, and if one of its members was a doctor. To test this theory, Atkov, a cardiologist from the Ministry of Health and a pilot who had developed the ultrasound cardiograph used by Lebedev and Berezovoi while in space, was added to the crew. Atkov would monitor the crew's health, review their exercise routines, and keep track of everyone's psychological state.

While Atkov did his medical work, Kizim and Solovyov would attempt the daring repair of the leaking propellent line, requiring a half dozen space walks during which they would have to cut open the station's hull, rebuild its propellent system, and then put everything back together. In addition, they would also install extension strips on another of the station's solar panels.

Like too many Russians, Leonid Kizim's childhood had been ripped apart by World War II. Born in the Ukraine in 1941, he and his parents, both railroad workers in the same coal-mining region where Vladimir Lyakhov's father had been killed, had to flee their home, sending mother and baby east beyond the Ural Mountains during the German occupation. His path to space was also very typical. When he was seven he suddenly announced he wanted to be a pilot. Graduating college with honors in 1963, he served in the air force in the Caucasus. Upon completing his two-year stint, he applied to join the cosmonaut program. Over the next decade he worked as a capcom, trained as a cosmonaut, and took correspondence courses leading to a degree in 1975 from the Gagarin Air Force Academy.[56] Finally, in 1980, Kizim made his first space flight, commanding the first three-man test flight of the refurbished *Soyuz-T* spacecraft.

Five years younger, Vladimir Solovyov's background was more like Alexander Alexandrov's. His father had been an aviation engineer, his mother a mathematics professor. Right from college Solovyov was hired at the Korolev design bureau, becoming a specialist in rocket engines and helping to design the fueling system for the first *Salyut* station. In 1977 he applied to the cosmonaut program, and was accepted a year later. By 1983 he and Kizim were teamed together, doing simulation space walks in the giant water tank in Star City to get ready for their space walks on *Salyut* 7.[57]

Before they could begin their daring repair effort, the three cosmonauts had to play host to the second noncommunist to visit a Soviet space station. This time the visitor was Indian Rakesh Sharma, who arrived in early April 1984 for an eight-day mission. Like Jean-Loup Chretien's French mission, this Indian mission had an impetus all its own. By April 1984, its architect, Leonid

Brezhnev, was dead, and former KGB boss, Yuri Andropov, was in
charge. Even with the end of detente, martial law in Poland, Soviet
troops still in Afghanistan, and the Soviet Union boycotting the
Olympic Games in Los Angeles, no one in either India or the Soviet
Union seemed willing to abandon the plan to send the first Indian
into space.

In addition to the typical political gestures (he spoke to Prime
Minister Indira Gandhi, posted her picture on the wall of the station, and exhibited the Indian flag on television), Sharma performed
the same kind of scientific research of past international missions,
taking multi-spectral photographs of his native land and doing a
variety of biological and psychological experiments. Possibly the
most interesting experiment he performed was his attempt to test
how yoga and weightlessness interacted. (It didn't seem to make
any difference.)

With Sharma back on Earth and the political mission over,
Kizim and Solovyov quickly got down to their real work. Over the
next few weeks the two men were to complete an amazing series of
aggressive space walks, spending more time outside their spacecraft than all previous Soviet space walks combined while completing repair tasks that until now had been inconceivable for either the Russian *or* the American space programs.

The leak repair was first on their agenda. The leak was located
in a part of *Salyut 7*—the aft service donut—where no space walks
had ever been planned and that had few railings or attachment
points. A specially refurbished *Progress* freighter was launched to
provide them with a working area. A hinged platform was attached
to its exterior, near the front port. When the time came for Kizim
and Solovyov to do their repairs, ground controllers were to unfold
this platform to create a deck where one cosmonaut could attach
his feet and position his head and arms over the aft service donut.
A second platform, inside the freighter, resembled a hinged ladder.
Kizim and Solovyov were to attach this to the service donut's exterior, thereby giving a second cosmonaut a place to attach *his* feet
and tools.

The leaking oxidizer line itself was buried beneath the service
donut's hull under several layers of insulation, all of which had to be
removed and later replaced once repairs were completed. To make
that possible, the cargo ship brought with it two dozen specially
designed tools, including a pneumatic punch to poke holes in metal,

special scissors that could be inserted into the holes and used to slice the metal hull, and a customized wrench designed to unscrew various caps and pipes already existing on the oxidizer line.

On April 23 the space walks began. With Atkov inside *Salyut 7* coordinating their activity with mission control, Kizim and Solovyov spent more than four hours in space preparing the work area at the aft end of *Salyut 7*, almost 50 feet from the airlock hatch at the station's bow. The climb for both men along the full length of the station was slow and careful. The ladder was bulky, and the toolboxes awkward.

For Solovyov, the experience was natural and exhilarating. An emotional and philosophical man with a commanding personality, it seemed to him that hanging in empty space was almost second nature. "I must confess that I had such a great urge to dive into that limitless space," he later explained. "Regardless of all safety precautions, I just wanted to face the ship and push away from it with both hands." Unlike Vitaly Zholobov from *Salyut 5*, who had seen the blackness as a terrifying emptiness, Solovyov reveled in the vastness of space. Or, to use the words of Robert Heinlein, outer space seemed to Solovyov as ". . . an enormous room, furnished in splendor, though not yet fully inhabited. It was [his] own room, to live in, to do with as [he] liked."[58] As Solovyov described years later, "You want to break away from your home, from that umbilical cord, though it seems unhuman, against our physiology. But who actually knows where our home is? Perhaps it is in space. And perhaps that is why you feel pulled into it."[59]

At the aft service donut, the two men maneuvered and fitted the hinged ladder into position, then unfolded it and locked it into position using guy wires. To fasten the wires to the station they used the pneumatic punch, piercing the hull eight times. Through these holes they strung the wires, pulling the ladder tightly into position. They then attached their several toolboxes to the work site, and headed back inside after being outside more than four hours.

Three days later, they were back. First they attached a television camera to a bracket near the airlock hatch, aimed to look aft along *Salyut 7*'s hull so that flight controller Ryumin and ground controllers could watch what was happening. Then, they positioned themselves at their respective posts—Solovyov attached to the *Progress* platform and Kizim strapped to the hull ladder—and be-

gan cutting away thermal blankets and tearing the station's hull
open to expose the propellent system inside. To do so, they once
again used the pneumatic punch to poke a hole in the hull, then
inserted the special scissors in this hole and cut the hull open.
They next cut aside the layers of insulation underneath that protected the propellent lines.

The whole system before them was a complex web of pipes and
valves. The lines fed oxidizer and fuel not only to the two main
engines, but also to the 32 small attitude thrusters and a handful of
tanks. Moreover, the lines had been designed with a certain amount
of redundancy, which further increased the system's complexity.

To find the leak would involve repeated gas-line tests. Throughout the system were dead-end pipes, installed for ground tests and
then capped and sealed. The cosmonauts were to attach valves to
these and use them to shut off gas flow to various parts of the
oxidizer line. Ground engineers would then pump nitrogen through
the system and Kizim and Solovyov would look for leaking nitrogen, closing different valves, one by one, to isolate different parts of
system. Simple in concept, the actual work was tediously complex.
The caps had been sealed with epoxy. One in particular took almost two hours to unscrew. With their spacesuit provisions fast
running out, the men finally located what they thought was the
leak. Eager to install the hose that would bypass it, Kizim and
Solovyov asked Ryumin if they could keep working. "No," he told
them curtly.

On their third space walk three days later the men installed the
bypass hose. Then they closed the valves and Ryumin had ground
controllers pump nitrogen through the new hose to test its seal. To
everyone's dismay, a second gas leak appeared in a different place
in the system. Already outside for more than two hours, Kizim and
Solovyov quickly began unscrewing two more caps, hoping to install another hose to bypass this second leak. The first cap came off
quickly, but the second resisted their effort. Once again, Ryumin
ordered them to stop. They put away their tools and returned to
Salyut 7's interior.

Three days later they were back for the fourth time, the journey along Salyut 7's length becoming almost routine.[60] For three
more hours the two men worked, installing the second bypass hose.
When they finished, ground engineers once again pumped nitrogen
through the system, and once again, another leak appeared, this

time in a place where a bypass hose would not work. To eliminate
this leak, the oxidizer line would have to be pinched closed, isolating it from the system. Unfortunately, the men did not have the
right tool for pinching the line. Moreover, the leak was in a place
where the proximity of the cargo ship made work impossible. To
finish the repair, a new tool would have to be manufactured and
brought to them, and the *Progress* freighter would have to be
undocked.

Kizim and Solovyov finished up. To protect their repairs from
the harshness of space, they fitted a special frame over the opening
they had made, and stretched insulation over this. They then folded
the pried-open section of the metal hull back into place, and covered everything with thermal blankets.

Their space walks weren't over, however. On May 18, Kizim
and Solovyov went out for the fifth time and spent just over three
hours installing extension strips to the station's port-side solar
panel.[61] Unlike Lyakhov and Alexandrov, Kizim and Solovyov had
trained to do this work, and had Atkov inside *Salyut 7* to help them.
Moreover, their experience working on the engine leak made this
installation seem almost routine. While it had taken Lyakhov and
Alexandrov two space walks lasting a total of 10 hours to install
two extension strips, Kizim and Solovyov did both strips on one
space walk lasting only three hours. They even had time to float
across to the starboard solar panel and cut out a section for later
analysis on Earth.

By this time ground engineers were convinced that the remaining leak was the last. They fabricated a tool with jaws able to apply
five tons of pressure, gave it to Johnny Dzhanibekov to test in the
giant water tank in Star City, and videotaped his every action.
Then, with Dzhanibekov in command, *Soyuz-T 12* lifted off on July
17, 1984. To allow the mission time to complete its work while
also giving Dzhanibekov time to properly train Kizim and Solovyov
for their last space walk, the flight was extended to 12 days, four
longer than normal.

Dzhanibekov's crewmates were Svetlana Savitskaya, making
her second fight in space, and rookie Igor Volk. A skilled test pilot
and part of the Soviet program to build a reusable space shuttle,
Volk had been moved into the rotation so that he could get some
experience in space before flying the shuttle.

Savitskaya's flight was once again inspired by the desire of

Glushko and the Soviet leadership to score firsts in space. NASA
planned to have Kathryn Sullivan do a space walk in the shuttle's
cargo bay late in 1984, to test tools for attaching a fuel pump to a
tank and then use the pumps to transfer fuel from one tank to another. Glushko suggested that Savitskaya do a space walk in July,
scooping Sullivan and making the first woman to walk in space
Russian instead of American.[62]

On July 25, 1984, Savitskaya and Dzhanibekov exited *Salyut* 7
for a four-hour space walk, during which she performed a number
of tasks whose purpose was to demonstrate and prove in-space construction techniques. In one case she used a newly developed, portable, electron-beam welding tool to cut, weld, and solder a variety
of metal plates, while Dzhanibekov photographed and described
her actions. Dzhanibekov then traded places with her so that he
could also test the equipment. Just before ending their space walk,
the two cosmonauts retrieved several exterior experiments testing
the effect of space on sample materials.

With the success of Savitskaya's space walk, the training for
the all-woman mission to *Salyut* 7 moved forward. If all went well,
sometime late in 1985 three women would spend four months in
space together on *Salyut* 7.

For the next few days, Dzhanibekov prepared Kizim and
Solovyov for their space walk, showing them the videotape of his
underwater simulations while teaching them how to use the
pincher. Then, a week after Dzhanibekov and his crew had returned
to Earth, Kizim and Solovyov performed their sixth space walk,
pinching the line after removing the heat insulation that surrounded it.[63] This last repair completed the job. The line was sealed,
making *Salyut* 7's engines and thrusters completely operational
again.

The experience repairing this fuel system was invaluable to the
Soviet space program. By the time they were finished, Kizim and
Solovyov had spent more time in spacesuits than all other Soviet
space walkers combined. They had repeatedly traveled back and
forth from one end of the station to the other, covering hundreds of
feet. They had used tools custom-designed for their work. And they
had successfully completed maintenance work as sophisticated as
any on Earth. Their success laid the groundwork for the far more
sophisticated space walks soon to occur on *Mir*.

More importantly, Kizim and Solovyov had proved once again

that human beings in space were capable of fixing anything. Put humans on a spaceship, give them the tools and equipment they need, and they could conceivably make that spaceship run forever.

Furthermore, during their eight-month flight Kizim, Solovyov, and Atkov had few conflicts or serious psychological problems. Crew dynamics in the cramped confines of a *Salyut* space station seemed to depend mostly on individual personalities, not on how many were there. Some crews, like Lebedev-Berezovoi and Volynov-Zholobov, could not work together no matter their skills, training, or intelligence. Others, like Grechko-Romanenko, Lyakhov-Ryumin, Kizim-Solovyov-Atkov, and all of the American *Skylab* crews, figured out ways to make it work. It seemed that if the crew were mature, cool-headed, and disciplined, they could work their way through almost any difficulty. As Grechko noted, "We were professionals. We needed to do it right so the next mission could benefit."[64]

As was now routine, the three men increased their daily workouts as their return date approached. Then, on October 2, 1983, after 237 days in space, Kizim, Solovyov, and Atkov returned to Earth, each initially having the expected difficulties re-adapting to Earth gravity. Doctors equated their condition to that of patients after long periods of bed rest.

Atkov was also about 2 inches taller! In space, the lack of gravity had caused his spine to straighten, and then lengthen. In fact, by this flight, the accumulation of long-endurance space missions had shown that about two-thirds of all individuals experienced some form of temporary back pain once back on Earth. Doctors suspected that the re-shaping of the spine, as with Atkov, might be the cause.

Their loss of bone tissue was more significant. While overall their skeleton mineral density seemed the same, and had even increased in their upper body, they had lost 3 to 15 percent density in different bones in their lower body, loss rates ranging from inconsequential to almost 2 percent per month. After almost a decade of research, the rate of bone loss in the lower weight-bearing bones still ranged between 0.3 and 3 percent per month. Whether exercise reduced the loss, thereby explaining why some individuals lost bone at one-tenth the rate of others, remained unclear. Soviet doctors were inclined to believe that it did. American scientists had continuing doubts.[65]

Either way, after a few weeks all three men were generally back to normal. Their healthy recovery, plus the new endurance record of almost eight months, set the stage for the planned longer missions.

Unfortunately, those marathon missions would not happen on *Salyut 7*.

Like a Phoenix

As Dzhanibekov guided his *Soyuz-T* spacecraft toward *Salyut 7*, he could see that the station was dead. No lights. No power. Its three solar panels hung haphazardly, rather than angled in parallel to catch the maximum sunlight. The station was a hulking 20-ton lump, and could give Dzhanibekov no information about his distance and position for lining up the two spacecraft.

Without panic, Johnny calmly guided the *Soyuz-T 13* spacecraft toward the imposing station. To ease the buzzing in his right ear, acquired during his years as a pilot, he had inserted an earphone so that he could listen to soft music.[66] Every 30 seconds, he fired a laser range finder at the station, read off the distance number to his crewmate, Viktor Savinykh, who punched it into a calculator and announced their speed.

Somehow Dzhanibekov had to do a docking with a dead spacecraft, something no Soviet space pilot had ever managed to do. In all previous successful Soviet dockings both spacecraft had working radar systems, and both had been maneuverable and functioning. And even with everything working, numerous previous Soviet dockings had failed.

Launched on June 6, 1985, as a rescue mission, Dzhanibekov's *Soyuz-T 13* mission, though reminiscent of the *Skylab 2* flight of Conrad, Kerwin, and Weitz in 1973, was far more ambitious. Primed by experience with *Salyut 6* and two years of repair work on *Salyut 7*, the Soviet space program managers felt ready and able to do anything in space. The years of failure during the late 1960s and early 1970s were finally over.

At a distance of six miles the two men spotted *Salyut 7* as it came into the sunlight. "It was very bright, gleaming red," Savinykh remembered.

Dzhanibekov took manual control of *Soyuz-T 13*. To increase his chances for a successful docking, the *Soyuz-T*'s automatic docking system had been removed, lightening the ship so that its tanks could be packed with as much fuel as possible. The descent capsule's middle seat was also taken out, and the pilot's standard controls and the 3-inch by 3-inch split screen black-and-white video monitor were moved to one side of the *Soyuz-T* descent module so that Johnny could look out the window while he piloted the ship. Unlike any previous Soviet space pilot, Dzhanibekov could actually see his target as he moved in for a docking. Supplementing these changes was Dzhanibekov's laser gun to get distance and Savinykh's calculations to get speed.[67]

At about 200 yards distance Dzhanibekov stopped his approach. The sun was behind the station and blinding his view. Patiently he waited. With both spacecraft racing around the earth at 17,500 miles per hour, he knew it would not take long for the sun to move out of his line of sight. After 10 minutes the sun had dropped away, almost touching the horizon. In a few minutes it would set, and they would be in darkness.

Dzhanibekov eased *Soyuz-T 13* in again, circling *Salyut 7* so that he and Savinykh could inspect it. Neither man saw any significant damage other than the randomly hanging solar panels and its darkened state. ("The station wandered about drunkenly," Savinykh wrote later.) Whatever had caused *Salyut 7* to go dead several months earlier had not ruptured its hull.[68]

Dzhanibekov lined the *Soyuz-T 13* craft with the *Salyut 7* bow port, matching his rotation with the station's. Then, with the sun gone, both spacecraft in darkness and his approach lit only by the *Soyuz-T* running lights, Dzhanibekov carefully brought the two docking ports together. With surprising ease the latches caught and pulled the two ships into a hard dock.

For a few seconds, the two men floated speechless. The docking had gone so smoothly that they simply didn't know what to say.

Now began their real job: bringing a dead space station back to life.

When Kizim, Solovyov, and Atkov left *Salyut 7* in early October 1984, they left a functioning station ready for its next crew. On the main control panel they had taped a package of bread and salt tablets, graced by a welcoming letter. "We leave you a lived-in, warm home. Work in unity, be cheerful." The next mission, to be crewed by Savinykh, Vladimir Vasyutin, and Alexander Volkov, was

planned as a 10-month flight, breaking the 8-month record of
Kizim, Solovyov, and Atkov, and laying the groundwork for the
first year-long mission to follow. Moreover, that crew was to hand
over *Salyut* 7 to the 4-month, all-woman mission, set to begin in
March 1986, thus completing the first direct hand-off of a station
from one long-term crew to another.[69]

On February 12, 1985, however, all contact with *Salyut* 7 suddenly ceased. The station began to drift: Its solar panels slipped
into darkness, its electrical batteries lost their charge, its systems
shut down, and its internal temperature dropped to below freezing.

In Star City near Moscow, the next long-term mission was immediately postponed, as was the all-woman flight. Planning for a
rescue mission began. Five cosmonauts, including experienced pilots like Kizim, Romanenko, and Dzhanibekov, were chosen to
train as commanders.[70] Dzhanibekov, who had returned from the
station only six months earlier, quickly became the prime choice.
Johnny had trained extensively in repairing the station's cooling
loops. He had also done one space walk on the station. And when
he piloted the international mission with Jean-Loup Chretien, he
had shown incredible coolness when his computers crashed and
he prevented his tumbling *Soyuz-T* spacecraft from colliding with
Salyut 7. He handled the situation so smoothly he almost made it
seem easy.

Johnny's crewmate in the repair mission was Viktor Savinykh.
On his second flight, Savinykh's background was somewhat different from most other Soviet-era cosmonauts. Though space exploration intrigued him, he had never really been interested in flying in
space. After completing his three-year army service after college,
he got a job at the Korolev design bureau in 1969, designing optical
instruments, including some used as part of *Salyut* 6. Then his
bosses suggested he apply to become a cosmonaut. Though he
didn't expect to be chosen, an order was an order. He filled out the
application and went through the medical exams. To his astonishment he passed, and in 1978 was accepted into the program.
Though he had never planned on becoming a cosmonaut, he was
thrilled to grab the chance when offered.[71]

Savinykh's first flight, 74 days long, had been the last long mission to *Salyut* 6, intended simply to complete the Intercosmos program. During this two-and-a-half month flight he and his crewmate, Vladimir Kovalyonok, played host to a Mongolian and a

Romanian cosmonaut. His next flight was supposed to be the record-breaking 10-month mission. When *Salyut 7* broke down, he was instead teamed up with Dzhanibekov, and the two were specifically trained to dock with the crippled *Salyut 7* station and repair and reactivate it. If their repair mission was successful, the Soviets would launch Savinykh's original crewmates and attempt to complete the remainder of their planned long-term mission.

After Dzhanibekov made sure he had a solid, hermetically sealed docking, Savinykh carefully released a valve to equalize pressure between the two spaceships, prepared to stop the process instantly if he found *Salyut 7* to be in a vacuum and sucking the air from *Soyuz-T*. The atmospheres equalized normally, proving that the station's hull was still sound. Now the question was whether the air was breathable. Wearing gas masks, Dzhanibekov opened the hatch and floated in, followed by Savinykh.

The station was dark and silent. Savinykh wondered aloud if the air was good, so Dzhanibekov decided to lift his mask and try it. "The air was cool and stale, like a cellar," Dzhanibekov remembered. It was also bitterly cold. He could see his breath freeze into ice crystals, and everywhere he shined his flashlight there was ice encrusted on the walls and instrument panels. To Savinykh it felt like he was wandering through an old, abandoned house. He went to the nearest porthole and opened its cover, letting sunlight stream in. On the control panel he noticed the salt tablets and crackers that Kizim, Solovyov, and Atkov had left. To get an idea how cold it was, the men spat on the wall and watched their saliva freeze. From this crude method they estimated the temperature to be around 14° F.

For the next week Dzhanibekov and Savinykh "commuted" every 40 minutes from their *Soyuz-T 13* capsule to the freezing interior of *Salyut 7*, wearing fur-lined suits and hats. With no working ventilation system, carbon dioxide built up in the air around their heads as they worked, causing dizziness, headaches, sluggishness, and exhaustion. To keep from suffocating, they periodically waved their arms, trying to stir up the air around them. They also ran a hose, linked to a fan, from *Soyuz-T* into *Salyut 7*. To make repairs, they had to work bare handed, the cold and bare metal burning their fingers. And because of the cold, no work session could last more than 40 minutes.[72]

First they had to get the station powered up. Because the bat-

teries currently connected to the solar arrays were not holding their
charges, they switched them with spares. Making the connections
". . . was hard, not simple," Savinykh wrote later. "We had to join
sixteen wires by hand." In replacing the batteries they also discovered why the station had gone dead. The problem was simple: An
electrical sensor that determined whether a battery was charged
and ready to be used or needed charging from the solar arrays had
failed. It could no longer tell when a battery needed charging. One
by one, each of *Salyut 7*'s batteries had run down, the sensor each
time failing to switch them to charge.[73] If a human crew had been
on board when the sensor failed, they would have easily spotted
the problem and fixed it. The station would never have gone dead.

By the third day the men had five batteries charged and could
turn on the lights and heat some food. The station was still frigid,
however. Dzhanibekov's feet became so cold at one point that they
stuffed his boots with the heated food cans to warm him up.

By the fourth day they had one Rodnik water tank unfrozen,
and could make tea.

By the fifth day the temperature was up to 42° F, and the ice on
the walls had begun to melt, causing a flood. "There were globs of
water everywhere," Dzhanibekov recalled. In zero gravity the cold
water tended to adhere to the walls and instruments, and to keep
dry the men had to try to float in mid-air, a task that wasn't really
possible. "Our feet, our clothes, were always wet," remembered
Dzhanibekov. They used towels and dirty clothes to soak up the
water, which they then wrung out into bags.

By the sixth day, June 15, almost everything on the station was
working, though to Savinykh the place looked a mess. "All the
panels were removed, the plates and cables twisting about slowly
like snakes in the air."[74]

Less than two weeks after entering the station, things were
functioning well enough that a *Progress* freighter was launched,
carrying three more storage batteries, another set of solar-panel extension strips, more drinking water, underwear, and food. According to Dzhanibekov, it also brought with it "lots of towels" so that
they could wipe up more water. Also packed inside were several
different greenhouses, carrying wheat, peas, arabidopsis, cotton,
and a variety of orchids and tulips.[75] The freighter did not bring
them any fresh food or personal letters, however, a lack that upset
and annoyed both men.

By July 2, Dzhanibekov and Savinykh had things working so

well that they were actually able to begin doing normal research. They turned on the station's multi-spectral cameras and began photographing agricultural regions of the Soviet Union. Then in August they went outside and completed a successful five-hour space walk, installing the third set of extension strips to the station's last solar panel.

And though most of their plants died because of the cold temperatures, Dzhanibekov kept trying to grow cotton. Repeatedly he installed a soft pack of cotton seeds next to a light and tried to coax the seeds to life. Repeatedly they sprouted, then withered and died because the station was still too cold. "We've made a hothouse for [the cotton plants]," Dzhanibekov told Nechitailo. "But it doesn't work." By September, however, the success of his third attempt amazed even him. "The cotton is pushing the ceiling," he exulted. "We don't know what to do. It's still green, and it's too early to crop it in." In three days the plant had grown 6 inches.[76]

Finally, after a summer of intense repair work, *Salyut 7* seemed ready to resume normal operations. On September 18, 1985, *Soyuz-T 14* was launched, bringing Savinykh's crewmates. The spacecraft also brought Georgi Grechko, flying into space for the last time. Since his 96-day flight seven years earlier, he had worked first in mission control as a capcom, then as a back-up flight engineer for the Soviet-Indian mission. When he was chosen to fly this short mission, he wasn't very eager to do it, since he didn't consider eight days in space long enough to accomplish much. "Yet, I had been chosen," he noted, and like any good Russian, he "had to fulfill my assignment." And despite his initial lack of interest, he behaved just as he had on his previous two flights, becoming so involved in his work that he slept hardly at all during the entire eight days. Once, he was so tired he drifted off to sleep while doing observations of the earth, and was awakened by Dzhanibekov, who found him floating motionless in the middle of the station.[77]

Soyuz-T 14 not only switched spacecraft, as was now routine, but it also rotated crews for the first time. While Dzhanibekov returned to Earth with Georgi Grechko, Savinykh remained on *Salyut 7* with Alexander Volkov and Vladimir Vasyutin, reuniting the original crew. Then, one day after Dzhanibekov and Grechko had returned to Earth, another transport-support module, dubbed *Cosmos 1686*, was launched, bringing with it more than three tons of fuel and five-and-a-half tons of cargo. Unlike the earlier transport-support modules used with both *Salyut 6* and *Salyut 7*, *Cosmos*

1686's recoverable capsule had been refitted into a platform for a battery of research telescopes and could not return to Earth. This lack, however, allowed the super-freighter to bring additional supplies to the recovering station.

If all went well, the three men would try to complete the rest of their planned flight, with Savinykh spending a full 10 months in space. The Soviets were once again going to try to use *Salyut* 7 to set space longevity records. And once again, events did not turn out as planned.

During the next two months, Savinykh, Volkov, and Vasyutin were able to follow a pretty normal schedule, activating the transport-support module and conducting scientific research. Their attempts to grow plants in space continued, and they managed to grow onion sprouts. Vasyutin sampled one, finding its taste ". . . bitter, as it does on Earth."[78] They also did tests on several older pieces of *Salyut* 7's solar panels, removed from the laboratory's exterior by Kizim and Solovyov during their last space walk.

During these same two months, however, Vladimir Vasyutin began to feel ill. He had difficulty urinating, and developed a fever as high as 104° F. At first the men tried to downplay his problems. However, by the end of October, his illness had become too serious, especially with an impending space walk. The pain and fever had increased, and he was becoming tense and emotional. "He felt he was botching the program," Savinykh wrote later.[79]

Savinykh and Volkov insisted that they tell mission control everything, and on October 28 all three men had a heart-to-heart with Ryumin and ground doctors. Their space walk was canceled, and Vasyutin was ordered to cease all work in the hope that rest would help. By the middle of November, however, his condition had worsened. On November 15 he vomited. The next day, under the supervision of doctors, Savinykh gave him a thorough medical. Then, when the doctors once again asked Vasyutin how he felt and he started to complain again about his symptoms, Ryumin cut in. "That's it. We stop. We want you to replace the radio communications blocks and by evening we will probably decide to order a landing." When Savinykh said they'd need at least a week to mothball the station, Ryumin said fine, take a week, but no more. Ryumin had terminated the mission.[80]

On November 21, the crew undocked and returned to Earth.

Vasyutin was immediately taken to a hospital where he spent a month recovering from inflammation of the prostate. Though he had symptoms before launch, neither he nor his doctors had realized the seriousness of his illness. As one doctor in the program noted, "It can be hard to diagnose," especially because the lower temperatures in the recovering space station helped disguise the disease.[81]

All three men were disappointed, frustrated, and unhappy because of their early return. Vasyutin felt intense guilt, while Savinykh resented that the illness prevented them from completing their mission and setting a new space endurance record.

Their early return prevented much more than this. Upon their return to Earth both Dzhanibekov and Savinykh reported that the station's usefulness was significantly hampered by its period of deep freeze. After some discussion, mission control decided to scale down the later missions to *Salyut* 7. Construction of the core module of the next space station, to be named *Mir*, was nearing completion. Why risk the expense of a flight on the less-capable *Salyut* 7, when a new, completely redesigned, and more-sophisticated station was soon to launch?

In accordance with this decision, the addition of the almost-completed *Kvant* module to *Salyut* 7 was canceled. Instead, it was modified to be attached to the aft port of *Mir*. Also canceled was the all-woman flight that was supposed to follow, as well as the entire Soviet female cosmonaut program. Though the reason given was that Savitskaya, who was supposed to command it, had become pregnant, the situation was more complex. Leonid Brezhnev was dead, Gorbachev was now in power, and to him space stunts like this did not have the same appeal. Moreover, Savitskaya's two flights had not had the desired political effect. Unlike the 1960s, when Tereshkova had flown, no one seemed to care that a Soviet woman had once again gotten into space ahead of an American woman.

Rather than fly women, the more old-fashioned Russians preferred to cancel their entire female cosmonaut program. Thus, in the almost 20 years since Savitskaya's two flights, only one other Russian woman has flown in space (and she happened to be the wife of Valeri Ryumin, flight director for *Salyut* 7 and later head of the Russian side of the American-Russian Shuttle-*Mir* program).

Even today, there are practically no women cosmonauts in the Russian space program.

Ironically, though the Russians had failed in their effort to use *Salyut 7* to learn how to *build* space stations in orbit, that failure had instead provided them with priceless experience in learning how to *repair* space stations in orbit. The Soviet cosmonauts had confirmed what the nine American astronauts had demonstrated on *Skylab*, that humans could make the in-space construction of spaceships more efficient by their ability to analyze problems and do repairs. Furthermore, the intense workload involved in trying to save the station defused and eliminated the serious psychological problems experienced by the earlier Soviet crews on long missions. Focused on construction and repair, the men had little time to dwell on their loneliness and isolation.

Finally, the audacious work of men like Kizim, Solovyov, Dzhanibekov, and Savinykh had proved that the Soviet technology to build spaceships had matured. It was now reasonable for Russian engineers to consider building a ship that could travel to Mars and beyond. *Salyut 7*'s failures proved that spaceships could be repaired in space and almost certainly habitable. If humans could show that they could survive in weightlessness for the time necessary, a similarly designed vessel assembled from a number of separate modules and given sufficient supplies and spare parts should be able to maintain a crew long enough to voyage across the black expanses of the solar system and return safely to Earth.

Yet, few cared. Despite the Soviets' success at building and repairing structures in space, to most of the world it was *Salyut 7*'s propaganda efforts that grabbed the most attention. The headlines, not very large or pronounced anyway, instead went to Savitskaya, Chretien, and Sharma. Savitskaya's flights provided propaganda not for Cold War competition but for gender politics, an issue near and dear to the hearts of many postmodern American intellectuals. The French and Indian astronauts' flights revealed the steadily widening cracks in the Iron Curtain that had held the Soviet empire in thrall for decades.

Things had changed. And most importantly, a new leader had moved into the Kremlin, with interests and goals far different than those of any previous Soviet chief.

On March 11, 1985, three months before Dzhanibekov and Savinykh docked with a lifeless *Salyut 7*, the Politburo of the Communist Party chose Mikhail Gorbachev as its new General Secretary. In the 28 months since the death of Leonid Brezhnev, the leadership of the Soviet Union had been tentative and unstable. Andropov had lasted only 15 months, dying from kidney failure on February 9, 1984. His successor, 73-year-old Konstantin Chernenko, had been ill with emphysema when he took over, and lasted only 13 months himself, dying on March 10, 1985.

Gorbachev, however, was only 54 years old when he took power that March, by far the youngest Soviet leader in history. Born in 1931 in Stavropol, that strip of Russia between the Black and Caspian Seas, he did not belong to the older generation of communist leaders who, like Krushchev, Brezhnev, Andropov, and Chernenko, had made their mark under Stalin. Unlike these earlier leaders, Gorbachev had been a child during these horrific purges, and had never participated in them. In fact, rather than instigating the purges, Gorbachev could even be considered one of their victims. In 1937, when he was six, his grandfather was arrested, and spent almost two years in prison.

Moreover, Gorbachev had lived through World War II, not as a political leader dictating troop movements, as had Krushchev and Brezhnev, but as a child like Grechko, Dobrovolsky, Lyakhov, and Kizim, with no control over what might happen to him. His father was drafted, the family's village was occupied by the Germans, and his grandparents and mother were threatened with arrest and execution. To protect the boy, Gorbachev was hidden on a farm on the outskirts of the village. Then, when the town was liberated and famine broke out, the family survived only because Gorbachev's mother sold her husband's civilian clothes for about 100 pounds of corn.[82]

Thus, Gorbachev's hands were clean of the worst totalitarian abuses, and he could act with some freedom and flexibility. Young, with fresh ideas and a different perspective, Gorbachev believed that the Soviet system was in serious trouble, and required drastic measures to fix. Like *Salyut 7*, the Soviet Union no longer functioned, and needed a radical rescue mission. As he noted in his book *Perestroika*, "At some administrative levels there emerged a disrespect for the law and encouragement of eyewash and bribery, servility, and glorification. [Many leaders] abused power, suppressed criticism, made fortunes and, in some cases, even became accomplices in—if not organizers of—criminal acts."[83]

During the two years following Brezhnev's death, relations with the rest of the world had also deteriorated. In Poland, where martial law had been imposed at Brezhnev's instigation, the Solidarity union movement remained banned, with many of its leaders in prison. In September 1983, Soviet fighter jets shot down a South Korean commercial airliner, killing all 269 people on board and inciting worldwide outrage. That same year Soviet diplomats were expelled from the United States, Great Britain, Iran, Ireland, Canada, and Belgium, even as the Moscow Helsinki Watch group disbanded because most of its members were in prison or exile. In May 1984, the Soviets pulled out of the Los Angeles Olympics. By 1985, Soviet embassies in the West were being subjected to repeated protests, some involving bomb threats and violence.[84]

Gorbachev came to power with a mandate from the Communist Party to change things. With his remarkably charismatic personality, Gorbachev quickly rallied the fading power of the Communist Party to his calls for "perestroika," or restructuring, and "glasnost," or openness. The entire corrupt system would be reshaped, and the whole process was to be done openly, with the whole world watching.

How these changes would affect the Soviet space program, however, no one yet knew.

7

Freedom: "You've Got to Put on Your Management Hat . . ."

Shuttle

On January 25, 1984, Ronald Reagan stood before a joint session of Congress for his annual State of the Union address. At the time Reagan's political situation was complicated. The campaign for his second term was just revving up. His chief rival, former Vice President Walter Mondale, had already been campaigning heavily for almost a year. Other Democrats, including former astronaut Senator John Glenn, were gearing up their own campaigns. Reagan was also battling with the Democratic Party in Congress, including the Democrat-controlled House of Representatives, over budget issues. He had gotten Congress to slash income taxes three years before, but had been unable to get them to trim spending. In those three years, the annual deficit had skyrocketed to around $200 billion per year, with every indication that it would remain out of control for years to come. The Democrats, including Mondale, had repeatedly attacked him on these deficits, claiming that his tax cuts were the cause, and that the only way to get the deficit under control was their repeal.

At the same time, the high interest rates and out-of-control inflation that had been left over from President Carter's presidency and had led to a crippling recession in the first two years of Reagan's administration, had begun to let up. Unemployment had dropped, interest rates had shrunk from 20 percent to less than 9 percent, and the gross national product had risen sharply throughout 1983. There were signs that a strong economic recovery was in the works.[1]

During his speech, Reagan went through the typical laundry-list of presidential proposals. In an effort to gain some control over the deficits, Reagan proposed a $100 billion reduction in spending over the next three years, while also asking for a line item veto—the ability to veto individual line items on any bill sent to his desk. On environmental issues, he announced, despite his proposed cuts, that he wanted to give the Environmental Protection Agency one of the largest budget increases of any agency. He also proposed extending the Superfund law, as well as doubling the funding to research the question of acid rain. As a conservative, he once again called for a return to prayer in school. He also pressed for tuition tax credits to "soften the double payment for those paying public-school taxes and private-school tuition." He called for an end to abortion. He demanded a greater effort to crack down on child pornography, drugs, and violent crimes, calling for tougher sentencing.

He also called for a more stable relationship with the Soviet Union, under the premise of freedom and human rights. "Governments which rest upon the consent of the governed do not wage war on their neighbors," Reagan said. "People of the Soviet Union, there is only one sane policy, for your country and mine, to preserve our civilization in this modern age. . . . If your government wants peace there will be peace. We can come together in faith and friendship to build a safer and far better world for our children and our children's children. And the whole world will rejoice."

The centerpiece of this speech, however, was an additional announcement. "Tonight, I am directing NASA to develop a permanently manned space station and to do it within a decade." Like John Kennedy more than 20 years earlier, Reagan wished to galvanize the country around a spectacular space achievement. As he enthused, "Just as the oceans opened up a new world for Clipper ships and Yankee Traders, space holds enormous potential for commerce today."[2]

For $8 billion, NASA was going to build the first permanently manned space station, and have it in orbit by 1994.

A decade had passed since the flight of *Skylab*. In those 10 years, while the Soviets had been building and flying increasingly sophisticated and successful *Salyut* space stations, under an increasingly well-organized and purposeful program, the United States' space program had stagnated. First came the difficult effort to design, build, and fly the space shuttle. Proposed during Richard Nixon's administration—not long after he had also agreed to the joint *Apollo-Soyuz* project—the shuttle's creation had been technically challenging. More importantly, every step had been hindered by poor management decisions, in both NASA and Congress.

To begin with, NASA had trouble getting sufficient funds from Congress to build it. The agency's initial and preferred design had called for two completely reusable, winged stages mated together at launch. The first stage would return to Earth after separation while the second would then ignite its engines to lift a cargo of approximately 7 to 15 tons into orbit, depending on the altitude and inclination. Both would land like an airplane on a runway. Furthermore, NASA wanted to incorporate this shuttle as part of an overall long-term vision that included a space station, lunar bases, and interplanetary exploration.[3]

Though this wide-ranging concept had the most potential for making space exploration practical and widespread, its development was expensive, costing in the many tens of billions of dollars. NASA figured the total cost of the two-stage, reusable shuttle alone at between $10 billion and $13 billion.[4] Neither Congress nor President Nixon was willing to cough up the dough.

Rather than get nothing at all, James Fletcher, NASA's administrator in 1971, focused on at least getting some form of space shuttle approved. Because he knew that Congress was not willing to provide NASA with enough funds to design and build a completely reusable spaceship, at least not with the technology available at the time, he brought the military into the equation. If the shuttle could become the sole system for launching all U.S. satellites, including military surveillance craft, its cost could be paid for by many different programs. The large, covert, military space budget could also be tapped to pay for the shuttle's development.[5]

The military, though not opposed to a completely reusable shuttle, did not want the two-stage shuttle design. They wanted something bigger, including a cargo bay at least 15 wide and 60 feet long and capable of carrying a payload from 18 to 30 tons into low Earth orbit, depending on altitude and inclination. In addition, the military wanted to be able to land the shuttle anywhere within a strip 1,500 miles wide. To be able to maneuver like this within the atmosphere at orbital speeds required a much heavier spacecraft with much thicker thermal insulation than proposed for the two-stage shuttle.[6]

Congress, meanwhile, was not willing to fund the research and development cost of a completely reusable shuttle. Instead, they offered Fletcher $5.15 billion, enough to build a small fleet of simpler, less-capable shuttles.[7]

To make everyone happy and get the project built under these limitations, NASA's managers compromised. They agreed to increase the shuttle's size and weight. Then, to save development costs, they eliminated the reusable first stage and replaced it with two powerful, solid-rocket boosters, while also making the shuttle's main fuel tank expendable.[8] With these changes, the military was willing to join hands with NASA and lobby Congress for a joint program. Congress, in turn, was willing to fund the program, because many congressmen were much more willing to spend money for military purposes than for space development. In July 1972, 10 months before the launch of *Skylab*, North American Rockwell was awarded the contract to build the space shuttle fleet, with the first scheduled launch sometime in 1978.[9]

For the first few years, the shuttle's development seemed to proceed without setback. For example, improvements in computer technology meant that the agency could buy off-the-shelf computers with enough processing power to guide the spaceship to a soft runway landing. Moreover, after a number of flight and wind-tunnel tests, engineers decided that they could save significant weight by eliminating the shuttle's landing jet engines and letting either the pilot or the spacecraft's computers guide the shuttle to an unpowered landing.

To commit to such a landing took incredible daring. If something went wrong as the shuttle approached the ground, there would be no way for it to power up, swing around, and try again. Every shuttle flight would have to get its landing right the first time. To prove the concept, NASA in 1976 unveiled a prototype full-scale shuttle, to be used for flight and configuration tests. Originally named *Constitution* in honor of the country's 200th anniversary, it was renamed *Enterprise* after a loud and boisterous public campaign by *Star Trek* fans.

Throughout 1977 a team of four astronauts, Gordon Fullerton, Fred Haise, Joe Engle, and Richard Truly, completed 16 aerodynamic tests of *Enterprise*, including five flights where it was released from the back of a 747 at more than 19,000 feet altitude and flown back unpowered, gliding safely to the dry lakebed at Edwards Air Force Base in California. Despite some minor problems (on one landing *Enterprise* bounced three times before slowing to a stop), these successful approach and landing tests proved that a 75-ton vehicle—more like a flying brick than an aircraft—could safely land powerless.[10]

Other engineering problems were harder to solve. To get the shuttle back to Earth, NASA had to develop some form of reusable protective skin, able to withstand the heat of re-entry—ranging from 700°F to 2,300°F over most of the spaceship's surface and reaching 3,000°F on the nose and leading edges of its wings.[11] Until the shuttle, all manned space capsules, American and Soviet, used some form of ablative material during re-entry. As the material heated up, its surface peeled away, taking with it the heat. While simple, easy to build, and reliable, ablative material is, by definition, *not* reusable.

The design NASA settled on was a form of silica fibers, made from sand and mixed with clay to stiffen it. Extremely light (about a fifth the weight of an equal volume of water), the silica material could absorb high temperatures without distortion, and then cool quickly without damage.[12] This material, however, was extremely brittle. If large sheets were used to drape the shuttle's outer surface, they would shatter like glass as the spacecraft underwent the stresses of dropping from more than 100 miles altitude and speeds of 17,500 miles per hour. Silica could work only if its protective layer was first cut into lots of little pieces and fitted into place like many differently-shaped bricks. The result: 31,000 ceramic tiles, either 6 or 8 inches square and 1 to 5 inches thick, carefully glued into position, one by one.

The third major design challenge for NASA was the shuttle's main engines. Using liquid oxygen and hydrogen, they were to be among the most powerful rocket engines ever built, able to produce more than 375,000 pounds of thrust at sea level. They also had to be completely reusable, able to fire for more than eight minutes, then return to Earth and be used again, at least 50 more times. Moreover, the engines had to be far smaller than previous engine designs, small enough so that three could fit inside the shuttle's rear housing.[13]

These design constraints made construction of the shuttle's main engines far more challenging that expected. The first firing tests didn't begin until mid-1975, 15 months late. Then, the cash restrictions placed on NASA by Congress meant that the agency didn't have the money to test and develop the engine in the most efficient manner. To save money, the management at NASA and at its contractor Rocketdyne decided not to test each component separately before assembling them together into an engine. Furthermore, NASA had fewer test engines built.

When testing began, these shortsighted cost-saving measures ended up being very expensive, in terms of both time *and* money. When Rocketdyne engineers attempted to fire the engines at full power, they leaked, exploded, or burst into flames. Seals broke, the fans in a fuel pump cracked, and bearings froze. With a shortage of test engines, the testing was often delayed while the damaged engines were repaired or redesigned.[14]

By 1977, the 1978 inaugural launch date was considered impossible, and was delayed one year until 1979. Then 1978 rolled by and the 1979 launch date became impossible. By the end of 1978 NASA admitted that the launch would probably have to be delayed until 1980. Worse, NASA had originally hoped to get the shuttle into operation early enough so that it could dock with *Skylab* before the station's orbit decayed. Though many doubted whether *Skylab* would still be habitable after years of abandonment, others hoped that it could be reactivated and put to use. Unfortunately, the delays in completing the space shuttle made these plans impossible. On July 11, 1979, *Skylab* came crashing down on Australia.

Throughout that same year the difficulties with the main engines continued. NASA had to go back to Congress twice, asking for an extra $400 million to complete construction. Then there were the problems with the shuttle's ceramic tiles. Getting each glued to the surface of the shuttle was far more time-consuming than expected. By March 1979, the first shuttle, *Columbia*, still lacked 7,800 of its 31,000 tiles. In an effort to hurry completion, NASA managers decided to ship the shuttle from the factory in California to Cape Canaveral, figuring that they could save time by preparing the shuttle for launch while at the same time finishing the tiling. To protect the shuttle's skin during shipment, dummy tiles filled the gaps, fastened in place with adhesive tape.

To the embarrassment of everyone, as the Boeing 747 pulled off the runway with *Columbia* riding piggyback, the dummy tiles showered from the shuttle's skin like confetti. Before the 747 could return to base so that the dummy tiles could be refastened, this time with glue, the pilot had to request that a crew with brooms get out on the runway to sweep it clear.[15]

Once the shuttle was in Florida, tiling continued, though much slower than planned. New people had to be trained, and the facilities were cramped and inappropriate for the work. A worker could take as long as 40 hours to attach one tile. The glue itself, applied in two steps, required a total of more than 30 hours to dry.[16]

Then, the shuttle's main contractor began doing pull tests on the tiles that had already been attached, and was horrified to discover that about one-quarter of the tiles, about seven thousand, cracked horizontally under stress. These had to be removed, replaced with tiles that had been strengthened, and then re-glued into position.

By the summer of 1979, an inaugural launch in 1980 was unlikely. Not only were there thousands of tiles still to install, there were further problems with the main engines. During a simultaneous test of three engines in November, a hydrogen fuel line burst, causing a fire and serious damage to both the test stand and engine #1. The damage to the test stand was so severe that further tests were delayed for four months. To NASA's chagrin, the investigation into the fire revealed that the wrong welding wire had been used when the hydrogen fuel lines on all the engines had been built. Rocketdyne had to inspect and repair every engine in its stock.[17]

For the American space program, 1978 and 1979 were very bad years. While Soviet cosmonauts were cavorting on *Salyut 6* with men from Czechoslovakia, East Germany, Poland, and other communist bloc nations, the manned space program of the United States had almost sputtered to a halt.

Nonetheless, the program did grind forward. The engine problems were fixed, one by one. Slowly and inexorably the ceramic tiles were tested, replaced, and glued into position. And despite NASA's requests for more money, the overall budget for developing and launching the shuttle fleet ended up costing just 15 percent more than originally planned, growing from $5.15 to $6 billion (in 1971 dollars), an increase that was surprisingly reasonable considering the difficulty of the task.[18]

Finally, on April 12, 1981, after three years of delays, *Columbia* lifted off from Cape Canaveral, crewed by John Young and Robert Crippen. For two days these two men put the spacecraft through its paces. Then, with more than a quarter of a million people crowding the hills that surrounded the dry lakebed at Edwards Air Force base to watch, they safely brought the 100-ton spacecraft back from space, softly touching down. Immediately afterward, as Young and Crippen enthusiastically inspected the shuttle's exterior, finding it to be in reasonably good shape after its first journey to and from space, Young couldn't help expressing his uncensored hopes for the future. "We're really not too far—the human race isn't—from going to the stars!"[19]

For the next 15 months, while Soviet cosmonauts completed the last flights to *Salyut 6*, and Lebedev and Berezovoi were growing the first seeds in space on *Salyut 7*, *Columbia* completed four test flights, proving its systems and working out the final kinks in its design. By 1983, the American space shuttle was in full operation. The second shuttle, *Challenger*, had been launched, and the third and fourth were nearing completion. By October 1985, all four ships were in use, and by the beginning of 1986 had made a total of 24 flights.

The shuttle missions during the mid-1980s were in many ways the vehicles' most ambitious, even compared to today's space flights. In addition to numerous scientific experiments, the flights repeatedly tested and proved a wide spectrum of daring in-space construction techniques. On two missions, astronauts tested an American-designed refueling system, similar to that devised for *Progress-Salyut* refuelings. The system was able to pump up to 550 pounds of fuel into the tanks of a satellite that the shuttle's robot arm had grabbed and retrieved. On the second test, Kathryn Sullivan became the first American woman do to a space walk (three months after Savitskaya) when she attached a portable fuel pump to a dummy tank to see if fuel could be pumped from tank to tank. During another flight, two astronauts, Bruce McCandless and Robert Stewart, did the first untethered tests of a jetpack, called a Manned Maneuvering Unit. They flew as far as 320 feet free of the shuttle, independent human *Sputnik*s orbiting the earth all by themselves.

In another mission, which took place as Vladimir Solovyov and Leonid Kizim were repairing the fuel line on *Salyut 7*, *Challenger* was used to capture, repair, and release the *Solar Max* research satellite, which had failed four years earlier. Then, during the inaugural flight of *Discovery*, the third shuttle, an experimental solar panel, 102 feet long and 13 feet wide, was unfurled from the shuttle's cargo bay, proving that the fleet could be used to lift such large structures into space. Later that year, *Discovery* was used to capture two disabled satellites, bringing them back to Earth to be repaired and relaunched. Meanwhile, the fourth shuttle, *Atlantis*, completed one mission in which two astronauts assembled a 45-foot-high girder extending from the shuttle's cargo bay.

By the mid-1980s, the United States undoubtedly had the world's first fleet of spaceships, able to build and assemble almost anything in orbit.

Unfortunately, the fleet had nowhere to go.

Station

When the space shuttle was first proposed in the early 1970s, NASA conceived it as a ferry for getting to and from space cheaply. Even if Congress and President Nixon refused to approve the grand plans for building space stations and exploring the planets, NASA believed that the shuttle would make the construction of space stations and manned planetary spaceships so inexpensive and easy that private investors would become eager to invest in space, allowing the space industry to bypass Congress for funding.

By 1983, it was obvious that the shuttle fleet would not fulfill this prediction. Forced to build the shuttle to meet the hodge-podge of specifications from both the military and Congress, it was no longer completely reusable, thereby making it much more expensive to use. Instead of bringing down the cost of getting into space, shuttle launches were costlier, estimated in the mid-1980s to range between $100 to $300 million *each*.[20]

More importantly, NASA seriously underestimated the difficulty of building a reusable spacecraft. The shuttle couldn't be used like a space truck because it simply wasn't a space truck. Instead, it was the first prototype and test vehicle for honing the engineering to build completely reusable spaceships. For example, over the next decade most of the troublesome ceramic tiles were replaced with silica thermal blankets that, unlike the tiles, could be cut in larger sheets and draped over wider areas. Other areas of the shuttle were coated with plates of reinforced carbon-carbon, able to withstand very high temperatures and much less prone than the tiles to damage or loss.[21]

Unable to get private funding for space construction, NASA

began lobbying for government funds to build a space station as soon as the shuttle was in operation. Able to launch from 10 to 28 tons of cargo into space, depending on altitude and inclination,[22] the shuttle gave the United States the chance to regain what was lost when the *Saturn 5* was abandoned in the 1970s—the chance to build towering cities in space. In 1982, NASA's administrator at the time, James Beggs, put together a task force to rally support in Congress, the administration, and among the public for an American space station.

The task force attacked the problem on all fronts. It awarded eight $1-million-dollar contracts to space contractors like Boeing, McDonnell-Douglas, Lockheed, Grumman, and TRW to propose their own station designs. It coordinated the many space-station proposals within NASA itself, trying to convince the different bureaucracies to work together. It lobbied Congress. It promoted the idea to the public, the task force's members putting on literally hundreds of slide presentations and lectures.

The space-station proposals from the private companies were generally simple and could be built quickly to support crews of four or fewer and involving the launch of five to seven modules, each lifted into orbit inside the shuttle's cargo bay. Several were remarkably similar to *Salyut 7*. Grumman's proposal, for example, used a core module with solar panels attached to a truss, and a small additional cargo module to bring supplies to and from the station.[23]

For Beggs's task force, trying to herd the many factions within the government was sometimes as difficult as designing the station. Men like David Stockman, Director of the White House Office of Management and Budget, and Caspar Weinberger, the Secretary of Defense, opposed the station, saying that the money was not available, that building a station was inappropriate for the government, and that it would place a strain on other, more important, government programs.[24]

Reagan, however, saw otherwise. Like Brezhnev, the station to him was a symbol, a demonstration of American capability both to the world and to the citizenry of the U.S. As he noted during his 1984 State of the Union speech, "America has always been greatest when we dared to be great. We can reach for greatness again. We can follow our dreams to distant stars, living and working in space for peaceful, economic, and scientific gain."

Reagan's optimistic and hopeful vision of the future came from his personal view of America. Reagan grew up in the flat, open prairies of the American Midwest. His father was an itinerant shoe salesman who struggled with alcoholism. His mother was a hard-working housewife with strong ties to her local church.

Yet for Reagan it was a sunny childhood, despite the family's poverty amid the Depression. Unlike the horrors of Soviet Russia, with its wars and genocide and disease and famine, the America of Reagan's youth was a remarkably benign place to grow up. "As I look back on those days in Dixon," Reagan later wrote, "I think my life was as sweet and idyllic as it could be."[25] His worst memory might have been of the night he came home, as an 11-year-old, to find his father drunk and unconscious in the snow in front of their house. Writing in his autobiography, Reagan described what happened. "When I tried to wake him he just snored. So I grabbed a piece of his overcoat, pulled it, and dragged him into the house, then put him to bed and never mentioned the incident to my mother."

And yet, even this hardship had a sunny ring to it. Reagan's father was never violent, and was apparently devoted to his wife and family. "It was prosperity that [my father] couldn't stand," Reagan remembered. "When everything was going perfectly, that's when he'd [drink], especially if during a holiday or family get-together that gave him a reason to do it."[26]

Though Reagan generally opposed too much government, he believed that space exploration was one area where the government should be involved. He thought that government, by paying the very high initial-investment cost, would jumpstart the juggernaut of private industry, enabling almost unlimited possibilities of commerce and growth. "Opportunities and jobs will multiply as we cross new thresholds of knowledge and reach deeper into the unknown," he declared.[27]

Despite his enthusiasm, Reagan's support of the space station was remarkably tentative when compared to that of men like Brezhnev or Kennedy. It took three years of politicking by NASA before he was finally willing to back the idea and put it to the nation in his State of the Union speech. Moreover, the distrust and dislike Reagan and the members of his administration had for new, large, government projects made them less willing to back the project as strongly as had Kennedy. The space-station project as proposed by Reagan was also too ambiguous. He hadn't really given it a clear goal, a situation made even more pronounced by NASA's lobbying effort, which tried, as it had with the space shuttle, to make the station all things to all people in order to garner as much support as possible.[28]

With such lukewarm backing, it is not surprising that the station quickly turned into a political football—the victim of battles not unlike those that had compromised the shuttle. Initially, to get approval from Congress, the administration had agreed to limit any spending on the program during its first two years to low-level inexpensive studies. Unlike the aggressive 1960s space program, which started strong and only got stronger, the station program in the 1980s was specifically designed to begin with a lot of fanfare, with little substance behind it.[29]

Meanwhile, the bureaucracy at NASA seemed more interested in drawing grandiose gold-plated blueprints than in building something simple but concrete. NASA was no longer the lean, trim, and hardnosed agency that had sent astronauts to the moon and built *Skylab*. Just like the stagnant Soviet bureaucracy under Brezhnev, there were layers of bureaucracy at every NASA center, made up of people more interested in protecting their jobs and building office empires than in building risky and daring spacecraft that would conquer the stars.

Even more ironically, NASA's management structure as created by Congress during the 1960s—various centers located in different parts of the country in order to spread the wealth to different congressional districts—had slowly evolved into the American version of the competing Soviet design bureaus. During the 1960s space race, the different centers had been given very distinct and separate tasks, thereby avoiding direct competition. The Marshall Space Flight Center in Huntsville, Alabama, where Wernher von Braun was in charge, designed and supervised the construction of the rockets that sent men to the moon. The Kennedy Space Center at Cape Canaveral, Florida, took those rockets, assembled them on the launchpad, and sent them into space. And the Johnson Space Center (called the Manned Space Center until 1973) in Houston, Texas, handled mission control, taking charge of directing the missions themselves.

By the 1980s, these different centers, especially the Marshall and Johnson centers, had evolved into competing institutions, each with its own idea about how the space station should be built, with each center hoping to dominate the next great American space project. While Marshall engineers focused on designing a small but growing platform for science and research, the Johnson center proposed a station that was a "transportation node," where more complicated satellites and orbiting facilities were assembled and sent to geosynchronous orbit, or beyond.[30]

Just as Korolev's design bureau had fought with Chelomey's in the early 1960s, a battle that ended up wasting billions of rubles, each of NASA's different centers campaigned to win as much control of the station as possible. For example, in the summer of 1986, while NASA was still reeling from the *Challenger* accident, the Johnson Space Center commissioned several contractors to do studies on a new, more compact station design—and kept NASA headquarters in the dark about the studies. At the same time, NASA chief, James Fletcher, who had returned to head the agency when Beggs stepped down (because of a Justice Department indictment that was later withdrawn),[31] testified before Congress that no new alternative designs were being studied.

That same summer, a feud broke out between the Marshall and Johnson centers. NASA headquarters wanted to shift responsibility for building the station's habitable modules from Johnson to Marshall. Johnson officials protested, and settling the feud required long and complicated negotiations that weren't completed for months. Simultaneously, the director of NASA's Lewis Research Center in Cleveland, Ohio, which had developed some of NASA's rocket engines in the 1960s, began lobbying to build the station's solar-power systems, fighting the Marshall center for the work.[32]

While the different NASA centers fought over the station, contractors like Martin-Marietta, Boeing, and Lockheed stopped proposing their own ideas. Instead, they began aping the ideas suggested by the NASA centers in order to curry favor.

The focus had shifted from building the best space facility possible for the lowest cost to a competition for building the biggest governmental empire. Like Soviet design bureaus, NASA's Byzantine management structure was making the construction of the station difficult, very expensive, and not very likely to get off the ground. As one task force member noted, "The goddamn centers won't sit there and work for another center." On paper at least, the station began growing, in both size and complexity.[33]

As bad as these bureaucratic battles were, they were far less of a problem than the micromanaging of the station's design by Congress. Besides the typical local pork-barrel spending used by some congressmen to spread money to their districts, Congress often made specific demands inspired by a whole range of lobbying groups. For example, in 1987, scientists persuaded Congress that the station had to have a second set of solar arrays or it wouldn't serve their needs. Congress agreed, and directed NASA to add the array.[34]

Congressional micromanaging sometimes took the form of outright opposition. For example, in 1988, William Proxmire got all spending on the station suspended until after the presidential elections, on the premise that the next president should decide whether the program should continue. George Bush, Sr., won and announced his support of the station. Amid these battles, Congress routinely cut the station's budget while stretching out its construction. While these changes reduced the money NASA spent each year on the station, they ended up increasing the total overall cost.[35]

Faced with trying to satisfy these different constituents, the station went through a plethora of proposed configurations throughout the 1980s. Six months after Reagan's 1984 speech, NASA chose its first preliminary design, dubbed the "power-tower." Flying in a gravity-gradient orientation to save fuel, five pressurized modules would be linked together into a "b"-shaped configuration and fixed to the base of a truss 450 feet long. On a cross-piece three-quarters of the way up the truss would be four giant solar panels providing about 75 kilowatts of power. NASA engineers figured that in this gravity-gradient attitude, the station's crew of six would be best positioned to do both Earth and astronomical observations. Scientists, however, complained to Congress and to NASA. To do microgravity research, they wanted the station's laboratories situated as close to the station's center of gravity as possible. Otherwise, the subtle stresses from being on the outside fringes of the station's minuscule gravity field would ruin their research. Moreover, the researchers complained that the power-tower design didn't provide them enough electrical power to do their work. They also worried that a crew of six was insufficient to do construction, maintenance, and science.[36]

In 1985, to satisfy the demands of both scientists and Congress, NASA junked the power-tower and came up with the "dual-keel" design. Unlike the power-tower, this placed the modules in the center of a 508-foot-long main truss, with two sets of two large solar panels on each end for balance. Also affixed to the ends of this truss were two revolutionary electrical-power generators that used mirrors to focus sunlight to boil water into steam which, in turn, ran turbines to generate power. Surrounding the module complex was an additional rectangular truss—the dual-keel—345 feet high and 147 feet wide, which added rigidity and room to attach later modules and equipment. The U.S. modules were reduced from five to three, with the Japanese and Europeans joining the station's construction by each adding one module of their own. The configuration was also made more compact, with the modules packed side by side and linked by small nodes at their ends. Overall, this new arrangement was about 30 percent larger than previous proposals and could support a crew of eight.[37]

Then, in 1987, Congress was shocked to learn that the dual-keel design it had forced NASA to adopt was going to cost at least $14.5 billion, almost double the $8 billion to which Congress had first agreed. To get Congress's approval and still keep the scientists on board, NASA once again reconfigured the station. Though the main 508-foot truss was retained, the rectangular dual-keel was eliminated, along with the solar turbines, reducing the cost to about $12.2 billion.[38]

This configuration, which seemed to please both scientists and engineers, was dubbed *Freedom*, in 1988. As explained by Marlin Fitzwater, Reagan's press secretary, at the announcement, "The name *Freedom* was recommended by a team of NASA representatives and international partners." He continued,

> The yearning for freedom is a basic human emotion, and freedom of the individual is a value shared by all the nations that will work together to build and use the space station. In a literal sense, the space station will provide freedom from the confines of Earth's gravity, enabling scientific and technological research, new commercial uses of space, and opening the way for continued exploration of space.[39]

For some Congressmen, *Freedom*'s newly reduced configuration was too expensive. Both Congressman Edward Boland, Democratic chairman of the House Appropriations Subcommittee, and Congressman Bill Green, the ranking Republican on the same subcommittee, opposed a continuously manned station, preferring instead a smaller "man-tended" facility that would be visited only periodically by astronauts to upgrade or swap the automatically running experiments.[40]

At one point these Congressmen tried to replace NASA's station entirely. Just like Chelomey had done back in the early 1960s when he convinced Khrushchev to fund a space program completely independent of and parallel to Korolev's, an independent group of former NASA engineers had successfully pitched their own station proposal to the two congressmen. Called the *Industrial Space Facility*, it was smaller and cheaper, could be launched quickly for about half the cost, and would be visited only about three times a year to swap experiments and do maintenance. In the 1987 appropriations bill Boland and Green had $25 million appropriated for this facility, while cutting NASA's station project by half. Only after intensive lobbying and backroom deal-making was NASA able to get Congress to quash this independent effort.[41]

Costs kept rising. And *Freedom*'s design kept changing. In 1989, the station went through another redesign, this time to trim its cost and streamline its construction in order to once again meet the demands of Congress. The main truss was shortened to 353 feet. One large solar panel was eliminated, reducing the station's power from 45 to 30 kilowatts. The attitude-control thrusters were changed from a hydrogen-oxygen system, designed to be self-sufficient by being fueled with excess water from the station's atmosphere, to more conventional hydrazine thrusters. Though requiring refueling from Earth, they would require little development cost. Also eliminated to save development cost was a completely closed-loop environmental system. Rather than have the station's water and atmosphere recycled, as the Soviets had increasingly succeeded in doing in their *Salyut* stations, the shuttle would ship up new supplies from Earth. The U.S. laboratory and habitation modules were shortened, and the module layout was rearranged so that several interconnecting node modules could be eliminated. The simplified redesign, unveiled in March 1991, was mockingly called "Fred" by many at NASA.[42]

By 1990 NASA had spent approximately $3.8 billion, almost half of Reagan's original budget goal, on the station over six years and had built literally no space hardware.[43] As space historian Martin Lindroos has noted, "The space station *Freedom* project finally collapsed under its own weight in 1990, when the design was found to be 23 percent overweight, over budget, too complicated to assemble while providing 34 percent too little power for its users."[44]

Part of the problem was that Reagan had not given the station program a focus. Though he wanted it built to initiate a renaissance of private enterprise in space, his proposal was never clear enough to make that dream possible. Instead, *Freedom* became a nebulous "laboratory" in space, a place where serendipitous discoveries would occur that would produce wealth for America and the world. Since a facility this ill-defined could serve any number of functions, no one had any clear idea of what it should look like. All those involved took this blank slate and painted their own fantasies upon it.

Compounding Reagan's lack of clarity was an overall lack of strong public support for the project. While Americans were generally in favor of space exploration, they had very mixed feelings about having the government pay for it. This lukewarm support—and the strong opposition by many in Congress and by most late-twentieth-century American intellectuals to human exploration of space—forced NASA administrators to try (as they had with the space shuttle) to keep the station a blank slate, able to fulfill anyone's dreams in the hope that they could rally support for the project.[45]

Finally, the bureaucratic in-fighting at NASA shifted the focus from letting the engineering determine the station's design, as had been the approach during the *Apollo* program, to letting managers use the station to build personal empires. The goal was no longer trying to learn how to live and survive in space, which is *always* the primary reason for putting humans there, but how to keep as many government workers as possible employed.

On January 28, 1986, the over-managed post-lunar American space program reached its nadir. Under pressure to prove the shuttle capable of frequent launches—operating like a truck—management at both NASA and its contractors chose to ignore the advice of their engineers.

The shuttle is lifted into space using a combination of rockets: the liquid oxygen-hydrogen engines at the tail of the orbiter and the two solid rocket boosters attached to either side of the giant external tank that holds the liquid fuel. The two solid rocket boosters used until 1986 were stacks of reusable segments filled with solid fuels. After each flight, the stacks were stripped down, re-filled, and reassembled, the joints between segments sealed with two rubber gaskets, a primary and backup, called O-rings. For the O-rings to work properly, they had to be pliant and flexible, able to fill the gap tightly when the fuel began burning and the gas pres-sure pushed them into the joint.

During the seventh of nine shuttle missions in 1985, engineers at Morton Thiokol, the manufacturer of the shuttle's solid rocket boosters, found significant erosion and damage to the primary O-rings. On another flight, in April 1985, they found a section of a primary O-ring completely burned through. The evidence suggested that when the outside ground temperature was low the O-rings did not work as engineered. Despite these engineers' increasing con-cerns, the shuttle continued to fly. As Morton Thiokol engineer Roger Boisjoly wrote in a memo in July 1985, "It is my honest and very real fear that if we do not take immediate action to dedicate a team to solve the problem . . . than we stand in jeopardy of losing a flight along with all the launchpad facilities."[46]

For the morning of January 28, 1986, the temperature at Cape Canaveral was predicted to be 36° F, close to freezing. The shuttle *Challenger*, about to begin its 10th flight and 25th shuttle mis-sion, carried seven astronauts, including schoolteacher Christa McAuliffe, the first American citizen who was not part of NASA's astronaut program chosen to fly into space. The Morton Thiokol engineers once again warned both their bosses and NASA offi-cials from the Marshall Space Flight Center who managed the contractor that the cold winter temperatures in Florida could have a negative effect on the O-rings.

The managers, at both Marshall and Morton Thiokol, were skeptical. They believed the data were inconclusive. More impor-tantly, they felt an inescapable but unstated pressure to maintain the schedule. Each time previously that NASA had delayed a shuttle launch, the agency had faced a chorus of ridicule and criti-cism from their opponents in the media and Congress about how the shuttle was not living up to its promise of frequent and cheap launches. This public pressure placed everyone at NASA in an awk-ward position: If they took a more cautious position and scrubbed questionable launches, they would be accused of failure and in-competence, and would give ammunition to those who wished to defund the agency. Rather than give their enemies proof that the shuttle program was a failure, the managers increasingly tried to tough it out, hoping to prove that even under difficult conditions the shuttle could function like a cargo airline company.

It was the same story all over again. Just as political concerns had caused NASA administrators to compromise the design of both the shuttle and the space station—making both vehicles inadequate to accomplish the tasks for which they had originally been conceived—the managers at both Morton Thiokol and NASA once again compromised their judgment in a futile attempt to please the politicians, press, and public that either didn't understand what they were trying to accomplish, or opposed it outright.

At one point the night before launch, as Thiokol engineers desperately tried to convince their managers that a launch the next day was too risky, a Thiokol executive looked at Bob Lund, manager of the engineering department, and said, "Bob, you've got to put on your management hat, not your engineering hat."[47]

For Lund, the pressure from above, the doubts about the data, and this insistence that he think like a manager, not an engineer, were enough to cause him to change his mind. He joined the other managers at Morton Thiokol to overrule his engineers, certifying the solid rocket booster as safe to launch.

Challenger lifted off at 11:38 A.M. on January 28, 1986. As feared by Boisjoly and the other engineers, however, at least one O-ring in the right solid rocket booster had been rendered stiff and non-pliant by the cold. With the O-ring unable to provide a good seal, black smoke began to leak from the joint almost immediately after ignition.

At 58 seconds, when the shuttle had gained more than four miles in altitude and was moving faster than the speed of sound, flame from the leak began impinging against the giant external tank. At launch, this tank had carried 143,000 gallons of liquid oxygen and 383,000 gallons of liquid hydrogen, most of it still unburnt.[48]

At 74 seconds, the liquid fuels in the tank ignited, and the shuttle, tanks, and rockets were ripped apart. The two solid rocket boosters shot wildly away like punctured balloons, forcing ground controllers to hit their self-destruct buttons. *Challenger*, torn into pieces, rained down into the Atlantic. Its crew cabin, designed to resist such forces, at first held together, some of the crew still alive inside. As it fell, however, the g forces became murderous, and upon impact with the ocean it shattered.[49]

Two weeks later, the Soviet Union launched the first module of its new space station, *Mir*, intended to establish the permanent occupancy of space.

8

Mir: A Year in Space

Restructuring

For Musa Manarov, the job at least was in communications, and at least it put him closer to the action. For five years he had trained to fly as a cosmonaut on the Soviet space shuttle. Then, in 1983, without explanation, he was dropped from the program, and given no other assignment. He went to flight director Valeri Ryumin for some explanation. Ryumin couldn't explain it. He simply said, wait, something will happen.

Nothing happened. Even though he had done everything they asked, passed all the exams, Musa Manarov was a cosmonaut without a mission.

He heard rumors. There were stories circulating that the reason he had been dropped and refused any further mission assignments was his ethnic heritage. Though his father had been a Soviet army officer who had fought bravely in World War II, and though the boy had been raised on military bases throughout the Soviet Union ("In the Soviet way," noted Manarov), the family were ethnically members of the Lakets people, a small tribe of around 100,000 to 150,000 people located in the Dagastan republic, a small province nestled between Georgia, Azerbaijan, and the Caspian Sea. The rumors said that certain higher political officials opposed Manarov's assignment to any flight because of his ethnic background. Such bigotry was not unusual in the Soviet Union.

Manarov could quit the program and do something else. He could go back to the design departments at Energia where he had worked as an engineer and designer and do that again. But he wasn't a quitter. After all those years of training, he wanted to fly into space at least once.

For the next two years he did as Ryumin suggested. He waited, working at Energia to help redesign the *Soyuz* spacecraft. Then, in frustration, he went back to Ryumin and asked if he could work in mission control as a capcom, handling communications between the ground and the orbiting space stations. He hoped that direct involvement in manned operations might increase his chances of getting assigned to a flight.

Moreover, Manarov had always loved working with radios. Because his father was an army officer, the family moved frequently. During his four years of high school he attended three different schools scattered hundreds of miles apart. Unable to make any steady friendships, the boy did his schoolwork and played with radios. He put together his own crystal sets, then began taking apart and fixing broken radios. "I always took my tools with me, and soon I was even doing repairs for my neighbors."

Graduating high school at top of his class, he decided to go to the Moscow Aviation Institute, not to study aviation like other future cosmonauts, but to study communications. His passion for fixing radios as a child had made him want to build them as an adult. Following his graduation in 1974, however, he was sent to Energia to work, where he was given the job of analyzing the telemetry being downloaded from various spacecraft. The work wasn't in communications, and it hadn't been his first choice, but he accepted it. It was an interesting job, and far better than many others he could have been stuck with.

"I had never imagined myself as a cosmonaut," he remembered. "I never flew airplanes. I never skydived. I wasn't a particularly fearless man. I just worked at my desk, like any other office worker." Nor did he consider himself a particularly healthy man. As a child he had been sickly with numerous colds and even pneumonia.

Nonetheless, when Energia hired him in 1975, his superiors strongly suggested, just as they had done to Savinykh, that he apply to the cosmonaut program. Energia wanted and needed its engineers to become cosmonauts. Though he hadn't gotten a direct order, he realized that it was in his best interest to go through the motions. At the time, all young men like him were required to be members of Komsomol, the Youth Communist League, and everyone was expected to do their utmost for the communist cause. If he didn't apply he might appear disloyal. Besides, the health rules were incredibly strict. Only 1 in 400 was accepted. Manarov, once a sickly child, would almost certainly be rejected.

To his amazement he passed every exam. It seemed that the doctors weren't looking for men who were unusually healthy, but ordinary men who simply had nothing wrong with them. Because Manarov's childhood sickliness had disappeared in adulthood, and he had no other physical problems, he passed with flying colors. The next thing he knew, he was in training, flying jets and skydiving, followed soon after, in 1978, by assignment to the shuttle program. To him, it all seemed like a dream.[1]

When he asked Ryumin for a job in mission control he had been a cosmonaut candidate for eight years, and had not yet flown in space. If Ryumin said no, he probably would never get into orbit. To his relief, Ryumin said yes, and in the spring of 1986, Musa Manarov began working as a capcom.

At the time, the Soviet space-station program was in transition. The last mission to *Salyut 7* was about to begin, and the next Soviet station, with a radically different design, had just been launched. On February 20, 1986, just three weeks after the *Challenger* accident, *Mir*'s first module, what the engineers uninspiredly called the base block or core module, had been placed in a 200-mile-high orbit.

Glushko, sensitive to the political winds of his time, had given this new station the name *Mir*, which in Russian can be translated as "peace," "union," "village," "community," or "new world," depending on context. Aware that Gorbachev was abandoning the strident, pro-communist point of view of previous leaders, Glushko hoped this name might resonate with the new leadership, and thus encourage continued financial support.[2]

At first glance, the new space station did not appear very different from previous Soviet stations. Designed around the limitations of the Proton rocket that sent it into space, the core module was a wedding cake tower about 43 feet long, weighing about 20 tons, with engines and a docking port at its aft end. Closer inspection revealed significant differences, all of which influenced the station's

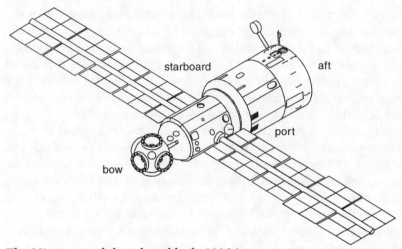

The *Mir* core module or base block. *NASA*

future, for both good and bad. For example, *Mir* carried a much more sophisticated attitude-control system, run by seven computers. Cosmonauts no longer had to periodically monitor and maintain the station's orientation, leaving this task to the computers. If the computers crashed, however (something which happened repeatedly in later years), the station's attitude would quickly drift, with the solar panels no longer facing the sun and therefore unable to charge *Mir*'s batteries.

Other changes: While the core module had a service donut incorporated into its aft end like the previous two *Salyuts*, with two main engines and 32 smaller attitude engines all using the same consolidated fuel system, the long-range plans did not call for the main engines to be refueled or used for routine orbital maintenance. Instead, their primary function was to lift the station from its initial orbit of around 125 miles elevation, placed there by the Proton rocket, to its permanent altitude of around 200 miles. Most later orbital changes were to be handled by the engines of the *Progress* and *Soyuz* spacecraft that docked to the station's aft and bow ports.

Much of *Mir*'s air and water supply was to be recycled. Using the Rodnik system first tested on *Salyut 3* and used by both *Salyut 6* and *Salyut 7*, water vapor in the atmosphere was captured, cleansed, and turned into potable water. In the bathroom, a cubicle located next to the aft exit tunnel, the washbasin water came from a closed-loop system that filtered and purified the water after use. Urine was processed so that the impurities were removed and then dumped into a tank. Initially this purified urine was dumped out into space. Later, an electrolysis unit, dubbed Elektron, was added to extract the oxygen from the urine to supplement and recycle the station's oxygen supplies. Until then, the station's atmosphere, a standard mix of oxygen-nitrogen maintained at sea level pressure, was replenished by *Progress* freighters.

To cleanse the atmosphere of carbon dioxide the core module used lithium hydroxide canisters, similar to the lithium perchlorate canisters of the *Salyut*s. Each canister, which operated passively, could clean the air of the carbon dioxide produced by two cosmonauts for several days. This system was later replaced by a permanent recycling unit, dubbed Vozdukh, included on a later module.

Unlike the *Salyut* stations, *Mir* was designed to grow over time, so that if it were completed as planned, the Soviets would have finally built a space station larger than *Skylab*. For this reason, the interior of the station was significantly changed. Because the core module was designed as the main living quarters for the yet-to-be-completed station, its interior space no longer contained the giant cone-shaped housing for research telescopes or surveillance cameras. Instead, the core module's main room was open and unobstructed, though narrower because its side walls were filled with collapsible equipment, storage compartments, tanks, an enclosed bathroom, and—for the first time on a Soviet space station—two small phone-booth-sized private cabins for long-term occupants. Placed on opposite walls, each cabin had its own window, fold-out writing surface, and sleeping bag.

The most obvious design change from the *Salyut* stations was also the one that made it possible for *Mir* to grow in size. Unlike the docking-transfer compartment that on the *Salyut* stations had one port, the multiple-docking adapter at the core module's bow was a spherical hub with five ports, one port at its bow and a ring of four radial ports on the hub's sides. *Mir*'s planned configuration, as stated in 1986 and scheduled for completion by 1989, called for five additional modules to be added to the core module, one at its aft port and one at each of the four multiple-docking adapter's radial ports.

In many ways *Mir* turned out to be a never-ending construction project. Its design called for, in addition to new modules, space-walking cosmonauts to add a third dorsal solar panel on the core module's top, and to supplement each solar panel with side strips, just had been done on *Salyut 7*. In later years, cosmonauts also assembled two large cranes for hauling men and equipment, and a 46-foot-long boom with a thruster engine at its end. Each of these new additions inspired engineers to rethink what they'd built and come up with new ideas. Equipment was constantly assembled, reassembled, repositioned, and disassembled. [3]

All these innovations aimed at accomplishing *Mir*'s primary goal: to achieve the first permanent occupancy of space. No longer would there be distinct expeditions to the station, as there had been with the *Salyut* stations. Instead—while the United States dithered in its own attempt to accomplish the same thing—one crew would pass control to the next, so that someone was on board and in space continuously. At the same time, the crew shifts would be lengthened, eventually reaching a year or longer. In this sense at least, *Mir* was the true prototype of an interplanetary spacecraft. If its components could operate in space for more than five years, it would prove that such a spacefaring vessel could be built.

The trouble was that there were real questions whether the new leadership under Gorbachev really cared to accomplish any of these space goals. Just as the Soviet space program was going through a transition when Manarov stepped into mission control as a capcom, so was the entire Soviet bureaucracy. In the 11 months since Gorbachev had taken power, he had started an aggressive public-relations campaign to reshape the communist empire.

Very little in the Soviet Union worked. There were endless shortages—of food, consumer goods, and heavy industries. Economic growth was stagnant, even falling in many industries. The harvests were so abysmal, reaching a 40-year low in 1980, that the Soviet Union was forced each year to import increasing quantities of grain to feed its people. The entire society was rife with corruption. To get anything accomplished required a kickback or a bribe. As Gorbachev himself noted in May 1985, "Try to get your flat [apartment] repaired: you will definitely have to find a moonlighter to do it for you—and he will have to steal the materials from a building site."[4]

Meanwhile, top officials in the Communist Party doled out

privileges and perks to themselves unimagined by the general population. For example, when Boris Yeltsin was named a candidate member of the Politburo in 1986, he was given a domestic staff of three cooks, three waitresses, a housemaid, and a gardener with his own staff. His family had access to special doctors and hospitals and to a separate telephone system. They could shop in special groceries stocked full with goods unavailable to the public. Their transportation was free, their clothes were custom-tailored, and they got free vacations at private resorts closed to everyone else. Nor was Yeltsin's experience unusual. Every Communist Party apparatchik traded loyalty for similar privileges. As they moved up in the ranks, the perks increased, thus insuring their loyalty and obsequiousness.[5]

Gorbachev, a child of this system, wanted to change it, to make the Soviet Union a more civilized and just society. As Gorbachev later wrote in his memoirs, "We longed for freedom so much that we thought that if we just gave society a breath of fresh air it would revive. We understood freedom in a broad sense. . . . We were aware—although we did not formulate this idea very specifically— of the need for democratization of society and the state, and for the development of people's self-government."[6]

Within weeks of becoming General Secretary, he proposed to make the Soviet Union more efficient and to get people focused on doing their jobs better. Perestroika demanded that everyone, from the lowest worker to the highest government official, ". . . change [their] attitudes. Anyone who is not prepared to do so must simply get out of the way!"[7]

Unfortunately Gorbachev had no idea how to do all this. To democratize the Soviet Union would require eliminating the privileges of his fellow Party apparatchiks. To make the country more efficient would require destroying the bankrupt, centralized, state-run bureaucracy and the Communist Party that ran it. Gorbachev, a passionate and proud communist, would have to introduce capitalism and private ownership, the very ideas the Communist Party had opposed for 70 years.

Rather than make such fundamental changes, Gorbachev's perestroika instead focused on superficial public-relations ploys. He announced a campaign to stamp out alcoholism and drunkenness. Vineyards were destroyed, the production of vodka was cut, the legal drinking age was raised from 18 to 21, fines for drunken-

ness were increased, and most painfully, vodka was no longer
served at official Party events. He demanded better quality control
in the factories and hired managers whose sole job was to weed out
badly produced goods. On the same note, in order to make the state-
run economy more efficient, Gorbachev reorganized the bureaucracy into a number of "superministries," each dedicated to a particular industry. For space, the numerous design bureaus were
placed under the umbrella of Glavkosmos, with Energia retaining
charge of the manned space program.[8]

He proclaimed a crusade against bribery and theft from public
institutions. In Moscow, for example, he fired the apparatchik in
charge of running the city and brought in Boris Yeltsin. Yeltsin had
been a hard-driving engineer who for more than a decade had managed the province of Sverdlovsk, located just east of the Ural Mountains and in many ways the Russian equivalent of the American
Midwest.

More daring than Gorbachev, Yeltsin was willing to test both
the limits of perestroika and the corrupt communist system of
which he was a part. Within 18 months Yeltsin had replaced almost 70 percent of the top officials in the Moscow Communist
Party structure. He continually browbeat the inefficient and lazy
apparatchiks in the Moscow bureaucracy. He rode the subway and
buses to see how people lived. He showed up at factories and stores
without warning.

In the West, Yeltsin's best-known innovation was the open fairs
he initiated in many empty lots, events that at least superficially
resembled what we in the urban United States call a flea market or
a farmers' market. At these fairs shipments of food were made available in booths, bypassing the warehouses and Moscow stores
where, Yeltsin believed, much of the best material was skimmed
off for sale in the black market. These fairs became very popular,
attracting musicians, performers, and large crowds and eventually
bringing Yeltsin's name to the attention of the Western press.[9]

Finally, Gorbachev's campaign insisted that the Soviet bureaucracy be held accountable to the public. This new policy, which he
named glasnost, or "openness," became evident almost immediately in the Soviet space program. From 1985 onward, Gorbachev
declared, the program's traditional secrecy was to end: Launches
were to be announced in advance, and failures were to be investi-
gated in public. As a prime example, *Mir*'s launch date was publi-

cized beforehand, and hundreds of television cameras were allowed to broadcast the event. (That the launch was rushed by several months so that it could be timed to just precede the opening of the February 1986 Communist Party Congress was an early sign that Gorbachev's reforms were not as deep or as sincere as publicly indicated. Like Brezhnev and Reagan, he still found space exploration a useful political tool.)[10]

Nonetheless, Gorbachev was far less interested than earlier leaders in the space program, either for exploration *or* as a political weapon. Because the focus of Gorbachev's government had shifted from establishing a universal communist utopia to reforming the bureaucratic mess, the pace of the Soviet space program slowed. Construction of the station's additional modules was delayed, and after the launch of the core module in February 1986, no long-term or record-setting mission was sent to the station for more than a year.

During that long gap, the only Soviet manned space flight was a four-month mission that spent almost as much time on *Salyut 7* as it did on *Mir*. On March 13, 1986, one month after *Mir*'s launch, the unique flight of *Soyuz-T 15* lifted off from Baikonur. Crewed by Leonid Kizim and Vladimir Solovyov, the same two men who fixed *Salyut 7*'s engine leaks during their marathon space walks two years earlier, their 125-day flight remains the only manned mission to have visited two different orbiting space stations.

Their mission was mostly one of mop-up and check-out. Many experiments on *Salyut 7* had been left unfinished or unresolved after the hurried exit due to Vasyutin's illness. A few could even be transferred to *Mir*. First, the two men docked with *Mir*, where they spent six weeks verifying its systems. They did a resonance test, shaking the station to verify its stability. They tested the use of a new Soviet geosynchronous satellite, *Altair 1*, which more than doubled the time the station remained in direct communication with mission control. They repeatedly evaluated *Mir*'s atmosphere to make sure its carbon dioxide scrubbers were working properly. And during all this, two *Progress* freighters arrived, testing the station's docking and refueling systems. Then, Kizim and Solovyov undocked and maneuvered to *Salyut 7*, where they spent another seven weeks.

While they were on *Salyut 7*, *Soyuz-TM 1* was launched and sent to *Mir*, the first unmanned test of a next-generation *Soyuz* spacecraft. That Manarov had helped do the redesign made it all the more appropriate for him to be in mission control, helping to put the spacecraft through its paces. Though quite similar to the *Soyuz-T*, *Soyuz-TM* was trimmer and fitter. Its weight had been cut, allowing it to launch an extra 500 pounds and bring back an extra 150 to 200 pounds. It also used a new radar rendezvous system, dubbed Kurs, which replaced the Igla system. Kurs, built by a factory in the Ukraine, had been developed in response to the crisis when *Salyut 7* went dead. The new rendezvous system could work at much greater distances than Igla, and its antennas were omnidirectional, able to lock onto a target even if that target was drifting. With Kurs, a spacecraft could approach *Mir* from any angle, with the station kept passive while the approaching craft did all the maneuvers. Igla had required both the station and the rendezvous vehicle to continually reorient themselves, something that was wasteful of fuel and impossible for the growing *Mir* complex to do.[11]

Soyuz-TM's fuel tanks had also been modified, somewhat belatedly, as a response to the gas-line leak on *Salyut 6*. The flexible bladder that had been used as a pump and had shown a tendency to leak was replaced with a metal membrane.[12]

Guided into *Mir*'s bow port, *Soyuz-TM* remained there for seven days while its systems were tested. It was then de-orbited and its descent capsule returned successfully to Earth.

Meanwhile on *Salyut 7*, Kizim and Solovyov performed two space walks during which they assembled and evaluated a 50-foot truss structure, similar to the giant structures being tested at the same time on American space-shuttle flights. On the first space walk they tested unfolding and folding the truss. On the second, they again assembled the frame and this time attached a detector at its far end to measure the truss's rigidity. They then proceeded to use the welding tool first tried out by Savistskaya and Dzhanibekov during their space walk in 1984.

For another three weeks Kizim and Solovyov alternated between making Earth observations and packing experiment samples, film, and equipment in their *Soyuz* spacecraft. Then, on June 25, they undocked from *Salyut 7* and returned to *Mir* for another three weeks in space, bringing with them 20 experiments from the older laboratory, including 6 different plant experiments and a variety of seeds.[13]

Since 1982, when Lebedev and Berezovoi had produced *arabi-*

dopsis seeds in space, the entire Soviet plant research program had stalled. *Salyut 7*'s technical problems kept interfering with the experiments. Worst of all, the failed docking of *Soyuz-T 8* resulted in the loss of most of the 200 space-born seeds produced by Lebedev and Berezovoi. As the first follow-up mission, the biology team had packed it with the seeds, hoping to see if the space-born seeds could produce a second generation of *arabidopsis* in space. When the docking failed, however, those seeds, along with everything else in *Soyuz-T 8*'s orbital module, had been destroyed, burning up in the atmosphere. "None of us imagined that such a thing would happen," Nechitailo noted. Then, *Salyut 7* kept breaking down, preventing them from sending up another set of seeds and starting over.

Thus, to find out if a second generation of plants could be grown in space, they had to wait for *Mir*. Even more frustrating, they also had to wait until *Mir*'s additional modules were added, because the core module was not designed to carry extensive plant experiments. For the moment at least, Nechitailo, Mashinsky, and the rest of the biology team at Energia had to content themselves with studying the few plants grown in the six greenhouses that Kizim and Solovyov brought back from *Salyut 7*. And like most of the research on *Salyut 7* since its systems had started to fail, none of these plant experiments were of much value scientifically.[14]

After Kizim and Solovyov left, *Salyut 7* voyaged on, never to be visited again. It was to remain in orbit for another four-and-a-half years, its fate undecided. While some wanted it maintained and used in conjunction with *Mir*, others in Gorbachev's government said that there was no longer money for such extravagances. Finally, on February 7, 1991, after years of indecision, the station was de-orbited, burning up in the atmosphere over Argentina.[15]

Meanwhile, when Kizim and Solovyov came home in July 1986, *Mir* was left unmanned, and would remain that way for the rest of the year.

For Musa Manarov, the second half of 1986 was extremely frustrating. He had joined mission control so that he could work directly with cosmonauts in space. Instead, he had become a capsule communicator with no one to communicate with. As far as he could tell, his chances at flying in space seemed dimmer than ever.

"You Are Distracting Us with Unnecessary Talk!"

Finally, on February 5, 1987, almost a year after *Mir*'s core module was launched, *Soyuz-TM 2* lifted off from Baikonur, the first manned mission to use the third-generation *Soyuz* spacecraft. Crewed by Yuri Romanenko and Alexander Laveikin, the mission's goal, announced publicly before liftoff, was to begin the permanent occupancy of space. Gorbachev's glasnost policy also insisted that the launch be covered live—as the launches of *Mir* and its first manned mission the previous year had been—beginning with the moment Romanenko and Laveikin climbed into their spacesuits.

In the decade that had passed since he lived on *Salyut 6* with Georgi Grechko, Yuri Romanenko had flown in space once, taking a Cuban into space during the string of international flights to *Salyut 6*. He then obtained a graduate degree at the Gagarin Air Force Academy and became the youngest colonel in the cosmonaut corps. In the early 1980s he worked as backup for several missions, then moved into mission control, supervising space walks. By 1984 he had the urge to go back into space. He wanted to participate in a long-term mission again, and do a space walk of his own, not just a short poke out the hatch like he had done with Georgi Grechko. "Seeing the stars floating under you, you feel as if you are the master of the universe," he remembered. He convinced Vladimir Shatalov, chief of the Gagarin Cosmonaut Training Center, to let him back into active duty to train for a long-term mission on *Mir*. By 1986, he and Laveikin were assigned as back-up crew for the first long mission to *Mir*.[16]

The prime crew was Titov and Serebrov, both of whom had been on the ill-fated *Soyuz-T 8* mission that lost its rendezvous antenna and couldn't dock with *Salyut 7*. While Serebrov had already flown a successful flight, joining Savitskaya in her first mission to *Salyut 7*, Titov had been unlucky. His only other mission had been the terrifying 1983 launch abort, when his rocket had exploded in flames on the launchpad.

Titov's bad luck continued. Only a week before he and Serebrov were scheduled to take off for *Mir*, the medical commission that had been set up after Vladimir Vasyutin's prostate illness on *Salyut 7* finally came back with its recommendations. They announced that every cosmonaut had to undergo further, more-stringent tests. Serebrov took the tests and failed. Both he and Titov were pulled from the first mission and replaced by Romanenko and Laveikin.[17]

For Titov, a good-natured man with roly-poly face, this grounding was incredibly galling. Twice he had been on a rocket, and twice the missions had failed. Then, his third assignment was grounded only days before launch. Some people were beginning to whisper that he was jinxed, like Petr Kolodin had been years before.

The primary goal for Romanenko and Laveikin was to extend the in-space record from eight to eleven months, thereby setting the stage for the first year-long mission, the absolute minimum time required for any spaceship to get to and from Mars. If nothing went wrong and all went well, they might even be allowed to set the year-long record themselves. Their secondary goal was to get *Mir* operational. Besides the normal routine of unloading *Progress* freighters and performing some science experiments, they were to greet the station's first new module and install *Mir*'s third solar panel during two space walks.

The new module, dubbed *Kvant*, had originally been designed to dock with *Salyut* 7. Delays in its design and construction, as well as the endless problems on *Salyut* 7, had caused it to be shifted to *Mir*.[18] Once docked with *Mir*'s aft port, the 11-ton *Kvant* would extend the station's 43-foot-long base block by 19 feet, adding 1,400 cubic feet of habitable space to the station.

Kvant, a shortened version of Chelomey's transport-support module, was mainly designed as an astrophysical facility, carrying two X-ray telescopes and one ultra-violet telescope. It also carried six large gyroscopes, what the Soviets called gyrodynes, to keep the entire complex oriented properly, as well as two recycling systems for maintaining the station's atmosphere. The Elektron oxygen electrolysis system generated breathable oxygen, replacing the lithium perchlorate canisters (which still functioned as a backup system). Linked to the station's bathroom in the base block, Elektron separated the oxygen out of the cosmonauts' urine (the hydrogen was vented from the station) to supply oxygen continuously for a crew of three. The Vozdukh carbon dioxide scrubber, similar to the scrubbers used on *Skylab*, replaced in turn the disposable lithium hydroxide canisters. While one molecular sieve flushed carbon dioxide out of the air, a second was exposed to the vacuum of space to let the gathered carbon dioxide naturally vent.

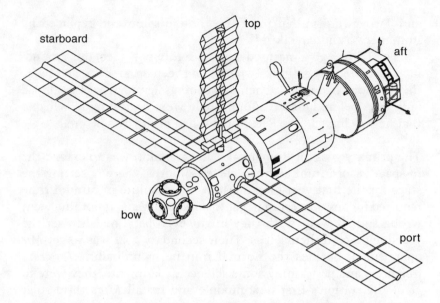

starboard top aft

bow port

Mir with *Kvant* docked to aft port and the dorsal solar panel installed. *NASA*

The system regenerated itself by switching sieves.[19] If these atmospheric recycling systems proved workable, they would then become standard components on any interplanetary spaceship.

Instead of the *Merkur* capsule carried by earlier transport-support modules, *Kvant* had docking ports at each end. Once docked to *Mir*'s aft port, *Kvant*'s aft port became the main port for all future *Progress* freighters. *Kvant* carried neither engines nor propulsion system. Instead, a space tug, dubbed the Functional Auxiliary Block, was attached to its aft port. *Kvant* would use the tug's engines and fuel to rendezvous and dock with *Mir*, after which the tug would undock and be abandoned.

On March 31, seven weeks into their flight, *Kvant* was launched. For five days its orbit slowly crept into sync with *Mir*'s. Then, on April 5, when the two spacecraft were still about 10 miles apart, the module's Igla radar system spotted *Mir*, and began guiding *Kvant* toward the station.* Everything seemed to be going well.

---

*Since *Kvant* had been built to dock with *Salyut* 7, it did not have the more advanced Kurs docking system.

Then, with the two spacecraft less than 700 feet apart and *Kvant* only minutes from docking, the Igla system lost its lock with the station. Ryumin immediately radioed the crew from mission control, telling them to "stay calm" but to quickly look out their windows to see what the module was doing.

To Romanenko's astonishment, the large module glided past his window only 30 feet away. If it had hit the station, its 11-ton mass would have destroyed it.

The next day Ryumin explained that the problem was that controllers had made the tolerances in the rendezvous software too tight or, as he said to the press, they had been "unduly cautious."[20] Two days later a second docking was attempted, and this time the maneuvers went smoothly and *Kvant* glided into *Mir's* aft docking port. However, when controllers commanded the spacecraft to hard dock, something prevented the docking latches from closing. Some obstacle was keeping the two spacecraft about an inch apart.

Peering through small portholes next to the docking collar in the base block, Romanenko and Laveikin tried to spot the obstruction. They could not. Then, after some discussion with the ground, Ryumin decided that both men would do an impromptu space walk to identify the problem and try to solve it.

For Romanenko, it was "deja-vu all over again." On his first mission, flight engineer Georgi Grechko, with Romanenko's support, had done a space walk to find out why the first docking attempt on *Salyut 6* had failed. Now flight engineer Alexander Laveikin, with Romanenko's support, was to do the same thing on *Mir*. This time, however, Romanenko would have permission beforehand to go outside.

On April 11, the two men sealed themselves inside *Mir's* multiple-docking adapter. Because this compartment was not intended as *Mir's* permanent airlock (to be launched as part of a future module), it was too small for both men to don their spacesuits at the same time. To give them more room, they also sealed the adjacent orbital module of their *Soyuz-TM* spacecraft, making the airlock a two-room chamber. After climbing into their suits and sealing them, they evacuated the air from both chambers.

Laveikin, an enthusiastic and talkative man who had always been eager to get into space, was suddenly horrified to see his spacesuit's air pressure dropping. If he didn't find the problem immediately he would suffocate. Hastily he ran through his suit checklist, checking switch after switch. To his relief, he quickly found that he had set one switch on the suit incorrectly, and immediately switched it back. At the same time on the ground, doctors noticed that for these few terrifying moments Laveikin's heartbeat began beating with a furious irregularity.

Laveikin felt nothing. While, unbeknownst to him, doctors were mulling over his heart data, he and Romanenko opened one of the adapter's radial docking ports and climbed outside. With Laveikin in the lead, they carefully worked their way along *Mir*'s length to its aft end. There, Laveikin peered over the station's edge at the docking port where *Kvant* was inserted. He radioed that he could see "some white object" sticking out between the two spacecraft. Could ground controllers back *Kvant* out of the port so that he could pull it free?

Kvant was eased back about a foot. With Romanenko holding a box filled with tools, the two men began working like a surgical team. Laveikin asked for a tool, and Romanenko passed it to him. After five attempts, Laveikin was able to use a pole as a jimmy to extract the white object from the port. It was a small refuse bag, filled with tissues and used toilet paper, that had gotten caught in the docking cone. When the last *Progress* freighter had been docked to the aft port and the docking cone had been open, the bag had somehow floated into it unseen. When the men swung the cone closed to be ready to receive *Kvant*, the bag had gone unnoticed and was, therefore, in the way when the port's latches tried to close.[21]

The obstruction gone, ground controllers ordered *Kvant* to move forward, and with a heavy clunk the latches closed and two spacecraft linked together in a hard dock. With *Kvant* attached to *Mir*, the station's aft main engines in the base block became useless, because *Kvant* blocked their nozzles. For the rest of *Mir*'s long life, the station's orbit was boosted by firing the rockets of a docked *Progress* freighter, a *Soyuz* spacecraft, or, much later, the American space shuttle.

Now that *Kvant* was docked, Romanenko and Laveikin could unload its main cargo, *Mir*'s third solar panel, and begin preparations for the two space walks they would do to install it on the outside of the station. The unassembled panel was folded like a pleated strip, packed into four bundles, each about 6 feet long. Two other packages held two lattice mast structures that, once installed in the mooring on *Mir*'s outside, would unfold and raise the panel sections upward like a flag. Over the next two months the men carefully checked out the packed arrays, and moved them from *Kvant*, through the base block, and into the multiple-docking adapter.

As they did this work, relations between the crew and ground control were far from pleasant. For one thing, because of the irregularities in Laveikin's heart that had occurred while putting on his suit, doctors had ordered him to do a medical check-up, including extensive exercises on the bicycle to test his heart's condition. Time-consuming and tedious, the tests ended up explaining little. As far as the doctors could tell, Laveikin seemed fine.

At the same time, Romanenko and mission control were having their own problems. Based on the emotional difficulties of earlier crews, ground controllers had been advised by the psychologists to handle Romanenko and Laveikin carefully. "At that time, we treated the men in space like children," said Manarov. "You couldn't make them angry, or worried, or anything." Or as Alexander Alexandrov noted, "No one knew what it was like to fly 300 days."[22]

Romanenko, however, was a tough, action-oriented man who didn't need or want coddling. "He was a very strict man," Manarov remembered, "Especially to mission control people." When he asked a question, he expected an answer soon thereafter. As time passed and he wasn't getting the answers he wanted, he became short-tempered, and even more demanding. Manarov found himself cringing at Romanenko's increasing rudeness. "I heard how he conducted himself and wondered." People on the ground began questioning Romanenko's emotional stability.

Despite what seemed to be building physical and emotional tensions, when in early June it came time to do the planned space walks, the doctors gave their approval. The two space walks took place on June 12 and 16. Laveikin and Romanenko slowly manhandled the mast packages outside through one of the docking adapter's ports to the top of the core module where the dorsal panel would reside. There, they fitted the two lattice masts together, one on top of the other, and attached the four pleated array bundles, two on each side. Then they raised the top lattice, the two sets of top panels unfolding as it rose, followed by the bottom lattice, which opened the whole array out. Thirty feet long, the

new panel added about 2.4 kilowatts to the station's power supply, bringing the total to about 11.4 kilowatts, exceeding *Skylab*
by several kilowatts.[23]

During the second space walk, doctors once again noticed
Laveikin's heart beating irregularly. The condition, which they finally identified as arrhythmia, occurs in athletes and is often temporary and treatable. Though Laveikin's heart once again settled
down, the doctors were concerned that he was having serious heart
problems, and they had no way of treating him while he was in
space. With the first international mission to *Mir* scheduled to arrive in only four weeks (carrying a Syrian cosmonaut), the decision
was made to bring Laveikin home. Alexander Alexandrov, who was
preparing for a short visiting flight and hadn't flown since his five-
month mission to *Salyut 7* in 1983, was chosen as Laveikin's replacement and given a fast month of extra training.

For Laveikin, the decision was a shock and disappointment. "I
certainly wanted to carry on with the work," he noted later. "I felt
well during the flight, I felt great throughout. But you can't argue
with the doctors."[24]

After landing, which in the spirit of glasnost was covered live on
radio, Laveikin appeared pale and weak, with one reporter comment-
ing that he looked like a man whose time in space had placed a
"heavy burden" upon him. A medical check-up revealed that the
arrhythmia had merely been his body's way of adapting to space.
There was no damage to his cardiovascular system, and after three
months he was completely back to normal. Based on these results,
some doctors on the medical team believed that the decision to bring
Laveikin home so quickly had been timid and overly cautious.[25]

Meanwhile, Romanenko's mission carried on. For both him and
Alexandrov, the situation was somewhat awkward. They had not
trained together. Moreover, Alexandrov was not as familiar with
Mir's systems as he, the flight engineer, would have liked. He actually needed Romanenko to give him some training. "He did something, I watched him, and then I did it myself." For Romanenko,
this extra burden was entirely unwelcome. Though he and Alexandrov worked well together, the situation helped to increase Romanenko's resentment toward mission control.[26]

The months stretched on and on. By September, soon after
breaking the 237-day record set by Kzim, Solovyov, and Atkov three
years earlier, the boredom and isolation, like on previous missions,

became numbing. Mission deputy director Viktor Blagov announced that the cosmonaut's workday had been reduced to five-and-a-half hours a day, with the workload for those hours reduced even more. In addition, Alexandrov had taken over all communications with the ground. Romanenko had simply become too testy to do it any longer.[27]

In fact, Romanenko's apparent distrust of ground operations resulted in the destruction of several plant experiments. Noticing in July that the internal temperature in the Fiton greenhouse was uneven, causing condensation on one wall, he decided, without consulting Nechitailo, to dismantle Fiton, clean it out, and water the plants manually. When Romanenko told Nechitailo what he had done, she winced. Fiton had been purposely sealed before launch to keep the integrity of its environment separate from the station's. By opening the greenhouse, Romanenko had ruined the experiment. Moreover, by shutting down the automatic watering system and watering the *arabidopsis* manually, he prevented the biology team from finding out what was wrong with their automatic system.

Nechitailo, however, was forbidden to say anything to Romanenko, ". . . especially with medical specialists standing by." She, like everyone else, had been ordered to treat him with kid gloves.

The result was that Romanenko's first set of plants, some quite healthy, was unusable for research. He had to replant everything in September.[28]

During the mission's last few months, Alexandrov did most of the heavy work, unloading three *Progress* cargo ships, one of which, in order to test new software designed to save fuel, undocked, backed off a mile and a half and then did a second automatic docking two hours later. Romanenko merely rested, devoting his time to writing poems and songs, and doing two to two-and-a-half hours of exercise per day, while wearing for eight hours a day a flight suit similar to the Penguin suit from earlier missions. Instead of having elastic bands, it had weights distributed throughout its lining.[29]

To try to ease Romanenko's languor and tedium as well as make Alexandrov's tour of duty more tolerable, the men spent much of Friday and Saturday nights talking to family and friends. Then in October, they did a number of television link-ups with a variety of personalities in France, Hungary, Japan, and England. They also

participated in a Moscow international space conference celebrating the 30th anniversary of Sputnik, chatting with many Soviet
officials as well as several American astronauts, including *Skylab*
astronaut Owen Garriott.

During the hook-up, Bulgarian cosmonaut Georgi Ivanov asked
Romanenko if he knew when he was returning home. Romanenko
told Ivanov to ask Blagov. "He never tells us when our flight will be
over."[30] Romanenko's resentment of Blagov partly stemmed from
the spaceman's strong desire to complete a full year in space, and
his inability to get anyone on the ground to tell him whether he
had permission to do it or not.

On November 7, 1987, the two men watched the parade in Red
Square, celebrating the 70th anniversary of the October Revolution. Romanenko had been in space for more than nine months,
more than 270 days, longer than any human in history. Even now,
he did not know whether he would go home in late December or
remain in space until February, completing a full year. What he did
know was that the decision had to be made soon.

In Red Square, Mikhail Gorbachev watched the parade from
the top of Lenin's Tomb, smiling occasionally at the large, carefully
chosen, supportive crowd below him. Only days before, he had
marked the occasion with a major speech, in which he tried (though
mostly failed) to give a more accurate appraisal of Soviet history.
The closest he came to complete honesty was when he called
Stalin's "wholesale repressive measures and acts of lawlessness . . .
enormous and unforgivable."[31]

Gorbachev had now been in power for more than two-and-a-
half years. Though he had made a big deal about perestroika and
glasnost, not much had really changed, and much had gotten worse.
Moreover, he had shifted his direction several times, issued contradictory policy statements, and in general accomplished little. On
one hand, in 1986 Gorbachev had called for reform in the electoral
process. On the other hand, by 1987 he had done nothing to accomplish this goal. On one hand, he had called for glasnost and free
speech. On the other hand, he had freed no political prisoners and
had kept the borders of the communist bloc closed, with more than
100,000 Soviet troops still occupying Afghanistan. On one hand, he
had called for the elimination of corruption and a rebuilding of the
nation's economic system. On the other hand, he had done little to
change the system of privilege, power, and corruption enjoyed by
the apparatchiks of the Communist Party.[32]

Gorbachev's contradictory and indecisive rule during these 32 months can best be seen by his interaction with Boris Yeltsin. For 70 years the idea of openly criticizing the communist leaders of the Soviet Union had been considered blasphemous. Even with the rise of Gorbachev and his demand for glasnost, the idea of standing up in front of the Central Committee of the Communist Party and telling Mikhail Gorbachev to his face that many of his policies did not work seemed impossible, dangerous, and downright stupid. On October 21, 1987, just two weeks before the 70th Anniversary celebrations, Boris Yeltsin took Gorbachev at his word and did just that. As head of the Moscow Communist Party and assigned the job of getting the city to function again, Yeltsin had become increasingly dissatisfied by the slow pace of change under Gorbachev's rule.

Like Gorbachev, Yeltsin was part of the younger generation of Soviet communists. He had been born in 1931, the same year as Gorbachev, in the Sverdlosk region about 250 miles east of the Ural Mountains amid the vast open steppes of Russia. His father had been a private farmer, exiled from his land and later sent to prison for three years for "anti-Soviet propaganda." After his release, he got work in a potassium processing plant. The family lived in the worst of poverty, sharing a primitive, wooden, one-story, 20-room hut with 19 other families.

During the war years starvation was never more than one meal away. Yeltsin and his mother cut hay at nearby collective farms and kept half, which they then sold for bread. "It was a fairly joyless time," Yeltsin wrote 50 years later in his memoirs. "We had only one aim in life—to survive."[33]

A hard-driving workaholic, Yeltsin did well at school. After graduation from college he moved steadily up the ladder, becoming an engineer, then a project manager. Wherever he went, people "were impressed by his ability to get to the heart of the matter."[34] Doing everything with an incredible ferociousness, he expected the same from everyone else. He took failing projects that had lingered for decades and got them finished, on time and under budget. By 1976, he had risen to be First Secretary of the Sverdlosk Region, which he ruled for almost 10 years, until Gorbachev called him in 1985 and asked him to take over the Moscow Communist Party operation and clean house.

Yeltsin soon discovered that, even with Gorbachev's public support, cleaning house was not as easy as it seemed. He found himself repeatedly stymied by many of the old-time communist officials still in power.[35] Like a bull in a china shop, Yeltsin attacked all their special privileges, only to find that he couldn't make any real changes without their cooperation. Because the changes would threaten their power and privilege, their cooperation was not forthcoming. Instead, they criticized or lambasted him, undercutting his power at every opportunity.

By September 1987 Yeltsin had had enough, and wrote a letter of resignation to Gorbachev, whose only response was to say they'd discuss it later. By the time of the October Central Committee meeting—the very meeting in which Gorbachev and the Communist Party were planning the November 1987 70th Anniversary celebration of the October Revolution—Yeltsin decided to mount the rostrum and announce both his criticisms and his resignation. The response of the communist delegates was an outpouring of contempt, insults, and outright hatred. Gorbachev himself led the way, encouraging others to step forward and attack Yeltsin. To Yeltsin's surprise, not one member of the communist leadership seemed willing to consider even the slightest possibility that anything he had said might be true.[36]

"A breach of Party ethics," was how Gorbachev later described it, also stating that Yeltsin's speech was "on the whole politically immature, extremely confusing, and contradictory." Less than three weeks later and only days after the parade in Red Square, Yeltsin was unceremoniously dumped from his position as head of the Moscow Communist Party organization.[37]

While his dismissal at first seemed a defeat for the reform movement, in the end it made Yeltsin a hero, and revealed Gorbachev's efforts to be halfhearted and resistant to change. In addition, Yeltsin's action served more than anything else to break the ice, to make everyone realize that they could speak out and not suffer serious harm. Less than two weeks after his dismissal, Gorbachev offered Yeltsin the job of First Deputy Chairman of the Soviet State Committee for Construction.[38] While not very important politically, the position demonstrated that the consequences of free speech in the Soviet Union no longer carried much negative weight. From this moment on Gorbachev's glasnost program began to take on a life of its own, no longer under his control.

In space, the New Year was approaching, and mission control had finally decided that it was time for Yuri Romanenko to come home.

His irritability and apparent tiredness made everyone, from administrators like Blagov to the many doctors on the medical team to former cosmonauts like Ryumin, reluctant to push him further than necessary. His return was scheduled for just before New Year's Eve after he had completed approximately 11 months in space. The job of completing a full year in space would fall to Romanenko's replacement crew.

The real surprise, however, was the names of that replacement crew: Vladimir Titov and Musa Manarov. For both men the assignment seemed strange and ironic. "They chose two men who had bitter memories," Manarov noted.[39] Titov's career had been increasingly stigmatized with allegations of bad luck, while Manarov had spent many years in a kind of ethnic exile, denied a mission merely because of his ethnicity.

Titov's background was in many ways similar to Manarov's. His father had been an army serviceman. After completing high school in 1965, Titov wanted to be a radio technician, but when he tried to enter the Kiev Institute of Civil Engineering, he failed the entrance exams. Instead, he got a job working on an oilrig. By 1966 Titov was faced with the certainty that if he didn't get into a college or university, he would be drafted into the military to serve two years as a conscript. That year he tried again, applying and getting accepted to the Chernigov Higher Air Force College in the Ukraine. After graduation he worked at the school as a pilot and flight instructor. He also worked with cosmonauts as they practiced flying. Like Manarov, he had never imagined himself a cosmonaut. When he was accepted into the program in 1976, it was a great surprise.[40]

The decision to pick Manarov as Titov's partner had actually begun with Vladimir Vasyutin's prostate illness on *Salyut* 7. After the medical commission's decision, there had been a major shake-up of crew assignments, with a number of men being grounded until they could pass the additional medical exams. Not only was Serebrov grounded, causing a crew change for the first *Mir* mission, but many other men also found their plans changed abruptly. Ryumin, who had been flight director during the *Salyut* 7 missions, was promoted to head the entire manned program, while Vladimir Solovyov, who had acquitted himself so well during his space walks on *Salyut* 7 but was now grounded, was promoted to flight director, a position he held through the rest of the *Mir* program and to the present.[41]

Titov, scheduled to fly the next mission, found himself without a partner. And because of all the problems on his previous
flights, no one wanted to fly with him.

Manarov had no such qualms. When others said they thought
Titov was unlucky, he disagreed. "It's crap," he said. "The man
survived having his rocket blow up under him. If anything, he's the
luckiest man in the program."

Manarov wanted an assignment and Titov needed a partner.
The match-up was obvious. Not only were he and Titov chosen to
fly together, they were told that their mission would last a year,
whether or not Romanenko managed to stick it out that long himself. After nine years of waiting, Manarov was finally going into
space. His decision to work capcom had paid off.[42]

Their third crewmate, Anatoly Levchenko, was to remain in
space for only nine days. He was trained as part of the Soviet
Union's space-shuttle program and Soviet air force doctors had decided, after watching the difficulties of men returning from space
after only a week, that Levchenko required at least one short space
flight before piloting their shuttle.[43]

These three men spent the last week of 1987 in *Mir* with
Romanenko and Alexandrov. Before blast-off, Ryumin had warned
them to avoid showing any "excessive emotion" in dealing with
Romanenko. Based on his conversations with the ground, the man
was thought to be emotionally fragile and very sensitive, requiring
careful treatment.[44]

"Even before Ryumin spoke to me I had these thoughts,"
Manarov remembered. "I was a bit afraid when I arrived on *Mir*,
not knowing what I'd find." When they opened the hatch, the
first thing Manarov noticed was that one of Romanenko's eyes
was slightly inflamed, making him look crazy, even somewhat
unbalanced.

The truth, however, was quite different. To Manarov's surprise,
Romanenko was far more normal and far less exhausted than implied by his conversations with the ground. "He was a military
man," Manarov explained. "It had all been a ruse on his part." Irritated with how ground controllers insisted on pampering him—
treating him like delicate, fine-cut glass—Romanenko had decided
to turn that attitude to his advantage and manipulate it to get what
he wanted. He believed that treating ground controllers harshly
would keep them on their toes and help maintain discipline.[45]

Nonetheless, Romanenko's physical and mental state was certainly strained. "He really was tired," Alexandrov remembered, noting how disappointed Romanenko was that his mission was ending after 11 months. "He had very much wanted to fly a full year in space."[46] On December 28, the day before his return, Nechitailo got on the radio to confirm that Romanenko had packed the entire four kilograms of biological samples for return to Earth.

Romanenko responded by breaking out in a torrent of anger and annoyance, demanding that "all superfluous personnel" be banned from mission control. "We have our instructions," he fumed. "Everything will be delivered in the best possible condition. Since yesterday we've been rushing around like squirrels in a wheel. Experiments, comments on experiences, loading, you name it, we've been doing it. We've still not had time to pack our personal things. And you are distracting us with unnecessary talk!"

Worried that Romanenko might leave some of her precious samples behind so that he could take personal items, Nechitailo appealed directly to Glushko. The head of Energia, who was by this time an old man in his 70s and partially paralyzed by illness, came to the microphone and bluntly told Romanenko that he had better load everything as ordered. Romanenko meekly agreed.[47]

On December 29, 1987, after 326 days in space, Romanenko came home. He, Alexandrov, and Levchenko landed in a snowstorm on the empty and barren northern deserts of Kazakhstan, where 40 to 70 mile-per-hour wind gusts blew the capsule sideways, dragging it by its parachute several hundred feet before it stopped. Romanenko and Alexandrov were carried from the capsule and placed in chairs. Because the wind and snow prevented the recovery team from erecting a medical tent, it was decided to transport them immediately by helicopter to the nearby town of Arkalyk, then by plane to Baikonur.[48]

Romanenko wanted to walk to the helicopter. The doctors said no. He had been in space three months longer than any other person in history, and no one wanted to take any chances.

Once on the helicopter Romanenko again insisted he be allowed to stand. After some consultation, the doctors finally agreed. He stood up. "Look at me," he said. "Look at me, I'm standing. How's my pulse?" In Arkalyk, Romanenko walked, with aid, from the helicopter to the plane. When he arrived at Baikonur he walked unaided from the plane and into the arms of his wife and 12-year-old son Roman.[49]

Romanenko's muscles had withered somewhat, but recovered with no problem. His hipbones had lost 5.6 percent of their calcium, a rate of about 0.5 percent per month, rather low compared to men flying much shorter missions. Based on this result, the Soviet doctors concluded that ". . . the degree of demineralization of the tibia in flight depends, to a large extent, on the volume and schedule of physical conditioning."[50] To the Russians, the evidence increasingly suggested that journeys of one to two years to the other planets could be accomplished as long as the crew religiously followed their exercise routines. Their bones might come back weakened, but not irreparably so. All that was needed to prove this hypothesis was to fly several missions for that long.

"No Time for Silliness."

With the departure of Romanenko, Alexandrov, and Levchenko, Titov and Manarov settled down to what had become the normal work routine of a long-term space flight. They exercised about two hours a day, did maintenance and repairs, performed a variety of experiments, and in general, tried to stay composed and relaxed.

However, Titov and Manarov knew far better than crews on all previous long-term missions what was required to make it through the long months. The many earlier flights, each increasingly longer—from 30 to 63 to 96 to 211 to 237 to 326 days—had illustrated the right and wrong way to do to things. Both men had watched it all happen. Both knew what they had to do. Like Grechko, they were professionals, cool-headed, calm men who understood that to survive for a whole year in a small metal room no larger than the coach section of an ordinary commercial jet required that they restrain their emotions.

Ground crews were more knowledgeable too. To keep the men occupied and stimulated, mission control had scheduled a series of milestone events scattered throughout the year. In June a visiting crew would arrive with a Bulgarian cosmonaut. In February, late June, and October, Titov and Manarov would do space walks. And in between these events, five separate *Progress* freighters would arrive, each loaded down with letters and newspapers and videos from home, as well as fresh fruits and vegetables.

The psychologists, too, were more aware of what to expect. Though some ground controllers and capcoms still treated the crew far too gingerly, their experience during Romanenko's mission taught everyone not to overreact to minor incidents. Moreover, Manarov's experience as a capcom had taught him how to handle communications with the ground. "People who haven't worked in mission control don't know how things are done," he explained. "There's a chain of command. . . . Both cosmonauts and mission controllers who don't understand sometimes get nervous. They ask questions, and can't get answers. They think something is wrong. What they don't understand is that the man they ask doesn't have the answer and needs to find out. This can take time."[51]

For everyone involved—the two men in space, everyone in mission control, and even their families at home—the mission was best understood as a very long marathon, requiring a careful, deliberate pace to reach the finish line. If anyone tried to sprint, they'd run out of steam long before the final straightaway.

Reinforcing all this knowledge and experience, Manarov and Titov had their own strong personal motives to succeed. For Manarov, failure would prove the bigots right, that he wasn't capable of flying in space because he wasn't a Russian. He had waited almost a decade to get assigned to a flight. He wasn't going to fail now. For Titov, failure would once again prove how unlucky he was, and would certainly end his career. The last thing he wanted was to end up like Petr Kolodin, the victim of one unfortunate disappointment after another.

So, month by month, their marathon unfolded. In January, they welcomed their first *Progress* freighter, bringing the usual supplies as well as the equipment for their first space walk. In February they did their first space walk to upgrade the dorsal solar panel installed by Romanenko and Laveikin by replacing one of its four solar panel sections. The new section used a carbon-plastic composite frame instead of metal, making it lighter and hopefully more efficient.[52]

In March came the second *Progress* freighter, this time loaded with almost 900 pounds of fresh food and vegetables. By now the men were entering the second stage of a long mission, experiencing lethargy, boredom, and general dissatisfaction with their circumstances. They complained about the noisy fans in the core module, and its artificial smell of metal, equipment, and plastics.[53] Yet they slept well, and managed to stay focused. "No time for silliness," noted Manarov. They had too much work to do. There were cables to repair, experiments to fix, and parts to replace. For example, the color television monitor that showed them images of their families on Earth kept breaking, and required constant maintenance.

In May the third *Progress* freighter arrived, carrying a load of equipment, special tools, and replacement parts for a failed X-ray telescope in the *Kvant* module. The TTM X-ray telescope, built jointly by Britain, the Netherlands, and the Soviet Union, had not been designed for in-orbit maintenance. Not long after launch, the astronomers had found it unusable because of electrical interference in its main detector. A new detector unit, a cylinder 16 inches wide and weighing 90 pounds, was built and sent up on the cargo ship. During their second space walk, Manarov and Titov would try to install it.

In June, a Bulgarian cosmonaut spent eight days on the station, completing the original Intercosmos agreement signed by Brezhnev more than a decade before.

In late June, after the visitors were gone, Titov and Manarov climbed out across the surface of the base block in their first attempt to fix the X-ray telescope, lugging the new detector, their tools, and a 3-foot by 6-foot platform to attach to *Kvant* to give them a place to stand. However, when they reached *Kvant*, they could not attach the platform to the module and had to take turns holding each other in place in order to work. Then, the gyros in *Kvant* shut down, causing the orientation of the *Mir* complex to drift. Originally the plan had been to hold the station so that the sun was at their backs while they worked. Instead, with the station drifting, the sun ended up in their eyes.

Struggling to keep himself in place, Manarov, as flight engineer, took the lead. He cut through 20 layers of insulation to get at the panel covering the failed detector. Removing the screws that held the panel in place was very difficult, even with special pliers. "My gloves were very thick, and the screws were very small." In the end, he had to snap off the screw heads.

Then they were stymied. The screws that locked the detector's electrical connections in place were sealed with an epoxy glue. "You couldn't get to anything," remembered Manarov. Taking turns trying to hold each other down, they slowly sawed through the epoxy, using a small power saw.

Finally it was time to remove the old detector. Shaped like a short cylinder, it had been fixed in place with a strap, which in turn was held tight by a clamp, which in turn was locked closed with a simple lock and key. Manarov fitted the key into the lock and turned. "The first time I tried it I didn't think it worked. I turned it again, and the key snapped off in the lock."

There he was, floating in frustrated disbelief 200 miles above the earth, holding a broken key in his hand. At that point, the space walk was already almost five hours long. Exhausted and unable to complete their job, the two men put the insulation back and returned inside, lugging the new detector back with them. The unusable work platform they tied to the outside of *Mir*, out of the way.[54]

This unsuccessful repair job was probably the only real failure during the entire year-long mission. It mattered not. The men put their tools and spacesuits away, and went on to other work. The telescope's repair would simply have to await the arrival of new tools.

In July came the fourth *Progress* freighter, loaded with replacement parts for the station's various life-support systems. It also carried a new television monitor, the old one having finally failed entirely. By this time the flight had lasted more than six months. For a half-year the two men had managed to keep themselves on an even keel, avoiding the kind of conflicts that had made other missions so unpleasant. "We respected each other," said Manarov. "It wasn't a game. You had to hold yourself in, control yourself."

Not that they never disagreed, or argued. In fact, one dispute became so heated that they didn't speak to each other for three days. Rather than let the disagreement get out of control, however, they focused on their work, speaking only when they had a work-related question. "We simply avoided each other for a while," noted Manarov. Time passed, they cooled down, and realized that whatever they were arguing over was not that important.[55] In fact, the mature manner in which the two men handled their circumstances ended up making their long mission seem almost boring. Very little went wrong. They had few problems or complaints. The station in general continued to function well.

In August the men received their second visiting crew, including an unexpected international visitor and a third crewman who would keep them company for the rest of the flight.

Almost a decade after the initial invasion, the Soviet occupation of Afghanistan had proved to be an utter failure. More than two million Afghanis had fled the country, settling in refugee camps in nearby Pakistan. Neither the Soviet army nor its puppet

Afghan government had been able to take control of the country-
side, with the inefficient Soviet military experiencing corruption,
low morale, and heavy casualties. More than 13,000 troops had
been killed, and the cost to the Soviet economy, already in freefall,
had been immeasurable. From almost the moment Gorbachev took
power, he had hinted at a withdrawal by the end of 1989 at the
latest. Finally, on April 14, 1988, a treaty was signed in Geneva
between Afghanistan and Pakistan, with the United States and the
Soviet Union as guarantors. According to the treaty, all Soviet
troops (about 100,000) would be out of Afghanistan by February 15,
1989.[56]

However, the signing of this accord caused serious problems
within the Soviet space program. Eight months earlier, on July 20,
1987, Gorbachev had met with the Afghan leadership and signed
an agreement to send an Afghani to *Mir* sometime in 1989. By early
February 1988 two Afghanis had been chosen, and had just begun
the minimum 18 months of training.[57] With Soviet troops scheduled to withdraw from Afghanistan by February 1989, however,
there was a good chance that the Soviet-backed puppet government
would fall apart soon afterward, making a subsequent joint Soviet-
Afghan mission rather pointless.

To avoid this embarrassment, and maybe help the Soviet-
backed communists in Afghanistan, the joint mission launch date
was moved up six months to August 1988 to guarantee its launch
while the Afghan communists were still in power. Given less than
six months to train for their mission, the two Afghan cosmonauts,
prime crewman Mohammad Dauran Ghulam Masum and his back-
up Abdul Ahad Mohmand, struggled to get ready, working seven
days a week. Then, just six weeks before launch, Masum became
ill with appendicitis, and Mohmand had to take his place.[58]

Mohmand's crewmates were commander Vladimir Lyakhov,
returning to space for his third and last time, and Dr. Valeri
Polyakov, the third Russian doctor to fly in space, and the first
since Atkov flew on *Salyut 7* four years earlier, in 1984.

Polyakov's quest to become a cosmonaut, already more than
two decades old, was almost Homeric in nature. Born in 1942 in
Tula, a city 100 miles south of Moscow, he had enrolled at the
Sechenov First Moscow Medical School in 1961, intent on be-
coming a doctor. In his junior year however, he watched Dr. Boris
Yegorov become the first physician in space on *Voskhod 1*, the

first space flight carrying more than one person, and was so impressed that he decided to dedicate himself to the science of space medicine.

It wasn't that easy. In the Soviet Union in the 1960s you didn't have much freedom to choose your career. Upon graduation from Sechenov, Polyakov tried to enroll in the Institute of Medical-Biological Problems (IMBP), which was then becoming known as a major center for the study of space medicine. IMBP, however, had no room for him. Instead, he was send to the Martsinovski Institute of Medical Parasitology and Tropical Medicine, followed later by a tour at the Semashko Institute of Social Hygiene and Public Health.

Polyakov did not give up. He began working as a volunteer both at IMBP and at the Institute of Public Health, doing research on how to make space stations more habitable. There he met Nikolai Gurovski, whom even today Polyakov calls his "good father." Gurovski, who was the director at Public Health, was impressed by Polyakov. "You are strong," he told Polyakov. "You should become a cosmonaut." He sponsored Polyakov's efforts and, by 1971, was able to wangle a place for him at IMBP. He also signed Polyakov's cosmonaut application while lobbying heavily for its acceptance with people at Energia. By 1976 Polyakov was a member of the cosmonaut corps, training as a physician who could perform medical work in orbit, including surgery if necessary. In fact, he became the commander of a group of doctor-cosmonauts at IMBP.

For years he worked in the background, helping in both preflight and post-flight medical training. Often he was present at both the launch and the landing of crews. At the same time he participated in several expeditions to the remote Altai Mountains, testing survival techniques. By 1979 he had completed training in the operation of the *Soyuz* and *Salyut* spacecraft, and was qualified to fly in space. Getting his first mission, however, turned out to be almost as difficult as getting accepted into the program.

His first mission was supposed to be the second manned test of the *Soyuz-T* spacecraft. However, when the first manned mission in 1980 was an unqualified success, this flight was canceled. Then he spent years training to fly with Solovyov and Kizim in preparation for their eight-month repair mission to *Salyut* 7 in 1984. Three months before that launch, Yevgeni Chazov, Minister of Health and personal physician to the recently deceased Brezhnev, stepped in and insisted that the doctor on the mission be Oleg Atkov. Though Brezhnev was dead, the power politics of the Soviet era still ruled. Polyakov immediately became the backup, and Atkov the prime.[59]

"It was very difficult, disappointing," Polyakov remembered years later. A weaker man might have given up at that point. Polyakov, however, was single-minded. "Once I make a choice, I hold to it, without hesitation." Strengthening his resolve was a promise made to him by Glushko, guaranteeing that he would be the next doctor to fly in space. "And Glushko never broke a single promise," Polyakov noted.

Four years later, Polyakov was finally in space. Glushko had kept his promise.

According to the original plan, Polyakov's mission was to have begun after Titov and Manarov had already returned to Earth, as part of the March 1989 Afghan flight. Once in orbit, he would stay there for up to 16 months, and maybe longer. As Polyakov noted years later, "I was prepared to spend as much time in space as needed." When the Afghan flight was shifted forward from early 1989 to August 1988, Polyakov didn't mind. He would not only get into space quicker, he would get to fly on *Mir* during Manarov's and Titov's last four months in orbit, allowing him to get a very close look at what weightlessness did to two humans as they completed a year in space. On Romanenko's flight the cosmonaut's exhaustion seemed to worsen during the last few months. As a doctor and specialist focusing on the problems caused by long periods of confinement and isolation, Polyakov stood an excellent chance of alleviating any problems Titov and Manarov might have.[60]

And if all went well, Polyakov still intended to break Manarov's and Titov's year-long record. With this plan in mind, he and his backup at the institute, Dr. Gherman Arzamazov, underwent painful bone-marrow operations in order to provide preflight samples for later comparison.[61]

The eight-day Afghan mission was in many ways very similar to the earlier international propaganda missions. Much of the flight was devoted to reading political statements calling for peace and an end to the Afghan civil war. During one televised broadcast, Mohmand spoke with the communist leader of Afghanistan and both men once again called for peace.[62]

The propaganda, however, was in some ways profoundly different from Brezhnev-era missions. No one made proclamations about the superiority of communism. In fact, no one mentioned communism at all. Moreover, Mohmand brought a copy of the Koran with him, using it to pray both before and during the mission. In fact, soon after he entered *Mir* for the first time, he paused before Soviet television cameras to read a passage from the Koran to the Soviet public.

The mission was also different in that the news coverage was far more open. In Gorbachev's era of glasnost, Western journalists were allowed to view the spacecraft while it was being assembled, during its rollout to the launchpad, and during its launch. They were allowed to stand in mission control, ask questions, and watch as the mission unfolded before them. They were even able to watch as, during Lyakhov's and Mohmand's re-entry, the engine burn on the *Soyuz-TM* descent module cut off prematurely, leaving the two men temporarily stranded in orbit. On the second burn attempt, which also cut off early, reporters listened as Lyakhov called out "Emergency! The motor fired for 60 seconds, then switched off!" Lyakhov then tried to fire the engine manually, but had to terminate the burn prematurely when the spacecraft drifted off course.

While foreign headlines proclaimed the men "Stranded!" and "Lost in Space!" mission controllers simply rescheduled the burn for the next day, and brought the men home without further problems. The worst part of the delay, which did carry the possibility of stranding the men in orbit if the last burn did not work, was that Lyakhov and Mohmand were trapped in the cramped descent module in their spacesuits, with little food and no bathroom facilities, for 24 hours.[63] Later, glasnost meant that the investigation into the flight's problems could not be kept secret. Soviet space officials had to face repeated grilling by both Soviet and Western journalists, asking if the haste in which the mission had been scheduled had contributed in any way to its difficulties.

Meanwhile on board *Mir*, Titov and Manarov got back to work, and Polyakov luxuriated in the wonders of weightlessness. As his rocket had been roaring upward into space, all he could think about was the chant of American blacks after the Civil War: Free at last! Free at last! *Free* at last! "It was like being a bird in flight," he remembered. To his delight he experienced no space sickness or nausea.

While Titov and Manarov exercised, did repairs, and prepared for their last space walk, Polyakov studied the life-support systems so that he could recommend ways to improve them. He watched how the men did their exercises, and suggested changes that might help them when they got back to Earth. He experimented with special wrist cuffs to regulate blood flow. He worked on the station's kitchen and toilet, trying to improve their operation. And he took some of the workload off Titov and Manarov by conducting most of the medical experiments on himself rather than on them.[64]

Titov and Manarov, meanwhile, had that last space walk to do. The Afghan mission had brought them new supplies, new space-suits, and new tools to try to complete the repair to *Kvant*'s X-ray telescope. The new spacesuits, dubbed the Orlan-DMA, were more adjustable than the older suits. Their gloves had been improved to make it easier to do delicate work, with an emergency wrist cuff added in case of glove wear or leaks. They could supply oxygen for up to seven hours, two hours longer than the previous version. Most importantly, the suit was built so that, with the later addition of a supplementary package, it could operate without the electrical and communications umbilical cord, making it the first Soviet space-suit to allow a cosmonaut to work completely independent of his spacecraft (a capability that American astronauts had since the mid-1960s).[65]

On October 20, with live coverage on domestic Soviet television, the two men donned the new suits, grabbed the new tools and the replacement telescope detector that had been waiting inside the station for three months, and headed back onto the exterior of *Mir*. Polyakov, meanwhile, locked himself in the *Soyuz* descent module, in case something went wrong. Rather than struggle again with the broken key and lock, they came with a jack, and simply pried the strap off. Then they lifted the whole unit out and, with a push, tossed it away into space. The new detector was then fitted down, plugged in, and clamped into place using a new and simpler fastener. In all, the second space walk went so smoothly that they finished an hour ahead of schedule.[66]

In fact, the ease with which they completed their last space walk epitomized their entire year in space. Though they had been in orbit for more than 10 months, neither man was experiencing any of the problems of fatigue and mental exhaustion seen with

Romanenko, Lebedev, and others. Polyakov even confessed later to feeling a bit "disappointed" by how well Titov and Manarov were doing.[67] With only two months to go, the completion of a year by both men seemed certain.

On the ground, the Soviet space program seemed to be doing as well, if not better. On November 14, 1988, the Soviet version of a reusable space shuttle made its first space flight, lifting off from the dry desert plains of Kazakhstan. First proposed by Glushko when he took over Energia in the mid-1970s, *Buran* (which means "blizzard" in Russian) was intended to be the first in a fleet of reusable shuttles, each capable of lifting as much as 30 tons of cargo into orbit.[68] Looking remarkably like a slightly scaled-down version of the American space shuttle, *Buran*'s main fuselage was a blunt body with stubby wings protected from the heat of re-entry by ceramic tiles.

Unlike the American shuttle, however, *Buran* had no launch engines. Instead, its launch system was Glushko's expendable Energia rocket (pronounced the same as the design bureau, e-NER-gee-a with the "g" pronounced as in "get"), the giant heavy-lift rocket that he had proposed simultaneously with the shuttle. As spectacular as *Buran* was, Energia (which had already completed its first test launch 18 months earlier) was even more a triumph of Soviet space technology, able to lift 100 tons into orbit.[69] And ironically, Glushko had built it using cryogenic liquid oxygen-hydrogen engines, the kind of fuel that Korolev had tried and failed to get him to use back in the early 1960s. On *Buran*'s first flight, covered live by the Soviet media, the shuttle made one orbit around the earth and then landed safely on a runway at the Baikonur space center. The plans called for the Soviet shuttle fleet to begin regular operations soon thereafter.

Two weeks later, at the end of November, *Soyuz TM 7* took off, bringing *Mir* a new crew: Alexander Volkov (no relation to Vladislav Volkov who had died on *Salyut 1*), Sergei Krikalev, and Frenchman Jean-Loup Chretien on his second space flight. Chretien stayed in space for a month, during which he completed the first space walk by someone other than a Russian or American. He then returned to Earth with Titov and Manarov, while Volkov and Krikalev joined Valeri Polyakov for what was intended to be an even longer marathon mission.

Landing in a cold and wintry Kazakhstan, with temperatures a chilly 6°F, Titov and Manarov were carefully lifted from the descent module and placed on helicopters. As he was helped from the capsule, Titov looked at the men around him, the sky, the ground all around and said with a joyous smile, "It has been a long time since I've been here. A whole year!"

Taken directly back to Star City near Moscow, Titov and Manarov recovered much like previous cosmonauts. At first they were very weak and tired. When Manarov tried to urinate the first time, he needed help to stand. "I couldn't feel my feet. They were numb." However, both men managed to stand unaided within an hour-and-a-half of landing and within days they were carefully strolling about Star City with their families.[70] Nonetheless, it took several weeks for them to completely recover. "Even when you begin to feel normal you can't jump," Manarov said. "It is as if you are glued to the ground."

For the scientists, the cosmonauts' generally good condition was further proof that long space missions were possible. Titov had lost a few pounds in space, while Manarov had actually gained weight. Though they both experienced muscle atrophy, it was less than Romanenko's. Moreover, though they had lost bone calcium— from 6 to 10 percent depending on the weight-bearing bone—the rate of loss was close to Romanenko's, 0.5 percent and 0.9 percent per month, respectively, once again suggesting that with sufficient exercise the loss could be slowed, if not stopped.[71]

After more than a decade of space-station missions, involving seven different stations and dozens of different cosmonauts, the Soviet Union had finally kept men in space for more than a year. It was a moment of sublime achievement.

In fact, the entire Soviet space program seemed unstoppable. It had a space station under construction and in use. It had a space shuttle tested and ready to fly. It had a giant rocket able to lift 100 tons into orbit. Its cosmonauts had completed missions lasting as long as a year in space. And all involved in the program, from space-men to ground controllers, were seasoned and ready to do longer missions, even flights to other worlds.

Compared to the moribund American program, which had a crippled space shuttle program and a space station that was noth-ing more than paper, it appeared as if the dreams of Brezhnev were about to come true, and that outer space was certain to become part of the communist empire.

And yet . . . something was rotten at the heart of the Soviet empire.

Democracy

A year had passed since Boris Yeltsin had been fired as head of the Moscow Communist Party. Since then little had changed, despite Gorbachev's frequent proclamations that change was essential. His anti-alcohol campaign had ended a dismal failure, causing an increase in drug use and a drop in government revenues from the sale of liquor. His effort to eliminate shoddy workmanship did little but antagonize workers and lower output. Without a profit motive, no mechanism existed to encourage better-quality workmanship. And the superministries, intended to eliminate layers of bureaucracy, had merely served to add another layer. The campaign to end bribery and corruption never even got off the ground, mainly because the very officials who were to administer it were actually the embezzlers and the bribe-takers. They had no reason to punish themselves, and Gorbachev's program included no serious attempt at replacing them.

In fact, instead of improving the Soviet economy, Gorbachev's program had worked to cripple it. Although the Soviet command system was inefficient, corrupt, and badly run, it did manage to keep the nation operating, albeit poorly. Gorbachev's reforms undermined this system without introducing anything to replace it. The result was that there were more shortages, longer lines, and greater chaos.[72]

To put it simply, the money had finally run out.

For the Soviet space program, this lack of money soon became very evident. For example, after *Buran*'s first short flight, the entire Soviet shuttle program was mothballed because of lack of funds. Neither *Buran* nor Energia ever flew again. More fascinating were the space program's first efforts to raise cash. During the last few weeks of Titov and Manarov's year-long mission, the flight-control center in Moscow suddenly became plastered with advertisements. Soviet officials even tried to convince French officials to let them put ads on the spacesuits of Chretien and the Soviet cosmonauts. The French refused, not wanting their "spationaut" to resemble a race car driver.[73]

During the space mission of Volkov, Krikalev, and Polyakov, lack of money prevented them from doing almost anything they were sent into space to accomplish. They had been trained to commission *Mir*'s second and third modules. They had trained to do numerous space walks, installing sensors to improve the station's ability to orient itself and to test the Russian version of a manned maneuvering unit. Polyakov had arrived prepared to set a new endurance record in space.[74] And at the completion of their mission the three men were to hand *Mir* over to another long-term crew, who would again be visited periodically by foreign cosmonauts chosen from countries selected for purely political reasons.

None of the flight's goals was accomplished. Faced with a lack of money, space officials decided it was time to rethink. Glushko was dying (he passed away on January 10, 1989, only two months after *Buran*'s only flight), and the new head of Energia, Yuri Semenov, found space spectaculars and the colonization of Mars much less exciting than finding cash to maintain the program. The old space program, run for propaganda and pure scientific research, could no longer be supported.[75]

In February 1989, Semenov announced that the two modules could not be launched while Volkov, Krikalev, and Polyakov were in orbit. Energia didn't have the money to complete them in time. Without the modules the space walks were either unnecessary or impossible, and were thus canceled. Cash shortages also prevented the construction of rockets and of *Soyuz* and *Progress* spacecraft, which made it impossible to launch a replacement crew before Polyakov, Krikalev, and Volkov had to return.[76] Faced with these deficiencies, Semenov decided to have the crew mothball *Mir*, and come home sooner than planned, in April 1989.

Even as Semenov began reshaping the Soviet space program, Mikhail Gorbachev began rethinking his reform program. The Soviet leader had finally realized that his attempt to solve the country's stagnant economy and declining economic conditions by extolling hard work and greater responsibility was insufficient. Like Yeltsin, Gorbachev could not get the Party apparatchiks to institute the reforms that would put them out of power. In late June 1988, while Titov and Manarov were preparing for their second space walk, Gorbachev used the 19th Communist Party Conference in Moscow to force through proposals for the first general elections in Soviet history. The winning candidates of these elections, to occur 10 months hence in March 1989, would form a new legislative body, dubbed the Congress of People's Deputies.[77]

To protect Communist Party power, 750 of the 2,250 seats in the Congress were reserved for mainstream official organizations, most of which were under the control of the communist bureaucracy. Of these, the Communist Party was specifically guaranteed 100 seats.[78] Furthermore, the Congress's only real function according to the constitution that Gorbachev forced through was to elect by secret ballot a higher legislative body, called the Supreme Soviet, made up of 542 Deputies from the Congress.

Still, 1,500 seats to the Congress were to be open contests, with the two competing candidates chosen at public nominating meetings in late February.

Gorbachev's proposals shook the Communist Party to its very bones. As he wrote later, "Many delegates realized that what we were in fact discussing was the transfer of real power."[79]

Televised live, then shown again on tape each evening to the entire Soviet populace, and *then* published in full the next day in national newspapers, the 19th Communist Party Conference also changed the way the Soviet public perceived its leadership. Everyone watched. People formed long lines to buy the newspapers to read the transcripts. Said one person, "I could hardly believe what I was seeing."[80]

During the conference, Yeltsin stormed the podium and demanded rehabilitation, repeating again many of his criticisms and complaints about the Soviet bureaucracy that he had aired 20 months earlier.

> You know that my speech at the October Plenum was found to be 'politically mistaken.' But the issues brought up in it have [since] been raised by the press, the Party members . . . from this very podium, both in the report and in speeches. I think the only mistake I made in that speech was the inopportune timing of it: just before the 70th anniversary in October.[81]

When Yeltsin stepped down from the podium, he did it to applause. Then in the days that followed, an avalanche of letters, telegrams, and phone calls streamed into his home, expressing heartfelt and passionate support.[82] Yeltsin suddenly realized that it no longer mattered whether the Communist Party chose to "rehabilitate" him. In the public's eye, he was already a winner. Based on this amazing outpouring of support, Yeltsin decided to use Gorbachev's March 1989 elections to directly challenge the power structure of the Communist Party. He would try to run in District #1, representing the six million people of Moscow. If he could get on the ballot and win, the Party's power, prestige, and most importantly, control over the country, would receive a terrible blow.

Getting on the ballot was not easy, however. The 875 selectors at the final nominating meeting were mostly Communist Party apparatchiks. Of the 11 people (including Yeltsin) applying for nomination, only two could run in the final election in March, and the Communist Party had nominated two candidates: one powerful, one popular. The powerful candidate, and the Party's favorite choice, was Evgeniy Brakov, the director of the automobile factory that produced the Zil limousine. The popular candidate was, of all people, Georgi Grechko.

In the almost four years since his last space flight, Grechko had become head of a laboratory at the USSR Academy of Sciences' Institute of Atmospheric Physics, studying weather forecasting and the ozone layer. Now 57 years old, Grechko still dreamed of space flight. "I'm going in space a fourth time," he had recently announced, saying that he had gotten the okay from doctors. All he needed was permission from his wife who, he admitted, didn't seem interested in giving it. "Three times I went up without her permission, and now I don't know. Her health is poor and I'm well aware that this is partly due to my flights. She paid a dearer price for them than I did."[83]

Grechko's fame as a cosmonaut as well as his likable personality made him immensely popular. Moreover, as a longtime member of the Communist Party, he seemed a safe bet to support the Party organization if somehow he beat Brakov in the final election. With these two trustworthy candidates on the nominating ballot, the Party instructed the apparatchiks to vote for only them, and blackball the rest. As Yeltsin later wrote, "Only two names, Grechko and Brakov, had been hammered into the selectors' heads."[84]

Grechko, however, had other plans. He really didn't want to be a politician. He had already turned down one of the guaranteed seats offered to him by the Institute of Chemical Physics. On the day of the final nominating meeting, Grechko arrived an hour-and-a-half early. He sat down in the front row of the empty auditorium, stared at the floor, and wondered what to do. "Why am I here?" he asked himself. "Why am I doing it?" To him, a basically easygoing and friendly chap with honest ideas, politics was a dirty game. If he won, he knew he'd say and do things that would make many other, more-powerful, people his enemies. The risks and heartache weren't worth it. He wanted out, but didn't know how to do it gracefully. If he simply withdrew, he would look like a coward.

As he sat there slouched over, with his head down in thought, two well-polished shoes suddenly stepped into his line of sight. He looked up, to find himself staring into the face of Boris Yeltsin. Curious, Grechko asked Yeltsin why he was there. "Aren't you running in Sverdlosk?" Grechko knew that in Sverdlosk Yeltsin's candidacy would be a shoo-in.

Yeltsin scoffed, explaining that he'd rather run in Moscow. "The Muscovites tried to run me out of town. Let them *vote* for me now."

Grechko was immediately impressed with Yeltsin's spunk and determination. And he thought, "Here was a way to drop my candidacy and keep my dignity." He would make his speeches, trying to persuade the voters to support him. Then, at the last second, when he knew he had won their hearts, he would withdraw in favor of Yeltsin.

Yeltsin readily agreed to this plan. Both men knew that if Party operatives learned of Grechko's withdrawal too soon, they'd have time to pick a new candidate and arrange for the apparatchiks to vote for him.

The raucous meeting lasted more than 12 hours, past 2:00 AM. As it wound down, each candidate was given a chance to make a final statement and to answer written and oral questions. Yeltsin used his speech time to defiantly answer many of the most hostile questions, attempting to demonstrate his willingness to listen to criticism.

Then it was time for Grechko to make his last two-minute speech. He walked up to the podium and looked out across the assembly. "I felt myself a winner," he remembered. "I knew they'd vote for me."

He began his speech. Grechko knew that if he mentioned Yeltsin's name too soon, the Party officials in the crowd would shout him down, preventing anyone from hearing anything else he had to say. Instead, he told the audience how it was impossible for all 11 candidates to be the best choice for the job. "We need the best," he proclaimed. Since he didn't consider himself the best, he was withdrawing his candidacy. Would any other candidates do the same? "Not surprisingly, no one did," Grechko remembered with glee.

"Now," he added, "who do *I* want you to vote for?"

He paused. "We Russians have a saying: one defeated candidate is worth more than two who never ran. Therefore, I want you to give your votes to the most defeated man here tonight."

He paused again, holding the audience in his hands. Then he shouted, "Yeltsin!" causing the hall to erupt in jeers, yells, and applause. Grechko calmly stepped from the stage and took his seat.[85]

With Grechko out of the picture, and no other second candidate for the apparatchiks to vote for, Yeltsin became the obvious second choice. The final vote, by secret ballot, was Brakov 577 votes, Yeltsin 532.[86] Thus, on March 26, the citizens of Moscow would have their first real election choice since the fall of the czar almost 70 years earlier.

The election itself was a shocker, showing how much the Soviet public rejected the Communist Party that had ruled them for seven decades. In a humiliating blow, many Communist Party leaders found themselves defeated at the polls. In a few cases, these candidates had been running *unopposed*, and had still lost the election because the ballots gave voters the option of chosing "none of the above" and they thus had failed to get 50 percent of the votes. In Moscow, in Leningrad, in Kiev, the communist officials running for mayor were all defeated. In Lithuania and in Estonia, both of which had strongly organized opposition movements calling for greater autonomy, possibly even independence, communist candidates were resoundingly defeated. In Moldavia, six radical members of the local writers' union were all elected, while the chief of the Soviet Academy of Sciences, Aleksandr Zhuchenko, was not.

In one election in Moscow's northern suburbs, the only candidate to publicly advocate multiparty elections, Arkady Murashev, won easily, with the Communist Party's candidate, its District Party Secretary, coming in last. "The people understood everything," Murashev said. "People are very sensitive to any lie—that is what happened on election day."[87]

In Moscow, Boris Yeltsin, dismissed as head of the Moscow branch of the Communist Party 18 months earlier, won a resounding victory, getting an astonishing 89.6 percent of the vote.

A large number of cosmonauts ran and won, including Valeri Ryumin, Svetlana Savitskaya, and Viktor Savinykh.[88] For the three cosmonauts in *Mir*, voting was less than straightforward. The men had no way to secretly register their ballots. They were to give their choices to the capcom, whose job it was to register them officially. "It made me uncomfortable," said Krikalev. "My vote should be private. I didn't want it to be used by anyone for political purposes." While Volkov and Polyakov voted for men who represented their respective institutions, Krikalev decided to abstain. "I didn't have enough information to make a good choice."[89]

Meanwhile on Earth, Georgi Grechko voted for Boris Yeltsin. As far as he was concerned, it was time for a change, and Yeltsin seemed at the time the best person to make it happen.

In fact, Grechko was right. The elections were the first shot across the bow, a clear signal that the Soviet Union was actually changing. Many candidates had run and won on platforms demanding a cut in space spending. As proud as they were of their space program, the Soviet public now demanded that it prove profitable.[90]

In April 1989, only a few weeks later, Volkov, Krikalev, and Polyakov began mothballing *Mir*. They replaced several batteries so that the station's power supply would not fail, as *Salyut 7*'s had. They used a *Progress* freighter to lift the station's orbit to about 240 miles, unusually high, to extend its orbital life.[91] Then, on April 27, the three men entered their *Soyuz* spacecraft, undocked, and returned to Earth.

Perestroika had finally arrived in the Soviet space program.

9

Mir: The Road to Capitalism

False Profit

It was a bitterly cold 1990 February day on the Baikonur launch-pad, the temperature hovering at around $-22°F$. As cosmonauts Anatoly Solovyov and Alexander Balandin climbed the steps to the elevator that was to carry them to the top of the launchpad and their *Soyuz-TM 9* spacecraft, they turned to look down at people below them.

At the base of the tower were the typical news crews that glasnost now allowed. Soviet newspaper journalists, radio reporters, and television crews crowded forward, shouting both questions and cheers of encouragement to the two cosmonauts. Also pressing forward amid these official media was something new—a United States documentary crew from the NOVA television series, there to document the flight for the American public. As the cameraman aimed his camera upward at the two cosmonauts, the NOVA producer called for them to climb up one or two steps so that the camera angle was better.

Obediently, the two cosmonauts moved up two steps and then turned again to face the crowd. The NOVA producer then waved at them with a grin, calling up to them to wave back at him for the camera. At first, Solovyov, the ever-serious pilot and commander who at that moment really wanted to think only about the launch mere hours away, shook his finger sternly at the producer, trying

270

not to smile. The producer only became more insistent, waving both arms wildly and laughing. Solovyov finally grinned and waved back.

That same moment, almost to remind everyone of where they were, a cloud of steam engulfed them, blown from the fueled and very explosive rocket that towered less than 20 feet away. Making movies might be fine, but going into space was nonetheless a dangerous affair. Solovyov's smile faded, and he immediately turned from the camera and began climbing the steps again, herding Balandin ahead of him. It was time they boarded their spacecraft.

For Anatoli Solovyov, no relation to Vladimir Solovyov (the space-walking star of *Salyut 7* who was now flight director in mission control), this was his second space flight, his first having been a 10-day support flight during Titov's and Manarov's year-long mission. As a teenager he had attended the Soviet version of a vocational school, getting a job after graduation first as a factory worker, then as a locksmith. What he actually wanted was to become a military pilot. He went to night school, and in 1968 was accepted at the Chernigov Higher Air Force College that so many other cosmonauts had attended. He learned to fly, and was soon an active fighter pilot operating all kinds of airplanes.[1]

Words did not come easily to Anatoly Solovyov. He was an earnest, quiet man who constantly strived to control his emotions. For example, at one point during an interview for the NOVA documentary, he tried to explain how cosmonauts were no longer the media stars they once were. "Nowadays people don't know your name or your face." He paused. "We are not offended by this. It simply means that our profession has become more ordinary." He paused again. Suddenly he was at a loss for words. He looked at Balandin, sitting beside him, for help. The junior officer immediately stepped in, noting glibly how the novelty of space had worn off. While Balandin talked, Solovyov took a breath and gathered his thoughts to continue the interview.[2]

Unlike Solovyov, Alexander Balandin liked to talk. For him, describing the mission and his life as a cosmonaut was natural and easy. For example, at another time the NOVA crew gathered the two men and their wives and children in Solovyov's home, where they filmed them eating dinner. When Balandin was asked what he was doing to prepare for the mission, he grinned and joked, "I don't come home for weeks."[3]

For Balandin, the desire to fly in space had begun as a passion.
Born in 1953, he watched as a child as the first Soviet space missions were launched, and decided like so many other Soviet children at that time that he wanted to become a cosmonaut. He became an engineer, and was hired by Energia in 1976, where he
worked to design the Soviet space shuttle's instrument panel and
safety systems. At the same time he applied to become a cosmonaut, and was accepted in 1978 as part of the same class that included men like Manarov, Savinykh, and Alexandrov. For the next
12 years he trained as a cosmonaut, working first in the shuttle
program with Manarov, then in the *Mir* program. With time, the
passion to go into space was replaced by routine and hard work.
Balandin began to see being a cosmonaut as nothing more than a
job, not unlike that of a glorified factory worker.[4]

Only now, this so-called "factory worker" was finally making
his first flight into space. Solovyov and Balandin's mission was to
replace the first crew to occupy *Mir* since flights had resumed in
September 1989. However, their mission, unlike those of all previous Soviet crews, was not about science or exploration. Nor did
they have the goal of setting a new endurance record in space—
even before launch they knew that their flight would last about six
months. Instead, under the cash-hungry leadership of Energia boss
Yuri Semenov, their objective was to make space profitable, to
prove that *Mir* could function as a commercial enterprise. "We now
have to justify all the money that has been invested in this branch
of the economy," Solovyov carefully explained before the flight.[5]

Therefore, they had spent part of their last few months of training catering to the NOVA crew, doing interviews, staging shots,
and giving tours of the facilities at the Gagarin Cosmonaut Training Center.

Today, it was finally time to fly into space. After a smooth
countdown, *Soyuz-TM 9* blasted off and reached orbit seemingly
without problem. However, as they approached *Mir* two days later
for docking, ground controllers aimed *Mir*'s docking camera at the
Soyuz-TM spacecraft and were alarmed to see what looked like
three large flaps undulating like flower petals from the curved top
of the *Soyuz-TM* descent module.[6] Something appeared to be very
wrong. Solovyov proceeded with the docking, figuring that they
would be better able to deal with the problem once on board the
station. The two men then floated into *Mir*, joining Alexander
Viktorenko and Alexander Serebrov for six days of joint operations.

Viktorenko and Serebrov, the first crew to occupy *Mir* after its abandonment in April, had been in space since September 6, 1989, working to reactivate the station in this new era of commercialism. In that context, their launch had been covered live by television, radio, and news organizations, and their rocket emblazoned with advertisements, including a 150-foot-high banner for an Italian insurance company.[7]

Nonetheless, Viktorenko and Serebrov's mission was essentially a transitional one. Press releases said little about the need to make profits, and much of their work was identical to past space-station missions. For example, the crew spent most of their time in orbit inaugurating the use of new technology, including a redesigned *Progress* freighter, a Soviet version of a jetpack for space walking cosmonauts, and *Mir*'s next permanent full-sized module.

The first *Progress-M* ferry was launched on August 23, 1989, two weeks before the arrival of Viktorenko and Serebrov. With its own solar panels, identical to those on the *Soyuz-TM* spacecraft, *Progress-M* was no longer limited to only about two days of orbital life before its reserve of electrical power ran out. Furthermore, once docked with *Mir*, these panels could be used to replenish the station's batteries. The freighter also used the new Kurs rendezvous system, developed for *Mir* and already in use by the *Soyuz-TM*. With Kurs, *Progress-M* could dock with either the aft or the bow port on *Mir*'s core module.

The freighter's interior had also been enlarged so that it could carry approximately two-and-a-half tons of cargo into orbit. *Progress-M* could also carry a small recoverable capsule, dubbed *Raduga* ("rainbow" in Russian), which had been squeezed into the freighter's docking port and once ejected would return to Earth with about 330 pounds of materials. This capsule was created as part of the push to make money from *Mir*. Beforehand, Soviet space stations had no easy way to get their scientific samples and the materials they produced back to Earth. *Progress* freighters burned up in the atmosphere, and manned *Soyuz* spacecraft had little spare room. *Raduga* was the attempted solution, offering future customers a method for returning home more of their output, thereby increasing *Mir*'s profit margin.[8]

The new module, dubbed *Kvant-2*, was launched on November 26, 1989, after more than a year's delay. Derived from Chelomey's transport-support module, it became the first module to permanently dock to one of the radial ports on the multiple-docking

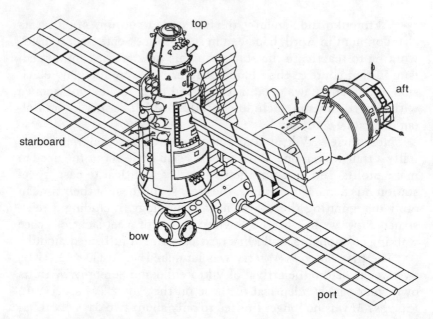

Mir, with *Kvant* in aft port and *Kvant-2* in top port. Note the airlock hatch at *Kvant-2*'s bow. *NASA*

adapter at *Mir*'s bow. Adding around 20 tons to the *Mir* complex, *Kvant-2*'s interior was divided into three hermetically sealed compartments. The main aft compartment contained equipment for making *Mir* more self-sufficient and livable, including a shower cubicle, the station's second bathroom, a second Elektron electrolysis unit, linked to the toilet, for turning urine into oxygen, and a Rodnik water-recycling unit for sucking humidity from the air and turning it into drinking water. Linked to these two systems were water and oxygen tanks, attached to *Kvant-2*'s exterior.

The middle compartment was reserved for scientific experiments, including a system for studying fluid behavior in zero gravity and a Japanese-built incubator to study the hatching and growth of quail eggs. This compartment, having airtight hatches on either end, could also be used as a back-up airlock.

The bow compartment was intended as *Mir*'s main airlock. It held two spacesuits and a roomy, three-foot-wide airlock hatch. It also carried the first Soviet version of a manned maneuvering unit for space walkers, officially called Icarus and originally intended for flight testing prior to *Mir*'s mothballing.

The wide airlock hatch, much larger than previous Soviet hatches, provided room for the larger cargo that the Soviet shuttles were supposed to haul into orbit. Furthermore, it opened outward so that the opened hatch was out of the way. This configuration had never been used before and carried some risks. When a hatch opens inward, the interior atmosphere presses against it, keeping the hatch sealed. (This explains why earlier crews had sometimes struggled to open their hatch: The last remaining bit of air in the station held the hatch closed.) Opening outward meant that nothing kept the hatch shut except its own bolts and clamps. In addition, opening outward meant that if the airlock was not completely depressurized when the hatch was opened, the gust of any leftover air could rip the hatch off its hinges once the hatch was unlocked.

To counter these concerns, *Kvant-2*'s hatch had 16 latches around its circumference, and was opened in two steps, using a hand wheel. First, the wheel was turned just enough to crack the hatch's seal, letting any residual air out. At this stage, protective, retaining hooks barred the hatch from opening more than one-hundredth of an inch, preventing its sudden release and subsequent damage. Once reassured by a sensor that the airlock atmosphere was gone, cosmonauts could turn the hand wheel the rest of the way, releasing the hooks and allowing them to push the hatch open.

For repressurizing the airlock's atmosphere after a spacewalk, *Kvant-2* carried a supply of sodium chlorate "candles." When ignited, these cartridges released a lot of oxygen quickly. Later freighters periodically restocked the airlock's supply.[9] The freighter also brought six large gyroscopes and a new, more-sophisticated computer system. When combined with the six gyros on *Kvant* and the 32 thrusters on the core module, the *Mir* complex's ability to maintain its orientation in space was significantly improved.

Overall, *Kvant-2* added about 2,000 cubic feet to *Mir*'s habitable space, bringing the station's total to approximately 7,000 cubic feet. The additional 20 tons brought the complex's total weight, excluding *Soyuz* and *Progress* spacecraft, to approximately 55 tons.[10]

However, joining the module to *Mir* was hardly simple. It was originally scheduled for launch in April while Polyakov, Krikalev, and Volkov was still in space, but funding shortages, the restructuring of the space program, and the failure of some of the module's computer chips delayed its completion for seven months. Then, after liftoff, one of the module's two solar panels refused to deploy entirely and its outer section swung freely instead of hanging rigid and straight. Ground controllers solved the problem by slowly rolling *Kvant-2* while simultaneously and repeatedly turning on and off the motor that was supposed to open the panel. The combination of repeated on-and-off commands and slight centrifugal force worked to lock the panel into its correct position.[11]

Then the docking itself had problems. With the cosmonauts waiting inside their *Soyuz* spacecraft in case something went wrong, the automatic Kurs docking system attempted to bring *Kvant-2* into *Mir*'s bow port. When the 20 ton module had moved to within 12 miles of *Mir*, however, the station's attitude-control computers inexplicably shut down, causing the six gyros on *Kvant* to shut down as well. The station began to drift, causing ground controllers to abort the docking.[12]

Though *Kvant-2* had a Kurs system and could dock with a drifting station, mission control decided to have Viktorenko and Serebrov manually control the station's orientation during docking instead. On December 6, Serebrov stationed himself at the main instrument panel in the core module, while Viktorenko positioned himself inside the *Soyuz-TM* spacecraft docked to the aft port. As *Kvant-2* approached, Serebrov watched it through his television monitor, shouting, "Right!" or "Left!" as necessary when the module drifted off his cross hairs. Viktorenko worked blind, using Serebrov's instructions to fire *Soyuz-TM*'s thrusters accordingly. "Both of us, as a team, piloted the station," Serebrov noted.[13]

Once the module docked, Viktorenko used a small robot arm on *Kvant-2* to grab the multiple-docking adapter, undock the module from the bow port, and rotate and dock it to the top radial port. Because only the bow port had radar equipment for automatic dockings, new modules could be shifted to the radial ports only like this. With the bow port once again open, Viktorenko and Serebrov climbed into their *Soyuz-TM* spacecraft, undocked from the station's aft port, and flew around to redock with *Mir*'s bow port.

The remainder of Viktorenko and Serebrov's mission was devoted, not to commercial endeavors, but to integrating *Kvant-2* into *Mir* and to testing the Icarus jetpack. To do these tasks, the two men did five space walks over a short, four-week period.

Their first three space walks laid the groundwork. They installed star sensors on the station's exterior to help orient the growing *Mir* complex. Before *Kvant-2*'s arrival, *Mir* had a shape like all previous American and Soviet space stations, a straight wedding-cake tower of docked modules. Once *Kvant-2* was attached to the top radial port, *Mir* formed an asymmetric "L" shape, which required much more-sophisticated calculations for keeping the station's solar panels facing the sun. The new star sensors aided these calculations. They also initiated *Kvant-2*'s airlock and made the first tests of the new add-on power pack that made the two Orlan-DMA spacesuits completely self-sufficient. For the first time, these space-walking Soviet cosmonauts used nothing but tethers and carabiners to hold them to their space vehicle. They assembled a winch and 200-foot tether and installed them on the exterior of *Kvant-2*. When they flew the jetpack, the winch and tether functioned as a back-up safety device, keeping them linked to *Mir* in case the jetpack failed.

Finally, on their last two space walks they tested Icarus. Unlike the American jetpack, which Bruce McCandless and Robert Stewart used to fly free of the space shuttle in 1984, Serebrov and Viktorenko never flew without being attached to *Mir*. Always linked by tether, on February 1, Serebrov used Icarus to soar more than 100 feet from the station where he did some turns and pirouettes. Four days later Viktorenko traveled almost 700 feet, cruising as much as 150 feet from the station.[14]

The men then parked Icarus inside the *Kvant-2* airlock, where it was to gather dust for the next six years, never to fly again. Built merely for political reasons to show that the Soviet Union could do anything the U.S. could, Icarus was too complicated, and certainly more dangerous, to use than simply allowing cosmonauts to climb along the station's exterior. Finally, in February 1996, to make room in the *Kvant-2*'s airlock, the jetpack was pushed outside and strapped to the outside of the station, out of the way.[15]

Even as Viktorenko and Serebrov were doing this space construction work, the pace of change initiated by Gorbachev continued to accelerate. In the year since the March 1989 elections and the April 1989 mothballing of *Mir*, much had changed in the Soviet Union and its communist empire. Month by month, the fall of communism rolled on. In May 1989, the Congress of People's Deputies held its first session. Though mostly controlled by Communist

Party apparatchiks, the three weeks of televised debate galvanized Soviet society. Suddenly, free speech became a normal and accepted way of life. Suddenly, opposing the ruling party was fitting and proper. As Yeltsin later wrote, the televised debates, ". . . in which almost the whole country watched . . . gave the people more of a political education than 70 years of stereotyped Marxist-Leninist lectures multiplied a millionfold."[16]

In June, elections were held in Poland, with the communists losing almost all their seats in the government. By October, a Solidarity-led government was in control. Officials of the formerly banned union began a detailed program to establish a market economy.

In July, the Soviet Union was hit with a devastating week-long miners' strike. More than 140,000 miners refused to work, demanding things that housewives in the Western world considered commonplace.

> Give each miner a towel and 800 grams [28 ounces] of soap a month for after-shift wash up. Provide miners with carbonated water. Give miners padded cotton jackets, because of high rates of sickness due to a strong stream of cold air on the way down to the pits during the winter season. Improve the content of salads and meat dishes.

That a communist regime, the so-called workers' state, couldn't provide even these most basic goods to its own workers, showed more than anything the complete bankruptcy of the system. After some initial resistance and half-hearted promises, which the strikers considered hollow, Gorbachev's government gave in, promising the delivery of 10,000 tons of sugar, 3,000 tons of washing powder, 3,000 tons of soap, more than 6,000 tons of meat, five million cans of dairy preserves, and 1,000 tons of tea.[17]

In September, the same month that Viktorenko and Serebrov reoccupied *Mir*, the Hungarian Parliament passed legislation allowing its citizens to travel abroad and emigrate. In October, that same parliament approved a new constitution, eliminating the special status of the Communist Party. The parliament also condemned the Soviet Union for its invasion in 1956, and declared itself an independent democratic republic.

In November, hundreds of thousands of Czechoslovakians peacefully marched through the streets of Prague, demanding an end to communist rule. By the end of the month this "Velvet Revolution" had thrown the communists out of power. The border to

Austria was flung open, removing all restrictions in travel to the West.

In East Germany, thousands of East German used the new rules in Hungary and Czechoslovakia (communist countries to which they were allowed to travel) to flee the Iron Curtain. By late October, the East German communist government, in power since the end of World War II, fell. Finally, on November 9, the Berlin Wall was opened. Within hours, millions of East Germans crossed that once deadly barrier to visit the West. Within hours, people were standing on that hated brick and barbed-wire fence, smashing it apart with sledgehammers.

In the Soviet Union, Gorbachev continued his stuttering reform campaign to restructure the Soviet economy. His effort had shifted from superficial public relations to passing laws that forced state-run industries to make believe they were private companies. In January 1988, a new comprehensive Enterprise Law took effect, requiring that over the next two years every government-run industry in the Soviet Union partly adopt free-market methods. Managers were to be given more freedom to choose their own customers. They were also to be given the freedom to sell 30 percent of their product on the open market.[18]

By January 1990, these new rules applied to the Soviet space program. Just before Solovyov and Balandin took off on *Soyuz-TM 9*, the administrators of the superministry Glavkosmos that managed the program announced it as the first Soviet mission whose primary goal was to make a profit from space. In the last year the space-program budget had been cut significantly, with many politicians calling for either more cuts or a return on every ruble spent in space. Furthermore, the collapse of the centrally controlled economy had left the Soviet budget in disarray, with huge deficits. To keep the program afloat, Glavkosmos's managers, apparatchiks who had no understanding of free-market capitalism and simply wanted to keep their perks and privileges, had to do whatever superficial thing they could think of to prove to their bosses that each *Mir* mission was profitable.[19]

Thus, mission managers calculated that launching *Soyuz-TM 9* cost about 80 million rubles. They also calculated that the products the crew planned to produce on the soon-to-be-launched *Kristall* module, *Mir*'s next module, would produce 105 million rubles of revenue when purchased by other Soviet industries. The difference was a "profit" of 25 million rubles, announced publicly with much fanfare before launch. That *Mir*'s Soviet customers were merely other Soviet government agencies, funded from the Soviet budget and required by government decree to buy *Mir*'s products, did not seem faze the men who ran Glavkosmos.[20]

Yet even this so-called "profit" depended on everything working smoothly. The strange "flower petals" flapping loose at the rear of Solovyov and Balandin's spacecraft threatened those profits. Protecting *Soyuz-TM*'s bell-shaped descent module were eight blankets of insulation material, each made of many layers of golden foil and an outer micrometeoroid layer. During launch, when the rocket's protective shroud was jettisoned, it apparently had pulled three blankets partly free as well. They now fluttered around the base of the descent module, exposing part of the heat shield to the hazards of space. The situation was not unlike the dance performed by American ground controllers years earlier when *Skylab*'s heat shield had ripped off. While parts of *Mir* had to be turned away from the sun to protect the *Soyuz-TM* spacecraft, the station also had to be turned so that its solar panels could get enough sunlight for electrical power. With these contradictory needs, ground controllers found themselves performing a constant balancing act, continually adjusting *Mir*'s orientation so that the heat shield and descent module would not be damaged by exposure to too much direct sunlight.

More worrisome was whether it was safe to use the *Soyuz-TM* descent module at all upon return. The loose blankets interfered with the infrared sensors used to orient the module in preparation for retro-rocket firing. Though the crew could probably steer the capsule manually, the torn insulation also left exposed to space three of the six explosive bolts used to separate the descent module from the service module at its rear. Without a closer inspection it was unclear whether the bolts would work when required. Furthermore, no one was sure whether the re-entry heat shield itself had been damaged. If either the shield was damaged or the service module could not be jettisoned, then *Soyuz-TM 9* was no longer a useful lifeboat. The situation resembled that faced by Ryumin and Lyakhov on *Salyut 6*—Solovyov and Balandin might need another *Soyuz* spacecraft to get back to Earth.

What was different now, however, was that Solovyov and Balandin's mission was driven, not by political concerns, but by an attempt to show a profit. To launch an empty replacement *Soyuz-TM* spacecraft, as had been done during Ryumin and Lyakhov's mission, would wipe that profit out.

Thus, Solovyov and Balandin began their work on *Mir* seemingly oblivious to the possibility that they might not have a way to get home. On the ground, four research bureaus were given the task of reviewing the problem. After much discussion, the general consensus, driven as much by the need to keep the program operating as by safety concerns, was that the damage was within acceptable limits and that the spacecraft should be safe to use.[21] Nonetheless, a space walk was planned to make a careful inspection of the heat shield and the explosive bolts. Ground engineers immediately started designing and building the necessary tools, including a special extension ladder that could be fastened to the exterior of *Mir* and allow the two men to get close to the descent module. At some point before the next *Soyuz-TM* launch, Solovyov and Balandin would climb into spacesuits and do an emergency space walk to see if they could fix the problem. If they could not, the next *Soyuz* would be launched unmanned.

While Solovyov and Balandin waited for these repair items, they also waited for the arrival of the *Kristall* module, originally scheduled for a March 30 launch. The launch was delayed until June, however, because of difficulties in getting *Mir*'s new computer system to coordinate the six gyroscopes on *Kvant* with the six gyroscopes on *Kvant-2*. Throughout March, the two men had struggled to configure this new computer system. One gyro had failed, and *Mir*'s asymmetric "L" shape made the computer calculations more difficult than expected. Moreover, the need to keep the damaged descent module from overheating made the orientation calculations even more complex. By the time they had finally configured the 11 working gyros, the tests had used up the fuel reserves of *Mir*'s small attitude thrusters. Ground controllers decided to hold off trying to dock *Kristall* to the station until the next *Progress* freighter had arrived and replenished these thruster tanks.[22]

Without *Kristall*, however, Solovyov and Balandin were unable to do the manufacturing work their mission had been touted for. *Izvestia* sadly announced that this delay trimmed mission's announced 25-million-ruble "profit" by 20 million.[23]

While they waited for *Kristall*, the two men unloaded the *Progress* freighter and performed a few scientific experiments, including activating the incubator in *Kvant-2*. On March 4 they placed 35 quail eggs inside. Then they spent 30 minutes each day monitoring the eggs, trying to coax them to hatch. Designed with a better ventilation system than the one that Ryumin had improvised on *Salyut 6*, the incubator hatched a healthy baby chick, the first living creature born in space, on March 17. By March 23, six chicks had popped from their eggs, and during one television broadcast the cosmonauts were shown hand-feeding the chicks.

As each chick appeared, the men transferred them to a customized birdcage. Leg bands fastened each chick into position, so that its beak was close to both a water supply and birdseeds. "It didn't work," Balandin remembered. Quail chicks have weak neck muscles. On Earth, this weakness works to their advantage: they can't lift their heads against gravity, so their beaks are always close to the ground where they find birdseed. In weightlessness, however, the weak muscles made it impossible for them to move their heads about and peck at the food. With their food only inches from their faces, the chicks were starving. "So we helped them feed," Balandin explained. "We put seeds in their mouths."

Despite their efforts, none of the chicks survived. Not only did they have trouble eating, they kept flapping their wings, unable to adjust to weightlessness. Within a few weeks they were all dead. Once again, the strange and alien environment of zero gravity had made it difficult for life to breed in space.[24]

On the ground, the political convulsions continued. On May Day, 1990, Gorbachev and the Soviet leadership once again gathered on top of Lenin's Tomb in Red Square to watch the traditional May Day parade.

This year, however, the parade was anything but traditional. While the official parade included its normal contingent of military hardware, highlighted with the usual giant posters showing the faces of Marx and Lenin, it also opened with an official demonstration of communist trade unions waving banners demanding a slowdown in Gorbachev's reform movement. Then the parade finished with an unofficial demonstration: thousands of protesters waving signs that were downright hostile to the leaders watching from above. Some carried signs saying things like "Seventy-Two Years on the Road to Nowhere!" "Down with Gorbachev!" and "Socialism? No Thanks!" The demonstrators also included some Hare Krishnas, and even a monk from the Russian Orthodox monastery at Zagorsk, holding a life-size representation of Jesus on the cross.

Faced with loud jeering and shouts, including chants like "Shame, Shame!" and "Gorbachev Resign!," Gorbachev and the other members of his ruling cabinet were visibly upset. After listening for almost a half-hour, Gorbachev slipped back into the Kremlin, while the square below was filled with mutilated Soviet flags, their hammer and sickle ripped out.

In a number of other Soviet cities, May Day celebrations had been canceled entirely. However, protests were still held. In Lvov, for example, tens of thousands of Ukrainians carried icons of the Virgin Mary, demanding independence for the Ukraine. In the Eastern European countries, now liberated from the communist bloc, there were neither protests nor official celebrations. Instead, the holiday became an excuse for shopping, block parties, and picnics. Nowhere were there communist banners, slogans, or public parades. In Budapest, Hungary, where a trade association ran a mass picnic, the communists were confined to running a single beer stand.[25]

Finally, on June 10, 1990, months (even years) late, *Kristall* arrived at *Mir*. Like *Kvant-2*, the docking had its problems. On the first attempt on June 6, an engine burn lasted too long, causing the computers to abort the docking. After a four-day delay, the docking was finally accomplished, using back-up thrusters.[26] The next day Solovyov used *Kristall*'s small robot arm to reposition the module to the bottom port on the multiple-docking adapter, turning the station's "L"-shape into a "T" and making *Mir* more symmetrical and thereby easing the complexity of attitude control.

Dubbed a factory in many Soviet press releases, *Kristall* carried five different furnaces for doing metallurgy and crystal-growth experiments, all supposedly aimed at producing profits for the space program. Like all the modules that were added to *Mir*'s base block, its basic hull was derived from Chelomey's transport-support module. At its bow was a spherical compartment built from a *Soyuz* spacecraft orbital module and very similar to the multiple-docking adaptor on *Mir*'s core unit. The bow docking port, however, was an updated version of the androgynous docking system first designed and used during the *Apollo-Soyuz* joint mission, and had been intended for use by the now-discontinued Soviet shuttle system.

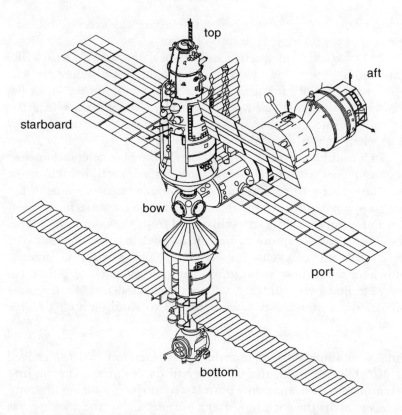

Mir, *Kvant* (aft port), *Kvant-2* (top port), and *Kristall* (bottom port). Note
the addition to *Kvant-2*'s airlock hatch of an attachment post for holding
the Icarus maneuvering unit. Also note the similarity between *Kristall*'s
bow and the core module's multiple docking module. *NASA*

Kristall's two solar panels increased *Mir*'s power grid by about
8 kilowatts to a maximum of 24 kilowatts, three-and-a-half times
the power of *Skylab* and five times the power of the most powerful
Salyut station. Because the location of the solar panels interfered
with a shuttle docking, they could be extended and retracted on
command. Their mounts were also detachable. Plans called for
them to be moved eventually and reattached to mounts on *Kvant*
where they would be permanently out of the way.[27]

Though *Kristall*'s main compartment carried a Bulgarian-designed greenhouse, dubbed Svet, the module did not house very

many plant experiments. Gorbachev's reforms had caused a serious curtailment of Energia's plant research program. The mission had to show a profit, and there was neither time nor money to devote to growing plants on *Mir*. Moreover, political concerns had insisted that the Bulgarian-designed greenhouse be used, even though Energia's well seasoned biology team had more advanced designs ready and waiting. "Technologically it was the same as Oasis 20 years before," Nechitailo remarked. "We couldn't learn anything from it." As a result, their quest to grow plants from space-born seeds remained a quest only. Balandin and Solovyov completed a few plant experiments, none of which taught the Energia biology team much that was new. The cosmonauts grew some onions (which they tasted), radishes (which remained too small to eat), several different orchids (none of which bloomed), and wheat (which produced no seeds).[28]

Thus ended the plant research that the Soviets had so aggressively pursued throughout the 1970s and 1980s. Except for one failed attempt to grow super-dwarf wheat, *Mir*'s greenhouse faded from use after 1991. It would take the arrival of American money to revive it.[29] Eight years after the first set of *arabidopsis* seeds had been grown on *Salyut 7* (and were subsequently lost after the *Soyuz-TM 8* docking failure), the question of whether space-born seeds could grow in space remained unanswered.

Overall, *Kristall* added about 2,100 cubic feet to the station complex, bringing its total habitable space to around 9,100 cubic feet, still several thousand cubic feet less than *Skylab*. *Kristall*'s 20-ton mass brought *Mir*'s weight up to around 70 tons, about a dozen tons less than *Skylab*.[30]

For the next five weeks Solovyov and Balandin raced to activate *Kristall*'s systems so that they could begin producing crystals and alloys in its furnaces. Their hope was to produce enough samples in the short period before they had to return to earth, which in turn could be "sold" by their bosses in Glavkosmos to other Soviet bureaucracies for a "profit" of three million rubles. In fact, in late June, these various Soviet industries requested that the mission be extended by 10 days to maximize the crew's so-called production. After asking and getting the crew's approval, mission controllers quickly agreed.[31]

During their last five weeks in orbit, the men also began their preparations for the space walk to inspect and maybe repair their

damaged *Soyuz-TM* spacecraft. With no specific pre-launch training for the space walk, they reviewed videotapes, beamed up to
them, of other cosmonauts doing simulations in the giant water
tank in Star City, talking at length with ground controllers about
what they had to do.

Finally, on July 17, 1990, after five months in orbit, Solovyov
and Balandin were ready to go outside. Though this was their first
space walk, both men were eager to get outside. They climbed into
their spacesuits, shut the interior hatch of *Kvant-2*'s airlock behind
them, and depressurized it.

As Balandin started the hatch-opening sequence, he turned the
hand wheel too far and released the safety hooks prematurely. The
last bit of air in the airlock pushed against the hatch, shoving it
open and banging it back against its hinges with almost 900 pounds
of force. Balandin himself was sucked outward and forced to grab
the hatch's rim to keep from being flung into space.

At that moment, both men were astonished by the intense,
bright sunlight that suddenly streamed in on them. For the first
time in five months, they were outside *Mir*'s tiny metal box.

Balandin was first out. He pulled himself out the hatch,
clipped himself into a handhold, and surveyed the terrain around
him. It was as if he were standing on the top of a four-story tower
that was flying hundreds of miles above the earth. Looking
"down" along *Kvant-2*'s length he could see on one side the *Soyuz-
TM* spacecraft more than 40 feet away, the golden flaps of its loose
insulation looking like petals on a flower. Looking "down" on
Kvant-2's other side, also more than 40 feet away, were the core
module and *Kvant*, stretching away almost 70 feet at right angles
to *Kvant-2* and silhouetted by the gigantic blue-white Earth. *Mir*'s
dorsal solar sail, assembled by the station's first two crews, rose
up beside *Kvant-2* so that Balandin could almost reach out and
touch its top.

"I could *feel* space all around me," he remembered. "It is not a
feeling you experience with your head. You experience it with your
whole body."

Balandin pulled himself out of the hatch and onto *Kvant-2*'s
exterior so that Solovyov could join him outside. As soon as he did
so, his perspective shifted. Instead of looking "down" the length of
a tower, he was lying on the side of a giant barrel, encrusted with
gold and silvery equipment. Like Kerwin on *Skylab* 17 years ear-

lier, Balandin found that he could choose his perspective entirely by the angle he placed his head. "You have to understand," he explained, "in space there is no up or down."

Using large carabiners and two short safety lines, the two men started their journey along the metallic, complicated surface of *Mir*. First they clipped themselves to the station with one line, unclipped the second line, moved forward, and then clipped the second line in so that they could shift the first line forward. In between, they had to also clip and unclip their large toolbox and two 6-foot packages holding the unassembled ladder that, once assembled, would give them a platform to work on. On one side the earth's surface raced by. On the other the sun raged, filling the sky.

For the first time, men were climbing on the surface of a space station shaped not like a long tower but having a complex and intricate structure, its surface strewn with sensors, antennas, windows, struts, tanks, cables, and handholds. At one point, when the two men described how complicated it was to navigate across *Mir's* encrusted surface, ground controllers joked that maybe they needed street signs to find their way.

The journey to *Kristall* took longer than expected. After a half-hour they were forced to stop when the sun set and they were plunged into darkness. For 45 minutes they could only creep forward slowly, waiting for the sun to return. After two hours, Solovyov and Balandin reached the spot on *Kristall* where they were to attach the ladder. As they assembled it, they discovered that it was much harder to unfold than suggested by the videotapes of the ground simulations. It took them another two hours to get the ladder stretched out to its full 13-foot length and fastened in position.

By this time they were more than an hour behind schedule. Their Orlan-DMA suits had only approximately two to three hours of atmospheric capacity left, depending on how much risk they and mission control were willing to take. Both men had to resist the urge to hurry.

Balandin was the first out on the ladder, with Solovyov at its base holding him to keep him and the ladder steady and stable. He felt no fear. The ladder was attached to *Kristall*, and backed up with a carabiner. He was attached to the ladder by a second carabiner, and to *Kristall* by a third. Solovyov, also attached by two carabiners, clutched him as well. And even if Balandin were

tempted to feel afraid (which he did not), his commander would
not let him. Nor would Solovyov let them waste time on sight-
seeing. With a few well-chosen but softly spoken words, Solovyov
kept the focus entirely on the work before them.

Hovering close to the damaged blankets, Balandin used a Japanese video camera attached to his shoulder to show ground controllers the damaged insulation as well as close-ups of the explosive bolts. Ground engineers asked him if he could see any white
patches or plastic foam around the bolts. As far as he could tell,
there was nothing unusual, noting that the bolt casings seemed
"very fresh."

Though flight engineer Balandin did most of the work, the two
men took turns over the next hour trying to reattach the three
loose blankets. To their dismay they discovered that all three blankets had shrunk, making it impossible to reattach them as planned.
Instead, the men folded up two blankets and pinned them down
and cut the third free with shears.

Time was running out. They had been outside almost five-and-
a-half hours, close to the safety limit of their Orlan-DMA suits. As
soon as the last blanket was cut free, ground controllers ordered them
to hurry back inside, even though they had to travel in the dark.
Leaving their tools behind and the ladder unfurled next to *Soyuz-
TM*, they raced back along *Kvant-2*, getting back to the airlock in
less than an hour. At the hatch, they struggled to coil their safety
lines, which in weightlessness refused to behave as they wanted.
Then they dove into the airlock and pulled on the hatch to close it.

To their horror it would not close. Even when both men tugged
at the door they could not get it to shut the last few millimeters.
For about 40 minutes they struggled with the hatch, tugging and
pulling at it without success.

By now, more than seven hours into their space walk, their problem was not running out of oxygen—they could tap into *Mir*'s oxygen supply—but that the lithium hydroxide canisters in their suits,
designed to absorb carbon dioxide, were almost certainly on the verge
of complete saturation. They knew that once the canisters were saturated, it would only take a minute or so for the levels of carbon
dioxide in their suits to become intolerable. If they didn't get inside
Mir soon, they would die from carbon dioxide poisoning, experiencing shortness of breath, headaches, dizziness, nausea, vomiting, convulsions, and eventually loss of consciousness.

It was finally "Do or die." Though it damaged the remains of the quail chick experiment, they vented the air from *Kristall*'s middle compartment and opened its hatch, using this small chamber as an emergency airlock.

For about 15 minutes they waited in the middle compartment, watching the air pressure slowly rise to normal. As far as they could tell, neither man felt any symptoms of carbon dioxide poisoning. Then, as soon as the air pressure was barely breathable, the men struggled to open their suits. "The handle to release the rear suit hatch wouldn't move," Balandin remembered. "It was too tight." Making things worse, in the tiny space of *Kristall*'s middle compartment, neither man could get much leverage on the handle. Desperate, Balandin opened an emergency valve in his suit, equalizing its pressure with *Kristall*'s. At the same moment, Solovyov grabbed Balandin's suit handle and pulled.

With a pop, the hatch on Balandin's suit opened. He stuck his head out. Less than a minute later, Solovyov emerged also. Both men floated silently for a few seconds, breathing deeply *Mir*'s mechanical-smelling but clean air.[32]

Inspection of the videotapes as well as detailed discussions between Solovyov and Balandin and mission control convinced everyone that the descent module was safe to use. However, the hurried end to their space walk required another. Before *Soyuz-TM 9* could undock from *Mir*, they had to close up the ladder and retrieve their tools. Furthermore, ground engineers wanted a close look at *Kvant-2*'s hatch, to try to figure out why it wouldn't close.

Eight days later, on July 25, the two men went back out. With Solovyov running the camera for mission control, Balandin tried to close the hatch. No matter how hard he tried, he couldn't get it to latch. Solovyov then videotaped the hatch's hinges, revealing one bent hinge plate and a partly unscrewed bolt. Ground engineers immediately guessed the cause of the damage: the force of opening the hatch the week before had bent the hinge.

While engineers reviewed Solovyov's videotape, the two men scrambled across *Kvant-2*'s length to the ladder and disassembled it, attaching both sections to *Kristall*'s exterior. Back at *Kvant-2*'s airlock, they tried again to close the hatch, without success. Then mission control asked a simple question: Had they ever removed the cover that they had placed on the hatch's rim gasket to protect it from the harsh environment of space? If the cover was still on, it would be impossible to close the hatch.

They looked at the gasket and, to both men's embarrassment, saw that the cover was still there and sheepishly removed it. Inside the airlock, Balandin took hold of the door, gave it a shake and, despite the damaged hinge, managed to pull it shut enough for its latches to grab hold. The men exited into *Kvant-2*'s middle compartment and closed the intervening hatch. Both compartments were repressurized, and for the next 24 hours ground controllers monitored the air pressure in the airlock, watching for any drop in pressure.

There was none. Though its usability was questionable (one engineer likened the hatch to a boarded-up window), mission control gave the crew permission to open the inside hatch, making the airlock accessible again. [33]

For the last two weeks of Solovyov and Balandin's mission they frantically used *Kristall*'s furnaces to try to make up for lost time, manufacturing semiconductor and crystal-growth samples for supposed sale to Soviet industries. They worked so hard that they neglected their exercises. Then, on August 1, the next *Mir* crew, Gennadi Manakov and Gennadi Strekalov, arrived. After six days of joint operations, Solovyov and Balandin boarded their damaged *Soyuz-TM 9* spacecraft and undocked it from *Mir*.

Standard re-entry practice called for them to first release the orbital module, followed by the service module, then fire the descent module's retro-rockets for return to Earth. This time, to reduce the chance that any of the loose insulation blankets might get snagged, as well as to give the release of the service module an added boost, both modules were released simultaneously. The tactic worked, and the return to Earth was completed without problems.

Though they came back healthy and alive, and had once again proved that humans could do difficult repair work in space, Solovyov and Balandin's overall mission failed in its primary goal. Even their so-called "profits" of 13 to 16 million rubles, trimmed by *Kristall*'s delay and the time spent repairing *Soyuz-TM 9*, were illusory.[34] Transferring money from one part of the Soviet budget to another to justify space expenditures simply did not work. To make space profitable, the space program would have to sell its services to someone other than itself.

And without that money, the assembly of *Mir* could not proceed. Semenov suspended construction of *Mir*'s last two modules, dubbed *Spektr* and *Priroda*. Until the space program could find the cash to launch them, they were to remain in storage on their factory floor.[35]

Even as Solovyov and Balandin had struggled to produce fictional profits in space, the slow crumbling of the Soviet Union and the Communist Party that ran it continued. Four days before the cosmonauts' death-defying space walk Boris Yeltsin walked up to the microphone at the 29th Communist Party Congress and announced that he was quitting the Party. Six weeks earlier, on May 19, at the first Congress of People's Deputies for the Russian Federation, Yeltsin had been elected its chairman. Then, on June 12, he had helped to pass a Russian law declaring the Russian Federation's "national sovereignty."

It was time to break his ties from the Soviet power structure, to quit the Communist Party. Yet, he agonized over his resignation speech for days. For the last 30 years the Party had been his life. Nonetheless, it was clear to Yeltsin that the Party was an obstacle to reform. He had to get out.

As he walked up to the podium there were jeers and boos from the audience of five thousand communist delegates. Yeltsin waited for quiet, then spoke softly, quickly. His speech was simple and to the point. He announced that as Chairman of the Russian Parliament, he had promised to sever all ties to other political organizations, and was therefore resigning from the Communist Party. He then pocketed his notes, stepped off the podium, and marched alone from the room. All around him were scattered boos and shouts of "Shame! Shame!"

Compared to three years earlier, however, when Yeltsin had resigned from the Politburo, the response was remarkably restrained. Gorbachev noted the change, and moved on to other subjects. He did not challenge Yeltsin's decision, nor call for others to attack or criticize him. In the past three years, Gorbachev had learned the political futility of such attacks.[36]

In fact, election politics, not party politics, now ruled. For Yeltsin, his resignation once again proved to the Russian people his dedication to their interests. His increasing popularity would soon become evident.

True Profit

The Soviet space program's quest to make money went on. Fake profits hadn't worked. Perhaps selling its launch services to foreign countries would. Glavkosmos began marketing its rockets to the West. They flew *Buran* to France for the 1989 Paris Air Show and stole the show. They also offered the various Soviet rockets as vehicles for launching public and private satellites. A public offer was even made to the United States to use the Energia booster to launch the space station *Freedom*. The offer was politely refused.[37]

Maybe joint missions with countries like France, Britain, Austria, West Germany, Spain, and Japan would bring in the rubles. Previous Soviet international missions had been provided at no charge because their purpose had been wholly political. Now, Glavkosmos decided that if, for example, a French astronaut wanted to work on *Mir*, the French government had to pay the cost. No longer would international visitors be flown to the station for free, just to satisfy the propaganda whims of political leaders. Now they had to pay for the privilege. Negotiations began with France, with Britain, and with Austria. The French, at first balking at having to pay for something they previously had gotten for free, agreed to pay about $12 million to fly their next cosmonaut to *Mir*. Similarly, the West German government agreed to pay around the same to fly a cosmonaut to *Mir* for a week in 1992. And in Britain, a private consortium agreed to fly a British cosmonaut to *Mir* for somewhere between $4 and $17 million.[38]

The most surprising of all the negotiations—and the most indicative of the Soviet need for cash—was the deal worked out with the Tokyo Broadcasting System (TBS). The network wanted to fly a journalist on *Mir*, and in mid-1989, Glavkosmos agreed. A Japanese reporter would fly to *Mir* for six days and send down daily reports, with TBS paying Glavkosmos about $12 to $14 million for the reporter's berth.[39]

For the first time ever, someone was going into space, not for scientific reasons, not for political reasons, not for propaganda reasons, but for reasons having to do with television ratings, profit, and the Japanese public's passionate curiosity about space exploration. And most remarkably, the flight would occur on a Soviet spaceship, using Soviet launch facilities in a Soviet Union famous for secrecy and hostility to capitalism.*

By late 1989 two Japanese reporters were training at the Gagarin Cosmonaut Training Center. Toyohiro Akiyama, 48 years old, was chief of TBS's foreign news division, while Ryoko Kikuchi was a camerawoman with the network. Akiyama was chosen as the primary launch candidate when, only a week before launch, Kikuchi went into the hospital to have her appendix removed.

On launch day, December 2, 1990, more than a hundred Japanese technicians swarmed throughout mission control in Moscow and the launch site at Baikonur. The rocket was painted with both the Japanese and Soviet flags, as well as TBS's advertising logo. Akiyama himself took six cameras, more than 100 rolls of film, and 40 hours of blank Sony Beta tapes into space with him.[40]

Assigned the job of baby-sitting Akiyama during his week in space was commander and pilot Viktor Afanasyev, making his first space flight, and flight engineer Musa Manarov, returning to space less than two years after the end of his record-setting year-long mission. When Manarov got back to Earth in 1988, his first thought was never to fly in space again. The mission had been exhausting and difficult, and it had left him far weaker than he liked. Then, after a long and welcome vacation with his family, Manarov felt the urge to go back into space. "When you come back from space you immediately decide it's your last flight," Manarov explained. "Never again, you say. Then a month passes and you want to fly again." He went to work in the Soviet space-shuttle program. *Buran* had just been launched, and he expected that the shuttle would soon be making regular flights. When that program was canceled, he asked to do another long flight on *Mir*, and was quickly assigned to a crew.

In between, he was picked as a candidate for the Russian Parliament, representing Dagastan, the republic of his birth and home of the Lakets people. Though he had spent very little time there

---

*Interestingly, the decision to fly a Japanese reporter caused a furor in the ranks of the Soviet press corps. The reporters, most of whom were Communist Party apparatchiks, were incensed that the first reporter to fly in space was not a Soviet citizen. To appease them, Glavcosmos initiated a program to choose a Soviet reporter for space flight, but without a financial backer, the flight never took place.

during his life, he was a famous cosmonaut who had spent a year in space. On March 4, 1990, he was elected. Still in office at liftoff, Manarov thus became the first elected Russian official to fly in space.[41]

His pilot and commander, Viktor Afanasyev, was a rough-hewn fighter pilot from the Soviet air force, flying MiGs at a variety of bases. Born in Bryansk, about 250 miles south of Moscow, he had dreamed of flying since childhood. Later he became a test pilot, then one of a handful of men chosen to pilot the Soviet space shuttle. In 1988, when the shuttle program was delayed due to lack of funds, he transferred to the main cosmonaut corps, acting as backup for the Solovyov and Balandin mission before finally getting his own flight.[42]

When Akiyama, Manarov, and Afanasyev arrived at Mir on December 4, they entered a space station about to celebrate its fifth year in orbit. Despite money shortages and the need for almost continuous maintenance, Mir's survival had proven that a vessel, stockpiled with the necessary supplies and given the necessary engines and fuel, could be built and launched from Earth and remain habitable long enough to conceivably take humans anywhere in the solar system.

Unfortunately, no one paid much attention to this achievement. The desperate search for money had changed the goals of the Soviet manned space program, much as the search for funding and support for Skylab had forced the NASA officials to change the American station's objectives in the late 1960s. Before Mir was left unmanned in 1989, Energia had been run by Valentin Glushko, a man who dreamed of sending interplanetary missions to the moon and Mars. Not only did Energia conceive and build its space stations as prototypes of interplanetary vessels, each subsequent manned mission had increased the flight length until crews were able to spend up to one year aloft. All that was needed to prove that men could get to Mars safely was for someone to fly an 18- to 24-month mission.

After Mir's 1989 abandonment and Glushko's death, these goals changed. To raise money, the Soviets shifted their focus to making a commercial profit. Longer, record-setting missions became unnecessary and, in fact, costly and detrimental.

So while Energia's new chief designer Yuri Semenov hunted for ways to make money from the program, a shift system was adopted.

Crews would go up, stay from four to six months (flight lengths that were now accepted as routine), and hand over *Mir*'s operation to a replacement crew. As noted by Vladimir Shatalov, head of the Gagarin Cosmonaut Training Center at the time, "That way, crews function better and don't lose interest in their work. They can go to space more often. But if you are there for 18 months or two years, the cosmonaut is like a disposable syringe, used once and thrown away. Who would go on to a[nother] two-year flight if they've already been up that long?"[43] Thus, Viktorenko and Serebrov spent five-and-a-half months on *Mir*, followed by Solovyov and Balandin for six months, and Strekalov and Manakov for four months.

Manarov and Afanasyev were to do the same. They would shepherd Akiyama through his week-long mission, then remain on *Mir* for another six months, maintaining and upgrading it for the arrival of the next crew shift and whatever paying customers the program could sign up.

Meanwhile, the Soviets' first paying customer, Toyohiro Akiyama, was struggling with space sickness. Before the mission he had candidly expressed surprise at being chosen, describing his health as that of "an average journalist," supplemented by four packs of cigarettes a day and shots of straight whiskey.[44] During his week in orbit, he was continuously sick, his head feeling stuffed and heavy from the body's redistribution of blood. Though he rallied to do one 10-minute television broadcast and two daily 20-minute radio broadcasts, when he came home on December 10 he was too weak to walk. His first words on air after landing were "I am hungry." He also noted that though the air smelled good, he wanted a beer and a cigarette.

Though business relations between the Japanese television network and the Soviets were generally good, there were several bumps in the road. For example, TBS thought, according to its negotiations with Glavkosmos, that Akiyama would get help 12 hours a day from Manarov. After launch, however, Energia demanded an additional one million dollars per hour for that assistance. TBS decided it could do without, which left Akiyama on his own for most of each day.

This confusion partly resulted from a political duel between the Soviet superministry Glavkosmos and Energia. Though Glavkosmos had negotiated the deals, Energia had the responsibility for running the missions and wanted a higher percentage of the

profits. Under Semenov's leadership, Energia used its leverage as
the organization that actually put men in orbit to push Glavkosmos
out. After the Japanese mission, all negotiations for buying time on
Mir were handled and controlled by Energia alone.[45]

On *Mir*, Afanasyev and Manarov set out to complete their six-
month tour. Though publicly Energia officials continued to claim
that Manarov and Afanasyev were going to produce semiconduc-
tors and other materials for profit, everyone in the space program
knew that this wasn't true, and that their real work was the maintenance and enhancement of *Mir* in preparation for later paying
customers.

Their main construction tasks all involved space walks: Try to
finish the hatch repair, then assemble and attach a 46-foot-long
crane to the outside of the base block. The crane, dubbed Strela
("arrow" in Russian), was needed to shift *Kristall*'s two solar panels
to the exterior of *Kvant*.

First came the hatch repair. The previous crew had attempted
and failed to repair the damaged hatch by trying to replace its bent
hinge pin. However, the pin was so twisted that it could not be
removed. Moreover, the ball bearings that surrounded the pin had
been crushed. Unable to complete repairs, all they could do was
attach a set of C-clamps to the hatch to act as a backup for the
hinge.[46] Rather than repair the hinge, Manarov and Afanasyev were
going to try to replace it entirely. They had done extensive simulations in the tank in Star City, using specialized tools for unscrewing and screwing the bolts that held the hinge in place. In fact,
Manarov had so much trouble with the bolts for attaching the replacement hinge that he had demanded better tools and gotten the
engineers to redesign them.

Once in space, Manarov still found it difficult to unscrew the
four bolts that held the old hinge. First, like so much else on *Mir*,
the designers had never thought that the hinge would need removal,
and sealed the screws with glue. Second, the washer behind one
bolt had been bent upward, making it impossible for Manarov's
wrench to grip the screw head. For four hours Manarov, with
Afanasyev holding him in place, struggled with the bolts. The last
bolt, with the bent washer, simply would not come free. Finally, in
desperation, Manarov took a screwdriver and managed to insert it
between the screw head and the washer and bend the washer flat.
The last bolt then unscrewed, and he pulled the hinge free.

The new hinge was quickly fitted in place and bolted down. Then the two men tested the hatch. To Manarov's delight and the disbelief of ground controllers in Moscow, it worked perfectly. "I was able to close the hatch with just one finger." Nonetheless, they left in place the C-clamps put there by the previous crew—just in case.

With the hatch fixed, next came the installation of the 46-foot-long Strela cargo crane. Still on the same space walk, they opened the hatch again, tightened the bolts, and headed down the length of *Kvant-2* to the port side of the base block until they reached its widest section, carrying with them the crane's mount. Unlike the hinge removal, fastening the mount to the base block's exterior went quickly. It and its attachment point had been designed so that men in spacesuits could fit them together easily. Within an hour they were finished and back inside the airlock.

Nine days later, a *Progress* freighter docked at *Mir*'s aft port, bringing with it the Strela crane. Strela was a telescoping series of five tubes, only 6 feet long when collapsed. When extended, using simple hand cranks, it stretched to 46 feet, long enough to reach *Kristall*, *Kvant-2*, and *Kvant* from its perch on the port side of the core module. Once attached to *Mir*'s exterior it would be used to not only move *Kristall*'s solar panels to *Kvant*, but also give space walkers a mobile handrail for moving themselves about the *Mir* complex.

The crane was packed in a 6-foot-long crate weighing 100 pounds. The two men spent a week carefully manhandling it from the freighter, through *Kvant*, through the base block, through the multiple-docking adapter, "up" into *Kvant-2*, and into the airlock. Then, on January 23, they donned their suits, and gingerly hauled it out the airlock, "down" *Kvant-2*'s exterior, and across to its mount on the base block.

Below them in the Persian Gulf, war raged. Both men described how they could see fires and the billowing smoke from the burning oil wells left behind by the Iraqis when they fled Kuwait.

It took an hour to reach the base block. They attached Strela to its mount, working far faster than planned. In fact, they were able to complete Strela's installation in one space walk, instead of the planned two. They finished so quickly that Manarov had time to take Strela for a ride. Before launch he had been given permission to climb up two sections of Strela. He knew that if he asked permission from flight director Vladimir Solovyov to go up all five sections the permission would be denied. "They'd be afraid to take the risk."

So, rather than ask, he simply waited until he was out of contact with the ground, and went ahead and did it. With Afanasyev at the controls to make sure the boom did not hit any part of the station by accident, Manarov climbed on and gingerly worked his way up each section, shaking the boom periodically to make sure it was solid. "I climbed a little bit by little bit to the very top."

Out there, holding on to the top of that narrow pole, he was like an ant on the end of a twig. Four stories below him was the core module. About 15 feet to his side was the top of its dorsal solar array, with *Kvant-2* 20 feet beyond that. Off to one side he could see the blue-white Earth glisten. And behind it all the velvet blackness of space was endless. "It is an unreal feeling," Manarov remembered, "hanging by your hand with this great space station above your head and the earth below you."

After a while, the men traded places. Then Manarov swiveled Strela so that its far end, with Afanasyev clinging to it, was parked against *Kvant-2*'s side about 15 feet from the airlock. Afanasyev climbed off and scrambled to the airlock. Manarov then locked Strela down and climbed out along the boom's length, using it as a thin highway back to *Kvant-2*.

Three days later, they were outside a third time, this time to attach the solar panel mounts on either side of *Kvant*, their journey across the growing *Mir* complex trimmed because they could use Strela instead of scrambling around the station's gear-encrusted surface. First Manarov climbed along Strela to the boom's controls. Then Afanasyev fastened the panel mounts to Strela's end and then clipped himself on as well. When he was ready, Manarov swung him around to *Kvant*, a hundred-foot-long arc across the vacuum of space. Then Manarov locked Strela down and climbed along its length to join him at *Kvant*, where the two men spent the next four hours attaching the two mounts to the module's port and starboard sides. To Manarov's annoyance, he had to make an extra trip to *Kvant*'s opposite side because they had brought the wrong mount with them. The mount for the port side would not work in the starboard position. "I had to take it back and get the other package."

Working on *Kvant*, it seemed as if they were at the bottom of a 70-foot tower, with two 45-foot-long modules at the tower's far end hanging over them at right angles. Rather than get distracted by the magnificence around them, both men found it best to focus on the work at hand.[47]

Their space walks completed, for the next two months they grew crystals and alloys, did Earth observations, and awaited the March 21 arrival of the next *Progress-M* freighter, bringing with it supplies, experimental equipment, and most importantly, letters from home. For a while at least *Mir* seemed to be operating as planned.

Then, for reasons that no one understood, as the cargo ship eased to within 500 yards of the aft port on *Kvant*, the Kurs rendezvous system sensed something wrong with its approach and aborted the rendezvous. *Progress* freighters had been docking with various Soviet space stations since *Salyut* 6 in 1978. While the large modules (*Kvant*, *Kvant-2*, and *Kristall*) had had docking problems, and while cosmonauts had sometimes been required to take manual control of their *Soyuz* capsule to finish its docking, this was the first of the more than 40 dockings attempted by unmanned *Progress* freighters to have difficulties.

After two days of consultation on the ground between men like Ryumin and Vladimir Solovyov, the decision was made to try again. On March 23, the freighter crept in toward the aft port. At first everything seemed fine. Then, when the 8-ton *Progress-M* eased to within 70 feet, only seconds from docking, the freighter's camera showed that it was not lined up properly with the docking port—even though the Kurs system insisted that all was well. Alarmed at the possibility of a collision, ground controllers immediately ordered the freighter to pull away. The order was too late to get *Progress* to simply back up, however. Instead, momentum carried the cargo ship just past *Mir*, missing the port solar panel on the base block by less than 15 feet.[48]

Ground engineers were baffled. The Kurs system insisted that the docking had been on target, even though the video of the docking clearly showed the freighter gliding off-line and to port. Furthermore, tests of the systems on both *Mir* and *Progress* indicated that all was in order. To find the problem, Solovyov decided that Manarov and Afanasyev would use their *Soyuz-TM* spacecraft to try a docking themselves. If an unmanned craft had docking problems, maybe a manned craft could find the cause.

Three days later, with the freighter hovering nearby in a parking orbit, Manarov and Afanasyev boarded their *Soyuz-TM* spacecraft and undocked from the station's bow port and flew around to *Mir*'s aft. As they lined up for docking, Afanasyev let Kurs do the work, automatically bringing the *Soyuz-TM* toward *Mir*. At several hundred feet it was clear again that the Kurs system on *Kvant* was the problem, somehow transmitting a false homing signal and sending the spacecraft off course. Afanasyev took control and completed the docking manually.

Two days later the cargo ship successfully docked with *Mir*'s bow port. It was now obvious that the problem was with the Kurs antenna attached to *Kvant* at *Mir*'s aft end. Engineers theorized that the antenna had been damaged or misaligned during the January 26 space walk, when Manarov had been climbing back and forth from one side of *Kvant* to another. Though the station could still function, without a repair to the docking system operations would be increasingly hindered.

Because Manarov and Afanasyev felt that they had probably caused the problem, they immediately began lobbying to extend their last scheduled space walk so that they could inspect the Kurs system. After some discussion, they were given permission, and on April 25, while Afanasyev did some routine exterior work near the airlock on *Kvant-2* (replacing a video camera that had been removed on an earlier space walk so that its lens could be replaced), Manarov took a solo journey out to the station's far regions. At *Kvant*, he discovered that the suspect Kurs antenna was not simply misaligned, it was completely gone. "There was no antenna dish," he remembered. In fact, the antenna mount was clean and unbroken, almost as if the antenna had been, on purpose, carefully unscrewed and removed. After videotaping the mount of the missing antenna dish for ground controllers, Manarov returned to *Kvant-2* where he and Afanasyev did some more routine work, installing "road signs" for later space walkers and removing test equipment for return to Earth.

They went back inside. Fixing the rendezvous system at the aft port would have to wait for the next crew, scheduled to arrive in only three weeks. More importantly, by this time the men were eager to come home. "I was tired of it," Manarov remembered. "Half a year of flight. My motivation wasn't strong. And my enthusiasm was less." On Earth it was spring. The flowers would be out. The air would be warm and scented. "I wanted to go home."

On May 20, 1991, their replacement crew arrived. The docking itself was tricky, because it had to use *Mir*'s aft port where the nonworking Kurs docking system was located. At first, the commander, Anatoli Artsebarski, tried to let the Kurs system complete the docking. When this didn't work properly, he took over, steering the spacecraft in manually. Even after the *Soyuz-TM* spacecraft was hermetically linked to *Mir*, the Kurs readouts insisted that the capsule was still more than 300 feet away.[49]

Arriving with the replacement crew of Artsebarski and Sergei Krikalev for a 1-week visit was Helen Sharman, the first British citizen to fly in space and the first woman to occupy *Mir*. Her flight had originally been an arrangement between Glavkosmos and a British business consortium. They would pay Glavkosmos for the ticket from money earned from the publicity, research, and entertainment produced by the first space flight of an ordinary British citizen.

Sharman herself was probably the least likely person to fly in space. Twenty-six years old, she had been a chemist working at a giant candy company where she helped design the equipment that produced candy bars. When one day, coming home from work, she heard the radio broadcast announcing that the consortium was looking for applicants to fly to *Mir*, she scribbled the phone number on a gas receipt. "It was a distracting and intriguing thought: by [the consortium's] fairly broad criteria . . . I was already . . . in a particular segment of the population any of whose members could become an astronaut."[50]

Sharman's eight days in space were far from unique. She was offered bread and salt upon arrival. She experienced no space sickness. She talked with school children by ham radio. She did some science experiments.

And to entertain herself and the men on *Mir* on her second night on board, she put on a ridiculous-looking, pink, frilly jumpsuit given to her by retired cosmonaut Alexei Leonov just before launch. "I got one of the ladies at the hotel to make it up for you." Leonov said, his sweet round face lighting up in an infectious grin. "Just for fun."

To her delight, her crewmate Krikalev responded by digging out a tie that he had smuggled aboard so that he could dress "formally" as well. The tie, of course, refused to hang down; instead it floated straight out throughout the evening.[51]

During her mission *Mir* began showing its first signs of age. Several times the computer system that oriented the station and its solar panels shut down, preventing the panels from producing electrical power from sunlight. Alarms rang, lights dimmed, and ventilation fans turned off as the station's systems automatically acted to preserve its limited battery power. Though at no point were they in any danger, each time the crew scrambled to reboot the computer and regain control of the station. After a few hours, when the sun had recharged the batteries, things went back to normal.[52]

Curiously, Sharman's mission once again demonstrated the difficulties the Soviets were having in establishing their space program as a viable and profitable business. The original deal had called for a British consortium to pay between $4 million and $17 million for an eight-day flight. Instead, the British partners dropped out, and other monies never appeared. Negotiations in 1990 and early 1991 with various television companies went nowhere because media interest was instead focused on the Persian Gulf War. Covering the launch of a British cosmonaut seemed less newsworthy. Rather than cancel Sharman's mission entirely and face the bad publicity that might ensue, Energia decided to send her up anyway, use her as a guinea pig to study medicine in space, and hope her flight would produce enough goodwill and positive public relations that other customers would soon follow. Though it was unclear where the money came from, Energia announced after the flight that the revenues from Sharman's flight totaled about $1.7 million.[53]

Despite these problems, Sharman's mission, and Akiyama's six months earlier, showed the way for Energia. Selling seats on *Mir*, unlike growing crystals or doing metallurgy, did raise hard cash, even if sometimes things didn't go smoothly. For men like Yuri Semenov, Energia's boss, and Valeri Ryumin, now one of the top managers of the *Mir* program, this was a lesson that they would not soon forget.

In space, the new crew settled in. As flight engineer, Krikalev was returning to *Mir* for his second mission in space. After having his first flight cut short in 1989 because of cash shortages, he hoped that this time everything would go as planned and he and Artsebarski would spend about five months in space, returning in mid-October.

Things did not go as planned. Ironically, the same political problems on the ground that shortened Krikalev's first journey into space were going to leave Krikalev stranded there. The Soviet system on Earth was finally about to come crashing down.

10

Mir: The Joys of Freedom

Top of the World

June 12, 1991: It was the moment of truth. After almost five years of bitter rivalry, Boris Yeltsin was about to move ahead of Mikhail Gorbachev. In the first direct national elections ever held in Russia, he was about to prove to the world that he had the support of the Russian people.

The West adored Gorbachev—and for good reason. In the six years since he had taken control of the Communist Party, he had moved the Soviet Union from a belligerent totalitarian state using force and barbed wire to keep its own people and half of Europe imprisoned to an open society able to give up its European empire and hold peaceful elections.

In the Soviet Union, however, Gorbachev's stock had plummeted. As freedom grew and became accepted, he was increasingly unable to keep pace with events. In late 1989, Gorbachev had considered several economic plans for rebuilding the Soviet economy and—after much consideration and the advice from the many Communist Party apparatchiks who surrounded him—rejected the most radical plan calling for a quick, step-by-step privatization of the economy. Instead, Gorbachev picked the plan designed to keep the Communist Party in power for as long as possible.[1]

Then, on November 7, 1990, during the annual celebrations to mark the anniversary of the communist revolution, Gorbachev, while standing on Lenin's Tomb less than 50 feet from the crowd in Red Square, had to duck two bullets aimed at his head, fired by a lone sniper who irrationally believed that killing Gorbachev would "show that there were people ready to die for democracy." Then, six days later, Gorbachev faced a five-hour confrontation with a thousand military deputies, who presented him with a long list of grievances, then heckled and badgered him when he took the floor to respond.[2]

After these events, Gorbachev's willingness to reshape the Soviet Union slowly ebbed. Over the next three months he began dismissing the most radical reformers in his government, replacing them with hard-line communist apparatchiks who opposed perestroika—all of whom eventually teamed up to try to overthrow Gorbachev later that summer.[3]

On New Years Eve, rather than be dismissed, Foreign Minister Eduard Shevardnadze, Gorbachev's closest adviser and ally during the early years of perestroika and glasnost, abruptly resigned, declaring publicly that, ". . . the reform has gone to hell. Dictatorship is coming; I state this with conviction. No one knows what kind of dictatorship it will be and who will come—what kind of dictator—nor what the regime will be like. . . . Let [my resignation] be my contribution—my protest if you will—against the onset of dictatorship.[4]

Two weeks later Shevardnadze's predictions seemed prescient. On January 13, 1991, six days after Manarov and Afanasyev had repaired *Kvant-2*'s airlock hatch and ten days before they began construction of the Strela crane, Soviet storm troops shot their way into a Lithuanian television tower in Vilnius, killing 13 people and wounding another 160. Gorbachev's reaction to the violence was lukewarm at best. He denied knowing anything about it beforehand ("I learned about what happened on Monday morning."), even though he had refused to come to the phone on Sunday when the Lithuanian president had tried to call him. Moreover, in the days before the attack Gorbachev had threatened the Baltic states with what he called "presidential rule" and a takeover by Russian troops if they didn't back off from their increasing demands for independence.[5]

In contrast, Boris Yeltsin seemed to increasingly understand the nature of freedom and democracy. His ouster from power in 1987 followed by his stunning election victory in 1989 had taught him that to rule justly a ruler needs the support of the electorate.

When he heard of the Vilnius attack, Yeltsin moved aggressively to put himself on the side of Lithuania and its citizens. He flew to the Baltics and immediately signed an agreement condemning the attack and recognizing the sovereignty of Lithuania, Latvia, Estonia, and Russia. The next day, he held a press conference where he once again condemned the attack and called on the soldiers of Russia to resist any orders to impose martial law in the Baltics.

> You could be given orders to move against legally constituted state bodies, against the peaceful civilian population which is defending its democratic achievements. You may be told that your help is needed to restore order in society. But can violation of the constitution and law be considered the restoration of order? . . . Before you storm non-military installations on Baltic soil, remember your own hearth, the present and future of your own Republic, your own people. Violence . . . against the peoples of the Baltics will engender new crises in Russia itself.[6]

Not just horrified by the use of force against innocent citizens, Yeltsin realized that by saying so publicly he was saying what most Russians believed, and would, therefore, increase his popularity.

His political instincts were accurate. On June 12, 1991, less than three weeks after Helen Sharman's return to Earth, the gigantic province of Russia, comprising about 70 percent of the territory of the Soviet Union, held direct elections for the first time in its thousand-year history. Previous elections, including the voting two years ago that had resurrected Yeltsin's career, had been only for a variety of different parliaments. This election was to directly choose the person to run Russia.

As noted by even the *New York Times*, "Should Mr. Yeltsin win, he would gain a popular legitimacy far surpassing that of Mr. Gorbachev."[7] Gorbachev had never been willing to submit to a direct public vote. Yeltsin had done it twice before, winning by incredibly high margins.

During the campaign the Communist Party tried aggressively to block Yeltsin's election. They used the media, still under their control, to slander him, even airing on television a panel of psychiatrists who questioned his mental stability.[8]

The results were unambiguous. Yeltsin garnered 57 percent of the vote—three times more than the closest runner-up. In the major cities he received more than 70 percent, with the candidates of the Communist Party coming in dead last. Moreover, though many thought the military would not back him, Yeltsin got more than 80 percent of the votes from sailors on ships in the Pacific and Indian Oceans. In other elections, reform mayors in Moscow and Leningrad also won easily. Furthermore, in a stunning albeit symbolic defeat for more than 70 years of communist rule, more than 55 percent of the voters in Leningrad stated that they wanted the city's name changed back to St. Petersburg.[9]

This victory gave Yeltsin the moral authority to demand a different political arrangement between the central Soviet authority and Russia. Though the negotiations had been dragging on for months, over the next six weeks Yeltsin was able to get Gorbachev to agree to a new All-Union Treaty, to be signed on August 20 by five of the fifteen republics that made up the Soviet Union, with four other republics soon to follow. This treaty would give sovereignty to each region. They would control their own taxes, their own laws, their own borders. The power of the central Soviet government, once practically immeasurable, would be limited to foreign policy and security.[10]

Up in space, Artsebarski and Krikalev had talked about the election and the various candidates, but had kept their preferences private. They understood that, unlike the sham Soviet elections of the past, this vote was for real, and to remain so, their votes had to be kept secret.

For Artsebarski, a young and idealistic lieutenant colonel on his first space flight, it was an exciting and historical moment. A thin and unimposing man with a ready smile, Artsebarski was born in 1956 in the southwestern Ukraine, in the province of Dnipropetrovsk, Brezhnev's home territory and the heart of his power. When Artsebarski was five, Yuri Gagarin became the first man to fly in space. That same night, the boy and his father went out to look at the stars. They saw a satellite drift by, and the father described how humans could now fly there. To Artsebarski's childish eyes, the satellite was large, and he imagined he could see the red banner of the Soviet Union draped across it as it flew over.

At that moment, he decided to become both a jet pilot, to defend his country, and a cosmonaut, to fly to the stars. He became a test pilot, flying all manner of new planes for the Soviet air force. When he was selected in November 1985 to fly the Soviet space shuttle, he had already accumulated more than 1,400 hours of flight time.

Though friendly, charming, and easygoing, Artsebarski was also a man who could be firm and to the point, saying only what needed to be said. Moreover, when faced with a challenge or difficulty, his face would harden, betraying a cold, quiet, and implacable intensity typical of many Russian cosmonauts.

His crewmate, Sergei Krikalev,* was no different. Tall and thin for a Russian, Krikalev had a young, friendly face with features suggestive of Tom Cruise. Like Artsebarski, he had an easy smile and a good-natured disposition. But like Artsebarski, when he wanted to focus on the job at hand, the smile faded, the eyes sharpened, and the task became the be-all and end-all of life. Over the next few years, this ability to focus would make Krikalev one of the most sought-after spacefarers, chosen repeatedly to fly some of the most important landmark missions to occur on either *Mir* or the *International Space Station*.

Born in Leningrad in August 1958, Krikalev had watched the 1960s space race as a child, and not surprisingly, had been inspired by it. By the time he was in high school he had decided that, of all careers, being a cosmonaut would be best of all. Krikalev's logical mind told him, even as a teenager, that his chances of getting into space were slim, if not impossible. Rather than aim directly at becoming a cosmonaut, he would keep space flight as a distant goal, while training himself in a wide range of activities, all applicable to going in space. "This way, at least, my work would be interesting."

He joined his high-school swim team, becoming one of its best athletes. He studied engineering, figuring that this training would be mandatory in space. He joined the local Leningrad flying club, learning to pilot and maintain airplanes and competing in airplane aerobatic events. Then, he graduated college first in his class. "In Soviet times, people were assigned to their job after college. If you had good grades, you had more choice." Having read a newspaper article describing how numerous cosmonauts got their start working at the Korolev design bureau, now renamed Energia, Krikalev chose Energia.

There, he was assigned the job of writing the procedures that cosmonauts follow when they use equipment, a job that at first disappointed him. He had wanted to build hardware, and instead he was doing paperwork. "I was luckier than I knew," he reflected.

---

*Unlike most Russian names, Krikalev's name in English is not spelled phonetically. The correct pronounication is kri-kal-YOFF.

If he had gotten a job designing bolts, he would have learned only how to design bolts. Writing procedures gave him hands-on experience with almost every piece of equipment used on manned vehicles. It gave him the opportunity to go to many launches. It got him involved in the efforts to solve almost every problem on *Salyut* 6 and *Salyut* 7, including the gas leaks on both vessels and the docking procedures used by Dzhanibekov to dock with *Salyut* 7 when it was dead in space.

By 1985 he was accepted into the cosmonaut program, passing his exams one year later and becoming a full time cosmonaut one year after that. Just 12 months later he was assigned his first flight, on the crew that relieved Musa Manarov and Vladimir Titov at the end of their year-long mission. His quick assignment occurred partly for the same reason Manarov had gotten his assignment—the shortage of cosmonauts caused by the medical commission decree that every man had to undergo more stringent testing after Vladimir Vasyutin's prostate problems on *Salyut* 7 in late 1985.[11]

It also occurred because Krikalev had made himself superbly qualified to fly in space. Though he had always kept his dream of space flight distant and removed, his wide-ranging campaign to train himself in every possible skill a cosmonaut could ever need had made him one of the world's best possible candidates. Energia officials would have been foolish to pass him by.

In orbit with Artsebarski on election day, June 1991, he listened as the capcom explained how they should submit their votes. Each was to submit his choice using a secret code letter. The capcom, the only other person to know what candidates the letters stood for, would then relay their votes to election officials.

Artsebarski voted for Yeltsin. "I liked Yeltsin's apparent openness," he remembered. Maybe he could bring some of that openness and honesty to Russian society.

Krikalev did what he had done in March 1989 when he was on *Mir* during the elections for the first Congress of People's Deputies. He abstained, choosing "none of the above." To his analytical engineer's mind, he simply didn't think he knew enough about any candidate to make a rational choice. Moreover, he was uncomfortable with the possibility that some people on the ground might use his vote for political purposes. "Making a decision without good information would either be guessing, or an attempt to gain some kind of political advantage." He didn't want to do either.

In space for almost a month, both men had begun preparing for an arduous series of space walks planned for the next five weeks, possibly the most ambitious construction project ever attempted in space. Their primary goal, after they had replaced the missing Kurs antenna dish, was to assemble a 46-foot-long girder (named Sofora after a fast-growing Asian shrub) on the top, or dorsal, side of *Kvant*. Unlike previous orbital installations (such as the Strela crane), this girder was not preassembled and ready to mount. Sofora had to be put together in space, on the outside of *Mir*, by men in spacesuits. If Sofora was built successfully and proved strong enough to withstand the rigors of space, a thruster engine would be installed on its peak. In this position, about 50 feet from the station's main axis, the engine would use about 85 percent less fuel in orienting *Mir* than the small attitude jets located on the circumference of the core module's service donut.[12]

First on the agenda was the damaged docking system. On June 25, Artsebarski and Krikalev climbed out of *Kvant-2*'s airlock and clambered across *Mir*'s growing structure to *Kvant*. At the damaged Kurs antenna, Krikalev looked at the base of the missing dish in puzzlement. "It looked like the dish had been carefully removed, not kicked off like Manarov thought." There were no broken pieces, no torn metal. Later, Krikalev theorized that the dish, which spins when in operation, had slowly worn its screw holes wider and wider until the whole unit simply spun itself off its mount and out into space.

Using a variety of tools, including a dentist's mirror, Krikalev removed the base of the missing dish and attached the new antenna in its stead. The work was slow. As usual, weightlessness made it difficult for Krikalev to get any leverage, even with Artsebarski giving him support. The screws were very tiny. They had been glued in place to keep them from loosening. Because the screws were so small and delicate, Krikalev could work only during daylight. And the screws were not the only problem. The electrical connections to the dish mount had to be unplugged, and they were very tiny, and sealed with both glue and wire. "Some engineers in mission control believed them impossible to work with." By the time Krikalev was done, his hands were very very tired. "It was very delicate work, with very bulky gloves."

This repair completed, Krikalev and Artsebarski spent the next three weeks preparing for the installation of Sofora, which was quite different from the girder assembled by Kizim and Solovyov during the last two space walks on *Salyut 7*. That girder, which had unfolded like an accordion and whose hinges then locked mechanically, had proved to be too weak and shaky. Sofora's design was almost science fiction in concept. It comprised 21 sections, each a 1.5-foot cube of pipes to be assembled in space by the two cosmonauts. The sections were to be attached to each other by sleeve joints made of an unusual alloy of nickel and titanium that somehow, when heated, "remembered" and reverted to its original size and shape. To link two pipes, a sleeve of this alloy was fitted around two pipe ends. Then an electrical current heated the sleeve, which then "recalled" that it originally had a smaller diameter, and shrank to that size, thereby tightening and locking the two pipe ends into place. It was hoped that these joints would be sturdier than the mechanical hinges tested by Kizim and Solovyov.[13]

To reduce their outside work, and to gain practice in assembling the sections, the two men assembled the first section inside *Mir*. Once outside, this section would become the top of Sofora, with each of the remaining 20 added to its base. Artsebarski, confident that Sofora's construction would go well, then added to this section his own personal embellishment. Months before launch, during the simulations in the Star City water tank, Artsebarski had gotten the idea that, like construction workers worldwide, they should celebrate the assembly of Sofora by hoisting the Soviet flag to its top. He immediately requested the creation of a flag made of space-hardened materials. No one objected, but no one did anything to make it happen. Despite repeated requests, when he and Krikalev arrived in Baikonur for launch there was no flag ready for them. In desperation, Artsebarski asked a friend to go into nearby Lenisk and buy an ordinary silk Soviet flag. Then, he snuck the 3-foot by 6-foot flag onto *Soyuz-TM*, since neither he nor Krikalev had permission to carry its extra weight into space.

With Sofora's first section complete, and destined to become the girder's peak, he and Krikalev fastened the flag to it. To keep the flag unfurled in the windless vacuum of space, they sewed poles along its edges.

On July 15, the space walks began. Like clockwork, they were scheduled to occur one after another, once every four days through the end of July. In the first space walk, the goal was simply to get their work platform attached to *Kvant*. For four hours they dragged the 150-pound platform across to *Kvant*. There, they fitted the platform to the fixtures originally used by the rocket shroud that had protected *Kvant* during launch.

As they worked, both ground controllers and Artsebarski noticed that his spacesuit was losing air at a higher than normal rate. The suits the two men were using had been in space since 1988, brought to *Mir* by the Afghan flight during Titov's and Manarov's year-long mission. In that time Artsebarski's suit had been used the most, 10 times, logging more than 60 hours total.[14]

Trying to pin down the source of the leak, Artsebarski looked at the suit's gloves, and noticed how abraded and worn they were. After all that use, his oxygen supply was literally dribbling away through his fingertips. To prevent further air loss, he activated the suit's special emergency wrist cuffs. Though uncomfortable, the cuffs prevented him from suffocating. He and Krikalev needed to finish their work. He could always replace the gloves later.

The next space walk took place as scheduled on July 19. Krikalev scrambled to Strela's controls and manually swung Artsebarski, the first two boxes of Sofora girder parts, and the already assembled first section over to *Kvant*. They then locked the first section into the work platform and began adding girder pipes to it. Once a new section was finished, they moved everything sideways, and attached a new section to the growing structure. Step by step, they built Sofora outward toward *Mir*'s aft, parallel to *Kvant* and the core module. Once completed, the almost five-story-long girder would be hinged upward so that it rose above *Kvant* almost vertically.

Both men discovered that the footholds on the work platform were useless, positioning the cosmonauts farther from the work area than during simulations in the water tank on Earth. They instead had to hold themselves in place by hand and foot, an awkward and difficult technique in weightlessness. To keep to their schedule the men continued to work even after the sun had set. When they finished this space walk they had managed to assemble three sections of Sofora, while also videotaping their work for ground engineers.[15]

Four days later, they were outside again, using Strela to move another 11 sections across from *Kvant-2* to the outside of *Kvant*. In the days between they had preassembled as much as they could of each section, and by the end of this space walk all 11 sections were added, bringing the girder's length to more than 32 feet long, its far end reaching almost past the service module on *Soyuz-TM 12* docked to *Kvant*'s port.

Finally, on July 27, 1991, they added the last six sections and hinged the entire 46-foot-long truss upward into position, hoisting the blood-red hammer and sickle flag of the Union of Soviet Socialist Republics upward as well. To test the girder's rigidity, Artsebarski started climbing it, rung by rung. At each section he stopped, shook the girder gingerly to satisfy himself that it was firm, and then went on. Soon he was at the peak. Held open by metal posts, the hammer and sickle flew beside him, high above both *Mir* and the earth. Magically, and quite unintentionally, at that moment

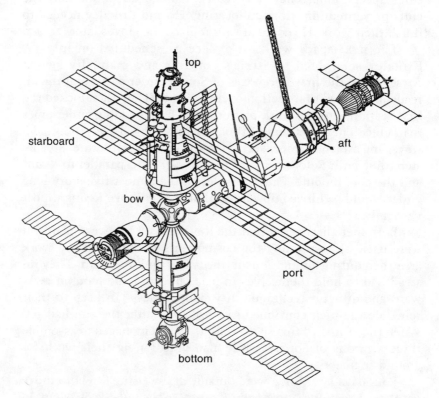

Mir, with the addition of the Sofora girder on *Kvant*, and the Strela crane to the core module's port side. A *Progress-M* freighter is docked to the aft port with a *Soyuz-TM* spacecraft docked at the bow. *NASA*

the dream of a five-year-old boy, imagining that the red Soviet flag was emblazoned on the side of an orbiting satellite, had come true.

Artsebarski looked down Sofora's five-story length, aiming his television camera at Krikalev floating near the girder's base. Spreading out below him hung the entire station, its erector-set modules and solar arrays silhouetted by the earth and the vast green and brown territory of the Soviet Union peeking out through white clouds.

Then, to Artsebarski's alarm, fog began forming on his visor. Soon he could barely see anything. He gripped the girder in momentary fear and radioed to Krikalev what was happening. Despite having a new spacesuit available, Artsebarski had gotten permission to use the old spacesuit one more time, "as an engineering test." The trip to the top of Sofora had overloaded its aging thermal system so that it could no longer absorb the moisture from his body, which instead condensed on his visor. In less than a few minutes, Artsebarski was blind, floating at the top of arguably the highest masthead ever built by man.

Very quickly the two men worked out a plan. By jutting his chin out and rubbing it against the visor's base Artsebarski was able to clear a tiny area. Slowly and carefully, his eyes angled downward to peek out this small opening, he worked his way down the girder while Krikalev climbed up. They met in the middle and Krikalev took the video camera from him, carefully untangling its electrical cord that had gotten wrapped around Artsebarski's body. Then, while Artsebarski waited in hope that his vision would clear, Krikalev climbed to the top of Sofora himself. "I wanted to test Sofora." Floating beside the flag, he shook the truss, and satisfied himself that it was solid and rigid. Then he came back down and joined Artsebarski. Together, they worked their way down Sofora, across Strela, and back inside *Mir*.

At the top of Sofora the hammer and sickle hung proudly in space. On the ground a Soviet television commentator noted that he could understand why the two cosmonauts had risked their lives to make this gesture. "After all, our country has not totally fallen apart yet. There are still things which we do better than anyone else in the world."[16] For this one last shining moment, the communist empire ruled all of outer space.

But only for a moment. Everything was about to change.

The Last Soviet Citizen

Three weeks later, at 6 A.M. on August 19, programming on
every radio station in the Soviet Union was interrupted by the voice
of Soviet Vice President Gennadiy Yanaev. "In connection with the
inability for health reasons of Mikhail Sergeevich Gorbachev to
perform the duties of the President of the U.S.S.R., effective 19
August 1991 I have assumed the duties of President." At the same
moment, long columns of tanks, trucks, and armored personnel
carriers, filled with machine-gun-toting troops, were converging on
Moscow.[17]

Yanaev was part of the self-styled "State Committee for the
State of Emergency." Its members were all handpicked officials chosen by Gorbachev in the last eight months to be part of his government. "I had promoted all of these people," Gorbachev wrote in his
memoirs. "And now they were betraying me!" In the Black Sea
resort of Sochi, where Gorbachev and his family were on vacation,
the Soviet leader was arrested. The resort's telephone service was
cut off, its mail and newspapers were confiscated, its radio and television service were disconnected, and guards were stationed to patrol its beaches and grounds.[18]

Yeltsin, meanwhile, awoke at his state dacha in the small village of Arkhangelskoye, about a dozen miles west of Moscow,
where things seemed oddly normal. Yeltsin could watch television
or listen to radio. No troops arrived. Friends and political allies, as
well as Yeltsin, drove in and out at will. And both the telephone
and fax machine continued to work.

In fact, that very morning, only a few minutes after Yanaev
made his public declaration, a construction crew arrived to asphalt
the driveway and grounds. After months of haranguing, the custo-
dian of the dacha had finally managed to schedule the paving.
Yeltsin went out and chatted with the workers, telling them to
keep working. "The hell with the coup," said the former engineer.
"Our asphalt's getting cold!"[19]

By 9 A.M. Yeltsin had written and transmitted an appeal to the
Russian people, exhorting them to defy the coup. Within an hour it
was being read worldwide. By 10 A.M. he was on his way to the
Russian Parliament building, dubbed the White House, located in
Moscow on a busy main thoroughfare. There he found the building
surrounded by tanks, as well as hundreds of Russian citizens.

People were bringing their cars into the square, using them to build barriers to protect the White House from attack. Others were talking with the tank crews, trying to get them to abandon their weapons. "Why are you here?" they were asked. The crews had no idea. They had been awakened in the wee hours of the morning and told to roll their tanks into the city. They had not been given a mission or objective.

Yeltsin immediately went inside the White House and held a press conference, declaring the coup illegal. Then he noticed that people were standing on a tank in front of the building. "I had to be out there right away, standing with those people," he thought. He wrote a quick statement and, surrounded by bodyguards, went down to the street and joined the crowd. Standing on the tank, with ordinary citizens around him and the tank commander sitting beside him with his face buried in his hands, Yeltsin declared

> We have found and continue to find strong-arm methods unacceptable. They compromise the U.S.S.R. before the entire world. . . . Accordingly, we declare illegal all the decisions and orders of this so-called Committee. . . . We call on the citizens of Russia to give the organizers of the coup an appropriate response and to demand the return of the country to normal, constitutional development. . . . We call on military personnel not to participate in the reactionary coup.

He then shouted, "Terror and dictatorship . . . must not be allowed to bring eternal night!"[20]

For the next two days Yeltsin refused to leave the White House. On the street outside, the crowds continued to grow, reaching a hundred thousand at one point. Construction workers built barricades of concrete blocks, while ordinary citizens added streetcars and more of their own vehicles to the barriers. Four tank crews changed sides, becoming defenders of the parliament building.

Just past midnight on August 21, a tank crew tried to force its way through. While the crowds around threw Molotov cocktails, three men climbed on top of the lead tank, trying to talk the crews into stopping. All three men were killed when the crews fired on them.

The barricades held, however. As the evening wore on, the crowds refused to disperse, despite the rain and the threat of death. As one man said while waving the Russian tricolor flag, "[The White House] is where the legitimate authorities that we elected are. So we *do* have something to defend."[21] By morning, the tanks were pulling out of Moscow, the Russian Parliament was meeting without opposition, and the coup had collapsed.

The coup failed for many reasons. For one, its instigators were not Stalin, who ordered the murder of millions to ensure the collectivization of Soviet agriculture, and later ordered purges that killed thousands of his most trusted Communist Party comrades because they might threaten his control.

Nor were they Khrushchev, who when threatened by the growing power of Beria and his secret police, coordinated a takeover that included walking into the conference room where Beria waited and shooting him with a pistol. Later, Khrushchev and his fellow communist allies had no compunction about destroying tens of thousands of churches and imprisoning their worshipers and priests, simply because these individuals did not believe in communism and atheism. Khrushchev even nonchalantly built a wall of barbed wire and brick down the center of a city, merely to guarantee control over his failing utopian society.

Neither were they Brezhnev, who calmly participated in a palace coup that overthrew Khrushchev, his patron and ally for more than 30 years. Brezhnev and the aging communist leadership left over from Stalin's time then organized their totalitarian rule into an ironclad corporate system, allowing no dissent, no disagreement, no debate, no freedom, only their own concepts of control and social order.

Instead, the coup's leaders were like Gorbachev. Children during or after World War II, they had not fought in a war as their predecessors had. In fact, they were part of a nation that had not seen a war for almost 40 years. And while they themselves were still scarred by the wars their parents had fought, the scar tissue from that war merely prevented them from engaging in the brutality required to maintain an iron-fisted rule. Like Gorbachev, they naively believed that bringing freedom to Russia would have absolutely no effect on their power and control. Surely there was no contradiction between giving the Soviet people their freedom while simultaneously denying them the right to pick their leaders. Surely they would continue to bow meekly to communist rule.

These men hadn't the slightest idea what freedom meant.

The coup also failed because the people of Russia *refused* to bow meekly to communist rule. They no longer were the citizens that Stalin, Khrushchev, and Brezhnev had faced down. As Yeltsin himself noted in his memoirs, ". . . much had changed in people's psychology during Gorbachev's perestroika. The public had grown accustomed to the fact that all of us, including the leaders, were individuals."[22] Yeltsin was himself one of those changed individuals. Raised in communism and witness to its failures for his entire life, he had come to realize that it was wrong. "We have oppressed the human spirit," he noted sadly during one press conference.[23] Seventy years of idealistic Soviet propaganda had finally had its effect. Yeltsin, along with the majority of the Russian population, at last knew what they had to do to make their country free and prosperous.

And so the public rallied to Yeltsin. Like the citizens of their former allies in Eastern Europe, the Russian people gathered to shout, "No more!"

On August 21, two planes left Moscow for Sochi. In one were some of the coup-plotters, hoping they could convince Gorbachev to help them in some manner. In the other was a delegation of the Russian Parliament, going to release Gorbachev and bring him back to Moscow, putting an end to the coup. In Sochi, Gorbachev refused to see the plotters unless all phone lines were reopened. Once that was done, he spent his time calling Yeltsin and President Bush, as well as the leaders of several other Soviet provinces. He also called the men in charge of security at the Kremlin, demanding that the phone lines belonging to the conspirators be disconnected and that if any conspirators were found they were to be arrested.

Then the Russian delegation arrived. "It was then that I really felt I was free," wrote Gorbachev later.[24]

In space, Krikalev and Artsebarski could only listen and wait. Ground controllers alternately relayed broadcasts from the pro-coup Soviet Central Television and the anti-coup Russian radio.[25] Neither man made any public comment about the events in Moscow, fearful of unknown consequences. Better, they both felt, to stay focused on the job at hand.

The entire Soviet space program rolled on as well with the same kind of tunnel vision. On August 21, even as the coup was dissolving and a Russian plane was flying to Sochi to rescue Gorbachev, a new *Progress-M* freighter was launched from Baikonur, carrying supplies as well as, of all things, two specially designed Coca-Cola soda dispensers. Though built partly to see if carbonated soda could

be poured in some fashion in weightlessness (a similar dispenser
had first been tested on the shuttle Challenger back in 1985), the
handheld units were as much an advertising ploy as anything, paid
for in cold cash by Coca-Cola to Energia.

Krikalev was not interested in appearing on tape doing a cheap
publicity stunt. He volunteered to be cameraman, leaving the dispenser testing to Artsebarski. It was Artsebarski's responsibility as
commander anyway. With a bemused grin, he spritzed Coke into
his mouth, first right-side up, then upside down. So, while the last
two Soviet cosmonauts drank Coca-Cola in space, the Soviet Union
dissolved on Earth—though it would take another five months before everyone realized it.

The two men on *Mir* dealt with the uncertainty of their situation and their nation in different ways. Artsebarski, the more ro-
mantic and philosophical of the two, found ways to relax by using
the indescribable gentleness of weightlessness. Late during their
sleep hours, when all was quiet, he would get out of his small cabin
in the core module and head to *Kvant-2*, which had the largest
window on *Mir*. There, nude, he'd float motionless, his arms floating limp before him, his head turned so that he could stare out into
the emptiness of space. If the station's orbit was in nighttime, he'd
gaze at stars in endless numbers, dwindling down, smaller and
fainter, until they seemed to fill even the sky's uttermost blackness like a fine haze. As he hung there in the darkness with only
Mir's fans whirring in the background, he could hear his heartbeat:
ba-bom . . . ba-bom . . . ba-bom. Soon, each beat would gently
shake his weightless body, and he'd begin to rock back and forth,
back and forth, back and forth. There was nothing but the quiet
fans, his heartbeat, and the endless innumerable stars. "Compared
to this universe," he thought, "A man is absolutely insignificant."

Krikalev, more pragmatic, tried to take the situation in stride,
joking about it with mission controllers. During the uncertain weeks
following the coup, strange rumors about the future of the space
program circulated through mission control and up into *Mir*. At one
point the capcom kidded Krikalev, telling him that, to raise cash,
Energia was going to sell *Mir* to NASA, lock, stock, and barrel.

Krikalev joked back, "Are they going to include *us* in the deal?"[26]

Ironically, and to Krikalev's misfortune, there was some truth
to these rumors. That summer, during the first Sofora assembly
space walk, Energia officials had announced that the next two

Soyuz missions had been combined because there was not enough money to build and launch both spacecraft. The original plans had called for flights in October and November, 1991. The first flight was to carry an Austrian, Franz Viehbock, his ticket paid for with $7 million in cash by the research firm Joanneum. Also on that flight were to be Artsebarski's and Krikalev's relief crew, Alexander Volkov and Alexander Kaleri.

The second flight was to carry a West German, also a cash-paying customer, and Toktar Aubakirov of Kazakhstan. The All-Union Treaty that Yeltsin had negotiated to replace the Soviet Union with the Commonwealth of Independent States, once signed, would make Kazakhstan independent, putting the Soviet spaceport Baikonur on foreign territory. To ease relations with Kazakhstan, the Russian government had offered it its own week-long space mission, similar to the Intercosmos missions to *Salyut* 6 back during the Brezhnev years.

The shortage of cash, however, forced Energia to rearrange this flight schedule. There simply wasn't enough money to send up a mission that merely visited the station for a week and then returned home. Each flight had to provide a crew change. Thus, the November mission was delayed to spring 1992, when the next crew change was scheduled. Furthermore, Aubakirov's assignment was shifted forward to the October 1991 mission, bumping Krikalev's crew relief, Alexander Kaleri. Since the Kazakh was not trained to stay in space for six months, Krikalev would have to remain in space for two consecutive crew shifts. [27]

Before these changes were announced, Valeri Ryumin, as head of the *Mir* program, got on the radio and asked Krikalev if he was willing to do it. In his typical analytical manner, Krikalev evaluated the situation. He realized that if he said yes to Ryumin, it would mean that he accepted the responsibility of finishing the mission. If he later asked to come home early, they would certainly oblige him, but consider it a failure on his part. He knew that the tough and blunt Ryumin, who had himself agreed to lengthen his first mission so that he could do a space walk and remove the antenna stuck on *Salyut* 6, would neither forgive nor forget.

Then, he considered his new crewmate. That Alexander Volkov had been his commander on his first space flight in 1989, and that they had gotten along well then, made staying seem somewhat less painful.

Finally, as a former long distance swimmer, he knew the limits of his stamina, how to pace himself for a long, hard, endurance test. The trouble was, his mission had become a long-distance race in which, halfway through, the distance had been almost doubled. "I had to estimate my remaining strength," he remembered. "I decided I had to slow down, to pace myself. As an athlete I could make that judgment."

So on October 4, 1991, *Soyuz-TM 13* arrived at *Mir*, carrying what was now an international crew, Alexander Volkov of the soon-to-disappear U.S.S.R. (though he noted in a prelaunch press conference that he had been born in the Ukraine, and could actually represent that country as well), Toktar Aubakirov of Kazakhstan (though he had been a Soviet air force pilot his whole working life), and Franz Viehbock of Austria.

For six days the five men worked together. Then Artsebarski said goodbye to Krikalev, giving him a high-five as he exited through the hatch, joining Aubakirov and Viehbock in their return to Earth. For Artsebarski, the separation was wrenching, as if he were abandoning his partner. Shortly after Krikalev had spoken to Ryumin, Artsebarski had tried to persuade his own boss, Shatalov, to let him stay longer. "It would be much better to keep the crew intact," he had said. Shatalov politely disagreed. As he descended to Earth, Artsebarski decided that his mission wouldn't be truly over until he saw his crewmate safely back on the ground.[28]

Five days after Artsebarski left, Volkov and Krikalev boarded *Soyuz-TM 13* to move it from the bow to the aft port. With Volkov doing the flying, they made repeated approaches, testing the new Kurs antenna at the aft port. Satisfied that it was working, Volkov then let the system complete an automatic docking.[29]

During the next few months, Krikalev and Volkov watched as the nation they had grown up in dissolved below them. Not only did 11 of the Soviet Union's 15 republics become independent nations linked together in a so-called Commonwealth of Independent States, 4 republics refused to join even this as Georgia and the Baltic states declared that they wanted nothing to do with their former oppressors.

On Christmas Eve 1991, the Soviet Union was removed from the United Nations, with Russia taking its seat on the Security Council. The next day, Mikhail Gorbachev sat before television cameras and announced his resignation, followed 24 hours later by a declaration by the upper chamber of the Soviet Union's Supreme

Soviet that the Soviet Union no longer existed. After almost three-quarters of a century, the utopian dreams of the communist revolution had dissolved into nothing.[30]

On board *Mir*, the two men continued the routine of previous long-term missions. They used one of the station's furnaces to produce semiconductor crystals of gallium arsenide, smelted samples of germanium and cadmium sulfide, made X-ray observations of the black hole candidate Cygnus X-1, and gathered data on the environment surrounding the orbiting laboratory. They also continued the tediously boring regimen of two hours of exercise per day.

They contended too with what was now an aging space station. Though *Mir* (with two new modules) had three times more habitable space than when Krikalev and Volkov were there together in 1989, miscellaneous equipment cluttered every module. Hoses and wires hung everywhere. The *Kvant* module, originally dubbed an "astrophysics laboratory" when launched in 1987, had instead become the station's storage "attic," located as it was at the aft end of the base block, or at the "top" of the station depending on your point of view. Many of its telescopes no longer worked, and since it was where crews unloaded *Progress* freighters and filled them with garbage, crews dumped old, unwanted equipment there, leaving only enough room for them to get to the aft docking port.

Worse, some of the station's equipment was ailing. Only seven of the original twelve gyroscopes on *Kvant* and *Kvant-2* still functioned. Several storage batteries had failed, reducing the station's power capacity. The station's solar arrays degraded about 5 percent per year, and now produced only between 10 and 20 kilowatts of power, depending on the orbit and the available sunlight. And after more than five years in orbit, the plastic windows on the base block had become blurry from the cumulative impacts of micrometeoroids.[31]

Moreover, the first of their two cargo ships, arriving in October and using the bow port, required two docking attempts, two days apart, before it managed to dock successfully. Then, when it came time for it to leave, it could not because some of *Mir*'s gyroscopes had failed, once again causing attitude-control problems. Krikalev and Volkov had to spend a day rewiring the gyros to get them to work so that *Progress* could undock.[32]

The second cargo ship, arriving at the end of January 1992, had no such docking problems, but its launch was almost delayed because of a threatened strike by unpaid employees at mission control. People in Baikonur had become so desperate that a riot broke out the next week, with four army barracks burned to the ground and three people killed when unpaid soldiers tried to ram a stolen car into their commandant's office.[33]

By early 1992, with the Soviet Union gone, it was unclear who owned the space program and its associated military operations. Throughout the autumn of 1991, Yeltsin and the Russian Parliament had been taking control of some agencies and refusing to fund others, letting more than 80 ministries and departments wither away. Both Glavkosmos and the Soviet Ministry of General Machine Building (which for 40 years had managed the space program) were abandoned, the administration of space taken over by the new Russian Space Agency. However, a system for paying salaries or expenses was not yet in place.[34]

For Krikalev and Volkov, the second cargo ship was important because it was supposed to bring tools and spare parts to repair the station's failed gyros, as well as fuel, food, letters from home, and, just as important, honey, requested by both men. (The station's supply of honey had run out early in Krikalev's mission.) The tools, fuel, and letters were there, along with fresh onions, horseradish, and lemons. No honey, however. Though Krikalev's wife had purposely donated a jar of honey for her husband, there hadn't been time or resources to test it to make sure it wouldn't contaminate the station's atmosphere. The chaos on the ground had made such tests difficult to accomplish. In fact, the chaos had made it difficult even to keep *Mir* supplied at all. In the last year or so the variety of meals on the station had dropped from about 300 to 100.[35]

The shortage of cash on the ground had also forced the program to change its repair philosophy. Rather than replace many items that, though functional, had reached the limits of their expected operating life, Energia officials decided that wherever *Mir* had sufficient redundancy, such as with batteries and its two Elektron recycling units, they would wait for parts to fail before replacing them. This approach, while increasing risks and long-term costs, would save the company money in the short run.[36]

The collapse of the Soviet Union caused other problems. The fleet of ships that were used to provide communications links with *Mir* when it was above the ocean could not be used; there was no money to launch them, and their berths in the now-independent Baltic states were denied them. Ground stations in Kazakhstan and other former Soviet provinces were also closed when their hosts demanded rent for their use.[37]

Moreover, the constellation of geosynchronous orbiting satellites, launched to replace the ships, were failing, and there was no money to replace them. And though one working satellite could still provide communications for most of the day, it belonged to the Soviet military, and they demanded rental money for its use, money that the Russian space program didn't have. Thus, for the first time in several years, cosmonauts could communicate with the ground only when their orbit took them over Russia. On some days and orbits, they were incommunicado for up to nine hours.[38]

Curiously, from the point of view of both the crew and ground control, this lack of communication had its advantages. It gave the crew more free time to work on their own, while saving money on the ground because not as many people were needed in mission control. "Every time you have to talk to mission control, you can't do your work," noted Krikalev. More importantly, too much communication with the ground made the station and its crew less autonomous. If humans were going to travel millions of miles to other planets, where communications with Earth would be rare and difficult, they had to learn how to manage without help from Earth. "The lack of communications helped make *Mir* more self-contained, more self-sufficient," noted Krikalev.[39]

Despite his confidence that he could pace himself to the end of his extended mission, by early 1992 Krikalev was struggling. He had gotten married only a year before, and had an infant daughter on Earth who was now beginning to speak. Early in the mission, before Artsebarski had left, his wife had even joked how he should "come back quickly or your own child will forget you." With all that had occurred, Krikalev wondered if that might actually happen. During a family session with his wife in January, he grumbled about how difficult it had become to do his work. In response she berated him gently, "Don't overreact. My God, what to do? You've got to live somehow, to adapt. If you're healthy, everything will be fine. You must take care of yourself, don't forget about exercises."[40]

On February 20, the two men did their only scheduled space walk together. However, just after they opened the airlock hatch the heat exchanger on Volkov's spacesuit failed. If he detached himself from *Mir*'s systems, the interior of his suit would fog up just as Artsebarski's had in July. Rather than abort the space walk, Volkov stayed near the airlock and did the scheduled tasks for that area while Krikalev climbed down Strela, across the base block and across to *Kvant*. There, he dismantled the work platforms he and Artsebarski had set up to assemble the Sofora girder and cleaned the lens of the docking television camera near the aft port.

Above him, Sofora rose five stories high, the flag that he and Artsebarski had raised still gleaming in the harsh sunlight. Wondering if Sofora was still as solid and rigid as it had been seven months earlier, Krikalev decided to go up to its top again. He was also curious to get a closer look at the flag. Pulling himself hand over hand along the girder, he quickly reached its peak, mere inches from the flag.

Already the silk fabric showed signs of decay, almost as if something were eating away at it. There were tears along its length, and large sections were gone, dissolved away. What was left, including the hammer and sickle in the upper-left corner, looked like it couldn't last many more months. Sofora, however, felt solid and sure. After more than a half-year in space, the joints of shape-memory alloy still remembered their shape. Krikalev took some pictures, then waved across at Volkov, whom he could see directly across from him at the *Kvant-2* airlock, about 60 feet away.

On his way back, Krikalev stopped at the dorsal solar panel on the core module and removed one of the test solar panels installed by Titov and Manarov four years earlier. By the time he and Volkov were back inside *Mir*, Krikalev was very cold. "It took me such a long time to warm up afterwards." Moreover, his fingers were once again badly irritated from working inside the thick spacesuit gloves. As he slowly warmed himself inside *Mir*, he consoled himself with the fact that he had accumulated more than 36 hours of space walk time, a record that would last for more than four years.[41]

Finally, on March 17, 1992, Krikalev's relief was launched into space. *Soyuz-TM 14* blasted off from Baikonur in Kazakhstan, sporting Russian and German flags on the side of the rocket, and carrying Russians Alexander Viktorenko and Alexander Kaleri as well as German Klaus-Dietrich Flade. For the first time, Russian cosmonauts were flying under a Russian instead of a Soviet flag. Flade was a paying customer, his airfare of about $12 million paid by the now unified German government. The former communist East

Germany had been absorbed by West Germany in October 1990, and one of Flade's trainers was Sigmund Jahn, a former communist cosmonaut who had flown to *Salyut 6* in August 1978.

After spending seven days handing *Mir* over, Krikalev, Volkov, and Flade returned to Earth. Krikalev had completed 313 days in space, a record exceeded only by the 366 days flown by Vladimir Titov and Musa Manarov, and the 326 days by Roman Romanenko.

On the ground, Anatoli Artsebarski, promoted to colonel in what was now the Russian military, was there to meet him, and to escort his former crewmate back to Star City. During the last five months he had gone to mission control every day, keeping tabs on his partner. When he was sent to the Crimea for post-flight recovery, he had taken his ham radio equipment with him so that he could talk with Krikalev every night. Only now, with his crewmate safely back on Earth, did Artsebarski consider his mission "finally over." On the flight home they talked of family and friends. Artsebarski tried to get Krikalev to drink some cognac. "It's medicinal," he explained. Like many Russians, he considered a little alcohol to be healthy. Krikalev, a pragmatic teetotaller who didn't really believe such tales, took a sip out of courtesy to his friend, and stretched out to rest. After 10 months in space, the weight of Earth was heavy on his muscles. It was good to be home.[42]

The country that Krikalev returned to, however, was not the one he had left. His hometown had changed its name from Leningrad back to St. Petersburg. His college was now the St. Petersburg Mechanical Institute, instead of the Leningrad Mechanical Institute. His employer was now Russia, not the Soviet Union. His income was no longer guaranteed. His family's apartment was unsure. His daughter's school was unknown.

Everything had changed. Everything was unknown. The joys of freedom had descended upon what had once been the Soviet Union.

11

Mir: Almost Touching

Trading Places

On December 11, 1993, representatives from all the major space programs worldwide gathered in New York City, along with press and television cameras from all the major networks and news organizations. The attendees also included cosmonauts Alexei Leonov and Vladimir Shatalov (who had flown the first successful *Soyuz* docking mission in 1969 and had from 1987 to 1991 been in charge of the Gagarin Cosmonaut Training Center in Moscow).[1] Even Yuri Gagarin's widow Valentina showed up that day in New York.

The event had nothing to do with some new achievement in space. Instead, these luminaries gathered at the Sotheby's auction house, where they and the Russian space program had consigned for sale more than 200 items from the remains of the Soviet space program. Many items—such as Yuri Gagarin's dress uniform and Oleg Makarov's flight jacket, gloves, and watch worn during his *Soyuz 18-1* launch abort—were owned by individuals. Other items—such as the *Soyuz-TM 10* capsule that had brought Japanese TV journalist Toyohiro Akiyama back from *Mir*—were owned by various Russian space organizations. In either case, the purpose of their sale was to earn money to pay bills.*

---

*In fact, a year earlier, Sergei Krikalev had criticized the program for abandoning the spacesuit that had failed when Alexander Volkov had done his

For sale was the *Almaz* reconnaissance capsule from *Salyut 5*. It had returned to Earth in February 1977, carrying spy pictures of U.S. military operations. Now it sold for $48,875. For sale were the work sheets used by Georgi Grechko and Yuri Romanenko during their 96-day mission on *Salyut 6*. Sold for $41,113. Also for sale were Grechko's spacesuit glove, the one used to pull Romanenko back into the airlock. Someone paid a cool $10,925 for it.

Before the auction, Sotheby's had estimated that the 200 Soviet-era items would sell for more than three million dollars. Not unexpectedly, the bidding was hot, with total proceeds easily doubling that number. Ross Perot was an anonymous telephone bidder, spending more than two million dollars to buy a *Soyuz* space capsule, three spacesuits, and other items. Another unnamed bidder, possibly Perot, even paid $68,500 for a piece of paper deeding ownership of an abandoned unmanned lunar scout ship, though the ship was stranded on the moon and there was no way of getting it back.[2]

The Russian space program and its employees needed cash. For the last seven years, they had struggled to learn how to be capitalists. Unfamiliar with the market and profits, naive managers had not known what prices to set for selling space on *Mir*. While the French, the Germans, and the Japanese paid $12 million to fly a cosmonaut in space, the Austrians paid only $7 million, while the British sent Helen Sharman into space for less than $2 million.[3]

These prices were simply too low, insufficient to keep the Russian space program afloat. The lack of money became so acute that some employees at mission control were not paid for months. More than a third found jobs elsewhere. By the end of 1992, launches had dropped to their lowest levels in decades, with less than half of all planned launches taking place. Furthermore, less than half the money the government had allocated to the space program was actually received.[4]

With the fall of the Soviet Union, the Yeltsin government set about reorganizing the space industries, focusing them more on private ownership. The Russian Space Agency, modeled somewhat like NASA in that it was supposed to manage all governmental

---

space walk. Krikalev, thought that the suit could have been brought home and sold for cash. Instead, the crew was ordered to eject it from *Mir* so that it would eventually burn up in the atmosphere. (Portree & Trevino, 87)

space efforts, was given the job of contracting the work to private and semiprivate companies like Energia. To the post of director, Yeltsin named Yuri Koptev, an outspoken and tough former space apparatchik from the now-dissolved Soviet defense bureaucracy.

Energia, meanwhile, underwent another reorganization. The design bureau that Glushko had originally headed and which had been merged with Energia back in 1974 was pulled from it to form a separate organization. Also pulled from Energia was the division that had formerly been Chelomey's design bureau to create the Khrunichev State Scientific-Production Center, a semiprivate government agency responsible for building the Proton rocket as well as space-station modules. Energia was left as a semiprivate company whose job it was to maintain *Mir* and the manned program, supervised by Koptev and the Russian Space Agency. The government would continue to subsidize it, but only partly. Private revenues would have to make up the difference.[5]

These changes did little at first to alleviate the cash shortages. The last two modules for *Mir*, originally to be flown in 1991 and 1992, remained unfinished on the ground. And construction on *Mir-2*, an upgraded replacement station for *Mir* that had originally been scheduled for launch in 1992, was also stalled.

Compounding the program's difficulties was the location of the Baikonur Cosmodrome, situated in what was once a province of the Soviet Union but now the independent country of Kazakhstan. To retain access to its primary launch facility, the Russian government and its various space industries had to work out a deal. By late 1993 the negotiations between Russia and Kazakhstan centered on defining what part of Baikonur's 11 million acres were to be leased, and the total amount of money to change hands.[6]

The cash shortage was not limited to the space program. Following the collapse of the Soviet Union came the collapse of the Russian economy. In 1991 the country's gross national product declined a shocking 12 percent, while prices rose 203 percent. Oil production dropped by half, forcing Aeroflot to cancel most flights and close half the nation's airports because of fuel shortages. By the end of 1991, production was only 21 percent of what it had been the year before, and the ruble—with Lenin's face soon to be removed—had lost 86 percent of its value against the dollar.

People were openly fearful of famine. The harvest, bad in previous years, was the worst in 15 years, producing 25 percent less than

1990's totals. By October 1991 only one-third of the needed grain and food necessary to feed the population had been produced. At one point Moscow had no bread in its stores, the first time this had happened since World War II.[7] Maintaining an independent space program merely for propaganda reasons, as Brezhnev and Gorbachev had, now seemed impossible, if not insane.

In November 1991, Yeltsin announced what he called "shock therapy" for the Russian economy. What followed was surely a shock. In January 1992 most price controls were lifted, government spending was slashed by 70 percent, and privatization begun. Funding for the military, including the space program, was cut by 85 percent. The government sold off half of all small and medium-sized enterprises. Land was to be privatized. And the ruble was allowed to float on the open market, available for the first time for exchange with foreign currencies.[8]

As Yeltsin said when he announced these draconian measures, "I have never looked for easy roads in life, but I understand very clearly that the next months will be most difficult. If I have your support and your trust, I am ready to travel this road to the end with you." Until this moment, he had had the support of the Russian people. Whether he could continue to depend on that support remained an open question.

Meanwhile, the American manned space program was experiencing a somewhat different set of problems. Superficially at least, the manned space programs of the United States had made a reasonable recovery from the *Challenger* accident. In the six years since *Challenger* was destroyed and its solid rockets redesigned, more than two dozen shuttle flights had lifted off from Cape Canaveral. Besides putting in orbit a number of NASA communication satellites, the shuttle had sent probes to Venus and Jupiter, and to circle the Sun. It deployed several Earth research satellites. It launched both the Hubble Space Telescope and the Compton Gamma Ray Observatory.

In addition, the shuttle completed a half dozen or so *Spacelab* missions, working in cooperation with Canada, Europe, Germany, and Japan to perform medical, biological, and metallurgical research, including a renewal of plant research, this time by American scientists. A series of American-designed greenhouses were tested as researchers attempted to duplicate the plant research begun by the Energia biology team in the 1970s. Once again, *arabidopsis* was grown in space, and once again the scientists had trouble getting the plants to produce seeds. Weightlessness in a cramped space shuttle made everything more complex. Getting water to the roots was difficult. Air circulation required filters and a carefully designed ventilation system. And lighting had to be continuous and bright, yet take up little room and mass and use very little electricity.

Despite the shuttle's apparent renewal, NASA was still plagued by sloppy management. The main mirror to the Hubble Space Telescope had been improperly ground, resulting in a $1.5 billion telescope whose images were all out of focus.[9] Though NASA eventually launched a successful shuttle repair mission at the end of 1993 to correct the focus problem, the error was a continuing embarrassment to the agency.

Meanwhile, space station *Freedom*, still unbuilt, went through redesign after redesign with its budget rising. By 1993, its total estimated construction cost had grown, according to some, to more than $31 billion, not including launch costs. In June of that year, the program barely escaped termination in Congress when a House motion to kill the program failed by one vote, 216 to 215. Five months earlier, in February 1993, the new President, Bill Clinton, ordered a complete reevaluation of *Freedom*'s design, demanding that the station's budget be cut by as much as two-thirds. Then, in April, he proposed that NASA combine its space station with the Russians', thereby saving money and construction costs while also fulfilling Clinton's foreign-policy goal of helping the Russians.[10]

At this time, the U.S. was already in the initial stages of a small astronaut-exchange program with the Russians. In the summer of 1992, amid the heated presidential campaign between George Bush, Sr., and Clinton, several NASA officials and Bush aides got together and lobbied to negotiate a joint space deal. Bush liked the idea. He had suggested something similar in a summit with Gorbachev the previous year, before the fall of the Soviet Union. To get such a joint program going would not only serve some of his foreign-policy goals, it might help generate votes for him in the election.

The final plan, agreed to in October 1992, called for two astronauts from each nation to be trained by the other nation. One Russian would then fly on a single shuttle mission in early 1995, followed by the launch of one American on a Russian rocket that same spring. The American would spend several months on *Mir*, after which he and his Russian crewmates would return to Earth on the American shuttle. Moreover, the U.S. would pay the Russians $10 million, much like the French, Germans, and Austrians had already done.[11]

For George Bush, the deal failed to help him win re-election. Less than a month later, the November 1992 elections removed him from office—in part because Ross Perot's third party candidacy split the vote—and placed Bill Clinton in the White House.

For Clinton, space exploration by itself had limited appeal. Moreover, Clinton was completely uninterested in using the space program to produce American jobs. Unlike Reagan or Bush, who saw the program as a way of inspiring U.S. commerce, Clinton would have nonchalantly gone along with those Democrats in Congress who wanted to trim the space program to the bone or eliminate it entirely, if the program's only goal was job creation. Instead, he saw other uses for the space program. While Kennedy, Nixon, Reagan, and Bush had all considered the foreign-policy aspects of space exploration important, Clinton accentuated this angle above all else. To Clinton, space was entirely a foreign-aid program, useful for helping the Russians while simultaneously encouraging international cooperation. As he said during the celebration of the 25th anniversary of the lunar landing in July 1994, "Our space explorations today are important models for cooperation in the new post-Cold War world."[12]

So, rather than eliminate Bush's Russian-American space deal, Clinton, for foreign-policy reasons, wanted to expand it, getting it incorporated into the combined space-station program he was trying to get NASA and Russia to work out. By combining the two programs, Clinton could reduce U.S. spending on space and at the same time increase foreign aid to the former Soviet Union.

Ordered by Clinton to rework the space-station program, by June 1993 NASA proposed three possible new configurations. Option A took the latest *Freedom* design and simplified it to try to reduce its construction cost to about $16.5 billion. Option B was the most expensive option, essentially *Freedom* with few changes. Option C was the simplest, quickest, and cheapest concept, decommissioning the space shuttle *Columbia* so that its engines could be used to launch the station's base block. Of the three, Option C required the least in-orbit assembly; its modules would be

launched as complete units, much like *Skylab* and the Soviet
Salyut and *Mir* modules.[13]

After two weeks of thought, Clinton chose Option A, though
he added to it some of the features included in Option B.[14] Then he
upped the stakes by once again insisting that the station be built in
cooperation with the Russians. The name *"Freedom"* was dropped,
informally replaced by the nondescript and bureaucratically neutral name *"Alpha,"* as in Option A[lpha]. Clinton then gave Dan
Goldin, NASA administrator since March 1992, 90 days to work
out the final specifics, including a deal with the Russians.

In Russia, meanwhile, despite the collapse of an entire country, the
space program had managed to keep a continuous human presence
in space since September 6, 1989. If they could keep that streak
going, they could claim that Russia began the permanent human
occupation of space.

At the Institute of Medical-Biological Problems, Valeri Polyakov still lobbied for another chance to prove that a human could
survive in weightlessness for more than a year. After his first mission ended prematurely, the change to a shift system on *Mir* had
stymied his efforts to get another long flight. Furthermore, in the
last few years it had become obvious that the psychological difficulties experienced by earlier long-term crews had more or less
faded. Unlike the *Salyut* stations, *Mir* provided crews with enough
distractions to keep them busy. If they weren't doing repairs or
space walks, they were unloading *Progress* freighters. And the
station's larger size, with its private cabins and multiple modules,
made each occupant's life significantly more tolerable, similar to
the experience of the astronauts on *Skylab*. Krikalev's 10-month
stay in space, for example, was only two weeks shorter than
Romanenko's. Yet, his mental state during his last few months in
space was far less stressed than Romanenko's had been. This success made it harder for Polyakov to convince anyone to let him fly
his 18-month mission.

Then the Americans gave him the ammunition he needed to
get his flight. In the spring of 1992, just before the return of Krikalev
and Volkov and several months before the Bush-Yeltsin joint space
agreement, a handful of U.S. astronauts arrived in Moscow to get a
first look at the post-Soviet space program. As Polyakov gave them
a tour of the Institute of Medical-Biological Problems, one Ameri-

can astronaut asked him some detailed questions about the problems of long space flights.

"Why do you ask?" Polyakov responded.

The American smiled good-naturedly, as Americans are wont to do. "Well, we expect to do year-long flights on our space station *Freedom*. Any information you can give us to make those flights more successful would be helpful."

That night, Polyakov couldn't sleep. Unlike most modern Russians and Americans, he didn't put much faith in the recent efforts to promote international space cooperation. Though he was a kind-hearted man, he liked the idea of competition—healthy, peaceful, good-natured competition. It was that kind of competition that had fueled the landing on the moon, the unmanned exploration of the planets, and an ambitious Soviet-Russian space-station program that had given Russia almost every manned space record. From what the Americans were telling him, however, those records might soon fall. It appeared that the Americans wanted to leapfrog the Russian space program.

The next day, Krikalev and Volkov returned to Earth. Amid the medical debriefings, press conferences, and general celebration, Polyakov had a chance to buttonhole Yuri Semenov, Energia's head since 1989. "You know that the Americans plan to fly astronauts for more than a year, breaking our records?"

"Yes, I know." Semenov noticed that Polyakov seemed upset by this. "It will be a shame to lose our lead in space."

"We shouldn't." Polyakov once again launched into his pitch to fly an 18-month mission on *Mir*. He pointed out that though Titov and Manarov had proved that a man could live in space for a year and recover, they had not proven that a man would then be strong enough to land on Mars and immediately begin work. Polyakov wanted Russia to be the nation to prove that point, and he believed that he was the man who could do it. Though Semenov was proud of his country and wanted it to hold all the manned space endurance records, this argument didn't carry much weight with him. He needed cash to run Energia, and a trip to Mars wouldn't get it for him.

Then Polyakov threw in the kicker. He reminded Semenov of his increasing difficulty attracting foreigners to buy more time on *Mir*. "The interest in space exploration is shifting from us to America," he said. "The Japanese, the Europeans, the Canadians, their idea of joint space flights is only with America. They won't talk to us anymore."

Polyakov had caught Semenov's interest. "Yes? So?"

"If you let me complete my 18-month mission on *Mir* and I am able to walk immediately thereafter, it will bring us prestige and attention. It will show the world the superb quality of our space program, its medical and life-support systems, its engineering, its station and rocket designs. It will get others to talk to us again. It will bring us business."

This time Polyakov had pressed the right buttons. Semenov was clearly intrigued by the idea of using the long mission to sell the Russian space program. He began to pepper Polyakov with questions. What were the risks? What would he do to keep healthy? Was it different than what cosmonauts did now?

Finally, Semenov had to leave, caught in the rush of post-flight activities. As he left, however, he told Polyakov that this time he'd give the idea of a long space flight some serious thought.[15]

As spring moved into summer, Polyakov kept hearing rumors about his proposal, how various high-level government officials besides Semenov were interested, and even approved. He even heard rumors about how Yeltsin himself had endorsed it. Then, in the summer of 1992, at the same time the Russians were negotiating the Bush-Yeltsin deal on the first shuttle-*Mir* crew exchange, word finally came down: Polyakov's mission was approved. The possibility that it could bring both national prestige *and* additional cash to Russia had convinced both the government and Energia to give it their sanction. Moreover, after much discussion, they had decided that only Polyakov was qualified to fly it.

Even so, accomplishing Polyakov's full 18-month mission was still touch and go, even years before launch. The just-signed Bush-Yeltsin joint space agreement essentially created a finish date for the long flight, determined not by science, space, or Polyakov's abilities, but by finances. According to the agreement, the American was to arrive on *Mir* sometime in the spring of 1995. Because the U.S. deal included a $10 million payment and the use of a shuttle to complete one *Mir* crew shift—eliminating the need for one Russian rocket and thereby saving a further fortune in cash—Polyakov, as the third Russian on board *Mir*, would have to vacate his spot when the paying American customer arrived.[16]

Thus, even if all went well, Polyakov's dream mission would have to end sometime in the spring of 1995. Only if it could start, as scheduled, in November 1993, would he have the time to complete a year and a half in space.

Polyakov immediately went into training. While there was little he could do to shape the politics of his space flight, he could certainly get himself ready to fly when the time came.

Above, *Mir* continued its unceasing journey around the earth. In the year and a half after the return of Sergei Krikalev, four crews came and went, including two Frenchmen flying on two separate missions. The station, now more than seven-and-a-half years old, had completed more than 40,000 orbits and traveled just less than a billion miles, far enough to have journeyed to and from Mars once. Though aging and requiring increasing maintenance, its systems still functioned reliably.

The first crew after Krikalev's return had begun the repair on *Kvant-2*'s four failed gyros by doing a space walk and slicing through the module's exterior thermal blankets so that future crews had access to the old gyros. They also successfully tested a new docking system, called TORU, that the Russians hoped could replace the Kurs system in use since *Mir*'s launch. The factory that built the Kurs system was located in the Ukraine, now an independent country. With the breakup of the Soviet Union, this factory had been demanding higher and higher fees for each new system. TORU, built in Russia, was essentially a remote-control system for firing the thrusters on the *Progress* freighter, using a joystick panel. Television cameras on *Mir* and the freighter gave the pilot visual cues, while his partner used a laser gun to obtain distance and speed. In cases where the Kurs system was also available, the pilot could use its radar to gather distance and speed information too.[17]

The second crew, including Anatoli Solovyov on his third space flight and second long-term occupancy of *Mir*, installed the thruster engine on the top of the Sofora girder. Brought on board by the first *Progress* freighter to dock at the aft port since its repair, the installation took three space walks. On the second excursion, Solovyov and his partner Sergei Avdeyev, under orders from the *Russian* government, removed the Soviet flag placed there by Artsebarski and Krikalev. After more than a year in space, the last flying flag of the Union of Soviet Socialist Republics was finally lowered—tattered and almost completely destroyed from exposure to the harsh space environment.

The third crew was the first to dock their *Soyuz-TM* spacecraft at the *Kristall* docking port originally intended for Soviet space shuttles. This crew installed the new gyros on *Kvant-2* and replaced the station's air-conditioning system. They also installed drive motors to the solar-panel mounts on *Kvant* that Manarov and Afanasyev had installed in 1991. When the panels were finally transferred, these motors would keep the panels facing the sun. However, after they installed the first motor, they discovered that one of Strela's hand cranks had come loose and floated away. Without the crank, they couldn't operate the crane, and could not install the second motor. They had to wait two months until the next *Progress* freighter arrived with a replacement hand crank to finish the installation.

By this time, the fiction that the Russian crews were there to make money from their *Kristall* furnaces had been completely abandoned. The men continued to do a few smelts and tests, but since there was no practical way of getting most of their results back to Earth (the *Raduga* capsule, while useful, had proved too small to be profitable), the experiments were done mostly for purely scientific reasons. Instead, the Russians more and more accepted the role of maintenance men, keeping the station running so that Energia could sell *Mir* rental time to researchers and politicians from other countries.

The fourth crew, Vasili Tsibliev and Alexander Serebrov, completed five space walks, during which they built a second test girder on *Kvant*, about a third the length of Sofora, and did a careful and complete inspection of the station's entire exterior.

These two men were a study in opposites, with Serebrov one of the program's most experienced cosmonauts and Tsibliev one of its newest. Serebrov, 49 years old, had completed his first space flight more than 10 years earlier, escorting Svetlana Savitskaya on her first mission to *Salyut 7*. A clever and passionate engineer, he had helped design *Salyut 6*, *Salyut 7*, and many of *Mir*'s components. On *Salyut 7* he redesigned the ventilation fans so that one could do the work of seven, thus reducing the station's noise and electricity use. He also redesigned *Salyut 7*'s cockpit seats, replacing the silly "couches" with bicycle-type seats devised for weightlessness. On *Mir* he designed its cabins and its kitchen table where food was heated. He also helped build the Icarus manned maneuvering unit, and when he and Viktorenko had reoccupied and reactivated *Mir* in

September 1989 after the station had been mothballed, he had been the first man to fly it.

An emotional and outspoken man, Serebrov liked to joke that he had really only been born in April 1983, when he had been the flight engineer on the near-collision between *Soyuz-T 8* and *Salyut 7*. Later, when crew assignments had been juggled because of the new medical restrictions after Vasyutin's prostate illness on *Salyut 7*, he had refused the job as flight director when Ryumin moved up to head the *Mir* program. "I am an engineer, a scientist. I don't want to be a boss, ordering people about and punishing them."[18]

Tsibliev was 10 years younger than Serebrov and a rookie. Tsibliev's determined effort to get into space was reminiscent of many other spacemen, in both Russia and America. Abandoned by his father at a young age and growing up in a tiny village on a remote collective farm, Tsibliev had dreamed of becoming a cosmonaut since the day Yuri Gagarin flew in space. Though a qualified fighter pilot in the Soviet air force, he was repeatedly rejected when he applied to the cosmonaut corps. His first application was turned down because he was too young and inexperienced. He then tried to become a test pilot and was rejected five straight times, with the last rejection telling him he was too old and should not bother applying again.

In 1984 he heard that the cosmonaut corps was looking for new applicants. He applied again, and was rejected again. Again he didn't take no for an answer and reapplied, finally getting accepted in 1987. Six years later, he was finally in orbit.[19]

With angular features and dimpled cheeks on a large, square face, Tsibliev was in many ways far less schooled than other air force cosmonauts. He had never been a test pilot or a flight instructor. Neither was he an engineer like Serebrov, nor a scientist like Polyakov. His experience was solely that of fighter pilot, flying jets on patrol near the East German border or above the Black Sea.

When Serebrov and Tsibliev lifted off on July 1, 1993, their mission called for them to be in space for a little longer than four months, returning in November. Polyakov would then arrive with their replacement crew to begin his marathon 18-month mission.

Even as these two men began their four-month mission, Dan Goldin and other NASA officials were holding extensive meetings with Yuri Koptev, the Director of the Russian Space Agency, and

Yuri Semenov, General Director-Designer of Energia. Goldin was under intense pressure to fulfill Clinton's demand to incorporate the Russians into the American space-station program, while Koptev and Semenov were faced with a serious lack of capital.

Taking a radical approach to save their space program, Koptev and Semenov proposed that NASA use the Russian space industry as its prime contractor for building space station *Alpha*. They provided Goldin with a 137-page English-language report, describing how NASA could save $2.5 billion by simply paying the Russians to launch *Mir-2*, using *Soyuz* spacecraft as ferries, *Progress* freighters as cargo ships, and the Baikonur Cosmodrome as the launch site. As the report bluntly noted, "The contribution of the Russian side consists of resources on the *Mir* orbital station, promising space technologies, as well as [their] resources on the international station. The contribution of the U.S. side consists of *monetary allocations to the Russian side*. [italics added]"[20] The Russians would build the hardware, while the Americans would pay for it.

For Goldin, the idea of using some Russia hardware instead of building everything anew in the United States was very appealing. However, replacing the American space station entirely with a Russian one was going just a bit too far. Instead, he proposed a more equal division of labor. The Russians and Americans would build different modules, each country contributing what it did best to the overall station. The Russians would provide the tankers, the lifeboat, and the initial habitable module (*Mir-2*), while the Americans would use the space shuttle to haul into space the larger American modules that would follow.*

Though many specifics still needed to be worked out, the basic outlines soon became clear, and on September 3, 1993 in Washington, D.C., Russian Prime Minister Viktor Chernomyrdin and Vice President Al Gore sat down together and signed the preliminary

---

*Interestingly, by giving the Russians the task of providing the tankers, lifeboats, and initial habitable space, Goldin essentially made their participation mandatory for building the station on time. Without them, the International Space Station simply could not be maintained and occupied. Goldin did not address this fact in his testimony before Congress, however, repeatedly stating instead that "we have an alternative plan if the Russians withdraw." (See United States Congress, 103-2, House, Hearing, May 17, 1994, 4, 47-48, 54, 59-60, 86; United States Congress, 103-2, Senate, Hearing, June 7, 1994, 729, 746-747, 762, 799.)

space agreement worked out by Goldin, Koptev, Semenov, and their subordinates.[21]

The merged American-Russian manned space-station program would follow a three-phase schedule. In Phase 1, the United States would pay Russia $400 million for the use of *Mir* as a training facility, using the shuttle to fly more American astronauts to the station to teach them the tricks and foibles of long-term space missions as well as showing both the Russians and Americans how to work together. The payment meanwhile would give the Russian space program an influx of desperately needed cash, while also funding the extension of *Mir*'s life by three to five years.

Phase 1 essentially incorporated and expanded upon the original Bush-Yeltsin agreement. The first American flight to *Mir*, scheduled for spring 1995, taking off on a Russian rocket and returning on the shuttle, would initiate the shuttle-*Mir* dockings, to be followed by several years of a continuous American presence on *Mir*. And instead of only one Russian flying on a single shuttle mission, two Russians would fly on two separate shuttle flights.

Phase 2 of the joint program would have the two nations combine their resources to build what was now dubbed the *International Space Station*, using the partly built components from the never-completed *Freedom* and *Mir-2* stations to assemble a larger space station for less money. To save money, the U.S. would initially use Russian *Soyuz-TM* spacecraft as lifeboats, and the Russian modules for living quarters. They would also depend on *Soyuz-TM* and *Progress-M* spacecraft to both refuel the station and boost it when its orbit decayed.

Phase 3 would have the station expand, incorporating the modules and components designed and built by Europe, Japan, Canada, and the other foreign partners.

While these negotiations took place on Earth, the mission of Tsibliev and Serebrov rolled on in space. On August 12 and 13 they rode out a particularly intense Perseid meteorite shower. Placed on 24-hour alert, they watched as about 240 meteors burned up in the atmosphere *below them*. Tsibliev reported "battle-wounds" to ground control, while Serebrov described how the solar panels changed color from dark to light blue wherever a micrometeorite hit them. They also spotted 10 new small impacts in the station's windows, including one crater about an inch across. During the height of the storm *Mir*'s sensors indicated a two-thousand-fold increase in particle flux in the atmosphere surrounding the station. Once, they watched one large meteor burn its way across the northern horizon for more than two seconds. When the storm ended on August 14 both men expressed "relief" in their radio conversations with Earth.[22]

Though the station survived without serious damage, a thorough inspection of its exterior seemed essential. Moreover, NASA engineers had requested this inspection out of concern that *Mir*'s aging condition was not up to the shuttle-*Mir* missions scheduled from 1995 to 1997. After the Perseid shower, mission controllers decided that instead of just one inspection space walk Tsibliev and Semenov would do two, carefully scrutinizing every inch of the station's exterior and solar panels.[23]

The ups and downs of the cash-poor Russian space program continued. One day after the shower, a *Progress* freighter docked with the station. When the men opened the hatch, they were surprised at how little was there. To Serebrov, it seemed that the cargo ship was less than two-thirds full, or as he told mission control, "Too empty."[24] The little-utilized cargo space held, amid the usual packets of letters, supplies, fresh food, the disassembled 16-foot-long test girder that the two men would build on the exterior of *Kvant* next to the Sofora girder. Made of the same nickel-titanium alloy used by Sofora that "remembered" its original shape, the girder was a test of an essential component planned for the *Mir-2* station. Over the next month they unloaded supplies and equipment and shifted the girder to the *Kvant-2* airlock. Then, in a three-week period beginning in the middle of September, they completed three space walks, the first two to install the girder and the third to begin the *Mir* inspection.

The girder assembled easily. Unlike Sofora, all the men had to do was activate the heaters in the new girder's joints, which then unfolded to their "remembered," straight position and hardened, expanding the girder to its full 16-foot length in less than three minutes.[25]

The first inspection space walk, on September 28, 1993, began less successfully. The cooling system in Tsibliev's spacesuit failed, forcing him to stay close to the *Kvant-2* airlock while Serebrov inspected the rest of *Kvant-2*'s surface. As Serebrov crawled across its surface, noting small meteorite impacts and worn insulation blankets, he complained that he, the first man to fly the Icarus jetpack, couldn't use it to facilitate the inspection. To his frustra-

tion, Semenov had refused to authorize the launching of new batteries for Icarus. Their weight would have precluded other cargo,
which in turn would have cost the program money it did not have.
For the budget-minded chief designer, the hands and feet of a space-
walking cosmonaut were an able and less expensive substitute.[26]

Missing batteries were not the only thing that Serebrov had to
complain about. Just ten days later, during their wake-up conversation with mission control, Semenov came on the radio to tell
the two men that their return to Earth and the start of Polyakov's
marathon mission had been postponed from November until early
January. There were no rockets available to launch their relief
crew. The Russian government, lacking funds, had failed to authorize money for their construction, while the factory that made
them was more than 14 billion rubles in debt and bankrupt.
Though one rocket was already stockpiled in Baikonur, it couldn't
be used for Polyakov's mission because it was configured for
launching a weather satellite and was able to carry only two men,
not three.[27]

Serebrov ruefully told the head of Energia that he and Tsibliev
understood the decision and were, of course, willing to remain on
Mir.[28] In truth, Serebrov had been expecting this news even before
he left Earth. "It wouldn't be the first time," he had thought cyni-
cally. In fact, two weeks before Semenov came on the radio, about
the time of the first inspection space walk, Serebrov warned
Tsibliev. "They are going to ask us to stay in space longer. Be prepared for it."

What convinced Serebrov were the events occurring that very
moment in Moscow. From what he could tell, listening to his ham
radio each night, the improvised Russian government set up after
the August 1991 coup was about to fall apart.

In the confused legal situation that existed in Russia after the
fall of the Soviet Union, no one was quite sure who was in control
of what. According to the constitution that had been written during communist rule, the Supreme Soviet of Russia was supposed to
be in charge, though in reality it had always been subordinate to
the Communist Party. When the Party disappeared, that commu-
nist-written document became the only guide for running the Russian government. The Supreme Soviet took over, acting as the de
facto Russian congress with Yeltsin, the elected President, acting
as chief administrator.

The Supreme Soviet, however, had been voted into power before the August 1991 coup under rigged rules set up by the Communist Party, and was made up mostly of former communist
apparatchiks. In the two years since the coup, these men had slowly
realized that they could use the power of the legislature to over-
turn any reforms or changes that Yeltsin proposed. They could even
use that power to reestablish the communist-ruled Soviet state.

A "War of Laws" ensued, in which every proclamation or ruling announced by Yeltsin was countermanded by new legislation
from the Supreme Soviet. From January to July, 1993, Yeltsin signed
more than 1,150 decrees, matched by the more than 1,200 orders
declared by the congress and its chairman. The conflict grew so
heated that at one point Yeltsin tried vainly to suspend all congressional laws, followed shortly thereafter by a failed Supreme Soviet
attempt to impeach him.[29] By the middle of September 1993, the
rift had become hopeless. In desperation, and with great doubt
about whether he was doing the right thing, Yeltsin decided to dis-
band the Supreme Soviet and call for new elections in December.

That same day, Viktor Chernomyrdin returned home from his
meetings with Al Gore and proudly reported to Boris Yeltsin that
he had gotten the Americans to finance the Russian space program.

Yeltsin listened silently, and then, as he later wrote, "I quickly
brought [Chernomyrdin] back to planet Earth." Though Yeltsin supported the new space deal, it ranked very low on his list of priorities. Far more important to him was to establish in Russia some
form of stable government, a "law-based state" as he liked to call
it. He described his decree disbanding the Supreme Soviet to his
prime minister, which Chernomyrdin signed without hesitation.[30]

The next day, September 21, 1993—the day after Tsibliev and
Serebrov's first space walk—Yeltsin made his announcement on
television, calling for new parliamentary elections in December to
replace the Supreme Soviet of the communist era with a Federal
Assembly. The leaders of the Supreme Soviet reacted in kind. They
declared his action a "coup d'etat" and stripped Yeltsin of power.
Then they appointed their own acting President, and called on "soldiers, sailors, militia, and state security agencies" to defend the
parliament from "the criminal encroachments of B.N. Yeltsin and
his clique!"[31]

For the next 10 days, a motley crew of 400 to 800 mercenary
soldiers, former KGB thugs, and communist apparatchiks longing

for the return of Soviet rule held their ground in the White House, refusing to leave. Periodically there was violence (reporters were beaten up, an old woman was accidentally killed by a stray bullet, a police officer died when he was pushed into traffic).

Then, on October 3, a mob of 10,000 to 15,000 people stormed the cordon that Yeltsin had established around the White House, shouting communist slogans and demanding that Yeltsin be hanged. Among them was Oleg Shenin, one of the leaders of the August 1991 coup, who told reporters "No more talk about elections or referendums. We need to establish a tough state power."[32]

Amid rifles, submachine guns, grenade launchers, and red flags, the last few members of the Supreme Soviet, which had dwindled to only about 150 hard-line communist members, sent several military trucks filled with armed men to take over the Ostankino television center in the northern part of Moscow. The trucks crashed through its glass lobby doors, grenade-launchers firing. Within hours the building was captured. By the next day, October 4, 1993, with more people dead, the television center under mob control, and with more violence threatening, Yeltsin decided that he had no choice. Taking the advice of his security officials, he ordered troops to re-secure the White House, using force if necessary. A dozen tanks were positioned in front of the building, while military commanders repeatedly asked the men inside to surrender. Periodically snipers inside the building exchanged fire with the troops on the ground.

By October 5, with the bottom two floors reoccupied by Yeltsin's forces but the remaining twenty-plus floors still held by snipers firing wildly out windows, the tanks fired at the upper floors, setting the top half-dozen stories on fire. Within a few hours, the last communist resisters surrendered, holding white flags as they exited the burning legislative building—the very same building that Yeltsin had defended two years earlier when he climbed onto a tank and shouted that "terror and dictatorship . . . must not be allowed to bring eternal night." All told, slightly fewer than two hundred people were killed and more than a thousand injured during the revolt of the Supreme Soviet.[33]

Above, on *Mir*, Tsibliev and Serebrov said little publicly about the violence in Moscow. Serebrov, though, was hit especially hard by the turmoil. As prepared as he had been for the postponement of his return, and as much as he agreed with Yeltsin's actions, he was not

prepared to find out by ham radio that one of his oldest friends had
been killed in the crossfire at the Ostankino television tower. "They
raced him to the hospital," he remembered, his voice breaking. "It
was too late." Even today the memory brings tears to his eyes.[34]

Like other cosmonauts before him he refused to give in to despair. He and Tsibliev proceeded with their work, preparing for their
last space walk. However, when they went out on October 22 to
spend five hours checking the surface of *Mir*, the oxygen system in
Serebrov's suit failed almost immediately, forcing them to cut short
their walk and flee back inside the station after only a half-hour.

Before they went inside, they had a conversation with Viktor
Chernomyrdin, who was visiting mission control at the time.
Chernomyrdin explained that he had just come from meetings with
Semenov and other Energia officials, promising them that, despite
the violence of the past month in Moscow, the government had not
forgotten them, and that the money would be allocated to keep the
program going. Tsibliev and Serebrov would not be stranded in
space, and a rocket would definitely be available to send a replacement crew, including Polyakov, in January.[35]

For Serebrov it was a bitter conversation. He had lost a friend
and, trapped in space, could do nothing about it. He listened, made
the appropriate sounds of agreement, but in his heart, all he wanted
was to get home to his wife, his family, and his friends. He would
take Chernomyrdin's promises seriously only when he was back
on Earth.

One week later, the two men tried another space walk, beginning their excursion by jettisoning Serebrov's failed spacesuit. They
outfitted its arms so that it appeared to be saluting and pushed it
from the station. Then they traveled across the face of *Mir*, filming
its solar panels for ground analysis and inspecting the base of the
Sofora girder to see how it was doing after two years in space. At
one point they were startled when an unidentified piece of space
debris flashed past them. Neither was able to get a good enough
look to identify the metal shard and figure out where it came from.

Their inspection showed that the station had suffered more
than 60 small impacts from micrometeoroids—the largest punch-
ing a 4-inch hole through one solar panel. Moreover, *Mir*'s thermal
blankets were coated with a fine soot, thought to be exhaust from
the station's attitude jets.[36] The solar panels, though aging, were
actually in reasonably good condition.

Two days later, November arrived. Instead of coming home, all the two men could do was pack experiment results into the *Raduga* capsule of their second *Progress* freighter, then watch on November 21 as both disappeared over the horizon, the *Raduga* capsule returning safely to Earth and the *Progress* freighter burning up in the atmosphere.

On the ground, Polyakov was just as helpless. Instead of going into space, he could do nothing but continue his training, the final stages of which were not very pleasant. His backup, Gherman Arzamazov, resented the fact that Polyakov was going to fly twice in a row. As the back-up doctor during Polyakov's 1989 mission, he should have been automatically assigned to the prime crew of the next medical flight, as was customary. In a sense, what had happened to Polyakov when Atkov replaced him in 1984 was happening to Arzamazov now. However, the head of Energia did not promise Arzamazov that he would fly next as Glushko had promised Polyakov then. After more than five years of training as Polyakov's backup, Arzamazov decided he couldn't accept the situation any longer. In November he complained to his bosses, claiming that Polyakov wasn't ready for the flight and that he should go instead. "He was tired," Polyakov remembered. "He really wanted to fly in space, and didn't want to wait any longer."

The problem was that the 18-month mission had been approved at the very highest levels of the Russian government, and *that* approval had hinged on having Polyakov, and only Polyakov, fly it. "It was an extremely important flight for the nation," noted Polyakov years later. "They did not want to trust it to a rookie." Though Arzamazov was officially Polyakov's backup, he would not have replaced Polyakov under any circumstances. If Polyakov got sick or was injured, the mission would have been canceled.

Arzamazov's complaint not only didn't get him into space, it got him ejected from the program. According to Polyakov, his protests revealed the same kind of "psychological unsteadiness" that the program had seen with men like Lebedev, Volynov, and Zholobov. Arzamazov was removed as Polyakov's backup, essentially ending his career as a cosmonaut.[37]

On December 17, 1993, Polyakov and his mission were officially revealed to the public with the launch set for January 8, 1994. If all went well, he would begin his journey in space on that date, remaining aloft for between 14 and 16 months, depending on when the first American arrived on *Mir*.

Four days earlier, the Russian space program earned $6.8 million dollars in the auction at Sotheby's.[38]

Ironically, by the end of 1993, the Russian space industry was in far better condition than the rest of the Russian economy—even if no one realized it yet. The low labor costs in Russia allowed the Russian space enterprises—among the few industries from the former Soviet Union to have an excellent reputation—to underbid any other space company in the world. It was this reputation that gave Semenov and Koptev a good and credible negotiating position when they offered to build the entire *International Space Station* for NASA.

By the end of 1993, privatized design bureaus like Energia and Khrunichev were earning hundreds of millions of dollars, despite import quotas limiting the number of U.S. satellites launched on Russian rockets to only eight per year. In December 1993, for example, Lockheed, Loral, Energia, and Khrunichev teamed up, contracting to build and launch five telecommunications satellites over the next few years, earning a combined $250 million on the deal. The U.S. companies would build the satellites and the Russian companies would launch them. This partnership has since matured into what is today known as International Launch Services, providing Proton, Atlas, and Titan rockets for anyone who wants to put a satellite into orbit.[39]

Moreover, during 1993 the Russian space industry had already earned $150 million by launching different foreign satellites. For example, in August the Russian Space Agency was paid a fee to put a small, experimental Italian-German weather satellite into low Earth orbit using a Ukrainian-built rocket.[40]

Then on December 16, 1993, five days after the Sotheby's auction and one day before Polyakov's mission was officially announced to the public, Dan Goldin and Yuri Koptev signed the final agreement outlining the details of the American-Russian joint manned space-station program. Phase 1 would begin in March 1995, bringing the first of at least five American astronauts to *Mir* for periods of four to six months.

The significance of these events was that, not only was the Russian space program finally beginning to see revenues but, unlike any space endeavor of any other country on Earth, Russian space activities were actually beginning *to make a profit from space*. While revenues in 1992 had been only $75 million, by the end of 1993 they had increased to more than $300 million, with even more possibilities on the horizon.[41]

As predicted by many American pundits, including Ronald Reagan, the future of space was in private and free enterprise. The irony was that as the twentieth century crept towards its conclusion, it was in the former Soviet Union that this prediction was proved true.

On January 8, 1994, *Soyuz-TM 18* lifted off from Baikonur. Its commander was Viktor Afanasyev, the tough-looking fighter pilot returning to space three years after his mission with Musa Manarov. The flight engineer was rookie Yuri Usachev, an Energia engineer like Krikalev, one of the young, new generation of Russian cosmonauts ready and willing to work with foreigners. Rounding out the crew was Dr. Valeri Polyakov.

Moments before launch, Polyakov's thoughts were far different than those on his first flight. Then he had felt eager, excited, and joyous about finally getting into space. Now he felt only fear.

He wasn't afraid of dying. Far from it. What he feared now more than anything was failure. "What if something goes wrong?" he asked himself. This mission had been his goal since the day he decided to go into space almost 30 years earlier. In fact, his obsession had marshaled the entire Russian government into financing the mission, during one of the worst crises in Russia's thousand-year history. Like a litany, Polyakov couldn't help repeating to himself what had been done to make the mission happen. "I had sacrificed so much time," he thought. "The government has spent so much, more than they can afford. And I've learned so much for myself, for them.

"Better to die if something went wrong," he thought. "Better if I had a gun to shoot myself."

Then the rocket ignited, and lifted smoothly into the air. For a moment he relaxed, recognizing the feeling of acceleration from his previous flight. Then the fear returned. "It's moving too smoothly," he worried. "Something's wrong."

By the time they reached orbit and were weightless, Polyakov had relaxed, aware once again of the ebullient, free sensation of floating in weightlessness. "Only when a cosmonaut is in space does he really know freedom," Polyakov explained, years later. Until then, everything is boring preparation, catering to the demands of others in order to keep them from kicking you off the program. In space, Polyakov was free, finally doing the thing he had prepared his entire life for. [42]

For six days he and Afanasyev worked with Vasili Tsibliev and Alexander Serebrov. Then Tsibliev and Serebrov waved good-bye and closed the hatch on their *Soyuz-TM 17* spacecraft. With Tsibliev the pilot and Serebrov handling the camera, they backed the spacecraft away from *Mir*'s bow port. Instead of doing a routine fly-around inspection of the space station, the plan was for Tsibliev to stop *Soyuz-TM 17* when it was about a dozen feet from the bow port, then slide the spacecraft sideways down the length of *Kristall* so that Serebrov could get photographs of its androgynous docking unit. Because this was the port that the American space shuttle would use in 1995, detailed photographs of the unit were needed to give American shuttle pilots an accurate idea of what it looked like before they attempted a docking.

For Serebrov, this was his last view of *Mir* and the culmination of his career in space. After 16 years as a cosmonaut and four space flights, he was retiring from active duty. For Tsibliev, this moment was the completion of a very successful maiden voyage, including six months in space and five space walks. After years of waiting, his future as a spaceman seemed bright and exciting.

As Tsibliev began guiding *Soyuz-TM 17* along *Kristall*'s length, with Serebrov in the orbital module snapping photograph after photograph, he suddenly discovered that one of his control joysticks, used to adjust his sideways drift, was not working. No matter how hard he pushed, *Soyuz-TM 17* was creeping closer and closer to *Mir*.

Worse, Tsibliev was having trouble reading his control panel instruments. The main cabin light bulb was out, and rather than fix it beforehand, Tsibliev had ordered Serebrov to open the window covers on the sunlit side of the station. Unfortunately, the bright sunlight made things worse—the contrast between light and shadow was blinding.

Inside *Mir*, Afanasyev, Polyakov, and Usachev waited in their own *Soyuz-TM* spacecraft, listening in disbelief as Tsibliev and Serebrov scurried to try to prevent a collision. While Tsibliev frantically tugged at the controls, Serebrov flew from orbital module to descent module, trying to find something on the instrument panels that might be causing the control failure.

Suddenly, the ship's speed seemed to increase, hurtling them directly at *Kristall*. "We heard a horrible noise," Serebrov remembered. "I was sure we had crashed against the station."

After a pause, the spacecraft slowed, and then moved away, as if it had collided with *Mir* and bounced off.

On *Mir*, however, the crew felt nothing, no jolt, no bang. All that changed was the abrupt shutdown of the station's attitude-control system, triggered when the system realized that the gyros could not handle the sudden change in attitude caused by the impact.

Polyakov was convinced that no contact had been made, that the jet exhaust from the *Soyuz-TM*'s nozzles had acted as a cushion, preventing the spacecraft from ever touching *Mir*, while pushing the station sideways and causing its gyros to shut down. Serebrov was convinced that the antennas on *Soyuz-TM 17* had touched the station, bent from the contact, and then acted as springs to push the spacecraft away.

As the spacecraft drifted away he rechecked one of the capsule's switches that controlled the thruster controls, and discovered that, though it looked as if it was in the correct position, it hadn't clicked into place. As soon as he flipped the switch off and then on, the steering system came to life, and Tsibliev was quickly able to take full control.

Both men strained to see if there was any damage to *Kristall*, but their position made it impossible to get a good look at *Mir*. With their fuel limited, they had no choice but to head back to Earth. A later crew would have to find out if the damage was serious.

On *Mir*, the men cautiously came out of their *Soyuz-TM* lifeboat. As far as they could tell, everything was working as it should. There were no air leaks, and the station's temperature seemed stable. After a few minutes the station's gyros took control again, and the station reoriented itself as if nothing had happened.[43]

Polyakov went to the windows and watched as *Soyuz-TM 17* disappeared below him into the backdrop of Earth's white clouds, blue ocean, and brown continents. For the next fourteen-and-a-half months, this would be the closest he would get to Earth.

Going to Mars

For Sergei Krikalev, the experience was far more alien than spending 10 months in space. Everywhere he went, people smiled at him. When he drove a car down the road, the worker who waved him past a road construction site gave him a smile. When he went into a store to buy food, the cashier grinned at him. Even when he passed a security checkpoint, the guard examined his ID card, then smiled, and waved him through.

In Russia, no one smiled—unless they had good cause. To smile nonchalantly for little reason was considered rude, superficial, and a put-on. But Krikalev was in America now, and in America everyone smiled, all the time, for the slightest reason. All his co-workers at NASA seemed to be laughing and grinning incessantly. It made him uncomfortable.

If, just three weeks earlier, someone had suggested to Krikalev that he would spend the next three years in the United States, he would have thought they were crazy. And yet here he was. And here he would stay, surrounded by relentlessly cheerful people, for the next three years.

In 1992, during the Bush-Yeltsin negotiations establishing an astronaut-cosmonaut exchange program, NASA officials had suddenly put the high-level Russian politicians across the table from them in an embarrassing position. According to Krikalev, ". . . the Americans had listed several names of who they wanted to fly on *Mir*, and asked who the Russians intended to send to the shuttle." No one in the Russian delegation had considered the issue that deeply. Moreover, the negotiators were politicians and diplomats, not members of the Russian space program, and really had no idea whom to propose. Frantic not to lose face, they quickly pulled out of thin air the names of two of their best-known cosmonauts: Krikalev, the so-called "last Soviet citizen," and Vladimir Titov, co-holder of the record for the world's longest space flight.

Being picked in this ad-hoc manner caused Krikalev some awkwardness. Not only were he and Titov given no say on whether they wanted to go, the decision had been made without input from anyone at Energia, not even Ryumin, the man who usually made such selections. Later, Ryumin called Krikalev, wondering if he had played political games to get picked. He had not, but Ryumin was still irritated that he had not been involved in the decision. Furthermore, Krikalev and Titov were given only two weeks to get ready. By November 13, 1992, they were in Houston, with their families following one week later.[44]

"Too fast," Krikalev remembered. "Too fast." While *his* English was rather limited, his wife Elena, an Energia engineer, spoke none at all. Moreover, they and their three-year-old daughter were first lodged in a suburban apartment, where the typical resident used a car for the simplest of errands. The Russians, however, were used to living in an urban setting in Moscow, where they walked to everything or could take the metro or bus. For the first few months at least, Krikalev's wife spent most of her time at home, isolated and alone. "We could not go anywhere," Krikalev recalled.[45]

Krikalev often found American life downright jarring. At one point, soon after arriving in Houston, he got on an elevator with another American. As the NASA engineer pressed the button (he was going only one floor), he turned to Krikalev, smiled, and said, "How ya doing?"

Krikalev, still struggling with a new language, felt sudden panic. "How am I going to answer his question fairly," he thought, "in a language I hardly know, in the short one or two minutes before we get to his floor?"

Nonetheless, Krikalev tried, stammering out sincerely how much he liked America, how much he appreciated everything everyone was doing for him. He started to try to describe his apartment. The American watched with a grin. Then the elevator stopped at his floor, and with a hearty, "See ya later," got out, leaving Krikalev alone and in mid-sentence.

"I soon realized that Americans weren't really expecting an answer," Krikalev explained. "In Russia, you only ask a question like that when you really want to know. Otherwise, you are considered rude. In America, however, it is merely a form of superficial greeting."

Krikalev noticed other things, both good and bad. He was amazed at the quantity and quality of food available in any supermarket. He was also appalled at Americans' eagerness to eat in cheap fast-food restaurants. "In Russia, it is hard to get good ingredients, but when we do, we try to make good food with it. I was astonished how Americans take good ingredients and combine them so badly."

To Krikalev, as well as the many other Russians who came to the United States as part of the joint space program, the most negative aspect of American life was what they considered its almost superficial and shallow friendships. While Russians considered friendship intensely important, and spent years developing trust before they were willing to call someone their friend, Americans could say howdy to each other, drink a beer, and consider themselves lifelong buddies.

The happy-go-lucky grins and good-natured and easy friendships seemed to Krikalev irritating, false, and artificial. "Are they putting on a front?" he asked himself. It also made it hard for the Russians to take Americans seriously, seeing them instead as ever-grinning clowns, to be laughed at.[46]

However, for engineer Krikalev the chance to fly on the American shuttle made all the cultural challenges worthwhile. Here was another kind of space vehicle, the first that was even partly reusable and the first able to come back from orbit and land on a runway. "I think every pilot who flies on some kind of plane wants to fly on another kind of plane," he explained.[47]

Krikalev adapted well. Training was remarkably similar to how things were done in Russia. It included the same kinds of simulations, the same kinds of technologies (albeit more sophisticated), the same kinds of exercises, and the same kinds of procedures.

Ironically, he found NASA's way of tightly scheduling every second of an astronaut's time far more bureaucratic than anything he had experienced in his own country. In Russia, ground controllers made major scheduling decisions, such as when space walks and *Progress* dockings would occur, but left the more detailed day-to-day planning to the cosmonauts themselves. Twenty years of running three-month to six-month missions had taught them, as had *Skylab*'s longest mission taught NASA in the 1970s, that it was unwise to try to plan a spaceman's day too tightly.

American shuttle missions, however, were short, rarely more than two weeks long. With so little time in space, mission controllers maximized efficiency by dictating the actions of every astronaut for every second of every mission. As Krikalev's own American shuttle commander admitted, "You might think that in Russia . . . [cosmonauts] are pretty rigidly controlled. Such is not the case. They have a lot more freedom than we do in deciding what goes on."[48]

Finally, after 15 months of hard training, Krikalev joined five Americans for an eight-day mission. On February 3, 1994, the space shuttle *Discovery* blasted off, its mission to release three different satellite packages: a small, experimental science satellite built by the University of Bremen in Germany, a recapturable test factory for manufacturing thin and very uniform films in weightlessness, and a cluster of six small spheres, ranging from two to six inches in diameter, so that ground radar stations could test and calibrate their equipment. As one of two mission specialists in charge of using the shuttle's robot arm, Krikalev's main technical assignments were to release the German satellite and recapture the test factory. Unfortunately, he was not able to do the second task. Technical problems prevented his crewmate Jan Davis from releasing the factory in the first place.[49]

As usual, he was also expected to participate in several political public-relations gestures. On the mission's sixth day the American television show *Good Morning America* put together a link between the ground, the shuttle, and *Mir*, so that the two crews could talk to each other and to the ground. Krikalev made contact. "I hear you loud and clear," he said in Russian. "Can you hear me?" As he spoke, a *Good Morning America* interpreter translated his words into English.

On *Mir*, the three Russians broke into laughter on hearing the English translation. Afanasyev opened his mike. "Sergei," he asked innocently in Russian. "Why are you speaking English to us? Have you forgotten Russian?"[50]

The next day, Krikalev and the shuttle got a telephone call from Russian Prime Minister Viktor Chernomyrdin. At one point, the Prime Minister also joked how the day before he had heard Krikalev speaking only English to the Russian cosmonauts on *Mir*, and was wondering if he had "forgotten" his native language.

Krikalev, ever cautious when speaking with or about politicians, carefully corrected the Russian Prime Minister, explaining that it was the English interpreter Chernomyrdin had heard. He then added that the Americans on *Discovery* were even learning a few Russian words. "In fact, they can pronounce many words without an accent," he explained with a straight face.[51]

Even as Krikalev was learning the American way of flying in space, Valeri Polyakov's 14-month-plus mission on *Mir* was getting up to speed. Following the collision on January 14, ground controllers had vacillated about whether they should have Afanasyev and Usachev do an unplanned space walk to examine *Kristall* for any damage.

Before they made a final decision, however, the crew had to transfer their own *Soyuz-TM* spacecraft from the aft to the bow port—a perfect opportunity to do a cursory scan of *Kristall*. On January 24 Afanasyev undocked the spacecraft, then waited while ground controllers used the Sofora thruster to flip the entire station over. As the station rotated, the three men used binoculars and cameras to study *Kristall*. As far as they could tell, other than some minor scratches near Kristall's docking port, there was no serious damage.[52] Based on these observations, flight director Solovyov decided to postpone the space walk. The next crew, still on the ground, had time to get the required space walk training. Having the present crew do it without training seemed unnecessarily risky.

The next six months were in many ways boringly unremarkable. After almost 20 years of doing long space missions, the Russians had the routine down pat. Three *Progress* freighters resupplied *Mir*, arriving every two months, each bringing about four tons of supplies, including fresh apples for the men to savor. The freighters were unloaded, filled with garbage from the station, and sent on their way, burning up upon re-entry to the atmosphere. There were few major on-board equipment breakdowns. Other than the failure of one of the Elektron regeneration systems in May, which caused Afanasyev, Usachev, and even Polyakov to spend an inordinate amount of time trying to fix it, little else went wrong.[53]

Polyakov, as a doctor, took over responsibility for maintaining the station's environmental and life-support equipment. He cleaned the toilet system. He maintained the Rodnik water-recycling equipment. He watched his crewmates' condition, and adjusted their exercise routines to keep them fit.

He also helped pay the bills. A German experiment, first brought to *Mir* during Klaus-Dietrich Flade's 10-day flight two years earlier, had remained unfinished after Flade's return. Germany, however, had paid Russia $700,000 to do the research, and the Russians felt obligated to fulfill their end of the bargain. When Polyakov went into space, he carried the last remaining pieces of the experiment's equipment in his personal baggage, and during the mission he periodically put on the headpiece, resembling a virtual reality helmet, so that German scientists on the ground could record his eye movements.[54]

And his willingness to use his body as a lab rat helped him survive the long, lonely months in space. He questioned everything he did or experienced. Should he do more running on the tread-mill? More bicycling? Should he wear more weights on his wrists and ankles when he did it? Less? What were his bowel movements like? Hard? Soft? Easy? Difficult? What should he eat? More? Less? Different things? How did the different foods affect him? How was his sleep changing? Was it deep or light? Was he getting enough rest?

Everything became subject to thoughtful examination and evaluation. For example, on March 8, three months into his mission, he impassively dictated to mission control his attempt the previous night to fall asleep.

> I put out the lights, lay down, felt a little sleepy. But just on the verge of falling asleep I suddenly switched on and that was it. I was lying half awake for many long hours.

Methodically, he analyzed the situation.

> Why couldn't I sleep? I felt hot. I was sweating. The cabin was too tight. Sweat ran down my chin. It seems I've reached the stage typical for long space flights when your sleep pattern undergoes changes. I'm in a state of alert. Sleep is normal but very shallow. I react to every little noise, every movement. I even woke up when something was hanging 2-3 feet from my head. Though it was just hanging [and] didn't disturb me, it woke me up.

Then, after a pause, he described how he finally fell asleep.

> About 1:30 A.M. I lost patience. I took two sleeping pills. After 10-15 minutes I was fast asleep. (pause) I will discuss the details with specialists after my return.

The same day, he described how his body was changing due to lack of gravity.

> The skin on my feet is peeling off. It comes off in big flakes, especially when I take off my socks. We catch that skin with the vacuum cleaner so it does not float around. The sight of my feet is not a very pleasant one. They say that after six months you get rid of all the calluses. You have feet like a baby though it's hard to say that of a 46-year-old man. No hardened skin at all. (with a laugh) It seems that weightlessness is a sure way to get rid of calluses.[55]

Every once in a while, Polyakov's fear returned, the fear that something would go wrong and force him to return to Earth prematurely. Each time, he took a deep breath, focused on his work, and tried not to think about it. As he remembered years later, "I became more careful, more determined to complete everything."

On Earth, the breathtaking changes in the former Soviet Union continued. Increasingly addicted to the idea of profit, the Yeltsin government in early 1994 decided it was time to make Energia a private company. By presidential decree, Yeltsin ordered 49 percent of Energia's equity to be offered for sale to the public, including foreign investors. The remaining 51 percent would be retained by the government, at least for the next three years.[56]

Meanwhile, friction between Russia and Kazakhstan over control of the Baikonur Cosmodrome continued. Negotiations on rent and ownership had dragged on for months. In early 1994 the President of Kazakhstan, Nursultan Nazarbayev, visited the United States, where he met with Clinton and various American businessmen. In a desire to improve his negotiating position with Russia, he proposed making Baikonur available for rental to others besides the Russians. Nazarbayev's negotiating ploy helped push his haggling with Yeltsin to conclusion, and on March 28, 1994, he and Yeltsin sat down together in Moscow and hammered out an agreement. For an annual fee of $115 million, Russia would lease Baikonur for the next 20 years.[57]

Yet, the tug-of-war between the two new nations was far from settled. In fact, it even caused the delay of the next launch. One crew member, Talgat Musabayev, had been born in Kazakhstan but was both a Russian air force and commercial pilot. For public-relations reasons, Kazakhstan wanted to claim him as a Kazakhstan cosmonaut. The Russians readily agreed, but then contended that, since Musabayev was a foreign cosmonaut on a Russian mission, Kazakhstan had to pay Russia to fly him, just as the French, Japanese, and others had done. The Russians estimated that the cost of Musabayev's four-month mission was around $150 million. Not surprisingly, Kazakhstan protested. Throughout the spring the dispute persisted. In May the launch date of the next mission was postponed from June 20 for 10 days. Then, a few days later the mission was delayed again, until the first week in July. Finally, in mid-June the now-semiprivate ITAR-TASS news agency announced that an agreement had been reached. Musabayev's flight was to be considered a commercial flight, with Russia and Kazakhstan splitting the cost. According to Musabayev, the Kazakhs had agreed to pay for equipment, food, and other expendables, while also giving the Russians credit on their rent for Baikonur.[58]

So, on July 3, 1994, *Soyuz TM 19* arrived at *Mir*, bringing rook-

ies Yuri Malenchenko of Russia and Talgat Musabayev of Kazakhstan to replace Afanasyev and Usachev. For the first time since
1977 and the first flight to *Salyut 6*, the Russians were flying an all-
rookie space mission. Leonid Brezhnev's senseless order that no
mission be crewed entirely by rookies no longer had any authority,
and the space program now had other, more important concerns—
almost all economic in nature.

Originally, Polyakov's backup, Arzamazov, had been scheduled
as the third crewman. When he was pulled from the program, mission control then considered putting Gennady Strekalov, an experienced cosmonaut who had already flown five times, on the flight.
After some discussion, Strekalov was dropped also. By leaving off
the third crewman, the mission required fewer supplies, so that
instead of launching two *Progress* freighters, one in early July and
a second in late August, Malenchenko, Musabayev, and Polyakov
would conserve their provisions and wait for a single, heavily
packed supply ship in late August.[59]

After Afanasyev and Usachev returned to Earth in early July,
having completed what by the 1990s seemed to be a completely
mundane six-month shift in orbit, Polyakov and his second crew
quickly settled in together, hoarding their food and supplies and
waiting for resupply. By late August, *Mir* had only about two weeks
of food and oxygen remaining.[60]

On August 25, *Progress-M 24* lifted off from Baikonur. Not only
was it packed with almost 3000 pounds of water, food, and supplies, it also included about 600 pounds of equipment for several
European Space Agency (ESA) missions—scheduled to begin in October—as well as for the first American mission four months after
that.[61]

In the years since the fall of the Soviet Union, Yuri Koptev of
the Russian Space Agency and Yuri Semenov of Energia had worked
out increasingly larger deals renting *Mir* to foreigners. Besides the
U.S. deal, the biggest, they had gotten the French to pay about $36
million to fly three separate missions, each lasting from two to
three weeks. In addition, ESA had paid $60 million for the right to
send two Germans to *Mir* and to conduct long-term experiments
on the station. The first German was to stay on *Mir* for a month,
followed by a second for a four-month stay in late 1995. The second
visit was to be fitted in after the first American visitor (negotiated
under Bush) and before the start of the two years of continuous

American occupancy (negotiated under Clinton).[62] *Progress-M 24* brought the equipment needed to complete these deals.

After the usual two days of maneuvers, the freighter homed in on *Mir*'s bow port. Then, at a distance of only 30 feet, the Kurs docking system inexplicably shut down. Like a boat on the ocean, the spacecraft could not back up instantly. Instead, it veered sideways to barely miss one of *Mir*'s solar panels.

Ground engineers figured that the problem might be a software error, similar to the problem they had back in 1991 when they first tried to dock *Kvant-2* to *Mir*. They reprogrammed the Kurs system, increasing its tolerances to make a shutdown less likely.

On August 30 they tried again. As the freighter approached, it appeared on the video screens to be aligned properly. At contact, however, the ship drifted sideways, tapping twice against the port's outer edge before bouncing off and floating away. Inside *Mir*, the impacts reverberated through the station. While Malenchenko scanned sensors, trying to find out if *Mir*'s hull had been breached, Musabayev and Polyakov scrambled from window to window, trying to see if the impact had caused any serious damage. They saw nothing, and Malenchenko reported no drop in air pressure.[63]

Even though the station was whole and working, the situation was critical. The *Progress-M* had only enough fuel left for one more docking attempt and if it failed, the station's supplies would quickly run out. Because the Russians had no spare *Progress* tankers available, *Mir* would have to be abandoned and the international flights canceled. The Russians had already been paid some $60 million for the international missions so there was real fear in Star City that this desperately needed and already spent cash would have to be returned. More importantly, the $400 million American deal might be canceled if the Russians failed to make good on the initial American mission.

After several more days of brainstorming they tried a third time. This time ground controllers used the Kurs system to bring the spacecraft to within 500 feet, after which Malenchenko was to take over with the TORU system. When *Progress-M 24* was only 15 feet away and moving in at a speed of less than a foot per second, silence descended on mission control. In essence, the Russian space program would live or die on what happened in the next few seconds.

Above, the giant television screens showed *Progress-M 24* creeping slowly toward *Mir*'s port. For a moment, the freighter seemed to veer to the left. Then Malenchenko made a correction, the freighter lined itself up properly, and made contact. A pause, and then the telemetry poured in, indicating a clean docking. The control room burst into cheers.[64]

Serious questions remained. Two collisions in less than nine months required a space walk by Malenchenko and Musabayev to inspect both *Mir*'s bow port and the *Kristall* module. However, when they went outside, the two men found nothing but minor scratches and a torn 12-inch by 16-inch piece of insulation on *Kristall*, which they easily fixed by simply tucking the fabric back in place and fastening it down.[65]

What had caused *Progress-M 24*'s docking problems remained a more worrisome mystery.

Despite this unknown, on October 4 Alexander Viktorenko, Elena Kondakova, and Ulf Merbold lifted off from Baikonur, heading for a docking with *Mir* two days later.

With *Soyuz-TM 20* 500 feet from *Mir* and Viktorenko standing by at the controls, the Kurs system took control of the spacecraft and began steering it to the bow port. It had barely eased the spacecraft in by 80 feet when warning lights began blinking on Viktorenko's instrument panel. Kurs was once again causing the spacecraft to yaw sideways.

At that moment, flight director Solovyov got on the radio. "You are three minutes from a communications blackout, when you'll lose telemetry. The automatic docking system is too slow. Perform a manual docking, and do it in less than three minutes."

Viktorenko quickly took control, deactivating the Kurs system and manually piloting his spacecraft to a flawless, albeit hard, docking.[66]

That the problem had occurred on both *Progress-M* and *Soyuz-TM* spacecraft suggested that the cause was inside *Mir*'s systems. Unfortunately, no more tests could be done until the next *Progress-M* freighter arrived in six weeks. In the interim, Solovyov suggested that Musabayev and Malenchenko could do further docking tests when they and Merbold were scheduled to return home in early November. Their return was originally set for November 3. If they did these tests on November 2—packing *Soyuz-TM 20* as if they were heading home—and everything worked, they could redock and then leave two days later, thereby giving Merbold time to complete his full mission in space. If they could not redock because the system failed, they could simply return to Earth, costing Merbold only one day in space.

Before this docking test, however, the month-long EuroMir mission had to be completed, with its six-person crew of one woman and five men from Russia, Germany, and Kazakhstan. Merbold, who at 19 had fled communist East Germany just before the Berlin Wall was built,[67] had a flight program that included twenty planned experiments studying space sickness, the shift of his body fluids in space, and the effect of weightlessness on the cardiovascular system, the muscles, the bones, and on the system of balance. The research also included five experiments studying how materials melted and solidified in space, as well as three experimental engineering tests.

More exciting to the world press, however, was the presence of 37-year-old Elena Kondakova, the first Russian woman to fly in space in the 10 years since Svetlana Savitskaya's last flight in 1984. Trained as an engineer, her first job out of college in 1980 was at Energia, working as one of *Mir*'s flight controllers.[68] There she met Valeri Ryumin. Ryumin, whose first wife had never been comfortable with space flight, was available and interested. Before long he and Kondakova were married. In 1989, Kondakova was selected for training at the Gagarin Cosmonaut Training Center, and one year later she was part of the cosmonaut team.

Ironically, Ryumin opposed the idea of female cosmonauts, and had even signed the order that stopped all female cosmonaut flights after Savitskaya's mission. As he bluntly told the Associated Press in 1997, "It's my opinion that a wife should stay at home for the most part, not at work and not in space . . . I think the majority of men will support me, because the majority of us would prefer that everything in our home is taken care of and everything is quiet and okay."

When asked in preflight press conferences about her husband's opposition, Kondakova shrugged and said that her husband "had to resign himself to my plans."[69] She had decided to go, and if he really wanted "quiet" at home, he had better accede to her wishes.

Her mission was to join Polyakov during his last five months in space, thereby setting the record for the longest female space flight. Up to then the longest any woman had spent in orbit was 15 days, accomplished by Chiaki Mukai on the space shuttle in July 1994.

By this time Polyakov was completing the ninth month of his endurance flight. He continued to sleep less than he would on Earth. He also seemed to eat less. To reduce the chances of overexposure to cosmic radiation, he slept in *Kristall*, where its array of nickel cadmium batteries acted as shielding against cosmic rays and the bright flashes they would cause on the retina, whether a person's eyes were open or closed. Overall, Polyakov was pleased by how well he was holding out. "You must understand," Polyakov explained. "It is a wonderfully healthy lifestyle in space." Each day he worked out for two hours. His diet was carefully prepared and balanced. And he was relaxed, doing what he loved most.[70]

Occupied by six people for a month, *Mir*'s systems were being pushed to their limits. The two Elektron air-regeneration systems attached to the bathrooms in the core module and *Kvant-2* could not generate enough oxygen from urine, and had to be supplemented with one lithium perchlorate candle per day—the same kind of cartridges used since *Salyut 1* to generate oxygen while removing carbon dioxide from the air as they burned. On top of that, the EuroMir program called for extensive television coverage of the station, at least 20 minutes each day, which, in turn, put a heavy strain on *Mir*'s aging solar panels to recharge the station's aging storage batteries, some of which had trouble holding their charges.[71]

On October 11, only a week into the month-long visit, the computers that ran the station's attitude-control system crashed. At the time the crew was asleep and out of radio contact with mission control. By the time the alarms went off, waking the crew, the station's batteries had drained, forcing its gyros, lights, and ventilation system to shut down. Its six inhabitants floated in a silent and dark station, as had happened when Helen Sharman had visited *Mir* three years earlier.

For the next two days the crew struggled to bring the station back to life. Malenchenko climbed into the *Soyuz-TM* spacecraft docked to the aft port and used its rockets to manually orient *Mir* so that its solar panels were aimed at the sun. Musabayev, meanwhile, searched for the cause of the computer crash, tracing it to a short-circuit in three of the storage batteries. He unplugged any nonessential equipment (such as video cameras), and began the process of isolating those batteries from the system.

Meanwhile, Viktorenko and Kondakova took turns at the controls, making sure someone was standing by 24 hours a day in case the computer crashed again. During one shift change Viktorenko, a good-natured man who had done pirouettes and figure-eights with the Icarus jetpack back in 1989, joked with Kondakova and ground controllers, noting that though the exchange rate for the U.S. dollar—the hard currency that cash-strapped Russians almost worshiped—remained stable on *Mir*, he couldn't say the same for the voltages of the station's power systems.

By October 13, seven of the remaining nine working batteries were recharged, and the attitude-control system was reactivated, using a back-up computer.[72]

Then, on October 15, Polyakov went into *Kvant* to light a lithium perchlorate candle. Kondakova came with him, curious to see how it was done. He inserted a new canister into its holder, locked it down, and pressed the ignition button to get it burning. They were startled by a bright flash as blinding flames shot from the canister.

Polyakov's mind raced. "Smother the fire, I've got to smother the fire!" Stored nearby, held in place by a bungee cord, was Malenchenko's Penguin suit. He grabbed it and whipped it over the canister, covering the white-red flames with both the cloth and his body. Deprived of oxygen, and lacking gravity to circulate the air, the flames died almost immediately, leaving nothing but scorched equipment and some quickly dissipating smoke.

After he and Kondakova were sure the fire was out, they contacted mission control and told them what happened. Other than asking the cosmonauts to make sure that the unit that held the canisters was still working (it was), no one on the ground had any suggestions for dealing with the problem. Faced with the looming American visits to *Mir*, and knowing how nervous Americans were about fire on spaceships, Semenov, Ryumin, and Solovyov all agreed to keep this small anomaly a secret. "We couldn't risk [the loss of] American financing by releasing any information about the fire," Polyakov recalled. On his own, Polyakov decided to keep a wet towel nearby whenever he had to fire up a candle.[73]

Though *Mir*'s electrical system was completely restored, power limitations made all of Merbold's furnace experiments impossible while he was on the station. With only nine working batteries and six people on board, the aging solar arrays simply couldn't generate and store enough power. In addition, several of the required furnaces had malfunctioned and needed repairs. Russian and European officials agreed that the work would be done later, when *Mir* was occupied by only three Russian cosmonauts, and after spare parts were shipped to the station so that the furnaces could be repaired.[74]

On November 2, after a hectic and frustrating month in space, Malenchenko, Musabayev, and Merbold boarded *Soyuz-TM 19* to do the improvised docking test of the Kurs docking system. Inexplicably, everything worked perfectly. Malenchenko backed *Soyuz-TM 19* 600 feet from *Mir*, where he let the Kurs system take over. The automatic system smoothly guided the spacecraft back into its pier on *Mir*. All told, the test took less than 35 minutes.[75] The situation was infuriating, resembling a trip to a car mechanic to fix an intermittent electrical problem: the only time the failure occurred was when no one was prepared to study it.

Finally, on November 4, 1994, *Soyuz-TM 19* came home. Malenchenko and Musabayev had spent more than four months in space, Merbold 32 days—six days longer than Frenchman Jean-Loup Chretien's flight in 1989, making him the first person from a country other than America or Russia to stay in space longer than a month.

Several-month-long missions had become commonplace. No one saw them as especially unusual or difficult (though hardly to be taken lightly). Though men came back to Earth weak and dizzy and with thinned bones, they very quickly returned to normal. In fact, *Mir* had now been occupied continuously for more than five years, and on November 18 completed its fifty-thousandth orbit, having traveled just less than 1.2 billion miles. Not only was that far enough for *Mir* to make a journey to and from Mars, it was enough for the station to have stayed in Mars orbit for at least a full year, doing research.[76]

On *Mir*, Viktorenko and Kondakova settled into their five-month mission, with Polyakov continuing his marathon flight. For Viktorenko, this flight was nothing more than a repeat of his previous three shifts on *Mir*. He was very used to *Mir*'s cluttered space, its noisy fans and pumps, the constant maintenance work, and the loneliness and tedium that came with a long mission. A cheerful man, Viktorenko had been called a "masterly" pilot by his commanding officer when he served as an air force pilot in the 1970s. Born in the northernmost parts of Kazakhstan, he had gone to the military air force academy in Orenburg, graduating in 1969. Like many who wanted to fly in space, Viktorenko waited many years for his first mission. His career started promisingly when he was plucked from the military by cosmonaut Andrian Nikolayev in 1977. Nikolayev had come to Viktorenko's military unit looking for new, young recruits, and within a year Viktorenko was training to be a cosmonaut.

Almost immediately thereafter Viktorenko's career, and life, nearly ended. During training he was placed in a chamber to see if he could tolerate isolation for long periods. At one point he received a bad electrical shock that burned both of his hands and knocked him to the ground unconscious. While he lay there, engineers outside were slow to react, thinking he had merely fallen from exhaustion. The accident ended his flying career. Though doctors wanted to ground him from space as well, Viktorenko refused to accept this verdict. After repeated tests and four years of training, he finally became a full-fledged cosmonaut in 1982. Even then, it took years to get on a flight. *Salyut 7*'s shutdown in February 1985 caused Viktorenko's first mission to be canceled. Finally, he flew to *Mir* in July 1987 on the Syrian international mission, discovering that he took to weightlessness instantly, like a fish to water. Over the next seven years he returned to *Mir* twice, once to reinitiate operations and test the Icarus manned maneuvering unit with Alexander Serebrov, and again to command the mission that relieved the stranded Sergei Krikalev after the August 1991 coup. [77]

Adjusting to life on *Mir* was harder for Kondakova. She was determined to prove to the male-chauvinistic Russians that a woman could perform as well as a man, and felt that pressure very strongly. To protect her interests she refused to back down in any way if she felt challenged. During training, for example, she even had a prolonged argument with Viktorenko while a German documentary crew filmed them.

Viktorenko was not reluctant to kid her publicly about her stubborn persistence. At the last preflight press conference, he was asked if it was difficult to command a woman. "Yes, it is not habitual," he said. Then he joked, "If she does not obey the captain's orders, I will complain to the crew's doctor who may remove her from the flight on [mental health] grounds."[78]

During the flight the stress of the long mission showed. At one point Kondakova complained, according to Polyakov, that the doctor was not paying enough attention to her. He, in turn, was offended because it seemed to him that she had ignored his previous medical advice. For about a week they did not speak to each other, keeping to their respective work to avoid escalating the conflict. Like others before them, they found it quite possible to live confined in a tiny space no bigger than a three-room apartment and never talk. Then, according to Polyakov, he decided to clear the air and work out their differences. She agreed, and from that point on their relations on board *Mir* became more cordial.* According to Polyakov, "We became friends again."[79]

Meanwhile, the aging state of *Mir*'s power systems placed limitations on the research Kondakova, or anyone, could do. Three of the station's batteries could no longer store power, while the others had limited capacities. Nor could Russia afford to build and launch replacements. Instead, *Mir* crews had to carefully ration their electrical use.

Moreover, the crew had increasing problems with the heating, water, and attitude-control systems. At least three times, the attitude-control computers failed, sending the station into a random spin. At least twice, parts in the thermal system failed, requiring replacement or repair. And a water leak in *Kristall* took several days to fix and clean up.[80]

On November 13, the three occupants of *Mir* got further disappointing news. Construction of *Mir*'s last two modules, delayed for years because of the collapse of the Soviet Union, was behind schedule, even though American money was now paying the bills. The first module, *Spektr*, had been scheduled for launch in December, and in conjunction with its arrival, Polyakov and Viktorenko had planned to do two space walks to reposition *Kristall*'s solar panels to *Kvant*. But with the delay of the modules, ground controllers decided to postpone the space walks for later crews. Ironically, the delay this time was not entirely the fault of Russian cash shortages. Russian customs agents had impounded some NASA equip-

---

*Though I tried repeatedly to get Polyakov to give me more details about this dispute, he refused, first claiming he had made it all up to satisfy a persistent reporter, and then admitting that the dispute occurred but that "it isn't worth mentioning." While it very well might be true that this dispute has been overblown, Polyakov's denial is curious. One wonders if he wishes to avoid offending Kondakova, the wife of the very powerful Valeri Ryumin.

ment intended to be installed on *Spektr* for use by American astronauts, and it took months to get the snafu straightened out.[81]

Not all the news was bad. When the next *Progress-M* freighter arrived, docking in the aft port (bringing with it the sweet smell of apples and lemons), the Kurs system worked perfectly. This success suggested to some Russian engineers that the extra cargo on the earlier ships might have caused the problems. The additional weight had skewed each ship's center of gravity, a change to which the docking software had not been properly adjusted.[82]

Finally, the New Year dawned, with *Mir* continuing its endless journey around the earth. On January 8, 1995, Polyakov, Kondakova, and Viktorenko performed a simple celebration in honor of the Russian Orthodox Christmas, eating some fresh fruit and drinking what the crew euphemistically called "tasty treats" that they had been saving. The next day, Polyakov broke the record of 366 days set by Vladimir Titov and Musa Manarov six years earlier. And like them, he felt fit and healthy, ready to go on as long as mission control would let him.

However, his fear of failure returned, stronger than ever. He pondered the possibilities. Though he had recorded his efforts extensively on computer and in notes dictated repeatedly to the ground, he knew that his most important scientific result was harbored within his body and his mind. To make his mission a success, he had to get home alive. More importantly, he had convinced everyone to approve the flight by vowing to walk almost immediately upon return. With only two-and-a-half months to go, he was feeling enormous pressure to fulfill that promise.

He fixed his mind on surviving, and got to work. He began wearing his Penguin suit all the time, doing subtle adjustments to its elastic bands. He modified his wrist and ankle weights. He wore the lower-body negative-pressure Chibis suit about an hour a day. He increased the treadmill's resistance, while doubling his treadmill workout time to two hours in an effort to strengthen his leg muscles, his circulation, and his heart. In turn, he stopped exercising on the bicycle (which was mostly for heart training).[83] As the weeks wore down, Polyakov became like a machine, his mind focused almost exclusively on that single moment when he got back to Earth and tried to take his first steps held down by the brutal, unsympathetic gravity of Earth.

On January 11, two days after Polyakov set the new record, he, Viktorenko and Kondakova climbed into their *Soyuz-TM* spacecraft and successfully repeated the docking test performed by Malenchenko, Musabayev, and Merbold, moving about 500 feet from the station, then allowing the Kurs docking system to automatically dock them. Once again, everything worked without a hitch. This last test convinced the engineers in mission control that the previous docking failures had occurred because of a programming error on both *Progress-TM 24* and *Soyuz-TM 20*.[84]

With this mystery seemingly solved, it was finally time for the American and Russian space programs to meet in space. On February 3, 1995, the space shuttle *Discovery* blasted off from Cape Canaveral, riding a column of super-hot white steam into a clear, dark, midnight sky.

Technically, the flight's purpose was to test the maneuvers necessary for future shuttle-*Mir* dockings while also proving that the shuttle's engine exhausts would not damage *Mir*. Politically, the flight served to give Bill Clinton a boost in popularity. He, like Nixon, Brezhnev, Gorbachev, Chernomyrdin, and other politicians had done before him, called the astronauts at a key moment in the mission and, by association, claimed some credit for their success.

On a less cynical note, the rendezvous also helped ease tensions between the former Cold War adversaries, forging ties and friendships where none had existed before. It also helped guarantee that American money would continue to pour into the Russian space program, keeping it viable and functioning. Above all, the flight brought to life, if only approximately, the dreams of Wernher von Braun, Willy Ley, and Sergei Korolev. The world's first true spaceship was rendezvousing with the world's first true port in space.

On this, its 20th flight into space, *Discovery* carried a crew of six. The commander was James Wetherbee, on his third flight, with his co-pilot Eileen Collins making her first journey into space. Like so many Russian air force pilots who became cosmonauts, Wetherbee had started his career as an American navy pilot, completing more than a hundred night landings on aircraft carriers. Later he became a test pilot, helping to design the avionics on a variety of jets.

Collins was the first woman to co-pilot a spaceship. Graduating college with a degree in math and economics, she had joined the air force to become a pilot in 1978. For years she was a flight instructor, then a test pilot, flying more than 30 kinds of planes. By 1990 she was an astronaut, training to pilot the highest flying aircraft ever built. If she maintained her skills and nothing went wrong, she stood the chance of becoming the first woman to command a space mission.

The rest of the crew included Bernard Harris, Michael Foale, Janice Voss, and, most significantly, Vladimir Titov. Titov, whose last mission with Musa Manarov had set the year-long record that Polyakov had just broken, was following up on Sergei Krikalev's earlier shuttle flight, completing the first part of the Russian half of the shuttle-*Mir* missions. Foale was on his third flight, and was soon to join the shuttle-*Mir* program, heading for Russia. During the rendezvous he was to be navigator, monitoring *Discovery*'s approach using his computer and his very thorough understanding of orbital mechanics. Captain Wetherbee called him a "maestro with his computer."[85]

For two days, *Discovery*'s orbit slowly crept toward *Mir*. On February 6, five hours before the shuttle was expected to reach the station and when the two spacecraft were still a hundred miles apart, Titov announced that he thought he spotted *Mir*. Wetherbee quickly picked up the bright glittering star on the horizon. Soon everyone could see it. Titov used a hand-held VHF radio to make radio contact with the station, talking to Viktorenko in Russian.

In *Mir*, Viktorenko, Kondakova, and Polyakov pressed their faces against the station's windows, peering out toward the horizon, searching for *Discovery*. Viktorenko had already re-oriented *Mir* so that the base block hung perpendicular to the earth, with the bow port pointed down, *Kvant* and the aft port pointed upward to the stars, and *Kristall* in front taking the lead position as the station circled the earth. The flight plan called for *Mir* to remain passive with *Discovery* doing the maneuvering. The shuttle would approach *Mir* from below and behind, passing just under the station, after which it would settle into position about 400 feet ahead of *Mir*, then slowly let the station catch up until *Discovery* was within 40 feet of *Kristall*'s docking port.

Slowly, inexorably, the winged, white, gleaming shuttle drew closer. It came on with its cargo bay facing the station, its nose pointed up toward the stars. Finally, at 12:20 P.M. Eastern Standard Time, with the North Pacific gleaming below them, the 105-ton *Discovery* eased into position, a mere 37 feet from the 70-ton *Mir* complex. Viktorenko, Kondakova, and Polyakov could be seen in

Mir's windows, waving and taking pictures. On *Discovery*, the shuttle astronauts returned the favor, taking truly spectacular video and film images of the *Mir* station.

To Michael Foale, the view of *Mir* was both striking and perplexing. He had trouble comprehending the size of the station, having nothing beside it to give perspective. It was "like seeing the Great Wall of China or something from a distance. . . . You don't relate to it."

To Titov, it was a thrill to see once again the station he had lived on for a year, "to look at my house again." During the entire rendezvous he chatted excitedly with his compatriots on *Mir*.

To Polyakov, the shuttle looked beautiful and amazing. He talked with Titov by radio, posing so that Titov could take his picture peering out the largest window in *Mir*'s base block "I told him to take good aim and hurry, because the sun was shining and the window lets in ultra-violet light." If Polyakov stayed there for more than a few minutes, he would get a very bad sunburn. Titov's spectacular picture of *Mir* captured Polyakov peeking out its window, a "Snoopy" cap on his head, his headphone in front of his mouth, and an impish grin on his face.[86]

Wetherbee, trained to mouth the boring, canned speeches that the NASA of the 1990s demanded during these public moments, gave a short speech in both English and Russian, describing how the flight was "bringing our two nations closer together."

In a burst of exuberance, Viktorenko responded, saying it far better. "We are all one. We are human!" he exclaimed. "This is almost like a fairy tale. It's too good to be true!"[87]

Unbeknownst to everyone, as Wetherbee was about halfway through his speech, he was also fine-tuning the shuttle's position, responding to Michael Foale, who had noticed the two vessels beginning to creep together and sharply told Wetherbee, off mike, to "Back up! Back up!"[88]

For 10 minutes Wetherbee held *Discovery*'s position, then he eased the shuttle away about 400 feet. For the next three hours, with *Discovery* taking the lead, the two spacecraft raced in formation twice around the earth. As they flew, Russian ground controllers used the thruster on Sofora to rotate *Mir*, allowing the shuttle astronauts to see the station's entire surface. Then Wetherbee fired *Discovery*'s engines, taking the shuttle into a higher orbit, above *Mir*. Within an hour the station had drifted by and away, shrinking to a star again on the far horizon.

The next day, in a station-to-shuttle chat with Viktorenko, Wetherbee spoke more naturally about the rendezvous. "It was like dancing in the cosmos," he said. "It was great!"

Viktorenko, in turn, complimented Wetherbee on the precision with which he piloted the giant 100-ton shuttle. "It was like a jeweler [at work]. We were surprised how accurate you were in station-keeping."[89]

For another three days *Discovery* remained in orbit so that its crew could do experiments. Titov used the robot arm to control and release a recoverable astronomical satellite. Foale joined Bernard Harris in a space walk, testing new spacesuit-glove designs.

Discovery also carried several American plant experiments. One experiment, called the Chromex greenhouse, had been flown repeatedly on shuttle missions, studying why plants seemed to have problems producing seed embryos in space. Just as the Soviets had learned on their *Salyut* stations, a 1994 shuttle mission had found that the problems seemed to come mostly from the poor circulation of air and water in weightlessness, not from weightlessness itself. Using a new air-exchange system, the Chromex facility on *Discovery* attempted to grow super-dwarf wheat seedlings.

Another plant experiment was the Astroculture facility, which had also flown on a number of previous shuttle missions, researching space-greenhouse design. This flight, incorporating what had been learned on earlier flights, was the first to include plants, and tested the ability of wheat seedlings to grow in an enclosed automatic chamber.

And once again, the Coca-Cola company was there. Just as Krikalev and Artsebarski had guzzled Coke the week of the August 1991 coup using hand-held can-like dispensers, now the American and Russian astronauts on *Discovery* used a soda-fountain-like system for dispensing both diet and regular Coke. According to company officials, this research was part of an "experiment" to study how the astronauts' taste changed in space. Not surprisingly, no one noticed a difference in taste, though everyone noticed *what* soft drink was flying in space.

With *Discovery* gone, *Mir* once again flew on alone. In mid-February, another *Progress-M* freighter arrived, docking with no problems. Its cargo included more than 200 pounds of American research equipment for use by Norman Thagard, in training at the Gagarin Cosmonaut Training Center and scheduled to be the first American to fly on a Russian rocket and the first to occupy *Mir*.

Polyakov's mission, now 13 months old, was finally winding down. With Thagard's arrival mere weeks away, Polyakov did nothing but prepare himself for his return to Earth. On March 14, Thagard's *Soyuz-TM* spacecraft took off from Baikonur, docking with *Mir* two days later. Its Russian crew, Vladimir Dezhurov and Gennady Strekalov, were to take over *Mir* from Polyakov, Kondakova, and Viktorenko, while providing support for Thagard.

The launch of this *Soyuz-TM* set another space record. Twelve days earlier the American space shuttle *Endeavour* had lifted off from Cape Canaveral, carrying a crew of seven on a 16-day science mission to do both astronomy and on-board medical research. With six people on *Mir* (having arrived from three different *Soyuz* spacecraft) and seven people on *Endeavour*, a total of 13 humans were in space, the most ever.

While *Endeavour*'s orbit made it impossible to dock with *Mir*, and its mission was not directly connected to the accelerating shuttle-*Mir* joint program, its goals (to study the stars while also doing medical research) were really not much different than those on *Mir*. As Polyakov and Kondakova spent their time trying to find out how the human body adapted to space, the crew of *Endeavour* were performing experiments to find out if the weightless environment of space could be used to cure diseases like AIDS, breast cancer, and diabetes. And with both missions, the eventual goal, rarely stated publicly by anyone, was to find out how to build a thriving human community in space.

For four days, while these crews remained in orbit, the first glimmer of such a community actually existed. On March 16, 1995, *Endeavour*'s commander Steve Oswald chatted with Thagard for a few minutes, each man complimenting the other's ability to make a complicated job seem simple.[90]

Meanwhile, *Mir*'s six occupants got down to business. The Russians gave Thagard a tour of the station, showing him the equipment he would use in his research. Then, at Dezhurov's insistence, he settled into one of the crew cabins in the core module, leaving the veteran Strekalov to fend for himself.

Thagard was astonished by Polyakov's healthy condition. "His legs were just as big as tree trunks. . . . If he did that well after fourteen-and-a-half months, I probably don't have much to worry about for just three months."[91]

On March 22 Viktorenko, Kondakova, and Polyakov boarded
their *Soyuz-TM* spacecraft and undocked. As Viktorenko pulled
them away from *Mir*, he tried to get Valeri Polyakov to take it easy.
"Shhh," he said. "Shhh." Polyakov was so enthusiastic at the pros-
pect of finishing his flight and going home that his commander had
to calm him down.[92]

Polyakov looked at *Mir* with mixed emotions. He felt deep
gratitude for the station's success at sustaining him for so long and
for so well. He also felt sorrow, knowing that he would probably
never see it again.

Slowly *Mir* moved away, shrinking to a tiny star on the horizon. Then, with a blast from *Soyuz-TM*'s engines, they began their
descent to Earth. Polyakov was coming home.

The ground was heavy. His arms felt heavy. Everything felt heavy.
Two men helped him from the capsule, and attempted to carry
him to a lawn chair. Polyakov refused. While he accepted their
help, he insisted on walking the scant few feet to the chair himself. All around him swarmed people. The blades of two helicopters chopped the air. Newsmen snapped photos or took video. As
Polyakov sat in his lawn chair, looking out across the flat empty
expanses of Kazakhstan, his thoughts drifted to cigarettes, alcohol, and his wife.

They had landed only 22 miles northeast of the small Kazakhstan city of Arkalyk and about 300 miles northeast of Baikonur.
The ground was covered by a light mantle of snow, softening their
impact. Despite the snow, to Polyakov it seemed a fine spring day.
The sun shone down upon him with a gentle warmth, not the
harsh, intense light visible from space. And at his feet he could see
several tiny flowers peeking up through the snow.

A friend nearby was smoking a cigarette. Polyakov eyed it hun-
grily, then reached out for it. The man smiled and put it in
Polyakov's mouth. Another friend poured a small shot glass of
brandy and handed to Polyakov. *How good it tastes!* Polyakov
thought, savoring the strong flavor as he swirled it about his mouth
and then swallowed.

After a few minutes they carried him, chair and all, into the
medical tent. There he tried again to walk, managing to stand and
take one or two tentative steps on his own. On the helicopter he
again insisted on walking. This time he managed more than a few

feet, slowly pacing back and forth in the tiny space. By the time he returned to Star City, he could almost walk normally, and so he stepped off the plane unassisted.[93]

Polyakov's space flight had lasted 438 days (bettering a year by more than two-and-a-half months). Yet upon return, his health was not much different than other cosmonauts' after a long flight. After those first steps, he completely readapted to gravity within two months. Moreover, his bone loss had been very low, only around 7 percent in some of his weight-bearing bones, a rate of 0.5 percent per month, confirming once again his belief, shared by other Russian doctors, that the exercise program had kept that loss low—low enough for him to survive a two-year trip to and from Mars.[94]

In fact, though his body felt tired, much of his exhaustion came not from Earth's gravity but from the endless medical tests he had to undergo upon his return. "I needed to walk," he recalled. "I wanted to walk, but they wouldn't let me." The doctors were not completely wrong. The bottoms of Polyakov's feet had become as soft "as those of an infant." On return to Earth, it took time for him to build up his natural calluses, and walk without pain.[95]

Nonetheless, Polyakov was convinced that he had proved that human beings could survive in weightlessness long enough to travel to Mars and work there immediately. "That I could walk from the capsule to the chair proved it!"

But had he? Could Polyakov have *worked* on Mars? Even today, the question remains unanswered. Within hours of landing, his body was strong and healthy enough to stand and walk on its own. Within a week he felt almost completely normal. Moreover, Mars has a much lighter gravitational field than Earth, making the adaptation upon arrival far less stressful than on Earth. Polyakov was convinced he could have done the minimum necessary: take some pictures, grab some rock samples, and then come home—as the *Apollo* astronauts had done on the moon.

Yet, Polyakov had come back to Earth very weak. For at least those first few hours, he needed help from those around him. Any spacefarer arriving at Mars after a year in space must be prepared to face that same challenge.

For Valeri Polyakov, however, meeting that particular challenge was no longer his concern. Though he would be glad to return to space again, he had no desire to extend his record. It seemed pointless. As he said in an interview a year later, "The goal [of my flight] was to prove that humans can reach Mars and be in sufficiently good shape to work on it. This goal has already been reached. To increase orbital flights further means to subject cosmonauts to unnecessary stress."[96]

More importantly, until a mission to Mars was actually being planned, the political will to fly longer flights did not exist in either the United States or Russia. Space exploration was entering an era of cooperation. To try to top someone else's records while also working with them hand-in-glove would be diplomatically awkward.

Above, in space, three men, from two countries that were once sworn enemies and fierce competitors, settled down to operate *Mir*. While Russians Vladimir Dezhurov and Gennady Strekalov were there to maintain and upgrade the station in preparation for the arrival of the first shuttle docking later that year, American Norman Thagard's mission was to set up and run a plethora of scientific experiments.

His primary goal, however, was to break the American in-space endurance record set by the last crew of *Skylab* 22 years earlier. Along the way he would also relearn many of the same lessons NASA had discovered and then forgotten from its only space station.

He, and his Russian colleagues, were also to discover that cooperation was not as straightforward as some politicians claimed.

12

Mir: Culture Shock

Picking Up Where It Left Off

After almost 20 years of independent space exploration, the world's two preeminent spacefaring nations were finally stuck in the same place.

Since the *Apollo-Soyuz* mission in 1975, the U.S. and Soviet-Russian manned space programs had followed divergent paths. The Soviets had focused on increasingly lengthy marathon missions, learning how to live and work in space for long periods. The Americans had dedicated their efforts to short hops, learning how to repeatedly and effectively launch very sophisticated and large cargoes into space.

Now, the two disparate programs had finally come together on *Mir*. However, unlike the *Apollo-Soyuz* joint mission 20 years earlier, the circumstances were hardly balanced. For the first time an American, Norm Thagard, was living in space according to the rules of another country—a fact that became starkly obvious to Thagard within hours of arrival.

His crewmates were Gennady Strekalov, a 54-year-old veteran of five space flights, and Vladimir Dezhurov, the commander and a rookie spaceman 22 years younger. Strekalov's space career had sometimes been harrowing, sometimes breathtaking, and sometimes legendary. The legendary part took place in September 1957, when as a 17-year-old apprentice coppersmith he watched the fabrication of *Sputnik*, the first artificial satellite, in the factory where he worked. Twenty-three years later Strekalov was on his first space flight, on the first three-man *Soyuz* mission after the deaths of Dobrovolsky, Patsayev, and Volkov, which docked with *Salyut 6*. Later, he almost died twice, once in the near-collision of *Soyuz-T 8* with *Salyut 7* in April 1983, and then in the aborted launch five months later when the rocket he and Titov were in erupted in flames on its launchpad. His last flight had been a 132-day mission to *Mir*, during which he and Gennady Manakov had made the first attempt to repair the *Kvant-2* airlock hinge, babysat the Japanese television reporter, and finished integrating *Kristall* into the station's computer system.[1]

Dezhurov was only 32 years old and was on his first mission in space. Born in 1962 in the Russian heartland several hundred miles east of Moscow, he was only nine when Dobrovolsky, Patsayev, and Volkov died in space. During Strekalov's first flight in 1980, Dezhurov was a freshman in college, attending the Kharkov Higher Air Force School. After four years as an air force pilot, he enrolled at the Gagarin Cosmonaut Training Center in 1987, graduating in 1989. For the last six years he had been in training, preparing for this first launch.[2]

Despite the contrast in their experience, Dezhurov was in charge. According to policies established during the Soviet era and still followed religiously, the pilot of the *Soyuz* spacecraft, enlisted from the armed forces, was always the mission's commander. Dezhurov, as an air-force officer, was that pilot and, therefore, commander.

To Thagard's astonishment, almost immediately after launch Dezhurov became bossy and imperious, issuing orders in a rude and blunt manner. During one of their first meals on *Mir*, with Viktorenko, Polyakov, Kondakova, and Thagard watching, Dezhurov, the youngest person on board, curtly ordered Strekalov to get him some trivial item, something so minor that today Thagard can't even remember what it was. Without comment, but clearly irritated, Strekalov went to fetch it.

Kondakova, wife of Ryumin, laughed. "Well, you're awfully bossy, aren't you?" she said to Dezhurov. "Why don't you get it yourself?"

Thagard was taken aback by Dezhurov's character change. On the ground he had been a young, enthusiastic, and friendly crewmate, their two families having dinner together almost weekly. In space he suddenly became a brusque taskmaster, barking orders and expecting instant obedience. Thagard was also astonished at how nonchalantly Strekalov accepted these orders. Though Strekalov eventually lost patience with Dezhurov and told him to back off, it took him several days to do it.[3]

Thagard should not have been surprised. Though Dezhurov was clearly trying to compensate for the insecurity of commanding others who were far more experienced than he, he was also behaving in exactly the same manner as most Russian bosses. Russian culture had been built on centuries of serfdom followed by 70 years of totalitarian communist rule. Russians grew up expecting all major decisions to spring from their bosses. Everyone believed, without question, that those in charge had the right to arbitrarily rule those below them.

An excellent illustration of the ingrained nature of Russia's authoritarian culture can be seen in the design of the Soviets' early manned spacecraft. Maneuvers, rendezvous, and dockings were all controlled from the ground, with on-board guidance systems either not operational or designed as back-ups to be used only in emergencies. Though cosmonauts could override this remote system, they could do so only if they received permission from mission control. It was as if Soviet designers were afraid of free-flying cosmonauts steering their spacecraft to the United States to defect.[4]

Raised in this top-down, bossy society, Russians rarely protested if their boss or leader was crude, arrogant, overbearing, or bossy—as Dezhurov was to Strekalov. For Russians, such vulgar and domineering behavior was the norm, not the exception. In his memoirs, Boris Yeltsin described the coarse, top-down Russian way of doing things:

> I was brought up in the system; everything was steeped in the methods of the command system, and I too, acted accordingly. Whether I was chairing a meeting, running my office, or delivering a report to a major meeting—everything one did was expressed in terms of pressure, threats, or coercion.[5]

Similarly, Gorbachev described his dealings with his bosses when he was a young party official as "having to endure abusive language and rudeness. . . . All this taxed the nerves."[6] From the top to the bottom of the chain of command, from the Soviet leader to the lowest office clerk, the bosses treated the workers below them with contempt and rudeness. They screamed at them. They insulted them. Employees were expected to do petty chores for their bosses, to kowtow to them, to brown-nose them. In turn, employees knew they could treat *their* subordinates as badly.

This crude Soviet culture was made even more acerbic because Russia is a far rougher and rawer society than most Americans realize. Unlike the United States, where literacy is common and every person *assumes* that at least three or more generations before them could read, Russia is a land that only a hundred years earlier was made up predominantly of illiterate serfs, peasant tribes, and a royal caste system. Consider: The cosmonauts of the 1960s and 1970s were often the *first* people in their families who could read. The few who came from educated roots were a distinct minority.

In addition, the ravages of Stalin's rule and World War II had ripped many families apart. Too many cosmonauts—Viktor Patsayev, Vladimir Lyakhov, Gennady Strekalov, to name a few— had lost a parent before or during World War II.[7] Others were either exiled by Stalin or refugees from the German invasion—watching as the German war machine marched through Russia, burning and destroying everything in its path. Some—such as Georgi Dobrovolsky and Georgi Grechko—witnessed the fighting firsthand as children.

The destruction from World War II, the oppression of Stalin, the leftover peasant culture of ancient Russia, and the top-down, bossy Russian society left over from the Czars and adopted by the Soviets all combined to make Russian culture caustic, tactless, and coarse, its people rough-hewn and emotional, prone to extreme passions, from the sentimental to the callous.

Even when they are being civil, Russians have a tendency to express themselves in far more passionate terms than Americans. For example, when Viktor Savinykh was on *Salyut* 7 he had a conversation with Vladimir Solovyov, who, with Leonid Kizim, had repaired the station's engine-line leak the year before.

Solovyov was working as capcom. He asked Savinykh what it had been like when he and Dzhanibekov first opened the hatch to the dead *Salyut* 7. "Did you notice the smell in the passageway? The smell of burnt iron?"

"Yes," Savinykh said. "It was an outer space smell."

This prosaic description did not satisfy Solovyov. He had to add an earthy touch. "Yes," he noted. "A *manly* smell."[8]

It is hard to imagine any American astronaut expressing himself in this manner. Americans, especially astronauts, tend to be more reserved, less poetic, and more concrete in their descriptions. Consider how Jerry Linenger described that smell. "An unusual odor, although not particularly unpleasant." Bill Shepherd, the commander of the first expedition to the *International Space Station*, described it as simply having a very faint "kind of burnt toast odor."[9]

For Americans, accustomed to cordial, calm, and respectful social manners, the passionate and sometimes harsh behavior of Russians was often hard to take. Moreover, while Krikalev had been bothered in Houston by how frequently Americans smiled, the Americans in Moscow were equally disturbed by how all Russians seemed to walk around with a perpetual scowl on their faces. To the Americans, Russians didn't simply wear the light, empty, neutral look of subway-goers in cities like New York or Boston. Instead, they seemed to glower, their faces forever frozen in what appeared to be an expression of almost discomfort or anger. This sea of unhappy-looking faces made Americans feel uneasy and unsafe.*

Nonetheless, when Dezhurov snapped his first order at Thagard, the American bit his lip, kept quiet, and did as he was told. He was a guest on *Mir*, on a program whose supposed purpose was to foster international goodwill. Getting into a fight with his commander only a week into his mission would hardly promote international peace. Only after a few weeks did he decide to copy Strekalov's tactics and tell Dezhurov to back off. After that, their relations steadily improved.

On the ground, the various NASA officials trying to coordinate the shuttle-*Mir* missions were going through some of this same culture shock. For them, the Russian they found most difficult to work with was Valeri Ryumin, former *Salyut 6* cosmonaut, former

---

*When I was in Moscow myself I could see this phenomenon quite obviously on the subway. If several Americans got on board, everyone immediately knew. While the Russians sat silent, scowls of neutrality imprinted on their faces, the Americans laughed and joked as they talked about the wife, the kids, the house, or whatever.

mission control flight director, and now Russian director of the shuttle-*Mir* program.

Befitting his tank-commander background, Ryumin could be brutal and cold-blooded. Practically every American who dealt with him came away exhausted and infuriated. "I really did want to crawl across the table and rip his face off," said Peggy Whitson, the project scientist for Thagard's flight, after one session in which Ryumin had harangued her and others about how inadequate American astronauts were compared to the Russians.

"Ryumin treated me like dirt," said Bryan O'Connor, former astronaut himself. "He would bad-mouth people and complain about everything—and drink vodka. He was just a jerk."

Even the Russians admitted that Ryumin's style was brutal. "The man is an animal," remarked Alexander Serebrov, who had been forced from the Russian program by infighting with Ryumin and others. "Everyone calls him the Russian pig."[10]

What the Americans didn't appreciate—and Serebrov did—was that Ryumin was behaving as Russians expected their administrators to behave. In order to get the best deal possible, intimidation was considered a normal Russian business tool. Furthermore, Ryumin was fiercely proud of his country's accomplishments in space. To him, a veteran of almost a year in space, astronauts like O'Connor (who had only 16 days in orbit) were inexperienced neophytes.

For Thagard, the foreignness of these circumstances was accentuated by his isolation. Though he generally had friendly and good relations with his Russian crewmates, both of whom acted like gentlemen compared to Ryumin, and though he had regular, weekly, hour-long conversations with his family, either by radio or by video, he soon discovered that days could pass without hearing an English-speaking voice. Moreover, the lack of satellite communications meant that there were times when he went days without speaking to *anyone* on Earth. *Mir*'s limited communications with the earth, at best only a few minutes every 90 minutes, also meant that he had no access to American news: no newspapers, no television, no radio, no internet. And though the Russians sometimes sent up Russian-language broadcasts, he simply didn't have the language skills, even after almost two years in Russia, to absorb this rapid-fire talk.[11]

Thagard wasn't particularly bothered by this cultural isolation. He had served in Vietnam, and had often spent months and even years away from home. What he found disturbing was NASA's star-tling lack of interest in and disregard for his mission. From the very beginning NASA's commitment to Thagard seemed "basically non-existent." When he first volunteered to go to Russia, it took more than a year for the NASA bureaucracy to officially name him to the flight. In the interim, he had to take Russian language lessons on his own. When funds were finally found to send him to the Depart-ment of Defense's Defense Language Institute in Monterey, Cali-fornia, the most NASA would pay him was $10 a day. And they refused to pay for a rental vehicle, forcing Thagard to drive there in his own 10-year-old car. As he crossed into Death Valley on his way from Houston, one wheel fell off. As the car skidded, Thagard was able to get it to slide safely to the side of the road. After getting the car towed to Las Vegas, he ended up having to trade it in for another used car to get to Monterey. Thagard figured that all told, going to Russia cost him about two grand out of pocket.[12]

In Russia, Thagard found NASA less than helpful. Though they provided him the airplane cargo space to bring over a great deal of personal goods, they gave him little else in support. The one laptop they provided him was broken, and they were never able to fix it. "NASA didn't do squat," he recalled. "I would have been far hap-pier if NASA had sent me over there and left me alone."

Later, when he was in orbit, he and his ground support team repeatedly had trouble getting information from NASA. He would ask questions, and sometimes wait weeks while his team struggled to get answers from a space agency that had no apparent interest in what he was doing.

Sometimes, even his own ground team seemed uninterested. Once, he called down for some assistance and was told he had to wait because the American support personnel were involved in a simulation. "Since when did a simulation have priority over a mission in orbit?" he wondered in exasperation. This lack of sup-port implied to him that "the orbital activities just didn't have any priority."[13]

Nor was Thagard's experience unusual. During the entire shuttle-*Mir* program NASA seemed oblivious to the needs of the astronauts it sent to Russia. The same management problems of arrogance and ignorance that had hampered the building of *Free-dom* and had caused the destruction of *Challenger* had once again reared their ugly heads.

The problem stemmed from politics. To fulfill his foreign-

policy objectives, Clinton had declared that Americans would fly
on *Mir*. Nothing, not even the ill-treatment of those Americans,
was going to stand in the way of those objectives. Just as Thagard
had backed off so that he wouldn't threaten that political goal,
NASA officials, in order to avoid any possibility of offending the
Russians, repeatedly backed off when challenged by them. This ti-
midity weakened their bargaining position during negotiations because of the Russian tendency to play hardball in any haggling.

For example, the final deal worked out by Goldin required the
Russians to use part of the American money to build new duplex
apartments in Star City for the American astronauts. However, when
Jerry Linenger arrived with his pregnant wife, the duplexes were not
ready, and wouldn't be for months. Instead, multi-million-dollar
dachas were being built at Star City for important Russian space
officials. After months of waiting and repeated demands, all ignored
by both NASA and Russian officials, Linenger finally threatened not
to go back to Russia when he and his wife returned to Houston for
her delivery. This threat worked, and his duplex was finished. As
Linenger later wrote, "NASA shuttle-*Mir* management seemed at
best reluctant to confront the Russians. Duplexes not built as paid
for? No problem. Training materials not provided? Not to worry, our
astronauts will figure out a way to get by. This approach by NASA
management made the astronauts training in Star City feel that they
had been abandoned. No one, not even *our* guys, seemed to want to
make our training or living conditions any better."[14]

When American reporters questioned NASA officials about the
dachas, the officials passed the buck, saying simply that the Russians could pay their officials however they wished. When astronauts like Linenger complained about bad training manuals, NASA
officials pooh-poohed the issue. Rather than face the facts and get
the problem resolved, they tried to make believe the problem
wasn't there. "I dreaded going to see Linenger," said Frank Culbertson, the American program manager for the shuttle-*Mir* program
and himself an astronaut. "He always complained about how things
were not right."[15]

Jerry Linenger, who flew on *Mir* 10 months after Thagard, said
it best. "The [shuttle-*Mir*] program was not primarily concerned
with doing good science or advancing our expertise in space operations, but rather was conceived and thrust down NASA's throat by
the Clinton administration as a form of foreign aid to Russia."[16]

Therefore, it wasn't surprising that Thagard had difficulty accomplishing most of his scientific goals during his stay on *Mir*. Sent to *Mir* according to a fixed schedule—determined three years earlier in political negotiations between Dan Goldin and Yuri Koptev—he arrived before *Spektr*, the module that contained most of his supplies and a crucial freezer for preserving his blood and urine samples. Then, *Spektr*'s launch was so delayed that it arrived near the end of his flight.

The lack of ground support was not the only thing that hampered Thagard's research efforts. The physical condition of the nine-year-old station hindered him also. When he first entered *Mir*, he was immediately struck by how much the place reminded him of "someone's utility room"—packed and cluttered. Though he found the station roomy, clean-smelling, and comfortable ("During the work day, it was not at all unusual to not see [my crewmates] all that much."), he also noted, during a televised press conference on his fourth day on board, that *Mir* had "obviously been lived in and worked in for nine years."[17]

Everywhere he looked he could see floating gear. As the years had passed, the storage space behind equipment panels had become stuffed to the brim. Any repair job required an hour or so of unpacking (while also trying to keep the loose gear from drifting off or getting in the way). Doing experiments was further hampered because many of the necessary tools were often hidden away behind a panel, forcing the researcher to waste hours searching for them. In fact, six months after Thagard returned from *Mir*, a centrifuge was "lost" for more than two months amid the clutter, preventing its use during several scheduled experiments.[18]

While waiting for *Spektr*'s arrival, Thagard attempted to make do by using an old and failing European Space Agency freezer. However, the unit needed defrosting, and there was no practical method for doing so. After less than a month, Thagard finally gave up, throwing away about a quarter of his blood, urine, and saliva samples and squeezing the rest in odd corners of *Mir*'s two small freezers and one food refrigerator.[19]

His research program crippled, Thagard had little to do but watch videos, or stare out the window at Earth. When asked, he willingly helped his Russian crewmates. For example, in April they unloaded a *Progress* freighter and dismantled the rarely used shower on *Kvant-2* so that new gyros could be installed there.

However, the Russian distrust of foreigners combined with the Russian habit of belittling subordinates made Dezhurov and Strekalov at first unwilling to let him do anything very complicated. Though Thagard was an experienced spacefarer (having flown four times previously) and repeatedly offered to help them, they treated him just like the first Czechoslovakian cosmonaut: Thagard had come down with "red hands disease." They let him do only those specific tasks he had trained extensively for on the ground.[20]

For Thagard, a man of action who liked to finish what he started, the inability to complete his scientific program was both humiliating and infuriating. Born during the war to a poor truck driver and a mother who left when Norm was a child, the boy was mostly raised by his paternal grandmother. In high school he worked 35 hours a week to help pay the bills. At the same time, he figured out how to build radios, and got his ham radio license. By age 14, he had decided he wanted to be an astronaut. He also decided that to do it required that he become an electrical engineer, a fighter pilot, and a doctor. Amazingly, over the next 30 years he became all three. Then, after joining NASA in 1978 as part of the first class of space-shuttle astronauts, Thagard became one of NASA's better-traveled astronauts.[21] To be stuck in space with no work of his own and begging for something to do from the Russians was simply maddening.

As frustrated as Thagard was, and as reluctant as the Russians were to call on his aid, the month of May was hardly an idle month for him or any of the men on *Mir*. After almost four years of delays, *Spektr* was finally scheduled to arrive on June 1, 1995, followed by the first shuttle docking three weeks later. For both to occur, the *Mir* complex had to undergo significant reconfiguration.

When *Mir* was designed, each of the four modules attached to the multiple-docking adapter was intended for a very specific position. *Kvant-2* was already where it was supposed to be, at the top port. *Kristall*, temporarily parked in the bottom port, was intended for the starboard port, with its solar panels removed and placed on *Kvant*. (If the panels were left attached when *Kristall* was in the starboard port, one would smash against the base block's solar panels, resulting in a very unpleasant-looking mess. Moreover, the module had been intended as the docking port for the Soviet shuttle, and the solar panels would interfere with dockings.)

Until *Spektr*'s arrival, however, placing *Kristall* in the starboard

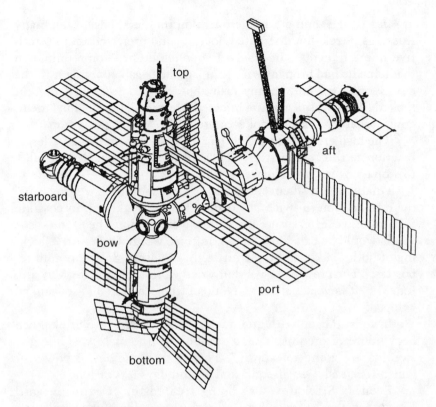

top

aft

starboard

bow

port

bottom

Mir, with *Kristall* moved to the starboard port (with one of its solar panels relocated to *Kvant*) and *Spektr* docked to the bottom port. Note also
the addition of the thruster to the top of Sofora and the second test girder
on *Kvant*. *NASA*

port would make the *Mir* complex asymmetrical and difficult to
steer. *Kristall* had instead been placed in the bottom port—*Spektr*'s
planned home—so that the entire station was T-shaped. Moreover,
though two crews had done space walks installing the mounts on
Kvant for *Kristall*'s solar panels, the panel relocation had been postponed until *Spektr*'s launch date was actually known. With *Spektr*'s
arrival pending, *Kristall* had to be shifted to the starboard port.
Dezhurov and Strekalov were to do at least four space walks, moving one solar panel to *Kvant* and partly retracting the second.

Before they even began, however, the whole program was
threatened by a rare injury. Strekalov had been cleaning behind one

of *Mir*'s panels, and cut his left arm on its metal edge. Like many Russians, Strekalov distrusted doctors, and preferred more-primitive home remedies. In his own supplies he had some unknown medicine he had brought with him, and he began applying it to the cut. Soon, a red and bumpy rash appeared. To Strekalov, the cut looked like it was becoming infected. He added more of his "medicine," what to Thagard looked merely like "some green stuff."

The rash grew. Convinced that the rash was merely an allergic reaction to the "green stuff," Thagard tried to persuade Strekalov to stop using it. Strekalov resisted.

Finally, the rash covered half his arm, with the swelling causing it to balloon to twice its normal size and making it impossible for him to wear his space suit. If the swelling didn't heal, the space walks would be canceled, which, in turn, would prevent the docking of both *Spektr* and the shuttle. As Vladimir Solovyov told reporters, "The crew has a very crowded work schedule for May and June. . . . It is very tense. . . . None of [the scheduled tasks can] be skipped."[22]

In consultation with ground doctors, Thagard gave Strekalov a heavy dose of prednisolone, a steroid. "If [the rash was] allergic, steroids are going to stop it," Thagard said. "They're powerful guns." The rash *was* allergic, and within days it was gone.[23]

Even as Strekalov was being treated, he, Dezhurov, *and* Thagard were sorting and installing a hundred feet of new electrical cable throughout the inside of *Mir*. Each *Mir* solar panel funneled the electricity it produced into storage batteries at the panel's base. When they relocated the *Kristall* solar panel, however, the batteries were going to remain on *Kristall*. To link the panels with the batteries, the men ran power lines from pre-planned plugs inside *Kvant*, through the core module, through the multiple-docking adapter, and into *Kristall*, where they were linked to the batteries. As he helped to rewire the station, Thagard couldn't help noticing that the electrical cables were being run through the hatches (alongside a number of previously installed wires and air ducts), a procedure that made him wonder how they would close them in an emergency.

Then, after weeks of preparation, and with Strekalov's rash gone, the two Russians opened the airlock hatch on May 12. As they crawled out along the Strela boom and out to *Kvant*, Thagard stayed inside *Mir*, moving back and forth from the core module to

Kristall, depending on what needed to be done. At *Kvant* the two Russians did some rewiring, preparing the mounts for the solar panels. Then they moved to *Kristall*, where they watched as Thagard remotely operated the panel's servomotors, folding three of its accordion sections back together and unfolding them again. The test showed that, after hanging extended in space for five years, the mechanism still worked—at least for the three sections that Thagard tested.

Five days later, the two Russians were outside again. First, Dezhurov climbed along Strela's length to its controls. Then Strekalov climbed onto Strela, and Dezhurov swung him over to the base of *Kristall*'s solar panel. Inside *Mir*, Thagard operated the servos that folded the panel up, with Strekalov at the panel to monitor the operation. Several times the servos froze. To get them to move, Thagard throttled them back and forth, as if he were rocking a car in snow. Often, Strekalov gave the sections a shake to help them along. Finally, after several hours of struggle and with the panel folded into a coffin-sized box with a mass of more than half a ton, Strekalov reached under it and released it from its mount. Very carefully, he strapped the folded array to the end of the Strela boom, then climbed on as Dezhurov swung him and the panel across to *Kvant*, Strekalov stabilizing the massive object as it moved.

At this point the two men had been outside for more than six hours, far longer than planned. Already they were using the emergency supplies in their Orlan-DMA spacesuits, which could last only another hour at most. Compounding the risk, as soon as they had retracted and disconnected the solar panel, the station's ability to accumulate electrical power was reduced. With only nine working batteries, *Mir* very quickly experienced power problems, and its computers began shutting down extraneous systems, including some lights, fans, and the Elektron oxygen regenerators.

With the panel floating just above its mount on *Kvant*, Strekalov tried to move it into position as Dezhurov guided it with the Strela crane. To their dismay, the job was infuriatingly difficult. Though they could easily move the massive panel with the mere touch of a fingertip, the inertia of its heavy mass made it almost impossible to stop once it was moving. It was like trying to get a thick thread into the eye of a needle—they kept missing the hole.

Finally, with time running out, Solovyov ordered the men to

quit and get back inside as quickly as possible. They hurriedly lashed the folded panel to *Kvant* using the various straps and lanyards used to keep their tools from floating away, then scrambled back to the *Kvant-2* airlock, completing an almost seven-hour space walk, the longest Russian space walk since Anatoli Solovyov and Alexander Balandin almost suffocated in 1990 when they couldn't shut the *Kvant-2* hatch. By the time Strekalov and Dezhurov got back, they had almost suffocated themselves—the carbon dioxide scrubbers in their suits were completely saturated. [24]

For the next six days, until they could get the solar panel plugged in, *Mir* operated on limited power, its lights dim, its food and beverage heaters using as little power as possible, its Elektron atmosphere-recycling system disabled and the men burning lithium perchlorate candles instead. To supplement the station's power, ground controllers decided to keep the *Progress* freighter docked to the aft port a few extra days so that the station could access the power from its two solar panels.

As the men on *Mir* prepared for their next space walk *Spektr* lifted off from Baikonur on May 20. Originally scheduled for launch before 1990, *Spektr* ("spectrum" in Russian) had been designed to test military technology for monitoring ballistic missile launches. The fall of the Soviet Union had prevented its completion. With the U.S.-Russian space agreement, Russia used American funds to redesign *Spektr* as a science module holding much of NASA's research equipment. In addition to the many remote-sensing instruments that festooned its exterior, *Spektr* also deployed four separate solar panels. (Because of *Mir*'s chronic power shortages, a second set of panels had been added late in *Spektr*'s construction.) When plugged into the station's electrical system, these arrays would boost *Mir*'s ailing power supply by a total of 6.9 kilowatts. The first set, attached to the module's main body, opened immediately on reaching orbit. The second, attached to the module's pointed bow, remained furled, to be opened on command after the module had docked with the station.

Spektr's interior was divided into two sections. The first 18 feet near the cone-shaped bow were unpressurized, containing the equipment for the module's scientific sensors. The module's pressurized, main section, 29 feet long, contained the instrument panels for operating these sensors. It also carried two freezers for storing the biological samples that Thagard was supposed to collect.

Also stored in this habitable space were two new storage batteries, intended to supplement *Mir*'s failing batteries.[25]

Overall, *Spektr* added about 2,200 cubic feet of habitable space to *Mir*, bringing this volume to about 10,600 cubic feet, still about 2,000 cubic feet less than *Skylab*. When docked with *Mir*, the complex's total mass rose to approximately 90 tons, finally making it heavier than *Skylab*, launched more than two decades earlier.

For the next 10 days *Spektr* was piloted along the most leisurely but fuel-efficient route available for rendezvousing with *Mir* while Dezhurov and Strekalov raced to get the station ready for the new module. On May 22, they space-walked to *Kvant* and, after some more struggle, finally attached the solar panel and plugged it in. Thagard then activated the servos and unfurled it. He next connected the power cables, and very quickly the panel began generating power, allowing the station's systems to come back to full life.

Meanwhile, Dezhurov and Strekalov climbed back to *Kristall* to monitor Thagard's partial retraction of *Kristall*'s other solar panel. He folded only 13 of its 28 sections, allowing the panel to produce about 60 percent of its electrical capacity and still be short enough to fit inside the shuttle's cargo bay during a docking.

On May 23, the three men finished packing garbage in the *Progress* freighter. Ground controllers then undocked it from the bow port, freeing that port for *Spektr*'s docking, and sent it into the atmosphere to burn up.

Having shifted *Kristall*'s solar panels, the men still had to shift the module itself from its temporary location at the bottom port, where *Spektr* would reside, to its permanent location at the starboard port. However, making this move required an additional series of complex steps. First, on May 26, Dezhurov used *Kristall*'s small robot arm to grab *Mir*'s multiple-docking adapter. The arm then lifted *Kristall* from the bottom port and swung it into the bow port.

To make the next shift to the starboard port required some internal reconfiguration. Each of the four radial ports in the multiple-docking adapter could be sealed with a flat, hinged hatch. However, for a module to be docked at any port, the flat hatch had to be removed and a cone-shaped, airtight, docking receptacle inserted instead. This cone was then used by the module's docking probe to find its way into position during docking.

To put receptacles in all four radial ports would waste too much space and fill the multiple-docking adapter. To save space, *Mir*'s engineers made the receptacle removable and sent two up to *Mir* when it was launched. One remained permanently in the bow port. The second was shifted around, depending on where a module was going to dock. Once a module was in place, the cone could be removed, and the flat hatch reattached.[26]

Since 1990, when *Kristall* had arrived, the two cones had sat in the bow and bottom ports. To shift *Kristall* from the bow to the starboard port required that the cone at the bottom port be transferred to the starboard port first. To make this shift, however, required what Russian officials labeled an "internal space walk." On May 29, with *Spektr*'s rendezvous only two days away, Dezhurov and Strekalov donned their spacesuits, climbed into the multiple-docking adapter, closed all its hatches and evacuated its atmosphere. They then repositioned the cone from the bottom to the starboard port.

The next day, with *Spektr* fast approaching, Dezhurov took control of *Kristall*'s small robot arm and attempted to shift the module from the bow to the starboard port. When he tried to bring *Kristall* into a hard dock, however, the latches to lock the module into the starboard port, not used in nine years, refused to catch. Dezhurov used the robot arm to pull the module back, reposition it, and try again. Again the latches would not lock.

If he couldn't get the module to dock, the entire shuttle-*Mir* joint program was threatened. With *Kristall* in the way, *Spektr* could not be docked, and the shuttle could not link up with *Mir*.

Dezhurov tried a third time, pulling the module back and readjusting its position. He then carefully and slowly guided it back into the receptacle cone. To everyone's relief, this time the latches locked. *Kristall* was finally docked in its intended port. The next day, June 1, the 20-ton *Spektr* module glided into *Mir*'s bow port and docked automatically on its first try, the first time one of *Mir*'s large-sized modules was able to do so.

That same day, a leak was discovered in *Mir*'s cabin, apparently coming from the starboard port where *Kristall* was docked. It appeared that the difficulties in docking had been related somehow to how the receptacle cone was fitted into place. Because it was a slow leak (taking approximately 5 percent of *Mir*'s atmosphere per day), ground controllers decided to see if it could be cured when the receptacle cone was next moved a day later.[27] On June 2 Dezhurov and Strekalov did another internal space walk, repositioning the receptacle cone back from the starboard to the bottom port so that *Spektr* could be moved there.

The gamble worked. Somehow, by removing the cone the leak disappeared. The next day Dezhurov used *Spektr*'s robot arm to move *Spektr* from the bow to the bottom port.

Even now the dance of modules was not complete. In only three weeks the space shuttle was scheduled to arrive and dock with the androgynous port at *Kristall*'s far end. To do so, *Kristall* had to be repositioned again by moving it temporarily from the starboard to the bow port so that its shuttle-docking port was as far from *Mir*'s solar panels as possible.

Before Dezhurov and Strekalov did this, however, they, with Thagard's help, began activating *Spektr*. For three days they unloaded the module's cargo while also plugging in and turning on its equipment. Then, on June 5, ground controllers tried to unfurl *Spektr*'s third and fourth solar panels, and mistakenly sent up the commands in the wrong order. While one panel unfolded completely, the other was held closed by an unreleased restraining strap.[28] As he had done three weeks earlier to get the *Kristall* array to fold up, Thagard tried cycling the array's servos back and forth, hoping to get the strap to come loose. This time the trick didn't work. Then Dezhurov tried firing several of *Mir*'s thrusters, hoping that the movement would shake the panel free. This didn't work either.

While Russian engineers pondered what to do about the stuck solar panel, Dezhurov and Strekalov unloaded the two new batteries from *Spektr* and installed them in *Kristall*. Then, on June 10, Dezhurov commanded the robot arm to shift *Kristall* from the starboard to the bow port, positioning it along the station's main axis, ready to receive its first shuttle. The same day, Thagard activated the American equipment on *Spektr*, including its freezers. He immediately began transferring his remaining samples there.

With surprisingly little fanfare, that same week Thagard surpassed the flight record of the last *Skylab* crew. Twenty-two years after Carr, Pogue, and Gibson had set the record for the longest space flight, long since eclipsed by many Soviet-Russian achievements, Norm Thagard became the first American to spend longer than three months in space.

The moment was filled with astounding irony. Thagard broke the record on Tuesday, June 6. Six days earlier *Spektr* had arrived, and immediately afterward Thagard's ground people asked him to do the check-out of *Spektr*, even though that required him to work on the weekend, on his days off. He agreed, and then discovered that no Americans were on the ground to answer questions because NASA had given "all the support people the weekend off." Once again, Thagard felt screwed by his own people.[29]

So, on that Tuesday when NASA officials brought Ed Gibson in from that last *Skylab* mission to congratulate Thagard on setting the new record, Thagard wasn't particularly interested in spending a lot of time doing a NASA public-relations event. As he noted years later, "I appreciated the moment, but I didn't want it to take that long." With *Spektr* there at last, he had a lot to do, work hampered by a lack of support from Earth.

Ironically, Thagard had no idea how similar his situation was to what Gibson's had been on *Skylab*. He imagined, wrongly, that the *Skylab* mission had gone like clockwork, made easier by having the full support of NASA. Instead, NASA had picked up exactly where it had left off 30 years earlier, with the people on the ground having no concept of what was going on in space. And that same lack of understanding was causing the same kinds friction and ill will.

Gibson understood however. "Norm, keep on trucking," he said, wrapping the conversation up quickly.[30]

"We Have Capture!"

Meanwhile, in Florida, the American shuttle *Atlantis* was standing ready on Launchpad 39-A, the same launchpad from which *Apollo 11* had departed to the Moon and *Skylab* to Earth orbit. Its crew was ready, its cargo was loaded. For the first time since *Columbia* was launched in 1981, the American space shuttle had somewhere to shuttle to. If all went well, *Atlantis*'s scheduled launch was set for June 23, with a docking for *Mir* two days later.

In Russia, space officials were still wrestling with several unresolved problems. In addition to the unfurled *Spektr* solar panel, a solar panel on *Kvant-2* refused to track the sun correctly. These problems, combined with the shortage of battery power, limited *Mir*'s electrical output. Moreover, the air leak, though seemingly gone, remained unexplained. Without knowing its exact cause, no one knew if it would reappear later, perhaps at the worst possible moment.

On June 13, Russian mission control decided, after consultation with NASA officials, to have Strekalov and Dezhurov do an additional, unplanned space walk before *Atlantis*'s arrival, inspecting the two solar panels and the starboard docking port to see if they could fix any of the problems.

On *Mir*, Strekalov listened to these plans, and then shocked everyone when he adamantly refused to do the space walk. "I was really yelling at them," he later remembered. Strekalov had a number of reasons for staying inside *Mir*. They didn't have the right tool to cut the strap that held *Spektr*'s panel—it was supposed to be delivered by the shuttle. Evacuating the airlock would put *Mir*'s air reserves dangerously low—especially considering the loss of atmosphere due to the still unexplained air leak. And according to Strekalov, if something catastrophic went wrong Thagard was unprepared to handle it. "We were trained for these emergency situations and Norm was not," Strekalov explained in that characteristic patronizing manner that Russians had for foreigners. "We had to take care of him."

Dezhurov desperately tried to change his crewmate's mind. For the younger man, Strekalov's refusal was horrifying. In the topdown culture of Russia, the refusal to follow an order could have serious consequences. At the least, Dezhurov feared that they might be blacklisted, ending any chance of flying in space again.

To Strekalov, none of this mattered. He was 54 years old and at the end of his cosmonaut career. He had flown five times previously—almost dying twice. He had nothing to gain by stepping outside *Mir*, and everything to lose.

In a sense, the changes that had been going on in Russian society on the ground had finally reached *Mir*. While past cosmonauts had exercised some discretion (such as when Musa Manarov had climbed to the top of the just-installed Strela crane without permission), none had ever before refused an order point-blank. The risks, including being blackballed or imprisoned, had been too great. After the fall of the Soviet Union these old rules no longer applied. Despite growing up in an authoritarian society, Strekalov now considered himself a free man, living in a free society. He knew that the worst his bosses could do was fire him. And in the free, capitalistic society that Russia had become, he knew that he could always get another job, as good or even better.

After two days of arguing, the Russian ground controllers finally decided to cancel the space walk. Though they eventually fined the Russian cosmonauts the equivalent of $9,000, 15 percent of the men's total salary, they simply could not make Strekalov do something he did not want to do. Though Russian culture was still fierce and overbearing, the apparatchiks in charge could no longer back that behavior up with violence. Strekalov understood, better than most, that his boss's bossiness was almost all bluff. (He even fought the fine and won, going to arbitration and getting it canceled.)[31]

For Dezhurov, the emotional turmoil of the situation was at that very moment compounded when, just as the argument between Strekalov and ground controllers ended, the mission's flight doctor decided to tell him that his mother had passed away unexpectedly.

The young man was crushed. Not only was he very close to his mother, he had been isolated for three months in the alien environment of a space station, circumstances that routinely made men and women overly sensitive to bad news from home. It was for this reason that Grechko had not been told of his father's death, back in 1978 on *Salyut 6*. Furthermore, the news came at a time when Dezhurov's budding space career seemed threatened by Strekalov's veto of the space walk. The deluge of bad news almost overwhelmed him. For the next few days he refused to speak to ground controllers, including his own doctor. He disappeared into his cabin and spurned all work. Strekalov even had to prod him to eat.

For Thagard, these events were baffling. Unfamiliar with Russian customs, he didn't know how to approach Dezhurov about his grief. And both he and his fellow Americans in Moscow were confused by Strekalov's actions. For them, it was inconceivable for such passionate arguments to occur during a mission.[32]

While these events were transpiring on *Mir*, the shuttle countdown in Florida began—despite predictions of violent thunderstorms on June 23. The night before launch, the predictions proved true, with lightning hitting the ground only five miles from the launchpad. After two hours of discussion, with heavy rain pounding Cape Canaveral, NASA officials postponed the launch for 24 hours.

The weather didn't improve. On June 24, a lightning bolt hit the launchpad just as the crew was being bussed there. Then, only

45 minutes before blastoff, the rain grew so thick that *Atlantis* became invisible from the public viewing stands only three miles away. Once again the launch was scrubbed and rescheduled for three days later.

Finally, on June 27, 1995, the weather cleared just enough. With fog hanging over the launchpad, *Atlantis* blasted off, beginning its two-day voyage toward *Mir*. For sentimental and publicity reasons, *Atlantis*'s flight—its 14th into space—had also been timed to be the 100th American manned mission into space, a string that had begun 35 years earlier with Alan Shepard's suborbital flight.

In those 35 years much had changed. On May 5, 1961, Alan Shepard took off in a tiny capsule named *Freedom 7* that weighed only 2,844 pounds. *Atlantis*, with cargo, weighed approximately 70 times that.

Shepard's flight was not intended as an orbital flight. The highest he ever got was 115 miles on a journey that traveled only 302 miles downrange from Cape Canaveral. His fastest speed was only 5,100 miles per hour. His space flight lasted only 15 minutes.[33] *Atlantis* was in orbit for just less than 10 days, orbiting the earth 152 times at altitudes ranging from 100 to 250 miles. In this roomy spaceship (which could have easily carried a half-dozen *Mercury* capsules in its cargo bay) was a crew of seven, six of whom had already accumulated more than 478 days in space, including one man (Russian cosmonaut Anatoli Solovyov) who had amassed more than a year in orbit.

Atlantis's captain was Bob "Hoot" Gibson, on his fifth flight. A skilled pilot who had been flying solo since he was a teenager, Gibson had already had a successful career as a shuttle pilot. His first flight in 1984 was the first to land the shuttle at the Kennedy Space Center at Cape Canaveral. His latest was the second flight of *Endeavour*, piloting the fifth spaceship to return to space more than once.[34] From 1991 to 1994 he headed NASA's Astronaut Office, where it supposedly was his job to assign crews to missions, based on skill, mission requirements, and fairness.

By the 1990s, however, NASA resembled the Byzantine design bureaus of the Soviet Union. Apparatchiks who knew how to curry favor with appointed political administrators or to make small-minded elected officials happy often ran the show. In the case of NASA, the apparatchik who held sway over crew assignments was a man named George Abbey, head of the Johnson Space Center.

Though he often exhibited fine common sense in choosing which astronauts should fly and which should not, he also played all the same petty power games that bosses in the Soviet Union played. You had to butter him up, drink beer with him. You had to be one of "his" guys. You had to watch what you said, or else his many spies would report back to him and get you in hot water. Most of all, you had to do *exactly* what he told you, or else you were frozen out of any space assignments.[35]

For the three years Gibson ran the astronaut office, he struggled mightily to wrest control of crew assignments from Abbey. Even though it was Gibson's responsibility to make the choices, none of his decisions mattered if Abbey or some other higher-up didn't approve them. Though the circumstances were hardly identical, NASA had become, like the centralized government bureaucracies of the Soviet Union, an organization whose natural order of operation was from the top down.

In his efforts, Gibson sometimes succeeded in bypassing Abbey. For example, after much maneuvering and back-room politics, Gibson managed to get his choice of crew approved for the first rendezvous with *Mir* in 1994. "We shoved that one right down ole George's throat," Gibson later exulted. Moreover, though Abbey opposed the selection of Thagard as the first man to fly on *Mir* (for reasons no one quite understood), Gibson accomplished an end-around and got Thagard assigned to the flight.[36]

Abbey, however, had the last word. Through intermediaries, he offered Gibson the command of the first *Mir* docking, an assignment that Gibson found difficult to refuse. By taking the mission, Gibson had no time to head the astronaut office, and stepped down.[37]

Among Gibson's seven-person crew were Bonnie Dunbar, who had been Thagard's backup in Russia, and two Russians, Nikolai Budarin and Anatoli Solovyov, who were to take over *Mir* from Dezhurov and Strekalov. Solovyov was no longer the young and taciturn commander from 1990—unsure of what to say or how to present himself to the public. In the interim years he had become a Russian hero, the king of space walks, the man who in 1990 and 1992 had saved *Mir*. Now he was confident and assured, aware that he was one of a special elite who was skilled at doing something that almost no one else on Earth could do. He also took great pride in his country's space achievements, and was eager to get back to its space station.

For two days *Atlantis* moved steadily closer to *Mir*, the two space vessels racing around the earth in slightly different orbits. *Atlantis* followed a flight path similar to that used by *Discovery* the year before, coming up from below and behind so that its own orbital speed, rather than its rocket engines, brought it to *Mir*. Unlike the previous rendezvous, *Atlantis* would make its final approach and docking coming up from below *Mir* rather than from in front.

On June 29, with *Atlantis* still 40 miles away, Dunbar announced to Thagard by hand-held radio, "We have you in sight." From her perspective, *Mir* appeared as a silvery, gear-festooned cross of girder-encrusted cylinders just above the earth's glowing horizon.

Thagard responded, "Isn't that the way it always is? You call for a taxi, and it takes weeks to get here."[38]

Slowly, *Atlantis* crept closer. On *Mir*, Dezhurov and Strekalov had set the station's solar panels so that they were fixed edge-on relative to *Atlantis*. In this position, the risk of damage from thruster firings was reduced.

When *Atlantis* was about a half-mile from *Mir*, Gibson took over the helm. From his perspective, *Mir* was above him, silhouetted by the black sky, with the earth under the shuttle. Carefully he guided his spaceship to within 250 feet, then held his position, awaiting permission from the flight directors in both Houston and Moscow to complete the docking.

At that moment *Atlantis* seemed almost to dwarf *Mir* in size. The shuttle was almost as long as the station, but its wingspan and overall width made it seem far larger. Any one of *Mir*'s modules could fit inside the shuttle's cargo bay. A collision between these two vessels, each weighing approximately 100 tons, would surely be deadly.

After a short consultation, permission was granted, and Gibson resumed his approach, using both the television screens on his console and the large shuttle windows over his head for visual guidance. In the shuttle's open cargo bay was a Russian-made androgynous docking port, its interior attached to *Atlantis*'s habitable space by a tunnel. His goal was to fit this port to the port on *Kristall* that had originally been intended for a Soviet-built shuttle.

When *Atlantis* was about 30 feet away, Dezhurov deactivated *Mir*'s attitude-control system, letting the 110-ton station go into free drift so that the shuttle could take over after docking. Seconds

later, Gibson brought the two space vessels together, moving at a
relative speed of less than an inch per second, slipping the shuttle
into the docking port on *Kristall*. As *Atlantis* made contact, the
ship's small thrusters fired, pushing it forward in order to guarantee a hard docking.

The firing caused a shudder to rumble through *Mir*. "We have
capture," Hoot Gibson announced gleefully. Though he could see
the docking ports come together, he felt and heard nothing. The
docking was to him so smooth it was as if the shuttle were still
orbiting the earth solo. On *Mir*, however, things were different.
"We felt it," Thagard later said. "One hundred tons hitting the station had some impact on us."[39]

For the next five days *Atlantis* remained docked to *Mir*, creating the largest man-made object ever placed in orbit. When all its
various pieces and the *Soyuz-TM* spacecraft were added, the
complex's mass was almost 250 tons. Moreover, the combined *Mir-
Atlantis* vessel carried 10 people, the most to occupy any single
human-built space-borne facility.

During those five days, about 3 tons of equipment and supplies
were hauled across to *Mir*, including water and air sampling equipment, hard-drive replacements for laptop computers, and the tools
needed by Solovyov and Budarin to fix *Spektr*'s stuck solar panel.
Water normally produced as a waste product of the shuttle's fuel
cells and ejected into space before re-entry was instead transferred
to *Mir*'s tanks to restock its supply. Oxygen and nitrogen were also
bled into *Mir*'s atmosphere.[40]

The crews removed a great deal from *Mir*, including one of the
failed batteries that prevented the station from storing power. Also
unloaded were more than a hundred experiment packages that the
Russians normally had no means of returning to Earth.

While transferring all this equipment, the difference between
the two nations once again asserted itself, illustrated by what, at
first, seemed a very trivial issue that quickly became significant.
New Russian cargo brought up by *Atlantis* had been labeled by the
Russians in Russian. When the astronauts tried to match these
items with their cargo, they used inventory lists that had been
translated into English, then translated back to Russian. In the process many names changed so that they no longer matched and the
astronauts had trouble figuring out which item was which.[41]

In between these practical matters, the crews performed for

the cameras, participating in the typical political ceremonies demanded by ground-based politicians. Vice President Al Gore and Prime Minister Viktor Chernomyrdin, who were meeting at the time in Moscow, got on the radio to talk with them. The crews posed, holding linked toy models of the shuttle and *Mir*. They also held up two halves of a commemorative medal, bringing them together for the camera. The commanders then signed a document "certifying" the time of the docking.[42]

Finally, on July 4, 1995 *Atlantis* undocked from *Mir*. Also undocking was the *Soyuz-TM* spacecraft, flown by Solovyov and Budarin. If all went as planned, for a short while both the shuttle and the *Soyuz* capsule would fly in formation around a temporarily unoccupied *Mir*. Several NASA engineers and management people had questioned this maneuver, worrying that if something on *Mir* failed at that time, there would be no one there to fix it. The Russians dismissed the worry, considering the excursion no different than a routine *Soyuz* fly-around from one port to another. As Ryumin bluntly noted, "It's our risk, not yours."[43]

At 6:55 A.M. (EST), Solovyov pushed the *Soyuz* spacecraft back from *Mir*, maneuvering it about 300 feet from *Mir*'s port side so that he and Budarin had a good view of the shuttle when it undocked. Fifteen minutes later, Hoot Gibson undocked *Atlantis*, using small springs in the docking unit to push the 100-ton shuttle away. Within minutes both spacecraft were several hundred feet from the station with both crews snapping pictures madly.

For Thagard, it was a moment of decidedly mixed feelings. Though he was thrilled to be going home, the thought that he might never fly in space again, never again float weightless above the earth, was saddening. "I wasn't sure I wanted it to end quite yet."[44]

At that moment, almost on cue, *Mir*'s attitude-control computers crashed. When *Atlantis* and *Mir* had been docked together, *Mir*'s control system had been turned off so that the shuttle could handle attitude control. Though the system was reactivated as soon as *Atlantis* pushed off, the force of the shuttle's separation was greater than the computer could handle, and it shut down, thereby causing the station's gyros to shut down as well. *Mir* began to drift out of control, quickly becoming skewed about 10 degrees from its correct orientation.[45]

Solovyov abandoned his camera and immediately swung his spacecraft around to perform a hurried docking with *Mir*'s aft port, where he and Budarin quickly scrambled to reactivate and replace computer-system software.

Three days later *Atlantis* returned to Earth, landing at Cape Canaveral with Norman Thagard, who had spent 116 days in space.

The cultural differences between Russia and the United States became evident once again on the ground. Experience had taught Russian doctors that it was best if a cosmonaut returning from three to six months in space did not walk immediately after landing. Cosmonauts (except for those setting records like Romanenko and Polyakov) usually accepted this command, waiting at least several hours before attempting their first steps. Thus, though his Russian crewmates meekly let themselves be carried off the shuttle in stretchers, Thagard strongly desired to walk off on his own two feet. Before landing, he checked with the NASA doctors, who said he could do it if he wanted, but, because some men had complained of back injuries by getting upright too quickly, they warned him to be careful.

After landing, he was the first to unstrap himself, forgetting to wait for the all clear from his shuttle commander. Because they were between him and the door, he had to wait for Dezhurov and Strekalov to be carried out before he could exit himself. Then, still wearing his pressure suit, he stood up gingerly, and walked into the bright summer air.

As expected, it took him several days to get back to normal; he felt heavier, weaker, and more awkward than he had after his shuttle flights. Controlling his movements was difficult. "If you had to make a left or right turn, you'd tend to overshoot. You'd tend to brush your shoulder on the opposite wall." If he leaned forward in his seat, he found it difficult to stop himself from falling forward on his face.[46]

Once on the ground, NASA wanted to follow the advice of the Russian doctors and have all three men kept together in isolation for several days before rejoining their families. When their plane landed in Houston, George Abbey came on board and told them that they were going to be sequestered. When they got to the bus, however, Thagard told Abbey thanks but no thanks. He turned to the bus driver and ordered him to take him home to his family.[47]

The return brought some culture shock for Dezhurov and Strekalov as well. For one thing, coming back on *Atlantis* posed an unexpected immigration problem. No one at NASA had thought of getting Dezhurov and Strekalov the proper visas for entering the

United States. (In a sense, both men, plus Thagard, were the first to use space travel to travel from one continent to another, taking off in Russia and landing in America.)

Dezhurov, already upset about Strekalov's revolt and the death of his mother, had been worrying about this situation for weeks, long before *Atlantis* arrived at *Mir*. A few days before *Atlantis*'s docking, he pulled Thagard aside. "We don't have passports," he explained. "I'm worried they're going to arrest us because we don't have passports or visas."

Thagard tried repeatedly to ease his fears. "I couldn't believe that in a million years they were going to arrest [them] because they arrived in the United States with no passport."

When they landed, the Russians discovered that the lack of passports was an issue after all. To make their arrival conform to customs regulations, the U.S. State Department had to go to the Immigration and Naturalization Services and apply for a "visa waiver for aliens from outer space." The visas were obtained without difficulty, though the effort caused more than a few jokes in the American press.[48]

Other cultural differences could be seen in small ways. For publicity reasons, Thagard was given a hot dog and a hot-fudge sundae during the initial press conference. The Russians insisted on, and got, vodka. And ironically, each side considered the other's traditions foolish and detrimental to medical research.[49]

For Thagard, that first post-flight press conference was a moment of truth. Though he was bothered by the lack of support and lack of scientific results, his biggest complaint was that NASA was trumpeting the supposedly great support they had given him and the results he had brought back. "They shouldn't be making claims that weren't true," he said later.[50]

Goldin, standing next to Thagard during this press conference, conceded that NASA had neglected the psychological aspects of Thagard's mission. "Dr. Thagard made it very clear to us. He called back and said 'Hey, you're not talking to me. I've gone days on end without news.' We really need to take a look at this."[51]

"Doesn't Anyone at NASA Even Care?"

High above, Budarin and Solovyov remained on *Mir*, the first Russian crew to come to a Russian orbiting station in a spacecraft other than a *Soyuz*. Over the next nine months they and two crews that followed them worked to get *Mir* in shape so that it could continue to operate through both the EuroMir and shuttle-*Mir* projects, scheduled to last through the next three years.

Once again, Solovyov proved the legitimacy of his reputation as a master of space walks, completing with Budarin three outside jaunts during their short two-and-a-half month mission. In the first, he and Budarin climbed out to *Spektr* where they repaired the balky *Spektr* solar panel using an American-built tool, manually unfurling all but one section (which remained half-opened and was considered an inconsequential loss of power). They also inspected the starboard port on the multiple-docking adapter, checking if any damage there could have caused the slow air leak. They found none.

On the second space walk Solovyov experienced a frightening moment of deja-vu. First, the cooling system in his Orlan-DMA spacesuit failed, forcing him to stay attached to the airlock while Budarin retrieved a variety of experiment packages. Then, when both men entered the airlock, they had trouble getting *Kvant-2's* troublesome hatch to close. Five years earlier Solovyov had struggled with this same hatch, having damaged it when he and Alexander Balandin opened it incorrectly. This time, however, the problem wasn't as serious. With a little extra effort, Solovyov and Budarin pulled it closed and repressurized the airlock.

After manually repairing Solovyov's spacesuit, the two men did their third space walk two days later to install a large European spectrometer for studying the earth's atmosphere, attaching it with clamps to the exterior of *Spektr*. When data from the instrument refused to flow to the ground, they traced the problem and fixed it. With this space walk, Solovyov became the true king of space walks, not just in word but in deed, exceeding Sergei Krikalev's record with eight space walks and a total of 41 hours 49 minutes outside.[52]

Space walking was not all that Solovyov and Budarin did. In between, they repositioned *Kristall* from the bow to the starboard port where it belonged, and did extensive repairs to the station's systems, replacing several ailing gyros on *Kvant* and *Kvant-2*.[53]

The next *Mir* crew included a German, Thomas Reiter, continuing the EuroMir project begun the previous year during Valeri Polyakov's long mission. Much of his research work turned out to be impossible because of *Mir's* limited electrical power due to the failure of many of its storage batteries and the degradation of its solar panels. Moreover, his mission was originally supposed to last about four-and-a-half months. The chronic cash shortages in Russia changed that. After the fall of the Soviet Union, the Russian space program had saved the cost of building new rockets by using its stockpile of unused military boosters. By the fall of 1995, when Reiter was supposed to come home, that reserve was exhausted. To launch the next replacement crew as scheduled, in October, would require building a new booster, something the Russians did not have the money to do immediately.[54]

So Thomas Reiter stayed in space for six extra weeks, completing a six-month mission, the first non-Russian to do so. When he finally came home, his medical data confirmed what all the Russian long-term missions before him had found—that the human body could easily handle weightlessness for at least this long. The only reservation remained the loss of bone tissue. His rate of bone loss, averaging between 0.5 percent and 1 percent per month, fit at the lower end of the scale of previous spacefarers, who had lost bone tissue in their lower extremities at rates ranging from 0.3 percent to 3 percent per month. The evidence that the exercise program the Russians had developed kept his bone loss low continued to encourage them that the problem was solvable. That he had still lost bone tissue suggested, however, that, no matter how much a person exercised, there might be a limit to how long a human could survive without gravity.[55]

Reiter's flight to *Mir* was plagued by new problems and bolstered by new equipment. In early November the Vozdukh carbon-dioxide-scrubbing system in *Kvant* inexplicably shut down. Simultaneously, *Mir*'s overall temperature began rising. Activating some lithium hydroxide canisters to make up for the lost Vozdukh system, the cosmonauts began a systematic search for the problem. Very quickly, they discovered that the station's primary cooling system had sprung a bad leak. Behind a panel in *Kvant* they found a half-gallon glob of floating ethylene glycol, the same kind of antifreeze found in car radiators.

Without the cooling system, the sieves in Vozdukh could not work and the station's temperature was difficult to control. The antifreeze now saturated *Mir*'s atmosphere and the long-term health consequences of exposure to it were unknown. In addition, the supply of lithium hydroxide canisters was running out. As a quick solution, the Americans offered 20 of the lithium hydroxide canisters used on the shuttle, along with an adapter to fit them to *Mir*'s system, all scheduled to arrive when the next shuttle docked with *Mir* in a few weeks.

After spending several days tracing the cooling system's pipes that ran throughout the station, sometimes hidden in panels behind years of accumulated junk, the cosmonauts located the leak and temporarily sealed it with a putty sealant. However, another two months passed before they could completely repressurize the system, using antifreeze shipped to them in a *Progress* freighter, and get Vozdukh running again.[56]

Though the thermal system had sprung leaks before, the failures had been intermittent and quickly sealed. This leak, however, was the first sign of what was to become a chronic problem over next few years. The pipes that distributed the antifreeze coolant throughout *Mir* were beginning to corrode after a decade in space. When water condensed at any point of contact between these aluminum pipes and their stainless steel grounding straps, corrosion occurred.[57]

Three weeks later, the American shuttle *Atlantis* arrived, bringing with it a Russian-built docking module for permanent installation on *Mir*. Without this module the shuttle could dock only if *Kristall* was repositioned from the starboard to the bow docking port, where a docked shuttle stood clear of the station's solar panels. Because the bow port was normally reserved for *Soyuz* spacecraft, *Kristall* could not be left there permanently. To use it for future shuttle dockings required that it be repeatedly repositioned, placing an increasing strain on the small robot arm that did the repositioning. In addition, each repositioning required that the *Soyuz* lifeboat docked to *Mir* be shifted to *Mir*'s aft port, thereby forcing the premature abandonment of any *Progress* freighter that happened to be docked there.

With the addition of the docking module on *Kristall*, *Kristall* could remain at its starboard port. The docking module extended *Kristall*'s length by about 15 feet, so that when the shuttle docked it was clear of *Mir*'s solar panels.

Two folded solar-panel arrays were attached to the docking module. One was entirely Russian-built, while the other was a joint U.S.-Russia project, with the U.S. building the panels and the Russians building the frame. These panels, installed during space walks over the next year, were part of the effort to increase *Mir*'s sagging electrical capacity.[58] *Atlantis* also brought *Mir* more than a ton of supplies, including the 20 lithium hydroxide canisters. The shuttle hauled more than 800 pounds of equipment and scientific results back to Earth, including another of the station's failed batteries and several hundred protein crystal samples brought there during the first shuttle docking five months earlier.

By this time, the supplies brought to *Mir* by the American shuttle had become essential to keeping the station alive and occupied. Despite getting paid large amounts of capital for launching foreign astronauts to *Mir*—exceeding $650 million in the last five years—the Russian government was reluctant to allocate that cash to its space program and instead put half of these profits into its general budget.[59] Thus, Energia didn't have the resources to launch enough *Progress* freighters. The American supply visits were desperately needed to keep the station in operation.

Despite these problems, on February 20, 1996, *Mir* celebrated its 10th year in orbit. Just as the growth of the American space program over the last 35 years was awe-inspiring, so were the Russian accomplishments on *Mir*. In those 10 years *Mir* had grown from 20 to almost 100 tons. It had proved its ability to maintain crews of six for weeks at a time. It had shown that a large structure of modules, assembled in orbit, could retain its integrity despite both collisions and dockings.

During that decade more than four dozen unmanned *Progress* freighters had come and gone, hauling about 80 tons of dry freight to *Mir*, almost equal to the station's actual mass. On top of that, the tankers had carried more than 40 tons of propellent, much of it pumped into the station's tanks.[60] Fifty-seven people from seven nations had visited the station, living and working there for months at a time.

Even the number of niggling problems over these years provided encouragement. The station had proved that humans could build, maintain, and *repair* a vessel in space long enough to travel to and from almost any planet in the Solar System.

All that was now required was the will to do it.

The next day, February 21, 1996, a new crew lifted off for *Mir*, signaling the start, at least for the foreseeable future, of the occupation of the station in low Earth orbit by mixed Russian and American crews. The commander, Yuri Onufrienko, was a 35-year-old air force pilot making his first flight. His flight engineer was Yuri Usachev, who had spent six months on *Mir* in 1994 during Valeri Polyakov's 14-month mission.

One month after these two Russians had settled into *Mir*, they were joined by American Shannon Lucid, who arrived when *Atlantis* docked with *Mir* for the third time, bringing with it a replacement gyroscope and three refurbished Russian batteries.

At the time of Lucid's arrival at *Mir*, the shuttle-*Mir* agreement called for 16 months of continuous American presence on *Mir*, with three more Americans following Lucid to *Mir*, rotating in and out just as the Russian crews did. Each astronaut would use the *Spektr* module as home and laboratory. Each was to perform experiments in both *Spektr* and the new *Priroda* module, due in April.

The launch of the last module to *Mir* had originally been planned for as early as 1989. For years after that, it sat unfinished in a warehouse. Only with the signing of the shuttle-*Mir* agreement was its completion and launch finally possible. On April 26, 1996, *Priroda* docked with *Mir*'s bow port, and the next day its robot arm repositioned it to *Priroda*'s permanent position on the portside docking port.

Finally, after 10 years of on-and-off construction, the *Mir* space station complex stood essentially complete, a 70-foot-long mushroom capped by a flower-petal cross of four-foot-long modules at its top.

Priroda ("nature" in Russian) added about 2,300 cubic feet to the *Mir* complex, bringing its total habitable volume to slightly less than 13,000 cubic feet, just exceeding the interior space of *Skylab*. The module's mass of 20 tons raised *Mir*'s overall mass to more than 110 tons, about the same mass as an average American space shuttle without cargo. When *Soyuz-TM* and *Progress-M* spacecraft were also docked to the station, the complex's total mass increased to more than 126 tons, making it the heaviest artificial satellite in orbit around the earth.

Priroda carried numerous Earth resource sensors, including three multi-spectral spectrometers, three multi-spectral radiometers, two multi-spectral scanners, and one synthetic aperture radar. Each was designed to look for different atmospheric and geological features in the Earth's environment. *Priroda* also carried American equipment and experiments for use by Lucid, including a spectrometer for analyzing medical specimens and a magnetic bottle for levitating samples during material-processing research.

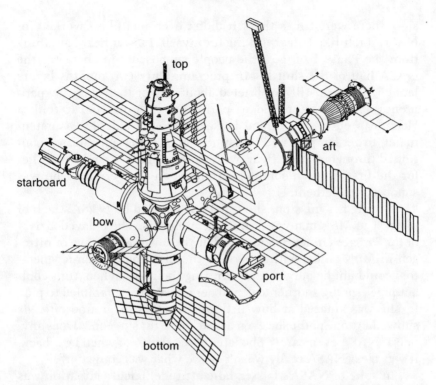

13. *Mir* complete, with all the core module's docking ports occupied, comprising *Kvant* (aft), *Kvant-2* (top), *Kristall* (starboard), *Spektr* (bottom), *Priroda* (port), and a *Soyuz-TM* (bow). A *Progress-M* freighter is also docked to *Kvant*'s aft port. *NASA*

The original plans had called for *Priroda*'s attachment to *Mir* before Lucid arrived, but the chronic cash shortages and a decision to bump the module's launch to give priority to a commercial satellite launch, caused Lucid to arrive first. Two *Progress* freighters were also delayed because of cash shortages. When the first freighter arrived on May 7, 1996 (the first time that all of *Mir*'s six ports were occupied), it brought many of the experiments and equipment that Lucid needed to do her research. It also brought fresh tomatoes and onions. "I never have had such a good lunch," Lucid exulted. "For the next week we had fresh tomatoes three times a day. It was a sad meal when we ate the last ones!"[61]

Just like Thagard's, Lucid's mission was handicapped most by NASA's seeming indifference to her. As she noted after the mission, there were times that "it didn't even feel like I worked for NASA. I felt like I was working for myself. I never heard anything from the Phase 1 office," the people supposedly in charge of the NASA half of the shuttle-*Mir* program. Only two months before launch, she was still not briefed about any of the science experiments she was supposed to perform while on *Mir*. It took a Herculean effort by her and her chief ground aide, Bill Gerstenmaier, to get her prepared for launch. In fact, that preparation continued throughout her flight, with Gerstenmaier often briefing her for the first time on some experiments on the very day she was supposed to do them.

During her mission, she often felt puzzled by NASA's lack of interest in *Mir*'s many intermittent problems. The delayed arrival of the *Progress* freighters meant that there was a shortage of nitrogen in *Mir*'s tanks, without which the Elektron oxygen-regenerators could not be purged, thus limiting their use. When *Mir*'s coolant pipes sprang another leak and her crewmates scrambled to plug it, she was amazed at how little interest NASA ground officials showed. At one point she even asked her flight surgeon, "Does anyone at NASA even care?" She later noted that ". . . when I got back, it was clear NASA really wasn't aware what was going on."[62]

In spite of NASA's laissez-faire attitude, Lucid's situation was not as isolated as Thagard's. CNN had arranged to provide her a one-hour weekly news update. She was also able to e-mail her family regularly, and both the Russians and NASA tried to arrange phone and video sessions with them for her. (Unfortunately, the video sessions rarely worked. As Lucid noted in her debriefing, "It was practically impossible to listen and hear what the other person was saying during the two-way video.)[63]

More importantly, Lucid's friendly and motherly persona clicked with her Russian crewmates. The Russian tendency to patronize foreigners never bothered her much. Onufrienko and Usachev called her "Miss Shannon," and the three got on famously. They didn't at first expect that much from her except emotional comfort, and she at first didn't demand much extra work from them, accepting their attitude so as to not rock the boat.[64]

While Lucid provided emotional support, Onufrienko and Usachev worked incessantly to keep *Mir* habitable. They struggled to plug the chronically leaking coolant loops so that the Vozdukh carbon-dioxide-scrubbing system would not shut down. They also battled with one of the Elektron oxygen-regenerators, which repeatedly broke down. With time, Lucid even joined them in their work, helping them wherever she could.[65]

In between, the two Russians performed five space walks, installing one of the two solar panels brought to *Mir* by *Atlantis* and fitting it to the exterior of *Kvant* to give it two working solar panels. They also installed a multi-spectral Earth-scanning camera on the outside of *Priroda*, deployed the antenna for *Priroda*'s synthetic aperture radar, and assembled and attached the second Strela crane to the base block's starboard side.[66]

During all these space walks, Lucid was left inside *Mir* to handle communications. To make sure she didn't do anything wrong, Onufrienko placed red tape across all communication controls. As Lucid noted at the time, she was given firm orders "absolutely not to touch [these controls] while they are outside." While Thagard would have cringed at being patronized like this, Lucid took it all with warm good humor. For example, during one long space walk, she was left inside with little to do but listen while Onufrienko and Usachev struggled to plug in one of the new solar panels. She was fascinated as the two men talked in Russian about "the mamas" and "the papas," the Russian version of what English-speakers call female and male electrical plugs. "It all feels so warm and homey," she later wrote in an e-mail to the general public. During another space walk, she took it upon herself to open a new food container and prepare their favorite foods so that the two men could eat as soon as they got back inside.[67]

Nonetheless, the cultural differences kept appearing in the strangest ways. During their first and last space walks Onufrienko and Usachev videotaped the first in-space television commercial, holding a 4-foot-high inflatable Pepsi can. For their work, Energia was paid more than a million dollars. Lucid, however, was forbidden by NASA from participating in this stunt. The agency's regulations, as well as a number of federal laws, specifically forbade her any activity that would appear to be motivated by financial gain.[68]

Thus, in her later writings about the mission she made no mention of the Pepsi commercial. She also had to be careful that her voice was not heard over the radio during the commercial's taping. God forbid that an American would do anything to promote commercial and profitable operations in space. Better to leave that task to the former citizens of a now-defunct communist empire.

Despite the lack of help from NASA, Lucid was able to get most of her research done while in orbit. Her most successful planned experiment involved burning 79 candles to see how their flames burned differently in space. With delight she noted that the flame consistently sat on the top of the wick like "a little blue igloo," rather than becoming a pointed flame as seen on Earth. In a statement that was to become very ironic in only a few short months, Lucid wrote how her research "on how flames propagate in microgravity may lead to improved procedures for fighting fires on the station."[69]

She also began the first American plant experiment on *Mir*. The July freighter had brought with it new super-dwarf wheat seeds as well as a set of American-made water sensors to be installed in the Russian-Bulgarian-built Svet greenhouse on *Kristall* that had so dissatisfied Energia's biology team. While they waited for her relief to arrive she and Onufrienko had time to install the sensors and plant 32 super-dwarf wheat seeds in Svet. For the first time in years, plants were once again growing on *Mir*.[70]

After Onufrienko and Usachev were replaced by the next Russian crew, Lucid nurtured the seeds, watering them and maintaining the greenhouse. Only days before *Atlantis* arrived to take her home, she looked into Svet's large glass window and to her delight saw that ". . . little baby seeds were growing." Excited, Lucid ". . . rushed into the base block. Valeri [Korzun] and Sasha [Kaleri] were there, and I said, 'Hey, you guys gotta come look quick, because the little baby seeds are coming!' So they came in there and looked, and they were real excited. It was just a real neat thing."[71]

At that moment it appeared as if Lucid might have become the first person since Valentin Lebedev on *Salyut* 7 in 1982 to successfully produce fertile seeds in space. After John Blaha replaced her on *Mir*, these plants continued to prosper. He would harvest them and bring them back to Earth.

Lucid's last weeks on *Mir* were commanded by the next Russian maintenance crew, Valeri Korzun and Alexander Kaleri. Lucid also shared *Mir* with another woman during this crew transition, Claudie Andre-Deshays of France, the first French female astronaut. Andre-Deshays' flight completed a three-mission joint French-Russian program, begun more than four years earlier. France had ended up paying Russia a total of about $33 million to place three astronauts on *Mir*. The funding shortages in Russia, however, caused Andre-Deshays' flight to be seriously delayed and as a result, many of the experiments left behind on *Mir* from the earlier missions were no longer viable. The Frenchwoman also noted later that she found a two-week stay on *Mir* insufficient to familiarize herself with the space laboratory and complete her research.*

Two weeks after Andre-DeShays returned to Earth, *Atlantis* arrived, bringing with it Lucid's replacement, John Blaha. *Atlantis* was originally supposed to arrive before the French mission, but problems with the shuttle's solid rocket boosters as well as the appearance in the Caribbean of Hurricanes Bertha and Fran forced a six-week delay. As a result, Lucid's final flight time of 188 days gave her both an American and a female space-endurance record, exceeding the 169 day record of Elena Kondakova. For the first time since *Skylab*, a space-endurance record was held by someone other than a Russian.

Once on Earth, Lucid also refused the stretcher, insisting on walking off the shuttle on her own two feet. And like all the other men and women before her who had stayed in space as long, she recovered quickly, returning to normal in a few weeks.[72]

The next American *Mir* astronaut, John Blaha, experienced the same culture shock and abandonment by NASA that Thagard and Lucid had. His isolation, however, was made even more explicit in that his Russian crewmates, Valeri Korzun and Alexander Kaleri, were not the men that Blaha had trained with. The commander of the original crew, Gennadi Manakov, had been hospitalized only a week before launch because of heart problems. As the Russians had done since *Salyut 1*, they promptly replaced the entire crew, inserting Korzun and Kaleri in place of Manakov and his partner Pavel Vinogradov.[73]

Like Thagard, Blaha struggled with the Russian habit of treating foreigners with condescension. Unlike Thagard, he refused to bite his tongue. He was a shuttle pilot, and curious about *Mir*'s design and operation. Repeatedly during his mission he took his commander Korzun to task for treating him like a child. "You

---

*Despite the delays and incomplete research, following Andre-DeShays' flight the French and Russians immediately signed a new cooperative agreement, calling for two more French astronauts to occupy Mir through 1999, the first for a period of several weeks and the second for four months, with France paying Russia approximately $40 million.

would have thought we were five-year-olds," Blaha later said, describing how Korzun sometimes talked to him.[74]

In response, Blaha's Russian crewmates treated him mostly with firm but distant courtesy, letting him participate in only a few maintenance tasks. As he watched, they installed a replacement pump for emptying *Mir*'s overflowing toilet tanks, performed two space walks, during which they added a Kurs antenna to the docking module and ran a new electrical cable from the new solar panel on *Kvant* to the batteries of the base block's dorsal solar panel (generally useless because its view of the sun was blocked by *Kvant-2*).[75]

For Blaha the worst aspect of his flight was the lack of support from his American colleagues. While Thagard had been irritated by it and had expressed his annoyance after the mission to NASA Administrator Dan Goldin, and Lucid had ignored it, managing to make her tour enjoyable despite it, Blaha simply felt betrayed. Once again, NASA had provided him no preparation at all prior to his launch. Blaha, who was trained as a pilot, was the first non-scientist to fly in the *Mir* program. Both Lucid, who was a chemical engineer, and Thagard, who was a doctor, were more familiar with the needs of the science work, and could therefore improvise more easily. In addition, the operations man that NASA assigned to Blaha knew less than he did, and was no help in getting Blaha's program organized. Then, when Blaha started to compile a flight program for himself, several NASA bureaucrats interfered. Incredible as it seems, they claimed that he was not allowed to bring unapproved personal notes such as this into space.

Then, while on *Mir*, Blaha found himself overworked. When he told his ground support team, which NASA kept changing, to ease up on the assignments, they ignored him. In a repeat of the experience of the third *Skylab* crew, the workers on the ground somehow thought they knew more about how Blaha should function than he did. No matter how much he protested, they refused to lighten their demands. Only when the *Mir* commander Korzun insisted they ease up did they finally listen.

By the time his replacement, Jerry Linenger, arrived on *Mir*, Blaha had had enough. His anger at everyone in NASA's management was so intense that it was almost inexpressible. He told Linenger as he was about to leave *Mir*, "Don't expect any help from the ground. You're on your own up here."[76]

And yet, despite the problems, Blaha persevered. During his four-month mission, he conscientiously nursed the wheat plants that Lucid and Onufrienko had planted, eventually harvesting 260 seed heads from about three dozen plants. The harvest, which he performed on December 6, 1996, took more than three hours to complete because of the number of plants and because of the care he took in retrieving every seed head. He then did a second planting, and by Christmas Day these new plants had grown 6 inches high, each plant vibrant, with multiple leaves.

Blaha's apparent success at growing wheat was further proof that past crop failures were not caused by some fundamental need of plants to grow in gravity, but by a technical failure of each greenhouse to provide light to the plants or water to the roots. Using the American-made moisture sensors near the roots, Lucid and Blaha were able to tell when the plants needed water. *Mir's* new solar panels and batteries also allowed them to light the plants 24 hours a day, instead of the 16 hours provided during the 1990 and 1991 Energia experiments.

When Blaha returned to Earth, NASA, the press, and the public got very excited about these wheat plants. Everyone, including Blaha and Lucid, thought that they had achieved a breakthrough, growing a plant in space from seed to seed.[77] Both were completely unaware that a Russian cosmonaut, Valentin Lebedev, had already done it 14 years earlier.

Unfortunately, when scientists opened the more than 260 wheat heads, they found that every single one was empty—the wheat plants had produced no seeds. After much testing, the scientists concluded that the levels of antifreeze in *Mir's* atmosphere (300 times higher than normal) had caused the sterility.* As one scientist noted, it was well known that "the use of ethylene induce[s] male sterility in cereals." To grow healthy wheat plants in *Mir's* greenhouse, they would have to add antifreeze scrubbers.[78]

Sometimes you need to look into a mirror to recognize yourself.

---

*Interestingly, the high levels of antifreeze were apparently not caused by the leaks in the station's cooling system. An analysis of *Mir's* atmosphere over time revealed that the ethylene mostly came from the periodic delivey of fresh fruits, which naturally secrete ethylene and which *Mir* had no system for removing from the atmosphere. (Bingham interview; Musgrave interview.)

During most of the 1980s and 1990s the dry rot that was settling in at NASA was ignored. Americans, and NASA officials, assumed that their behavior conformed to the American traditions that had sent men to the moon: free, bold, and innovative. Then, during the shuttle-*Mir* program, many in NASA, including men like Blaha and Thagard, were forced to compare themselves with another space program that in those same 20 years had accomplished incredible feats of bravery with far fewer resources. The contrasts were stark and disturbing.

In those two decades, the Russians had become progressively freer, their space program moving from government funding to private commercial free enterprise. At the same time, both Russia and its space program had become increasingly efficient and profitable, able to produce new technologies that could be built cheaply and used quickly.

The American space program, meanwhile, had degenerated, depending increasingly on government directives and less on private initiative. Larger and larger quantities of money were being spent to produce less and less. Bureaucracies flourished, accomplishing little. "We are becoming more bureaucratized, more rigid," noted Thagard. "With each passing year, the rules would tighten up." Except for the space-shuttle fleet, which had had a difficult and impractical birth, American manned space exploration had accomplished little in the decades since Apollo, shriveling down to a boring series of forgettable orbital missions. At the same time, NASA had spent billions of dollars and more than a decade drawing nothing but space-station blueprints.[79]

Americans tend to assume that our government acts as a servant of the people. Government officials are supposed to care about the citizens who work for them. They are supposed to use the taxpayers' money as the citizen wants, not as the bureaucrat desires. The shuttle-*Mir* program revealed how unconnected to reality this perception is. By the 1990s, NASA officials and American elected officials were more prone to act like Soviet-style bureaucrats, wielding power for their own ends, treating their employees more as political pawns than fellow human beings, and caring little about whom they hurt in the process.

The consequences of this reality were soon to become glaringly evident.

On January 14, 1997, *Atlantis* made its fifth docking with *Mir*, bringing with it Blaha's replacement, Jerry Linenger. Five days later, Blaha was back on Earth, struggling to readapt to the feeling of gravity. Though his recovery was normal, he was "absolutely stunned" at how weak and wobbly he felt.

On *Mir*, Linenger stepped into Blaha's slot, with Korzun and Kaleri his crewmates for his first four weeks in space. Then, on February 10, 1997, *Soyuz-TM 25* blasted off from Baikonur, carrying veteran Vasily Tsibliev, returning to *Mir* after four years, and rookies Alexander Lazutkin and German astronaut Reinhold Ewald. Tsibliev and Lazutkin were to replace Korzun and Kaleri.

The next six months, both in space and in the halls of government, were to become one of the most difficult periods for space exploration since the early 1970s. And sadly, the parallel nature of events on *Mir* and on the ground illustrated the cynical character of human space travel at the end of the second millennium.

13

Mir: Spin City

"Get a Fire Extinguisher!"

Congressman F. James Sensenbrenner was a very unhappy man. Chairman of the House Science Committee, he had spent the last four days wandering the halls of various Russian space facilities, trying to find out the status of the Russian share of the *International Space Station*. No one would give him a straight answer. In-orbit construction of the station had originally been scheduled to begin in late 1997 with the quick sequential launch of three modules, the Functional Cargo Block (*Zarya* as it was eventually named), Node 1 (or *Unity* as NASA finally dubbed it), and the service module (or *Zvezda* as its Russians builders named it).

Zarya ("sunrise" in Russian) was an updated version of *Mir*'s later modules, combining systems from both *Kristall* and *Kvant-2* to provide the station with guidance, communications, and power during the station's early assembly period. Built by the Russians, it had been financed by NASA so that the U.S. government could claim it as a jointly built project.

Unity, the first American-made module, had originally been planned as a node to connect *Freedom*'s different modules. It was a stubby 18-foot-long cylinder with six docking ports, one at each end and four around its circumference. Once linked with *Zarya*, it would provide a haven for later American modules.

Zvezda ("star" in Russian) was essentially *Mir-2*, an updated version of *Mir*'s base block. It was to provide the station's living quarters during the years of construction, as well as thrusters with refillable tanks to maintain the station's orbit. Without *Zvezda*, the other two modules could not stay in orbit longer than 15 months. After that, *Zarya*'s fuel would run out, and its orbit would decay. *Zvezda*, however, was unfinished. Despite repeated promises, the Russian government had consistently failed to fund its construction. Lacking an imminent and reliable launch date for *Zvezda*, the other two modules could not be launched, and the start of the station's in-orbit assembly was repeatedly delayed, pushed back to April 1998 and then to June 1998.

The Russian broken promises galled Sensenbrenner. As recently as two weeks earlier, in face-to-face meetings in Washington on February 7, 1997, Russian Prime Minister Viktor Chernomyrdin had promised Vice President Al Gore that $100 million would be transferred to the space station program by February 28. Only three days later, however, he issued a decree providing only loan guarantees totaling less than half that amount. When Sensenbrenner arrived in Russia, he asked for a copy of Chernomyrdin's decree. At first, no one could get him a copy, and when it was finally produced, it was missing one page, and the text mysteriously skipped from Article 3 to Article 9.

Everyone he met with passed the buck to someone else. Finally, Sensenbrenner had enough. During a walking tour through the cavernous floor of the Khrunichev design bureau factory (where the Proton rocket that launched Russian modules was built), Sensenbrenner let loose his anger. With a crowd of media people and television cameras watching, Sensenbrenner confronted Anatoli Kiselev, the portly director of Khrunichev, demanding some concrete information about when *Zvezda* would be finished. For 10 minutes, with the cameras recording every word, Sensenbrenner berated Kiselev about *Zvezda*'s unfinished state.

Sweating profusely, Kiselev nervously tried to sidestep Sensenbrenner's attack. He was only a factory manager. He didn't allocate funds, the Duma did. When the money arrived, the module would surely be done. "I am not the president of Russia. I am the president of my company," he said. "I'm not responsible for Russia."

Sensenbrenner didn't buy it. "No one is accepting responsibility for the delays," he told Kiselev. "The whole reason there is a crisis today is because your firm did not meet the schedule agreed to by Gore and Chernomyrdin in Moscow last July." As he told reporters later, "We're getting close to the point where the back of the camel is going to break. If Russia does not fulfill its promises here, its government should know that it will affect its relations with the rest of the world on matters far beyond space."[1]

What Sensenbrenner didn't know that day was that Chernomyrdin's decree was worth even less than the paper it was written on. February 28 passed, and no money was released, no cash, no loans, nothing. The Russian government was still bankrupt, and simply didn't have the funds for a new space project. *Zvezda* sat untouched on the factory floor, an empty hulk months behind schedule.

That same week, the six men on *Mir* almost died.

For two weeks the two crews had been living together on the station, with all six men almost entirely devoting their time to the experiments and television interviews required by German astronaut Ewald's flight.

On Sunday, February 23, the six men had just finished dinner together in the station's base block, sucking borscht and jellied perch-pike from tubes. While Linenger headed back to *Spektr* to do some computer work, Lazutkin floated into the adjacent *Kvant* module to light a lithium perchlorate candle. In the 13 days since Lazutkin's arrival, the presence of six people on *Mir* had caused them to light about three dozen such candles.[2]

Only seconds after activating the candle, Lazutkin was startled to see sparks shoot out from the unit. A second after that, a foot-long orange-pink flame ignited, boring its way through the canister's skin and burning with the intensity of a blowtorch.

For a few moments Lazutkin stared at it in disbelief. Because of the Russian tendency to sweep problems under the rug, no one had ever told him that an identical candle in exactly the same place had erupted in flames when Polyakov lit it in 1994. Almost timidly, he called out, "Guys, we have a fire."

Eight feet away in the base block, the three Russians and the German continued to chat, unaware of what was happening in *Kvant*. Then, almost simultaneously, Ewald and Tsibliev saw the flame and the smoke. "Fire!" they yelled.

Korzun immediately flung himself through the crowd and into

Kvant, stopping next to Lazutkin, who was trying to smother the fire with a towel—just as Polyakov had done two years earlier. The blowtorch flame burned through the towel as if it were paper.

"Get a fire extinguisher!" Korzun yelled, pushing Lazutkin back into the base block.

There, Tsibliev and Kaleri were already searching for extinguishers. Merbold, the least familiar with *Mir*'s operations, had pushed himself to a corner, out of the way.

With the smoke already thick and billowing, making breathing and seeing difficult, Kaleri handed a fire extinguisher to Lazutkin, who passed it to Korzun. The extinguisher would not activate. As Korzun struggled with it, he yelled "Get to the oxygen masks!"

The station's fire alarm blared.

Linenger, calmly typing on his computer in *Spektr* and unaware of the emergency, heard the alarm and assumed that the station's power system had once again failed. Though such failures could be dangerous, they took time to unfold. He coolly saved his files, shut down his computer, and pushed himself to the docking adapter to see what was happening. There, he almost collided with Tsibliev. "Fire!" Tsibliev yelled.

By this time the Russians had begun evacuation procedures. Tsibliev and Lazutkin had climbed into their *Soyuz-TM* spacecraft, docked to the bow port, and were activating its systems, while Kaleri was trying to download from *Mir*'s main computer the flight information for getting back to Earth.

Looking along the length of the base block to *Kvant*, all Linenger could see were flames and billowing smoke coming from *Kvant*'s hatch. He suddenly realized, as had everyone else, that the fire's location blocked access to the aft port where the second *Soyuz-TM* was docked. With only one *Soyuz-TM* lifeboat available, complete evacuation was really not possible—at least three men would die if the hull was breached.

As the men fought the flames, things kept going wrong. The first oxygen mask Linenger put on failed to work. One fire extinguisher failed, and several in *Priroda* were bolted to the wall so that they couldn't be removed without tools. (They had been secured that way for launch, and in the year and a half since no one had thought of releasing them.) Meanwhile, Korzun couldn't make a dent in the flames, despite dousing it with three working extinguishers. Linenger, who had slid into the docking tunnel between

Kvant and the base block to hand Korzun extinguishers and brace him when he discharged them, watched as blobs of molten metal drifted away from the flame.

Finally, after burning for 10 to 15 minutes, the fire seemed to die out on its own, as if it had used up all the nearby available fuel and suffocated on its own soot. Other than minor burns and lacerations on their hands and arms, no one had been seriously hurt.

For the next hour the men floated silently without moving, trying to stretch their limited supply of respirators while they waited for ventilation fans in *Mir* to clear the smoke. "Smoke filled the station with a density so thick that we could not count the fingers in front of our faces for nearly an hour," Linenger wrote later.[3] Unlike an Earthbound house fire, there wasn't much they could do until *Mir*'s ventilation system sucked the smoke from the atmosphere. They were stuck like Grechko and Romanenko in *Salyut* 6 20 years earlier; they couldn't open a window. Nor could they find a low point below the rising smoke. In zero gravity smoke permeated everywhere; there were no low points.

By early morning the air had cleared enough for them to stop using respirators—one man even joked that the smoke gave the station the smell of roasted turkey. They began cleaning up, wiping the soot from everything. They talked to Moscow mission control, describing their circumstances. They started the laborious task of bringing *Mir* back into full operation.[4]

Meanwhile, the events on the ground after the fire most exemplified what Sensenbrenner had been complaining about only a week before. The fire occurred at around 10 P.M. Moscow time. However, it was not until 10 A.M. the next morning, when the American contingent arrived for work at Moscow mission control, that the Russians told them about it. The Russians then tried to bury the problem, announcing that the fire had lasted only 90 seconds. They also wrapped up their investigation of it after only three days, concluding that the fire was an isolated event, due solely to an unspecified failure in the specific candle.

When Linenger tried to inquire about the health effects of the fire with the ground, the Russians simply wouldn't give him communication time. He and Korzun even got into a screaming argument when Linenger tried to interrupt a communication pass to discuss the fire.[5]

Telling his point of view to NASA was made even more difficult because the top-echelon people at NASA were not very interested in hearing about Russian space failures. In public, NASA
seemed willing to accept the Russian version of events, stating in
their own press release that "The fire, which began at 10:35 P.M.
Sunday, Moscow time, burned for about 90 seconds. The crew was
exposed to heavy smoke for five to seven minutes and donned
masks in response." In the same press release, Frank Culbertson,
Director of the Phase 1 shuttle-*Mir* program, was quoted as saying
that "there were no injuries," adding that "Russian management
and operations specialists have been very informative as to what
happened." Culbertson also tried to spin the facts to make the fire
seem as inconsequential as possible. He told the press that Linenger
was able to start his science work again "about a day or so later. . . .
He's back on track. All of that is proceeding normally. He says that
things are as they were before in that regard."[6]

Things had hardly returned to normal for Linenger. In fact, he
didn't even get to sleep until two days after the fire. During those
48 hours he repeatedly did medical checks on everyone, insisting
that the crew wear breathing masks at all times because he had no
way to test the toxicity of *Mir*'s atmosphere. "Sleep was pretty difficult for everybody," he reported.[7]

He also tried and failed repeatedly to report the seriousness of
the situation to his NASA cohorts. Linenger's take on the fire was
very inconvenient for the Clinton administration. Clinton had set
his sights on building an international space station in partnership
with the Russians, and a deadly fire on *Mir*—which the Russians
kept secret and then refused to investigate closely—was powerful
ammunition for the project's opponents.

Not that the opponents didn't already have ammunition. Despite many promises, Yeltsin's government simply didn't have the
cash on hand to finish *Zvezda*. In the chaotic but free Russian society established after the collapse of the Soviet Union, the Russian
government was probably society's least powerful and poorest component. In the half-decade since the August coup, the circumstances for ordinary citizens had improved enormously. Store goods
were plentiful. Work was abundant. Or as a Ukrainian friend told
me when I visited Kiev in 1995, "In 1990 things were cheap, but
nothing was available. Now things are expensive, but you can get
them when you need them. All you have to do is work hard and

make money."[8] For the authorities, however, things were not so good. Yeltsin's government found it difficult to collect taxes, partly because the collection system was so complex that everyone had an interest in dodging it. Government officials, appointed and elected, had no understanding of budgets, and routinely overspent their allocations.[9]

On February 24, the day after the fire and only a week after Sensenbrenner returned from Moscow, Yuri Koptev, head of the Russian Space Agency, held a press conference in Moscow to announce that, due to lack of money, completion of *Zvezda* was being delayed again until December 1998. Then, February 28 arrived, and the $50 million in loan guarantees to Energia that Chernomyrdin had promised did not show up.

In Washington, these broken promises brought furious condemnation. Sensenbrenner, who would shortly be named chairman of the House Subcommittee on Space and Aeronautics, called for congressional hearings, and soon thereafter, hearings in both the House and the Senate were scheduled for early April.

In space, the technical problems continued to pile up. Only days later, on March 3 and one day after Korzun, Kaleri, and Ewald had returned to Earth, both of *Mir*'s Elektron oxygen regenerators failed. After years of improvised repairs, the systems simply refused to work, and neither Tsibliev nor Lazutkin could figure out a way to fix them.[10] Plans were quickly whipped together to launch a replacement unit.

In the interim, the crew had no choice but to use lithium perchlorate candles, lighting about two or three per day in the one remaining working unit in the base block. Because no one knew what had caused the candle to erupt into flames on February 23, the crew was ordered to use only candles that were less than two years old. Moreover, during each lighting, someone was to stand by with a fire extinguisher, just in case.[11]

The next day, Tsibliev attempted a docking test, using only the TORU system to redock the *Progress* freighter that since early February had been flying nearby in a parking orbit. Until now, *Progress* freighters routinely used the Kurs system for rendezvous and docking. Only once, by Malanchenko in 1994, had TORU been used to dock a *Progress* freighter, and that had been a last-ditch effort to get the cargo ship into port. However, the price of Ukrainian-built Kurs systems had continued to rise. If Tsibliev's test worked, it would prove that *Progress* dockings could be handled entirely by TORU, and allow the Russians to stop buying the Kurs system.

On March 4, the *Progress* freighter was shifted into position so that Tsibliev could guide it into *Mir's* aft docking port. Then, for reasons that even today remain unclear, the camera on the TORU system malfunctioned. Tsibliev, who already had one *Mir-Soyuz* fender-bender on his resume, stared helplessly at his static-filled monitor as the freighter inexorably barreled toward them.

Despite being stationed at different *Mir* windows, neither Linenger nor Lazutkin could see *Progress*. Desperate for any guidance, Tsibliev flew to the nearest window himself, peering out into the blackness to see if he could spot it.

Lazutkin suddenly yelled, "It looks like it's coming right at us! It's coming way too fast! More braking, Vasili!"

Tsibliev flung himself back to the TORU controls and furiously fired *Progress's* thrusters, trying to get it to veer to the side. "Did that help?" he yelled to Lazutkin. "Is it slowing? Is it veering away?"

Simultaneously, all three men saw the freighter fly past *Mir*, "screaming by us," as Linenger later wrote. Thinking *Progress* was going to hit *Mir*, Tsibliev shouted for Lazutkin and Linenger to flee to their *Soyuz* spacecraft in preparation for an emergency evacuation. As Linenger darted to *Soyuz*, he expected at any moment to feel the impact of metal and cracking of *Mir's* hull.

The next few minutes passed silently, without a collision. To their relief, *Progress* had glided past, missing the station by what was officially estimated as several hundred feet.[12]

Once again Linenger tried to let NASA officials know what had happened. Once again, he found himself stonewalled. First, Russian controllers hid the docking failure by telling their NASA counterparts that the docking had simply been canceled. Then, when Linenger told his American support team what happened, they closed ranks with the Russians and tried to make the incident seem less dangerous than it was. To Linenger, NASA's nonchalant attitude could be roughly paraphrased: "Oh, we hear that there was some problem with the attempted docking, but we weren't given any details. Do you want us to look into it?"[13]

The problems on *Mir* were not over. At the end of March, a coolant loop in *Kvant-2* sprang a new leak, shutting down the Vozdukh carbon-dioxide-scrubbing system. To keep the atmosphere breathable, the crew had to use both lithium perchlorate candles and lithium hydroxide canisters, and their supplies of both were limited.[14]

The failure of the cooling system caused the station's temperature to rise to around 95°F. The failure of the Vozdukh system caused the carbon dioxide levels to climb to between 5 and 8 percent, a level that was almost toxic. The men had to cease exercising in order to keep the levels from rising further. Worse, the leaking antifreeze made the overheated station smell like a car repair shop.[15]

Amid all these technical difficulties, there was also human tension, both among the three men and with the ground. Tsibliev and Lazutkin were gregarious, friendly men who wore their emotions on their sleeves. In addition, Tsibliev was unlike most Russian commanders in that he disliked ordering people about. He preferred people to want to help him.

Linenger, however, preferred to keep to himself and do his own scientific work. Maintenance of the station was the Russians' responsibility, and Linenger was much more concerned about meeting the goals of his scientific research. Moreover, Linenger was disturbed with how Tsibliev and Lazutkin were allowing their work to mess up their sleep schedule, staying up until four in the morning trying to catch up and then sleeping into the afternoon hours.[16]

The repair workload was so overwhelming that by April the Russians began to desperately want Linenger's help—despite their knee-jerk disdain of outsiders. When ground controllers asked him to postpone his science work to focus only on station repair, however, he politely refused. Used to people obeying the requests of their superiors, the Russians found Linenger's refusal infuriating. How dare he not follow orders? Foreigners weren't supposed to be so independent and freethinking.

These tensions led to several arguments about minor issues. Once Linenger and Lazutkin avoided each other for two days after arguing about where to store some equipment. At another time, Tsibliev, in conversations with the Russian mission control out of Linenger's earshot, ridiculed the American's refusal to help.[17]

Linenger considered these crew conflicts minor and unimportant. As far as he was concerned, his relationship with the Russians, especially Tsibliev, was great. "We got along wonderfully,"

Linenger wrote later. "Tsibliev and I . . . had no major disagreements in the four months that we spent together on *Mir*."[18]

Relations with Earth, however, were much more rocky. Based on what John Blaha had told him about ground support, and confirmed by his own experience during the first few weeks of the mission, Linenger decided that he was not going to let the ground teams micromanage his activities, especially if he believed their advice was not helpful. Because radio communication was often difficult and accomplished less than nothing anyway, he simply announced that he would cease using it. On March 7 he told his ground team, "I think we should stop doing voice comm[unications]." From this point on, Linenger limited most of his contact with his U.S. ground controllers to e-mails. He hoped that by taking this action he could force mission controllers to fix the problem.[19]

Tsibliev, meanwhile, was getting increasingly infuriated with the demands that the Russian ground controllers were making on him. Not only did they expect him to keep the station operating, they also expected him to do research. Like a host of previous space dwellers, beginning with the last *Skylab* crew, followed later by Yuri Romanenko at the start of *Mir*'s operation, then by the Americans Blaha and Linenger, Tsibliev felt used and abused by the insensitive orders given to him from Earth.[20]

Repairing the station was becoming a 24-hours-a-day job. *Mir*'s aging systems on the base block, *Kvant*, and *Kvant-2* continued to malfunction. For example, in March the station's attitude-control computers crashed repeatedly, causing the usual chain reaction of failures. The gyros would shut down so that the station rotated into a gravity-gradient position with its solar panels no longer facing the sunlight. The station's batteries—many of which could no longer store much power—then ran down, causing the rest of the station's equipment to stop working.

Each time, the three men had to shut down the station so that every bit of electricity produced by the solar panels could be pumped into battery storage. Once the batteries were recharged—a process that could take several days because of the age of many of *Mir*'s solar panels—the men reactivated the systems and rebooted the computer so that it could once again maintain the station's orientation.

In between, Tsibliev and Lazutkin had endless repair work to

do, working 14, 16 hours a day. They replaced parts in the two
Rodnik water-recycling systems. They replaced more hoses to stop
leaks in the coolant loops. They switched dead batteries with live
ones to get as many powered up as possible.[21]

In Washington on April 9, congressional hearings began. Because
NASA officials had suppressed the worst reports of *Mir*'s malfunctions—for example, the near collision went completely unreported
in the American press—these hearings focused on the Russian
government's failure to keep its promises and build *Zvezda*. Chairman Sensenbrenner summed up the American perspective very suc-
cinctly in his opening statement.

> In January 1996, Russian Deputy Prime Minister Oleg Soskovets*
> promised Congressman Jerry Lewis and I that the Russian Govern-
> ment would pay its bills. It did not. In March 1996, NASA promised
> . . . that it would resolve this issue by mid-May 1996. It did not. In
> April 1996 the Vice President received more assurances that the
> Russian government would pay its bills. It did not. In July of 1996
> Russian Prime Minister Viktor Chernomyrdin promised Vice Presi-
> dent Gore in writing that the Russian government would pay its
> bills and meet several milestones. It did not. This February, Prime
> Minister Chernomyrdin promised Vice President Gore that it would
> give the Russian Space Agency $100 million by February 28th. It did
> not. On February 10th, the Russian government released a decree
> promising to provide a schedule of payments to the Russian Space
> Agency by March 10th. It did not.[22]

During these same House hearings, NASA official Wilbur
Trafton announced that, because of the Russian delays, the launch
of *Zarya* was officially delayed once again—by 11 months, to October 1998. Despite NASA's many promises that the Russian contribution to the station would not affect American plans, including
Dan Goldin's repeated testimony before Congress in which he
stated that NASA even had contingency plans if the Russians did
back out, the Russian failure to do what they promised was stalling
the launch of the American space station. Trafton also announced
that NASA was shifting $200 million from the space-shuttle budget
to finance further Russian construction, including *Zvezda*, a cost
that Russia had originally agreed to pay for.

---

*Soskovets was fired only a few weeks later, in April 1997, having been
linked to a scheme that embezzled millions of dollars from the Russian space
program.

Sensenbrenner and his fellow congressmen blustered a great deal about the delays, the funding shift, the misinformation and lies from both the Clinton administration and the Russian government. They threatened to cut off funding to the station. They threatened to cancel the program.

In the end the congressmen did little. They passed a resolution (which was never confirmed by either the House or the Senate) demanding that no further American astronaut go up to *Mir* unless NASA Administrator Dan Goldin personally vouched for the station's safety. They also set an August 1 deadline for Bill Clinton to once and for all decide whether to keep the Russians as partners in the *International Space Station* project.[23]

Interestingly, Sensenbrenner's perspective on the program was completely at odds with Clinton's. As the congressman said at the hearings, "I would remind [NASA and the White House] that we are building the space station to do science and open new commercial frontiers, not to provide foreign aid to the Russian aerospace industry or cash under the table to the Russian government."[24] Foreign aid, however, was Clinton's entire reason for supporting the space station. His administration would accept anything the Russians did to keep the joint program going, thus making the congressional resolutions worth about as much as Chernomyrdin's February 28 decree—less than nothing.

The day before the House committee hearings, on April 8, a new *Progress* freighter arrived at *Mir*, bringing several tons of supplies. The cargo included three fire extinguishers to replace those used during the fire, repair equipment to try to seal the station's coolant leaks, about 50 oxygen-generating lithium perchlorate candles, and a pair of new spacesuits for an upcoming space walk.

During the five weeks leading up to the arrival of the next shuttle, Tsibliev and Lazutkin worked themselves to exhaustion trying to repair *Mir*'s various atmosphere-recycling systems. By cannibalizing parts from one Elektron system, they managed to get the other operating, at least temporarily. To get a second functioning system, the Russians quickly built a new unit and shipped it to the United States so that it could be loaded into the cargo bay of *Atlantis* and taken to *Mir* during the next docking mission.

Then, on April 29, Jerry Linenger became the first American to do a space walk using a Russian spacesuit. He and Tsibliev were to install a U.S. built 500-pound optical sensor package and retrieve

experimental materials that had been left out on *Mir*'s exterior for
the last few years. As they suited up, Linenger got his first close
look at the *Kvant-2* hatch. He was astonished. As far as he could
tell, the hatch was held shut only by a set of ordinary C-clamps. No
one had briefed him on its earlier damage and repair—all Tsibliev
did was repeatedly warn him to "be very gentle" when he touched
the hatch. Linenger had no idea that the C-clamps were only a
backup, left over from before Manarov and Afanasyev replaced the
hinge in 1991.[25]

Tsibliev and Linenger then completed a five-hour space walk,
with Tsibliev using the starboard-side Strela crane to transport
Linenger and sthe optical sensor package from *Kvant-2* to *Kristall*.

It was the ride of Linenger's life. Initially, he was overwhelmed
by his sense of speed and exposure, hurtling at more than 17,500
miles per hour 200 miles above the earth, a tiny object at the end of
the 40-foot pole, being whipped from one *Mir* module to another.
"My heart raced. I wanted to close my eyes in an effort to escape
this dreadful and persistent sensation of falling. White-knuckled, I
gripped the handrail on the end of the pole, holding on for dear
life."

Then he got control of himself and got down to work. By the
time Tsibliev swung him back to *Kvant-2*, he was thoroughly en-
joying himself. "Yahoo!" he yelled, holding on to the end of Strela
as it bounced him across the emptiness of space "like a fishing
pole." If he could have waved a cowboy hat in the air like any
rodeo-hand, he would have.[26]

"My God! Here It Is Already!"

On May 17, 1997, *Atlantis* arrived at *Mir*, bringing with it
Linenger's replacement, Michael Foale. Notwithstanding the complaints of senators and congressmen, NASA was going to complete its shuttle-*Mir* program. The shuttle also brought more than
5,000 pounds of supplies to *Mir*, including the new 250-pound
Elektron oxygen regenerator as well as new seeds for the Svet
greenhouse. Going home with *Atlantis* were the remains of the
burned lithium perchlorate candle canister as well as the failed
Elektron regenerator.

Linenger's flight had lasted 132 days. When he got back home
he refused to be carried out of the shuttle on a stretcher. "I was
[either] going to walk off that shuttle or . . . crawl off." He, like

Thagard and Blaha, was startled by how weak he felt. Later, measurements of his bone mass revealed that he had lost about 12 percent, especially in his lower spine and hips, a high rate of around 3 percent per month, quite possibly caused by the many problems that had hampered his exercise program.[27] The Americans were learning what the Russians had already found out: Several months of weightlessness changed the human body, and exercise appeared to be the only way to mitigate that change.

On *Mir*, Michael Foale took Linenger's place. For him, *Mir* reminded him most of a male frat house, where guys had been living and working for a decade. The first thing he needed to do was scope out his own place to sleep. Because Tsibliev and Lazutkin had already claimed the only two cabins in the base block, Foale chose *Spektr* as his home. He moved his personal effects there and quickly settled down to a routine.

Unlike Linenger, who had been willing to shake things up to get what he needed, Foale was a more affable fellow. His father had been a RAF fighter pilot. His mother came from Minneapolis. Foale, the eldest of three, had grown up on air-force bases in Cyprus, West Germany, and Malta, watching his father go out on flight patrols over the Middle East and Europe.[28] Rather early on in his training in Russia, Foale decided to try to ingratiate himself with the Russians. He and his wife repeatedly invited Russians to come to their Star City apartment for dinner. They visited several Russian families in their own homes.

On *Mir* he immediately offered to do as much of the maintenance work as Tsibliev and Lazutkin would give him, volunteering to clean up the many globs of water that had condensed on equipment throughout the station. He also made it a point to share every meal with the two Russians, to spend time with them chatting over tea, and to participate in every orbital communication session with mission control. "It was a way to stay close to the crew," he explained later.[29] Foale's easy-going approach to the situation helped ease the tension left over from Linenger's flight.

Not that the situation improved that much. By the middle of June, four months into his mission, Tsibliev was burnt out. Unlike his first long-term mission with Serebrov in 1993, the problems on this flight had been hard and never-ending. His partner, Lazutkin, was an inexperienced man who didn't seem able to get things done as efficiently as Serebrov had.[30] And Tsibliev did not have Serebrov's experience to lean on. Frazzled from overwork, his ability to think independently was gone. Without guidance he no longer knew what to do.

Tsibliev didn't give in, though. Whatever his mission controllers asked him to do, he tried to do. They wanted him to do a debilitating sleep experiment. He did it. They wanted him to take blood samples so often that his fingers were covered with pinpricks. He did it. They demanded that he and Lazutkin try to repair the Elektron regenerators. He did it. Considering the deteriorating state of *Mir*'s atmospheric and electrical systems, Tsibliev's dedication and hard work surely kept the station alive during these difficult months.

Not that he didn't complain about the workload. In June, his exasperation with the demands placed on him and Lazutkin by mission control reached a boiling point. At one point he became so sarcastic and frustrated that ground controllers arranged to have him chewed out in public by Pyotr Klimuk, the hero of *Salyut 4* during the signing of the Helsinki Accords and now the general who ruled Star City. "A job is a job," he lectured Tsibliev. "You went up there to work, not to relax and have fun. It wasn't a vacation when I was up there in the 70s, you know. Be tough and hang in there."[31]

Tsibliev nodded agreement, but remained angry. How dare Klimuk imply that he wasn't working hard?

In the last week of June, mission control asked him to do another docking test of the TORU system. As he had always done, Tsibliev agreed though he had reservations. On June 24, while ground controllers undocked *Progress* from the aft port, he set up TORU.

Since the failed docking test in March, ground engineers had concluded that the malfunction of the television camera was caused by interference from the older Kurs docking system. To prevent this interference, they decided to turn the Kurs system off. Rather than using the radar data from Kurs, Tsibliev would judge the freighter's speed and distance by eye and by a hand-held laser rangefinder.

This kind of rendezvous and docking was very risky, at the least. American astronauts in the 1960s had learned the impossibility of judging distance solely by eye when they learned how to make the first rendezvous and dockings during the *Gemini* program. In 1983, Vladimir Titov had almost smashed his *Soyuz-TM 8* spacecraft into *Salyut 7* when he tried to eyeball distances. At least a half-a-dozen times since then, docking attempts had been aborted when either computers or men had had insufficient information to complete the maneuver. Tsibliev himself had had two bad experiences, once when his *Soyuz* craft had grazed *Kristall* during his 1993 undocking, and again in March during the first *Progress* docking test. On top of this, ground control had positioned *Progress* in a higher orbit than *Mir*'s. When Tsibliev's monitor showed him an image from *Progress* looking down at *Mir*, he had to try to find the station silhouetted amid the wash of Earth's clouds and landforms.

Once again, the commander positioned his crewmates at different portholes to search for the approaching freighter when he found TORU's television images too ambiguous to trust. Once again, they couldn't see it. Once again, Tsibliev guessed at the freighter's speed and distance. Once again, he had no idea where it was throughout most of the exercise. And once again, when the men thought *Progress* should be at least 1200 feet away, Lazutkin suddenly saw it bearing down on them only seconds from impact. "My God," he yelled. "Here it is already!"[32]

A moment later, the freighter's nose smashed into the gap between *Kvant* and the base block, then flipped over so that its aft end banged into *Spektr* and one of its solar panels . For a second *Progress* hung there. Then it bounced once more off of *Spektr*, after which it drifted away silently like an iceberg.

Inside, the three men instantly sensed that the impact had cracked the station's hull. They could feel their ears pop from the pressure change. Alarms went off. Foale thought the leak was in *Kvant*. Lazutkin, however, had seen the freighter hit *Spektr* and knew that was where the leak had to be. He quickly rushed past Foale, poking his head inside *Spektr* to hear the terrifying hiss of escaping air.

Tsibliev, overwhelmed by circumstances, remained unmoving in the base block, talking to Moscow mission control. "Everything was going on fine," he moaned. "But then, God knows why, [*Progress*] started to accelerate and run into [*Spektr*]."

"Can you close any hatch?"

"We can't close anything," Tsibliev said. "Here everything is so screwed up that we can't close anything."

In the two years since Norman Thagard had helped run electrical cables through the station, 18 different ducts and cables had been fed through *Spektr*'s hatch. Imagine you need to run telephone and extension cords from one room of your house to another and, rather than feed the wires within the walls, you run them through the doorways. The doors would no longer close tightly, but what does it matter?

On *Mir* it mattered. Sealing *Spektr*'s hatch was impossible with all those cables and ducts in the way.

To get Tsibliev out of his funk, flight director Vladimir Solovyov ordered him to immediately open the cocks on his emergency oxygen tanks, hoping to keep the station's air pressure up long enough for the crew to escape. Abandoning the radio, Tsibliev flung himself from the base block and up to *Kvant-2*, where the tanks were stored.

Meanwhile, Lazutkin scrambled frantically to disconnect the wires running through *Spektr*'s hatch. First, in a panic, he tried to cut them with a 4-inch knife. He cut a data cable and then sparks flew from a hot electrical cable when he put the knife to it. When he saw that cutting the cables wouldn't work, he flew into *Spektr*, tearing at the cable connections to unplug them one by one. Somewhere nearby he could hear the hiss of escaping air.

Then he and Foale struggled to get a hatch in place. At first they tried to close the hatch inside *Spektr*, but couldn't pull it shut because the air escaping from the station into *Spektr* kept yanking it open. Next, Lazutkin grabbed one of the simple flat hatches that was strapped to a wall of the docking adapter. As the two men shifted it into position, Foale could immediately feel it being sucked into place. "Truly, there is a leak on the other side of this," he remembered thinking.

Very quickly the pressure difference sealed the hatch lid in and, to their relief, the air pressure in *Mir* stabilized. The station was saved.[33]

However, disconnecting the cables from *Spektr* left *Mir* crippled. *Spektr*'s four new solar panels provided the station with half its electrical power. With the cables unplugged, that power was lost. Experiments couldn't run. The Elektron regenerator in *Kvant-2* had no power. Moreover, Foale lost all his personal belongings, from family photos to toothbrush. He didn't even have slippers for his feet, leaving him barefoot. Everything was in *Spektr*, out of reach.

The impact of *Progress* had also thrown the whole station into an uncontrolled spin, about one rotation every six minutes, too strong for the gyros and the attitude-control system to overcome.[34] With the working solar panels no longer facing the sun, the batteries were quickly drained, and *Mir*'s systems began to shut down.

For the next 30 hours *Mir* was practically dead in the water. During the first few night passes, literally nothing worked. No fans. No lights. No computers. Nothing.

During these dead periods, the men could do little but wait, floating together in the base block, drinking some leftover cognac that one of them had found hidden in a refrigerator, and talking about life and traveling in space. Foale recalled these moments quite clearly. "I will always remember being in total darkness, no power, no fans, and all of us in front of the big window, looking at incredibly complex, swirling auroras with the galaxy showering down on them, with nothing else for us to do."[35]

During the first few day passes, enough sunlight periodically hit the solar panels to get some systems running, but the uncontrolled spin made it impossible to collect enough electricity to power up the batteries.

At this point it was Michael Foale, the American—the foreigner on board—who saved the day. As an expert in plotting orbital corrections and maneuvers, Foale persuaded both his commander Tsibliev and the Russian ground controllers that the only way to get *Mir* back under control was to use the engines on the *Soyuz-TM*. Working together, like Viktorenko and Serebrov had back in 1989, Foale and Tsibliev got the station under control. With Foale gazing out one of *Mir*'s window to see how each blast affected their spin and then yelling instructions back to Tsibliev at his *Soyuz-TM* controls, the commander fired burst after burst, until they had the station facing the sun with the solar panels recharging its batteries. "A bit like sailing a large yacht," was how Foale described it to his wife in an e-mail later that day.[36]

Their work wasn't over yet. To get all the batteries charged required them to trade dead and live batteries from *Kvant-2* and the base block. It wasn't until 48 hours after the collision that they finally got the toilet in *Kvant-2* working. "Which was terribly important," Foale noted coolly later.[37] For the first time in days, the three men could get some sleep.

Though the station was under control and electricity was flowing to its storage batteries, the loss of *Spektr*'s four solar panels left *Mir* lame, with only 50-percent power. To keep things working, the men turned off as many systems as possible, cutting off all electrical systems in both *Kristall* and *Priroda*, as well as one working Elektron system in *Kvant-2*. Without power, the temperatures in the two modules dropped to almost freezing, which, in turn, caused water to condense on everything.[38]

Meanwhile, Russian ground engineers, among them Sergei Krikalev, were coming up with a plan to recover some power from *Spektr*'s solar panels. They built a special hatch, with plugs on both sides that would allow the cosmonauts to reconnect the module's electrical cables to the rest of *Mir*'s systems while still keeping its hatch closed. After they designed and built it, Krikalev took it into the simulation tank and tested it.

Even as the engineers and Krikalev worked out the details on the new hatch, *Mir*'s attitude-control computer crashed again, putting half the gyros out and throwing the station into free drift. For three days the crew scrambled to regain power and orientation, again using the thrusters on *Soyuz-TM*.

On July 5, the next *Progress* freighter arrived with the new hatch and some personal items for Foale, including a toothbrush. The plan called for an internal space walk, during which Tsibliev and Lazutkin were to seal off the station's multiple-docking adapter, depressurize it, replace *Spektr*'s flat hatch with the special hatch, and connect cables on both sides. While they did this Foale was to stand by in the *Soyuz-TM* spacecraft. If everything went well, the cables would restore about 30 percent of the station's power.

It was not to be, at least not while Vasili Tsibliev and Alexander Lazutkin were on *Mir*. On July 13, the overwork, tension, and circumstances finally took their toll on Tsibliev. Since the collision he had been fighting depression and burnout. The never-ending succession of failures and problems was almost more than he could handle. "My heart is breaking," he said on the day of the collision. "You wake up in the morning and look at all this, and it just looks so darn sad."[39]

During the daily transmission of medical telemetry, the data showed that he had developed a heart arrhythmia, similar to but more serious than the condition experienced by Alexander Laveikin in 1987 at the beginning of *Mir*'s operations. When Laveikin's heart first sputtered, the doctors had not known how to react, and ordered Laveikin to do a number of stressful exercises to measure his heart's condition, the worst possible activity for a man with an unsteady heart. As Polyakov noted years later, "If they had let him rest, he would have been fine." In 1997, the doctors knew better. They immediately ordered Tsibliev to cease all work.[40]

Then, before any decision had been made about whether Tsibliev could do the internal space walk, another disaster struck. The next evening, Lazutkin entered the multiple-docking adapter and began unplugging cables in preparation for the space walk. Before they could seal the adapter, they had to clear the dozens of cables and ducts that ran through its six hatches into the other modules.

As Lazutkin worked, he accidentally flipped to the wrong page in his procedures manual and unplugged the wrong cable, disconnecting *Mir's* guidance computer from sensors in *Kristall*. The system once again shut down, and the entire complex once again drifted in space with its solar panels no longer oriented toward the sun.[41]

This situation could actually have been far more serious than previous shutdowns. First of all, it took hours for the men on *Mir* to get ground controllers to take the situation seriously. Repeatedly they radioed that *Mir* was drifting out of control, and repeatedly the ground controllers told them not to worry, to just hang on and wait while they tried to decipher *Mir's* meager telemetry.

Second, the endless calamities had made both Tsibliev and Lazutkin gun-shy. Raised in Russia's top-down culture, they were naturally reluctant to do anything without orders from above. The disasters of the last few months made them even more fearful of taking independent action. Rather than shut the station's systems down to save battery power, they decided to wait for further instructions, thus allowing *Mir's* batteries to drain entirely.[42]

Third, a strange design flaw in the *Soyuz* spacecraft actually stranded all three men in space for about a dozen minutes. When a *Soyuz* spacecraft was docked to *Mir* for long periods, its internal batteries were disconnected from its solar panels and the power from the panels was fed directly into *Mir's* batteries. However, to operate *Soyuz's* thrusters required power from its internal batteries, and to reconnect these batteries to the solar panels so they could be charged required power from the station.

During previous power outages, the crew had reconnected *Soyuz's* internal batteries before all power was gone. This time they waited for orders from the ground, and by the time they tried to activate *Soyuz*, the station was dead. It wasn't until the second night pass, when Tsibliev entered *Soyuz* to use its radio to contact the ground, that he realized that his spacecraft was useless, and couldn't be turned on.

For almost a quarter of an hour the men waited for the station to reenter daylight. As *Mir* came out of darkness enough sunlight hit the solar panels to power *Mir*'s aged batteries to restart *Soyuz*. Tsibliev then flipped the switch, and got the *Soyuz-TM* batteries reconnected to its solar panels, recharging them.[43]

For mission control, Lazutkin's error and Tsibliev's heart problems were the straws that broke the camel's back. As Vladimir Solovyov shouted in despair when he heard that Lazutkin had accidentally pulled the plug on *Mir*, "This is not a kindergarten!"[44] The internal space walk was canceled. Moscow mission control decided to have it done by the next crew. On August 15 Tsibliev and Lazutkin returned to Earth, both men facing harsh criticism for the many problems during their flight. They, in turn, laid much of the blame on faulty equipment, poor planning, and lack of support from ground control.

Meanwhile, their replacements, Anatoly Solovyov and Pavel Vinogradov, began the repair effort on *Mir*. Solovyov, still considered the master of *Mir* maintenance and space walks, had been called upon to save the station. Over the next six months, the two Russian cosmonauts undertook seven internal and external space walks to repair the crippled space station. They began work on August 15 when Solovyov, Vinogradov, and Foale boarded the *Soyuz-TM* spacecraft and flew around the station, videotaping it to assess its condition. On August 22 Solovyov and Vinogradov did an internal space walk, entering *Spektr*'s crippled interior to retrieve Foale's laptop and some family photographs and connect *Spektr*'s three working solar panels to *Mir*'s systems by way of the special hatch. On September 6 Solovyov and Foale did a six-hour space walk to inspect the outside of *Spektr*. While Foale manned the Strela controls, Solovyov cut away several insulation blankets looking for the leak. He also used a boat hook pole to manually realign two of *Spektr*'s solar panels so that they were better aimed at the sun.

Meanwhile, Vinogradov figured out what was wrong with the old and frequently failing Elektron unit in *Kvant*. A "white-brown jelly-like" glob was clogging one of its pipes. After clearing it away, the unit started up immediately, turning water into oxygen. For the first time in months, the station had two working Elektron regenerators.[45]

After Foale had returned to Earth, Solovyov and Vinogradov did a second internal space walk to finish attaching power cables to the new hatch. Then in early November they did two space walks and removed one of *Kvant*'s old solar panels, replacing it with the second of the two solar panels brought to *Mir* by *Atlantis* in 1995. By the time he came home, Solovyov had completed the 16th space walk of his cosmonaut career, a record that will probably last for years.

Though the repair effort had included plans for recovering *Spektr* by removing the damaged solar panel and capping the leak at its base, the capping operation did not take place, partly because the leak's exact location could not be pinpointed and partly because other, more-urgent repairs (such as an airlock leak and the repair of the station's guidance computer) took precedence.

Nonetheless, when these two men came home in February 1998, *Mir* was probably in better shape than it had been in years. Its atmospheric systems were running more or less reliably and its electrical capacity was about 80 percent of what had been available before the collision—which was actually more than what had been available before *Spektr*'s arrival in 1995.

Ironically, despite all the tribulations during Michael Foale's stay on *Mir*, the greenhouse experiment in *Kristall* continued to operate. Power for the greenhouse didn't come from the solar panels on *Spektr*, but from the new solar panels that had been attached to *Kvant* by Strekalov, Dezhurov, and Thagard and then cabled through the base block to *Kristall* in 1995. Even when everything else on *Kristall* was shut down and the module's temperature dropped to less than 40°F, the greenhouse purred on, its fluorescent lights not only nourishing its plants but also illuminating the module itself.[46]

Before the collision, Foale had planted 52 seeds that he brought with him from Earth. Dubbed *brassica rapa* by botanists and field mustard by everyone else, the wild plant produces tasty light-green leaves that can be cooked or mixed in salads. Field mustard was chosen because, like *arabidopsis*, it has a short life cycle, flowering only 14 days after planting. Foale's plan was to try to coax these seeds to produce two generations while he was on *Mir*. Before planting he installed some new equipment in the Svet greenhouse. Sensors developed at the University of Utah gave him a much more precise idea about the amount of moisture reaching the roots. Fans kept the greenhouse atmosphere circulating and clean of toxins like antifreeze. He installed a new bed of artificial soil, using commercially available gardening products. The many previous attempts to grow plants by both the Russians and Americans had shown that the water reached the roots best if the soil grains varied in size, from 1.7 to 4.3 mm in diameter.

Planting the seeds required Foale to insert them into a hollow wick that was then inserted into the layers of soil. As the plants grew, he used tweezers to coax the shoots out from the wick. When leaves began to appear, Foale wrapped them in special polyethylene bags to filter the air reaching them.

After about four weeks—twice as long as it would take on Earth—the plants had grown 1 to 2 inches high and were ready to be pollinated. At that point, Foale became a human bee, using what gardeners call a bee stick to collect pollen from the plants and deposit it on the plants' stamen. "Bzzz, bzzz, bzzz—you go up and down the rows," said Foale.[47]

Then the collision happened, changing the entire nature of Foale's agricultural effort. The loss of *Spektr* meant that he did not have access to more polyethylene filter bags. It meant that for long periods, the remaining plants floated in darkness with no fans to circulate the air around them. It meant that they had to live in a much colder environment than originally planned, with temperatures sometimes dropping to almost freezing.

Nonetheless, Foale persisted. Just days after the collision, a number of the plants developed seed pods. "Just like pea pods," Foale remembered. "It was pretty clear that they were full of seeds."[48] In between reorienting *Mir*, wiping up globs of condensed water, and shifting batteries about, Foale carefully harvested these pods, reserving half to return to Earth for later study and preparing the rest for replanting in space.

For the next few weeks, when conditions on *Mir* were their worst, he set the harvested seedpods to dry. When he opened the dried pods to get the seeds for planting, he found them to be generally smaller than those grown on Earth. By August, he had a dozen or so seeds. Carefully, he inserted about six seeds in wicks and placed each in Svet. The rest he put aside for return to Earth for study.

For the next four weeks, Foale treated them like babies, carefully helping each seed find the light, carefully feeding them the precise amount of water. Amid the arrival of Solovyov and Vinogradov and their space walks to fix *Mir*, Michael Foale struggled to produce life from his six plantings.

By September, four of his six seeds had germinated, growing leaves and seedpods. From these, he was able to harvest about 15 to 20 seeds in total. Though it seemed to him that the seeds were bigger and healthier than the first space generation, all told, they had not prospered. "They were so weak and flimsy," he noted later in disappointment. "Only two or three of them were worth planting." After their return to Earth, six were planted, producing two viable plants, though smaller and less healthy than normal Earth-grown field mustard.[49]

For 26 years, since *Salyut 1* and the deaths of Dobrovolsky, Volkov, and Patsayev, humans had tried to grow plants in space. What was natural and easy on Earth had turned out to be artificial and difficult in space. The plants died. Or they withered after blooming. Or their seeds were sterile. Or they grew deformed.

Yet, as disappointed as Foale might have been, he had done something grand. For the second time in history an Earth-born form of plant life had given birth to life in space. Just as Valentin Lebedev had done in 1983 when he proved that *arabidopsis* could flower in space, Foale had shown that mustard plants could reproduce there as well. The technology required might be complex and subtle, but the chances of making a garden out of barren space had just doubled.

On Earth, few paid much attention to Foale's agricultural success. Instead, the focus and furor was on the collision, the problems that ensued, and Russia's continuing failure to fund its part in the *International Space Station* project. Many people both in and out of NASA questioned whether any more Americans should be sent to *Mir* after Michael Foale. Others wanted the United States to dump the Russians from the space partnership.

Congressman Sensenbrenner held more hearings. There, NASA's Inspector General, Roberta Gross, described how NASA had consistently underestimated the seriousness of *Mir*'s aging systems. The space agency seemed unable to analyze the situation impartially. "Various sources have voiced their concern about the objectivity and/or adequacy of NASA's risk/benefit assessment process in the face of stated national policy to maintain the Russian/American partnership."[50]

In other words, because it was President Clinton's policy to go to space with the Russians, no technical problem in space, *no matter how dangerous*, was going to stop the program. And in fact, though many in NASA had questions about continuing the program, no one in management seemed to have any doubt that the next astronaut in the shuttle-*Mir* program, David Wolf, would fly to *Mir*. Clinton demanded it. Dan Goldin wanted it. Everyone in the Phase 1 office believed in it.

To justify his decision, Goldin turned to a commission he had created in 1994. Led by former NASA astronauts Tom Stafford and Joe Engle, the Stafford Commission was made up mostly of NASA insiders. And Stafford, who had commanded the American half of the 1975 *Apollo-Soyuz* mission, was strongly in favor of the American-Russian partnership. Or as Hoot Gibson noted, "Tom Stafford is the equivalent of a lobbyist for the Russians."[51] Not surprisingly, Stafford's commission rubber-stamped the idea that the shuttle-*Mir* missions should continue.

On September 24, 24 hours before *Atlantis* was to blast off and take astronaut David Wolf to *Mir* to replace Foale, Dan Goldin announced his decision. The program would go on. "I approve the decision to continue with the next phase of the shuttle-*Mir* program," he said. "It's the right thing to do."[52]

Ironically, the relationship with the Russians had actually improved in the months after the February fire. The endless problems both on *Mir* and with funding *Zvezda* had forced the Russians to work with others. When the collision happened, they kept their American partners informed of events. "[The Russians] were a lot more forthcoming after the collision," remembered Keith Zimmerman, one of the American ground operators during Foale's mission. "After the fire they realized they goofed. They said 'Oops. Yeah, we should have told the Americans. We messed up.'"[53]

They even listened to the advice of NASA officials and postponed a three-week French mission, originally scheduled for launch in August 1997. "I have serious concerns that the life-support, power, and attitude-control systems will not support the load of six crew members for 21 days," wrote Frank Culbertson to Ryumin in July.[54] Shortly thereafter, the Russians agreed.

In addition, the infusion of American cash, to the tune of $273 million, made it possible to resume work on the stalled *Zvezda* module. As suspicious as the Russians wanted to be about Americans, the goodwill that cash engendered could not be ignored. By the summer of 1997, people in both Russia and NASA were increasingly confident the module would be ready for launch sometime in late 1999 or early 2000.

The partnership had also taught the Russians to accept, even depend on, the help of a foreigner. The first time Foale told Solovyov that they had to use *Soyuz* to reorient *Mir*, the Russian had doubts, refusing to do anything until he got instructions from the ground. Mission control, however, immediately told him that "Michael has had a lot of experience with this problem and knows the technique. Will you please follow what he suggests?"[55] As it had done in the late 1970s on *Salyut 6*, the Russian effort to explore space had forced them to put aside their natural suspicion of outsiders and join the rest of civilization.

During *Mir*'s last three years in space, the station was visited three more times by American shuttles. Though the original agreement had called for Foale to be the last American on *Mir*, the U.S. agreed to add two more shuttle flights and pay an additional $73 million dollars to Russia to finance *Mir*'s operation. This additional money, combined with the $200 million taken from the American shuttle program, made possible *Zvezda*'s completion.

During those additional flights, two more Americans lived in space for a total of 271 days. Though once again there was tension—resulting from the Russian insistence on running things and the American unwillingness to kowtow to them—the crews in general managed to get along.

Also living on *Mir* during its last years were two Frenchmen (one for six months), a Slovak astronaut, and a former political aide to President Yeltsin.

Even Valeri Ryumin, of all people, returned to space to live on *Mir* for about a week. After years of watching the never-ending problems from the ground, he had decided, just as he did many years earlier after *Salyut 1*, that he should go into space and see for himself what was going on. He proposed the idea to his bosses at Energia, and they approved. He then lost 55 pounds to get in shape for a 10-day flight on the space shuttle *Discovery*.[56]

His attitude during all the disasters of Tsibliev's and Lazutkin's flight had been that the crew must have screwed up. After all, he had lived on *Salyut 6* for almost a year and managed to solve its technical problems, and that had been a much smaller and less-capable station.

Once on board, however, Ryumin's tune changed. "I don't know how they live up here," he told Charles Precourt, *Discovery*'s commander. "This is awful. This is worse than I imagined. This is unbelievable. This is unsafe." What struck him most was the amount of accumulated junk that was packed into every inch of *Mir*'s storage space. "We lost control of stowage and inventory at about the three year point," he said in disgust.

And then, when he asked if he could discard some rubbish and unusable cables that he found stuffed behind a panel, mission control told him that nothing could be thrown out. Like everyone else who had flown in space, Ryumin once again discovered that there was a vast difference between what people on the ground perceived about space and its reality. "You know," he said, "when you don't see it yourself, you hear these stories, you just don't imagine how bad it really is."[57]

In its last years, malfunctions of and repairs to *Mir*'s systems continued. In February 1998, an overheated fan started smoking, requiring a fast shutdown and its subsequent replacement. For a short while *Mir*'s atmosphere was filled with smoke. Not surprisingly, the troublesome airlock hatch on *Kvant-2* had more problems, and required more repairs. During one mission a space walk was canceled when Nikolai Budarin and Talgat Musabayev broke three wrenches in a futile attempt to open the hatch. Their space walk had to wait two weeks until the next *Progress-M* freighter arrived with a new wrench.

Not everything went wrong, however. Both Americans and Russians did more space walks, assembling new equipment, swapping experiments, repairing and replacing solar panels. Several crews searched in vain for the source of the leak on *Spektr*. Musabayev and Budarin replaced the thruster engine on top of the Sofora girder, climbing the 46-foot-tall tower to stand high above the earth on a racing human space vessel.

The station's guidance computer continued to crash. That problem was finally solved when a single small ventilation fan was inserted to blow cold air on the computer. It seemed that the years of failures had been caused merely by an overheated computer motherboard. "In Russia we don't have a lot of experience with air conditioning," noted Alexander Serebrov with humor. "Heating systems we know." After this fix, the computer worked fine, with almost no failures during *Mir*'s final two years of operation.[58]

Mir's atmospheric systems continued to need repairs, but in general they worked more reliably. Having only two men on board put less strain on them. If the station had been traveling to Jupiter during these years, the crews could have survived and kept the station operating.

Despite all its failures in 1997, *Mir* had proved that it was possible to build an interplanetary spaceship. In fact, two months after Ryumin's return, the station was still able to keep a person alive in space for more than a year. Sergei Avdeyev arrived on *Mir* on August 15, 1998, and remained on board for the next 380 days, making his flight the second longest in history, after Valeri Polyakov's. And like Polyakov, Avdeyev had few serious problems when he returned to Earth. His return left *Mir* unoccupied for the first time in just under 10 years. In fact, when Avdeyev and his crew returned to Earth at the end of August 1999, it was the first time in nearly a decade that no human was in space.

A year and a half later, on March 23, 2001, *Mir* was finally deorbited, burning up in a huge fireball of 1500 pieces as it crashed into the remote South Pacific. Even so, during those last 18 months one more crew visited the station, staying for just less than two-and-a-half months, with most of their flight's cost coming from funds raised in the attempt to turn *Mir* into a commercial operation.

When *Mir* went down its total mass was more than 120 tons. It had been in orbit for more than 15 years. It had completed more than 86,000 orbits. During its lifespan, 104 men and women from a dozen countries visited it, and it was occupied for more than twelve-and-a-half years total, including 4 men who inhabited it for more than a year. As it circled the earth it had traveled a total distance of 2.2 billion miles, sufficient to have taken its occupants to and from Mars several times. With all the problems in its later years, no one ever died, and in fact, no one was ever even seriously injured while working on *Mir*. The station had proved unequivocally that the technology for going to other planets was available, and buildable.

Provide human beings with the necessary tools and supplies and they can go anywhere.

The problems on *Mir* were never a reason not to send men and women to the station. Exploring space will never be easy. Creating an entirely artificial, human-built environment is incredibly difficult. Things will go wrong. Machinery will break. Equipment will wear out. "Space flight is hard," Michael Foale said several years later. "This is a difficult thing we're doing."[59]

That it is hard, however, is not a reason to avoid doing it. If anything, the challenge *itself* is the reason to proceed, if only to prove that we can become something better than we are. To have canceled the later shuttle-*Mir* missions would not only have done nothing to make the space programs of both the United States and Russia better, it would have been a cowardly act, inflicting a terrible wound to human creativity and exploration that might have taken decades to heal.

Nonetheless, what was disturbing about the decision-making process by everyone in the Clinton administration, both in the White House and at NASA, was their willingness to ignore serious problems in their efforts to keep the program running. Just as management problems had caused *Challenger* to explode in 1986, individuals in the shuttle-*Mir* program were willing to lie, to spin, to fake it, to close their eyes to every failure in order to avoid the difficult realities.

Bill Clinton ignored the problems entirely. He didn't care if the American program was harmed, as long as he could send foreign aid to the Russians. After cutting the budget for *Freedom* by several billion dollars—money that could have gone to U.S. companies—he then funneled that same money to Russia and Energia so that they could harvest the rewards.

Dan Goldin tried to face the problems on *Mir*, but out of fear that the space program would be trimmed, he never addressed some essential points about the Russian participation in the space station program both to Congress and to the public. And though he voiced support for the program, he never gave the shuttle-*Mir* program the management support it truly needed. In his later years as head of NASA, the space-station budget went completely out of control, leaving behind such a mess that today no one is sure there is enough money to finish the American portion.

Others pretended the problems weren't there. When Jerry Linenger, a man who had been on *Mir* during one of its most critical periods, raised questions about the station's safety, NASA people were more willing to label him a complainer than to pay attention to his concerns. For example, when Jim van Laak, the number-two man in the Phase 1 program, heard Linenger's debriefing, his response was anger rather than concern. Instead of aggressively investigating Linenger's complaints, van Laak began to monitor Linenger's interviews with the press, while simultaneously trying to discredit him with reporters. "In my opinion, Jerry Linenger does not have the right stuff," he told reporters. He also squelched circulation of written transcripts or notes of any of Linenger's debriefing sessions.[60]

No one would stand up, face Linenger's concerns squarely, and say "This is hard. Things have gone wrong. We have made mistakes. We *still* intend to do it." Instead, everyone put on their manager's hat and made believe the problems did not exist.

Space is unforgiving, however. Human beings will not colonize the planets and stars by faking it. In the harsh, hostile, cold darkness of the greater universe beyond Earth's atmosphere, the only way humans can survive is by facing the truth.

On November 20, 1998, long before Avdeyev came home, and long before *Mir* was destroyed, sunrise came once again to Baikonur. *Zarya*, the first module in the *International Space Station*, was finally launched. The human urge to live in space had not yet ended.

14

International Space Station:
Ships Passing in the Night

"The Focus is Lost."

Beaming with childish excitement, 60-year-old Dennis Tito rocketed through the hatch and into the *Zarya* module of the *International Space Station*, moving so fast that he flew past everyone else and almost crashed into the far end of the module. Talgat Musabayev, his commander, and Yuri Usachev, the station's commander, had to grab him to slow him down.

For a few minutes, while his crewmates hugged and joked with each other, he floated motionless and speechless near the television camera, his eyes gleaming with joy. Once, he looked at the camera, grinned, and gave an elated thumbs up. Then he joined the others for the traditional group shot, during which he made only one excited comment. "I don't know about this adaptation [to space] they talk about," he bubbled, his enthusiasm reminiscent of Pete Conrad's almost 30 years earlier. "I'm already adapted. I love space!"

Getting to space for Tito, however, had not been easy. The wealthy businessman and former NASA engineer grew up in Queens, New York, the son of Italian immigrants. His father was a printer, his mother a seamstress. When, as a teenager in 1957, he watched *Sputnik* circle the globe, he decided that space was the place for him. He got an aerospace engineering degree and went to work for the Jet Propulsion Laboratory in Pasadena, California, calculating the flight paths for NASA's *Mariner* Mars probes.

By the late 1960s, he realized that he wasn't going to become an astronaut for NASA. More importantly, he wanted to earn more money. In 1970 he founded Wilshire Associates, using his mathematical engineering skills to chart the stock market instead. By the 1980s he was a millionaire. By the 1990s, he was a multimillionaire, with a net worth estimated at more than $200 million.

At the same time, he watched the space program of the former Soviet Union become increasingly capitalistic, selling seats on *Mir* to the Europeans, to the Japanese, to the British, and even to the Americans. Then, during *Mir*'s last years, when Energia was scrambling to find some way to keep the station afloat, Tito saw an opportunity and seized it. In April 2000 he struck a deal with the private American-Russian company MirCorp, which was trying to turn *Mir* into a commercial operation, and bought a seat to visit *Mir* for $20 million. Tito put the money in an escrow account, available only upon his arrival on *Mir*, and began eight months of intense training in Russia.

Unfortunately, Tito's money wasn't enough to save *Mir*. In December 2000, the Russian government finally decided to abandon the aging station and focus its efforts on the *International Space Station (ISS)*. Tito no longer had a place to go.

No matter. Energia simply rescheduled his flight, giving him a seat on one of their missions to *ISS* during a routine swap of *Soyuz-TM* lifeboats. His money would cover the flight's costs, and possibly encourage other wealthy individuals who wanted to fly in space to buy a seat on future *Soyuz* missions.[1]

The whole plan seemed straightforward. The Russians had been doing these kinds of deals for years. Who could object?

To their surprise, every single international partner they worked with did. The heads of both NASA and the European Space Agency were soon making noises about how "dangerous" Tito's presence on the station would be, and how Russia had to follow the "rules." "Would people visit a hotel when it's under construction?" Dan Goldin told *Space News* reporters. "We don't have time to hand hold tourists that don't have the proper training."[2]

"It is irresponsible to send amateurs to [the *International Space Station*]," said Jorg E. Reustel-Buecht, director of the European Space Agency's manned flight division. "Russia could do what it wanted with its own station, but this is an international facility and there are rules to respect. Maybe in 20 years, when we have fully checked out the station and flights are more common, we can consider this sort of thing. But not now."[3]

Even Bill Shepherd, commander of the first expedition to the station, had doubts about Tito's flight. "Our crew trained for over four years to get ready for our flight," he said. "The day will come when we'll have civilians and tourists up there, but it's not something you can enter into lightly."[4]

Then, only five weeks before launch, Tito arrived with his two Russian crewmates at the Johnson Manned Space Flight Center in Houston, Texas, for five days of training. When they showed up at the visitors center to get their security badges, NASA astronaut and space station manager Bob Cabana ushered them into a conference room and explained that while the Russians could be trained, Tito would instead be given a standard tourist tour, then spend his time in meetings with NASA lawyers. His Russian crewmates found this treatment offensive, and politely told Cabana that if Tito couldn't train with them, they couldn't train either. The three men got up and left, returning to their hotel rooms.[5]

The Russian response to all these protests was blunt, but polite. Tito would fly. After more than 20 years' experience in training foreigners to fly on their spacecraft, the Russians no longer feared their presence. Tito was medically fit and had successfully completed his training. Moreover, he had paid for his ticket and they needed the money to fly the mission.[6]

In the end, despite weeks of frantic effort to convince either Tito or the Russians to back down, it was NASA and the Europeans who blinked. On April 28, 2001, Tito took off from Baikonur, joyously entering the station two days later.

Despite this disagreement, in the more than four years since the station's first module was launched in November 1998, assembly of the station has proceeded with surprisingly few technical problems. The U.S. finally completed construction of actual space-station hardware. After more than 13 years of planning and billions of dollars, *Unity* was launched by the shuttle on December 4, 1998, two weeks after *Zarya*, and was subsequently linked to *Zarya*'s bow port using the shuttle's robot arm.

Russia, meanwhile, managed to get *Zvezda* into orbit, albeit 18 months late. Though its space program continued to be somewhat strapped for cash, Russia's terrible economic woes of the late 1980s and early 1990s had disappeared, replaced by one of the world's most prosperous economies, with growth rates exceeding 7 percent in both 2000 and 2001 (compared to an average U.S. growth rate of about 4 percent). Their creative energies unleashed by freedom and capitalism, the Russian people are quickly turning their nation into a boomtown, with new businesses springing up everywhere and the horrible decay and poverty of the communist era slowly evaporating into oblivion.

After *Zvezda*'s launch in July 2000, construction continued with remarkable speed and efficiency. In October the first crew arrived, beginning what is hoped to be the never-ending, permanent occupation of space by the human race. In February 2001, the American-built *Destiny* laboratory was docked to *Unity*'s bow port by the shuttle. In March the first crew rotation took place. In April, just before Tito's visit, the first of the station's two robot arms was installed. And in the years since Tito's visit, the station has continued to grow. As I write these words in December 2002, 20 months after Tito's flight, humans have been living continuously in *ISS* for more than two years, including two Americans who set a new record for the longest American flight, 196 days.

Rising from *Unity* are the beginnings of the gigantic horizontal truss first conceived by NASA engineers in the early 1980s. Attached to this are two of eight immense solar-panel arrays—each 115 feet long and 38 feet wide and able to produce 31 kilowatts of power, three times more than *Skylab* and 25 percent more than *Mir*, at its best. The station now has two airlocks, the American-built Quest airlock attached to *Unity*'s starboard docking port, and the Russian-built Pirs docking and airlock module, attached to *Zvezda*'s bottom bow port. Pirs's extra ports permit the Russians to dock one more module or one more *Soyuz* spacecraft on their half of the station. Moreover, the shuttle has also shuttled up and down from the station the two Italian-built *Leonardo* and *Raffaello* reusable cargo modules, each able to carry as much as 10 tons of equipment and supplies.[7]

During this very successful construction phase, *ISS* has also had its share of malfunctions and breakdowns, many remarkably reminiscent of *Mir*'s problems. Several times the computers crashed, causing the station's attitude-control system to shut down. The Elektron system in *Zvezda* has had intermittent failures, forcing the crew to once again depend on lithium perchlorate candles while they troubleshot the problem. One of the carbon dioxide scrubbers in the American *Destiny* module didn't work because of a stuck valve, and had to be replaced. A treadmill failed because a ball bearing seized up.[8]

None of these difficulties, however, has been very serious. All have been repairable. Though running somewhat behind schedule, the station's assembly in orbit has continued unabated, so that all told, as of December 2002, *ISS* weighs almost 200 tons with about 15,000 cubic feet of permanent habitable space, numbers that far exceed both *Skylab* and *Mir* at completion.[9] And the *International Space Station* remains less than half finished.

Inside the station, the first science work has begun, including several long-term plant experiments, using several different greenhouses. In the station's first plant experiment, *arabidopsis* seeds

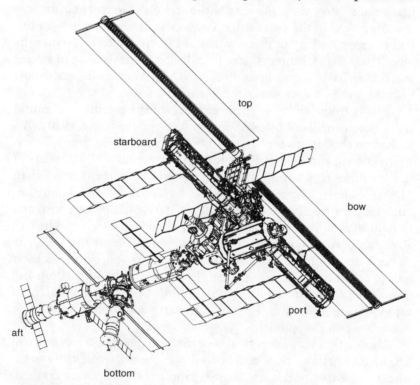

The *International Space Station*, as of December 2002. From aft to bow, the modules are *Zvezda* (with docked *Soyuz*), *Pirs* (with docked *Soyuz*), *Zarya*, *Unity* (with *Quest* and the beginnings of the large main truss attached), and *Destiny*. NASA.

were grown from seed to seed, once again confirming the achievement of the Energia scientists from 20 years earlier. More importantly, the space-born seeds were the healthiest ever grown. Ninety percent of the 91 seeds sent into space germinated. Seventy percent of these produced seedpods, with each plant producing about two dozen pods, each holding an average of three dozen healthy seeds.[10] If the station had devoted space for these seeds to grow, its astronauts could have turned the artificial environment of *ISS* into a garden, much as Ryumin did more than two decades earlier on *Salyut 6*.

For Sergei Krikalev, a member of the station's first crew, inaugurating the permanent occupancy of space merely continued his string of history-making flights. After his 1994 *Discovery* flight, he flew on the December 1998 shuttle flight that linked *Zarya* to *Unity*. On that same mission, he and American Bob Cabana activated the station and became the first two people to enter it, gliding in simultaneously as a gesture to the American-Russian partnership.

As one of *ISS*'s first full-time occupants, however, Krikalev was strangely dissatisfied by what he found there. Unexpectedly, the new station seemed far more overcrowded and cluttered than *Mir* had been in its last days. Because of the long delay between the launch of *Zarya* and *Unity* and the arrival of the first crew, one extra shuttle supply mission had brought with it more equipment than the not-yet-complete station was ready to handle. "People forget that the station is made of metal, not rubber. It doesn't expand," said Krikalev. "We were barely able to open the hatch."

In addition, *ISS*'s design, though far more sophisticated than *Mir*'s, disturbed Krikalev. Because the station was heavily computerized so that ground controllers could both monitor and maintain its operation, every piece of equipment required that computer system to operate. Krikalev couldn't even turn on the lights without booting up. This computerization, combined with the political deal that required both American and Russian mission control centers, meant that *ISS* needed *more* ground controllers than either *Mir* or *Skylab*. Instead of building a self-sufficient vessel able to sustain humans in space for long periods, the program had shifted its focus to creating jobs on the ground "to keep people busy," noted Krikalev.[11]

The station's sophistication had actually made it more dependent on Earth than were earlier stations. For example, during the station's fourth expedition in 2002, the crew reported a noxious odor coming from the Quest airlock, percolating out of a system that cleans the carbon dioxide scrubbers used by the American spacesuits. Ground controllers gave the crew permission to shut the system down, then ordered them to close the airlock's interior hatch to seal the smell in and retreat to the *Zvezda* module. While the crew slept, ground controllers shut down some of the station's ventilation fans and turned on an air-filtering system located in the *Destiny* module. When the crew woke up the next day, they reported that the air had cleared.[12]

As easy as this problem was for ground controllers to fix, one wonders at the long-range practicality of such an Earth-managed system. A spaceman can't very well ask for help from Earth when he is orbiting Mars. The message alone, traveling at the speed of light (186,282 miles per *second*), would take approximately 20 minutes to get back to Earth.

Interestingly, the station's stronger ties to the ground did nothing to eliminate the communication rifts between the Earth and the station. For example, NASA's extensive communication satellite network allowed Krikalev to speak to ground controllers 24 hours a day. On *Mir*, the lack of geosynchronous satellites dedicated to station communications meant that communication sessions were short, occurring only when the station passed over Russia. While this limitation was often inconvenient, it also left the crew large blocks of free time to work, unbothered by ground controllers who wanted to pester them with questions. On *ISS*, Krikalev found himself repeatedly distracted by questions from Earth.[13]

In another example, during the shuttle supply mission in June 1999, required because *Zvezda* was not yet ready, the crew found that if they stayed inside *ISS* too long, they experienced headaches, nausea, and eye irritation. They could alleviate these symptoms only by returning to the shuttle for breaks. However, because the astronauts were reluctant to describe the problem over the public airwaves, they told no one on Earth about it during the flight, making it impossible for ground controllers to pinpoint the problem. When, during post-flight debriefings, the crew finally revealed what had happened, NASA engineers—no longer having a crew on board the station to check—could only guess that the symptoms stemmed from stagnant air inside the station and a buildup of carbon dioxide.

To Krikalev, the annoyances and design questions reflected only a more fundamental lack of clear purpose. He knew that all the earlier Soviet stations were designed to learn how to build interplanetary, self-sufficient space vessels. Every change, every redesign, was done to make the next station more independent. "With *Mir* we almost had a closed loop," he remembered.

ISS, however, does not have this goal. In fact, Krikalev wasn't sure what *ISS*'s true goal was. In order to save development costs, the American portion of the station had long ago abandoned any effort to make its oxygen and water supplies self-sufficient. Instead of recycling these supplies, U.S. modules depend on the Russian recycling systems, its own lithium perchlorate candles, or the shuttle to haul up tons of supplies. "The focus is lost," said Sergei Krikalev. "We don't have a clear idea of what we are doing."[14]

This lack of focus can be seen by the simmering political, financial, and human problems that still surround the station on Earth—despite its relatively smooth in-space construction. In Russia, though the economy might be booming, the public no longer wishes to give its space program, or their government for that matter, a blank check. From personal experience they have learned the difference between freedom and tyranny, and now prefer their government agencies to be weak and poorly funded, thereby preventing them from again becoming too powerful. Thus, Energia and the Russian Space Agency have had to scramble to find revenues wherever they can.

Fortunately for these space agencies, such revenues do exist, and the heads of these organizations have learned, as good capitalists, how to compete to earn them. After Tito, they sold another tourist ticket on a subsequent *Soyuz-TM* lifeboat swap flight to Mark Shuttleworth, a South African internet entrepreneur, for approximately $20 million, while attempting and failing to sell a third ticket for the same amount (the American rock star who wanted to go unfortunately failed to raise the cash). The earned cash has, in turn, been plowed back into their program, keeping it solvent, at least for now. However, this focus on tourism and making a profit has distracted the Russians from their original goal of building interplanetary space vessels. Rather, they are focusing on turning their portion of *ISS* into a tourist and entertainment center in order to generate the funds to keep their operation afloat.

In the United States, the American contribution to the space station keeps shrinking. Conceived in the mid-1980s as an eight-person station, *Freedom* was supposed to include two laboratory modules, two habitation modules, a cargo module that the shuttle would haul up and down from Earth, two crew lifeboats, and the large truss to hold the eight gigantic solar panels.

By 2000, the American contribution to the *ISS* had shrunk to only one laboratory module, one habitation module, one crew lifeboat, and the truss and solar panels. Then, after the 2000 presidential elections, the Bush administration did an audit of the station budget left over from the Clinton administration, and discovered a chaotic mess. The budget to complete the station was estimated to be more than $4 billion over budget, a 50-percent overrun. As the audit bluntly noted, "The existing ISS Program Plan . . . is *not credible. . . .* The cost estimates for the U.S.-funded enhancements (e.g., permanent 7-person crew) are *not sufficiently developed* to assess credibility [their italics]."[15] In other words, there simply was no money to finish either the habitation module or the crew lifeboat.

Except for the already launched *Unity* and *Destiny* modules, and the truss and eight gigantic solar panels that will provide most of the station's power, the United States was forced to postpone indefinitely the construction of all remaining U.S. modules. The American contribution to its first station since *Skylab* has shrunk to practically nothing. And what little has been built has cost a staggering $24 billion, according to some estimates, three times more than the development and construction costs of the entire original shuttle fleet.[16]

This bad news might be counterbalanced if the private American space industry were forging ahead like Russian companies such as Energia. Unfortunately, it is not. Six months after the *Challenger* accident in 1986, the Reagan administration announced that the shuttle would no longer be available as a launch vehicle for private commercial ventures. NASA had been charging very low fees, maybe one-tenth of the actual launch cost, to encourage private enterprise to develop space industries. Moreover, the political agreements that got the shuttle funded in the 1970s required that *all* American launches use the shuttle. From Reagan's point of view, these conditions served only to choke off any private American initiatives to develop a private American launch industry. By forcing satellite companies to go elsewhere for their launch services, Reagan hoped to encourage the growth of a private launch industry. [17]

Unfortunately, this hope has not been fulfilled. Since 1986, only one new American aerospace company (Orbital Sciences) has been successful in entering the rocket industry, and the companies already in the business, such as Boeing and Lockheed-Martin, seem unable or unwilling to compete. While these companies continue to launch satellites for NASA and the military, commercial rocket sales have increasingly gone to either Russian companies, or the Ariane rocket—developed and owned by the European Space Agency.

Meanwhile, the American shift toward centralization and government control continues, as illustrated by the official response to Dennis Tito's flight. Poignantly, that response illustrated how much the opponents in the Cold War have traded places. Like ships passing in the night, the Russians have become freedom-loving capitalists, while the Americans have become xenophobic control-freaks.

Nor was that response an isolated event. Like the Soviet-era design bureaus, NASA's various centers in Houston, Huntsville, and elsewhere continue to focus on building empires for bureaucrats rather than spaceships that can go to the stars. As noted by the Bush audit, "The institutional needs of [NASA's] centers are driving the program, rather than the program requirements being served by the centers. . . . Deleting more hardware saves very little money since the bulk of the expenditures are in the 'people' category."[18] In other words, even as the American hardware on the station kept shrinking, the costs continued to rise in order to pay the salaries and benefits of NASA employees—*whether or not they built anything.*

This focus on maintaining the bureaucracy has also turned the American space program into one of the most regulated American activities imaginable. In fact, the autocratic nature of NASA is remarkably similar to the authoritarian work culture of Russia, a likeness that can be seen in many areas. For example, consider how tightly NASA schedules the work itinerary of its astronauts. When the two nations joined forces in space in the 1990s, astronauts from both countries were astonished at how much more independence and freedom Russian cosmonauts were given in space compared to the Americans. Based on their historical roots and their memories of the Cold War, everyone assumed the Russians would have little flexibility in their actions in space. Instead, it was the Americans who were expected to follow orders and a strict predetermined schedule, with little freedom to improvise if necessary.

And this inflexible scheduling policy continues to be NASA's preference. Consider how NASA manages science experiments on *ISS* compared to the experiments on past stations. With both the *Skylab* and *Salyut* stations, the crew was allowed a great deal of freedom to improvise. Garriott brought up a fish experiment in a plastic bag and hung it on a wall. Ryumin decorated *Salyut 6* with extra plants wherever he could. *ISS*, however, is supposed to be a government-funded laboratory, not a habitat for spacefaring humans. Thus, NASA must tightly control its use. Improvised experiments are strongly discouraged. Moreover, while plant research might be performed skillfully and with great success, it cannot be quickly adapted and put to practical use (such as using plants to decorate the station to improve morale while also helping to recycle the station's food and air) because such use doesn't fit with the rules that govern the use of federal facilities.

Consider also NASA's growing willingness to depend on psychologists rather than plain commonsense and good management for assigning and managing crews, a reliance far out of proportion to their ability to predict and eliminate crew conflicts. In one truly ominous example, a psychologist, with amazingly naive good intentions, proposed requiring every space-station crew member to wear a badge with an "infrared transponder in the form of a microchip that, on receipt of an infrared signal from another transponder on the wall, transmits a stream of 14 characters that identify the person or object to which the badge is attached." This psychologist then continued,

> With this equipment, it is possible to locate in 0.7 seconds any one of up to 65,000 persons. . . If desired, the phone nearest the person's current location can be rung in another 0.3 seconds. . . In a real space situation, use of monitoring would be of value in many ways. It could identify and report technological and human problems as they occurred. Badges would make it easy for each spacefarer to be found at all times. The officer of the watch would be able to see instantly the location of all crewmembers.[19]

Doesn't that sound just grand? Wouldn't we all like it if our boss could monitor our movements 24 hours a day?

Sadly few, inside or outside NASA, seem to see how oppressive such an idea is. In fact, the American public in the post-September

11, 2001, world seems increasingly eager to impose some form of similar identification-badge standards to the entire American populace.

Meanwhile, the Russians already know the simplest and most practical way to avoid the conflicts and miscommunications between those in the alien environment of space and those on the ground—a lesson that the Americans learned on *Skylab* but then forgot: Increase the experience of ground controllers and astronauts with each other's working conditions. Astronauts should work in mission control, and mission controllers should, as much as possible, be astronauts. This is what the Russians discovered during Romanenko's 10-month mission in 1987. Repeatedly, men like Manarov, Alexandrov, and Krikalev told me in interviews that once everyone understood what was going on in space, they relaxed, and could deal with the communication difficulties using basic commonsense. As Manarov noted from his own experience, "[These problems are] easy to understand if you've worked in mission control."[20]

Yet today, instead of commonsense, NASA seems willing to give increasingly favored status to their psychologists and the strict rules and requirements they have devised to restrict access to space.

Maybe the best way to see how small a part the concept of freedom plays in the American part of the *International Space Station* is to look at something as simple as the evolution of *ISS*'s name. Reagan proposed the station during the final years of the Cold War, eventually giving it the name *"Freedom"* to symbolize how freedom helps make all things possible. When Bill Clinton had the station reshaped as an international partnership, he dropped the name *Freedom*, a decision that was in at least one sense a convenient political move—he couldn't sell *his* space station project using a project name proposed by his predecessor. No official name was ever actually chosen for the new international station. The name "Alpha" came into use only because Clinton chose the "A" or "Alpha" option of the three station proposals presented to him. Later, *International Space Station* was adopted as a convenient and precise description, though hardly inspiring.

Clinton's abandonment of the name *Freedom* was also symbolic, considering how little freedom has to do with how things run on the station. The crew works under a military chain of command, with a commander. And everyone takes orders from the ground. Very little of what they do once they get into space has much to do with what they want to do.

Whether NASA's bureaucratic autocracy will dictate how things in space will run in the future remains unknown. The United States has the most money and power when it comes to running *ISS*. The Russians have their own agendas, and have demonstrated their willingness to push back if they think NASA is being unreasonable—as they did in the case of Dennis Tito. Yet, all signs continue to suggest that it will probably be a long time, maybe centuries, before the citizens in space will stand as free as the citizens of both the United States and Russia on Earth today. Like the early American colonists in the New World, the astronauts and cosmonauts in space are entirely dependent on their mother countries for survival. Everything, from their clothes to their food to even the very air they breathe, is given to them at the behest of Earth governments. And if they wish to stay in space, they must do the bidding of those governments, no questions asked.

In turn, just as the British in the 1700s looked down on their American colonists and the communists in the 1900s did the same to their citizens, the attitude of everyone on Earth is to treat these space explorers as something akin to indentured servants, hired to do whatever work their bosses on Earth demand. We have even passed international laws denying them the right to own property, declaring outer space as a kind of communist-style utopia where everything is reserved as "the province of all mankind."* No wonder many people were outraged when astronaut Dennis Tito, born an ordinary and free American but an astronaut nonetheless, had the audacity to pay for his space flight and go to *ISS*. How dare he!

And yet, the idea of freedom still lurks. During the effort by NASA to convince everyone that it would be all right to send another American to *Mir* when Michael Foale's mission ended, the shuttle-*Mir* office asked Foale to videotape and transmit to NASA an optimistic tour of *Mir* to show at committee hearings. Foale, who despite all his trials in space was entirely in favor of continuing the

---

*See the 1966 United Nations Outer Space Treaty, which came into force on October 10, 1967, and has been signed by more than a hundred countries, including the United States and the Soviet Union (now Russia).

missions, needed to broadcast when *Mir* was over Russia in order to get a strong enough communications signal to transmit the memory-heavy video signal.

Anatoli Solovyov objected, however. The commander, who in the 17 years since his first space flight had become the typically autocratic Russian boss, wanted Foale to do the transmission 20 minutes sooner, when *Mir* was over a much less capable American communications link. Bluntly, he ordered Foale to begin broadcasting then.

With his usual good humor, Foale refused. "Anatoli, I understand where you're coming from. But you Russians cannot order me to do this."

"You have to do it now!" shouted Solovyov. The quiet man who had once had trouble finding the right words to say was incensed that Foale would dare disobey his commands.

But Foale was a free man. Despite following orders for months, gladly doing some of the most menial tasks on *Mir*, he could not be *made* to do something he thought was wrong. With Solovyov fuming, Foale broadcast the tape while *Mir* was over Russia, as he needed to do.[21]

Once again, the willful nature common to Americans and Russians caused them to butt heads. The American was born to a culture that expected independence, while the Russian came from a society that had for centuries demanded diffidence and obedience from its citizens to their superiors.

And yet, Solovyov was a newly made Russian, forged in the last decades of the twentieth century when the first real glimmers of true freedom arrived in Russia. Ten months earlier, he had been selected as one of the prime candidates to fly with Krikalev as part of the first expedition to the *International Space Station*—and turned the mission down. The political deal between the U.S. and Russia provided for the first mission to be launched from Baikonur, using a *Soyuz* spacecraft on a Russian rocket, but it also stipulated that the only American on board, Bill Shepherd, would be the commander. Solovyov, one of the world's most seasoned space travelers, having done more space walks than any other man, couldn't stomach the deal. He believed that the mission should be commanded by a Russian.[22]

Free to choose, he chose to beg off. He could obey no one but his own conscience.

"Put Some More Engines on this Thing . . ."

A thousand years ago, a collection of shipbuilding tribal clans on the coasts of Norway, Denmark, and Sweden scattered outward, using their revolutionary ships to trade, conquer, and emigrate across a vast, worldwide expanse. The Vikings became bodyguards to the rulers of Constantinople. Their ships brought them to Greenland and the northern coasts of North America. In the Mediterranean, their sailing and shipbuilding techniques revolutionized ship design and trade.

They changed the world most, however, in how they shaped the culture of Europe—especially England and Russia. Vikings conquered and settled large regions in eastern England and the forests surrounding Moscow, bringing to both places their fiercely independent and freebooting culture of trade and exploration.

The consequences of that immigration was that, a thousand years later, the citizens of both the United States and Russia—coming from opposite sides of the globe with many social traditions that couldn't be more different—are in other ways remarkably similar. The ancestors of both peoples were pioneers. As the Americans worked their way westward, the Russians moved east. The land both groups settled was harsh, brutal, and unyielding. Death was omnipresent. Out of these two pioneer struggles have risen nations able to forge in the sky the first rockets, the first spacecraft, and the first tentative and grand attempts to colonize the stars.[23]

Now, both countries stand together in space, carving in the emptiness of space a new pioneer spirit, bull-headed and independent in its own way. Though their politicians still argue, and the bureaucrats and administrators on Earth still try to exert control over what these space explorers do, the men and women in space are nonetheless trailblazers, rough-hewn by the technically challenging and harsh environment of the universe they are trying to colonize.

And though they rarely say it out of fear that their incautious words might upset an Earth-based politician who might have the power to cancel their missions, they understand that the only real reason to build the *International Space Station* is to learn how to build interplanetary spaceships. As Wernher von Braun, Willy Ley, and Sergei Korolev noted a half-century ago, once in orbit it doesn't take much additional energy to propel a habitable vessel on its way to Mars, or anywhere else in the Solar System.

And spacefarers like Sergei Krikalev, Shannon Lucid, Valeri Polyakov, and Bill Shepherd want to go, very badly. As Shepherd said at one point during that first expedition to *ISS*, "Put some more engines on this thing and send up that Mars vector!"[24]

The trouble is, if they don't get the chance to put those engines on soon, the American and Russian effort to colonize the stars might simply peter out. Repetitive crew visits that don't seem to accomplish anything significant will only bore the public, destroying whatever excitement people might feel for space exploration. We can see this sad phenomenon already occurring with the first crews of the *ISS*. They go up, spend some time going around the Earth, and then come down. After the first few times, the whole trip begins to seem rather pointless.

Yet it isn't pointless. Or at least, it shouldn't be. If the focus were shifted to where it belongs, toward preparing the station for its eventual launch toward Mars, then everything these men and women did while in orbit would suddenly take on a greater intensity and significance. They would have a larger mission that would not only energize them but also excite the rest of the world.

Every technical question they face today would finally have a coherent purpose: to make the station as independent of Earth as possible. How reliable are the station's systems? To what degree can it grow its own food? How self-sufficient is its atmosphere and water supplies? If not self-sufficient, how long can its supply of water, food, and oxygen actually last? Instead of making the station's atmospheric systems less able to recycle its air and water (in order to save money), the goal would be to make them more self-sufficient. Instead of computerizing the station so that ground controllers could run it, the station would be operated entirely by the crew, and be made as simple as possible so that the crew could easily maintain and repair it themselves.

And instead of keeping the crew missions short, less than six months, the effort would be to lengthen the flights. Only by breaking Polyakov's fourteen-and-a-half month record and flying in space for eighteen months or longer (as he originally intended), will it be possible to find the likely candidates with a rate of bone loss low enough—less than half a percent per month—to travel in space for one to two years and still make it to the Martian surface without breaking their bones.

Suddenly, every action on board would have a purpose besides merely station-keeping. And the challenge (and greater danger) of lifting this huge hulking station out of Earth orbit and into the great beyond would become real and present, for all to see with every spin of the station.

Finally, and most importantly, by recognizing that *ISS* is really nothing more than a test version of an interplanetary spaceship, the very idea of journeying to Mars suddenly seems far less difficult and inconceivable. There is no need, as suggested repeatedly by many in NASA, to do decades of studies costing billions of dollars (while employing thousands of well-paid government workers) to figure out how to go. The first prototype of that vessel is already being built in orbit, and the extra cost and design work to "put some more engines on this thing" and send it to Mars is relatively insignificant.

It is now a half-century since the first spaceship builders dreamed of going to other planets. The twentieth century has ended. We now live in the third millennium, a time that for centuries was considered a symbol of the future.

Unfortunately, too many Americans today give a collective yawn to the idea of space travel. Too many pooh-pooh the idea of building an *International Space Station*, or of flying on the space shuttle. Too many get bored when traveling to Mars is mentioned. Their attitude is, "Been there, done that."

But we haven't been there, or done that. When Willy Ley wrote *The Conquest of Space* in 1949, he listed a number of mysterious places on the moon, and the persistent questions about those places. Of the crater Aristarchus, he wondered whether volcanoes might be the cause of the periodic sightings of clouds and patches of reddish color (which amateur astronomers still see periodically). Of the Straight Wall, on the east side of Mare Nubium, Ley wondered at the length (more than 60 miles) of this astonishingly straight shadow. Was it a gigantic fault, or a thousand-foot-high cliff face longer than anything on Earth? What caused it? Ley also wondered about the Moon's "Alpine" Valley, as it is dubbed, located near the crater Plato. This great valley, 80 miles long and 6 miles wide with a straight rill running along its center, slices at a right angle through the 10,000-15,000-foot-high Alps Mountains. Ley wondered why this valley's floor was so flat, and why it existed amidst such high mountains.

In other books, Ley wondered at the nature of distant stars, from red giants to strange and weird binary systems, and was baffled by their formation and structure. He also considered what was known of the Milky Way, trying to understand its size and shape.

In 1949 Willy Ley asked these questions, and noted that ". . . we'll never know [the answers] until we get there."[25]

Today, 50 years later, we still haven't answered any of these specific questions, and *we still haven't gone there*. Only a dozen humans have strolled on the lunar surface. Four of the six *Apollo* landing sites were chosen because of how boring and, therefore, safe they looked. In fact, the six lunar landings covered less territory than seen by a taxi driver in New York City on a typical afternoon. To this day, we, the human race, have yet to visit some of the most spectacular places on the lunar surface. To this day, all we really have is the imaginative musings of artists and writers and moviemakers. We still have not gone.

Even worse is the scarcity of people interested in such grand adventures. Too many instead cower in fear of a stupid, and harmless, computer-programming bug like Y2K, and seem unwilling to imagine heroic dreams, idealistic visions, or epic passions. Can we no longer imagine a future where humanity goes out and settles the far-flung stars? Have we become so small-minded that we cannot envision a tomorrow as idealistic and hopeful as that imagined by men like Ley, Korolev, and von Braun? I hope not.

In A.B. Guthrie's classic novel about the early American West, *The Big Sky*, there is a moment when great dreams meet reality, and the dreams win. Telling the story of the early explorers of the Rocky Mountains in the 1830s and 1840s, Guthrie describes how, after many years of difficult adventures in the wild, untamed wilderness of the Grand Teton Mountains, the book's main character, a mountain man, returns to his home in Kentucky. On his way east he meets all kinds of people with all kinds of foolish and silly ideas about what the American West is like, as well as what they can accomplish when they get there. A preacher wants to convert the Indians to Christianity. A man wants to go to Oregon with his family to farm. A young man has absurd visions of building a business empire.

The mountain man knows these dreams are wild and unrealistic. He knows what the West is *really* like, and, having seen the

West through his eyes, the reader tends to agree with him. The Rockies are savage and harsh, and the mountain man knows that the dreamers he meets are ridiculous, fooling themselves about how they can change that pitiless terrain. Yet today, at the beginning of the third millennium, these foolish and idiotic dreams are actually a closer description of today's West than the land known to the mountain man in 1840.

Our hopes and dreams are a definition of our lives. If we choose shallow and petty dreams, easy to accomplish but accomplishing little, we make ourselves small. But if we dream big, we make ourselves great, taking actions that raise us up from mere animals. I think of Don Quixote, who lived in a mad fantasy world of hope and honor, and who eventually brought the hard, cynical, and cold world to its knees, making it honor and accept his wild dreams.

Sometimes our dreams *are* foolish and mistaken. Sometimes they lead to terrible and tragic evils, as in the case of the communists in the Soviet Union or the terrorists on September 11, 2001. Sometimes they simply don't work and nothing comes of them, as in the case of Robert Goddard, the man who invented the liquid-fueled rocket but who died an unknown man who had never built a rocket able to fly more than a few miles into the air.

Sometimes, however, our dreams take flight, and we soar with angels. Sometimes, as they did in the American South in 1865 and in Russia in 1991, our dreams bring freedom and justice to slaves and the oppressed. Sometimes, as they did in 1969, our dreams take us to the moon, and possibly beyond.

It is now the twenty-first century. In a thousand years, just as today we look back at the Vikings, the human settlers on the Moon and Mars and Europa and the asteroids will look out across the blackness of space at the gleaming blue-white Earth and remember what our generation did to colonize the planets. The world these future space explorers shall live in will be shaped by what we do. Their language, their government, their society, everything about them, will be ordained by decisions we make now.

Will they speak English or Russian, or both? Or will another language, perhaps Chinese, dominate because the Americans and Russians got timid and didn't get the job done? More importantly, will they sing our praises, as we generally do of the Vikings, or wring their hands in misery and in anger because we did not act as wisely as we could? It is now in our power to shape that future.

We live in the world of tomorrow, and ironically, our technological capabilities are far better than anything von Braun or any of the engineers of the 1950s could imagine. For us to dream of settling the Moon or Mars is hardly a wild dream. It is clearly possible, and doable both cheaper *and* in a much grander scale than anything imagined half a century ago. It is time we did it. It is time we finally did what the better dreamers before us could only imagine, to have bold dreams, to be visionaries, to try a grand new adventure.

It is time we went to the stars.

Bibliography

I have tried to go to original sources wherever possible, using interviews with the people involved, flight transcripts where available, or original press reports in the United States and Russia. All quotes or specific details not footnoted come from these interviews.

For historical background a number of sources deserve special mention. Leon Aron's *Yeltsin, A Revolutionary Life* is by far the most complete history of the fall of the Soviet Union. Also helpful, though biased and incomplete, were the memoirs of Boris Yeltsin and Mikhail Gorbachev. Concerning the negotiation of the Helsinki Accords, John Maresca's *To Helsinki, The Conference on Security and Cooperation in Europe, 1973–1975* told the story from both a personal and a historical perspective. Another source for the Brezhnev years was Robin Edmonds's *Soviet Foreign Policy, The Brezhnev Years*.

The best source of general information about the *Salyut* space stations came from the regular reports produced by Gordon Hooper and Neville Kidger for *Spaceflight* magazine. Without their dedicated reporting, much of what happened on those stations would today be lost. Also helpful was David Harland's *The* Mir *Space Station*, which provided a good and thorough summary of these earlier reports. Furthermore, the biographical data gathered by both Gordon Hooper and Michael Cassutt has preserved much of what is known about the lives of the men and women who went to space.

For the history of the early Soviet space program, Asif Siddiqi's epic work, *Challenge to Apollo* provides the most information. I found the work of James Oberg, who has been reporting on Soviet space achievements for almost three decades, to be invaluable. In addition, Mark Wade's voluminous webpage, *Encyclopedia Astronautica*, continues to be one of the most useful sources of information about the history of space exploration. For plant research, *Space Biology, Studies at Orbital Stations* by Galina Nechitailo and Alexander Mashinsky describes the actual Russian research on these stations and includes flight transcripts, information available nowhere else. Dr. Nechitailo's personal help was also invaluable; she let me spend many hours in her home interviewing her, and gave me numerous heretofore unknown anecdotes of the 1970s and 1980s Soviet space program.

For the later years on *Mir*, Bryan Burrough's *Dragonfly* gives the most complete description, though he unfortunately does not footnote his material. Also useful were Chris van den Berg's regular *MirNews* Reports. For *Mir*'s entire existence, van den Berg listened to the ground-to-station communications and then provided summaries in English, which he posted on the internet. In addition, I found helpful David Portree's work in detailing the history of all Russian space stations and all human spacewalks. Finally, my interview with Alexander Cherniavsky, the engineer who designed the Sofora girder on *Mir*, was invaluable in helping to decipher the engineering work being done by the Russians in space.

Interviews

Alexander Alexandrov, 3/27/02, 4/3/02; Anatoli Artsebarski, 3/15/02, 4/4/02; Alexander Balandin, 3/25/02; Dr. Gail Bingham, 2/19/03, 2/21/03; Alexander Cherniavsky, 3/21/02; Marina Dobrovolsky, 4/3/02; Vladimir Dzhanibekov, 3/29/02; Georgi Grechko, 3/25/02, 4/3/02, 4/4/02; Sergei Krikalov, 3/31/02, 4/1/02; Alexei Leonov, 3/21/02; Jerry Linenger, 4/7/03; Oleg Makarov, 3/20/02; Musa Manarov, 3/28/02; Dr. Mary Musgrave, 2/13/03; Galina Nechitailo, 3/31/02, 5/14/02; Svetlana Patsayava, 4/2/02, 4/3/02; Valeri Polyakov, 3/18/02, 3/19/02, 3/25/02; Viktor Savinykh, 4/4/02; Alexander Serebrov, 3/26/02; Norman Thagard, 8/5/02, 8/6/02.

Other Sources

Academy of Sciences of the U.S.S.R. *Leonid I. Brezhnev, Pages from His Life.* New York: Simon & Schuster, 1978.

———. *Andrei Sakharov, Facets of a Life.* France: Editions Frontieres, 1991.

Adams, Carsbie C. *Space Flight: Satellites, Spaceships, Space Stations, and Space Travel Explained.* New York: McGraw-Hill, 1958.

Advisory Committee on the Redesign of the Space Station. *Final Report to the President.* Washington, D.C.: Government Printing Office, 1993.

Arnold, Anthony. *Afghanistan, the Soviet Invasion in Perspective, revised and enlarged edition.* Stanford, California: Hoover Institution Press, 1985.

Aron, Leon. *Yeltsin, a Revolutionary Life.* New York: St. Martin's Press, 2000.

Astronauts and Cosmonauts Biographical and Statistical Data, revised May 31, 1976. Washington: U.S. Government Printing Office, 1976.

Baker, David. *The History of Manned Spaceflight.* New York: Crown Publishers, 1981.

Barrett, Thomas M. *At the Edge of Empire, the Terek Cossacks and the North Caucasus Frontier, 1700–1860.* Boulder, Colorado: Westview Press, 1999.

Bekey, Ivan, and Daniel Herman, eds. *Space Stations and Space Platforms, Concepts, Design, Infrastructure, and Uses.* New York: American Institute of Aeronautics and Astronautics, Inc., 1985.

Belew, Leland F., ed. Skylab, *Our First Space Station.* Washington, D.C.: Scientific and Technical Information Office, NASA, 1977.

Berezovoi, Anatoli. *Cosmonaut Berezovoi's Memoirs on 211-Day Spaceflight.* Originally published in Moscow in *"Aviatsiya I Kosmonavtika, #7, 7/83, #8, 8/83, #9, 9/83,* translated by JPRS, USSR Report Space (JPRS-USP-84-004), 51–67.

Blagonravov, A.A., ed. *Transactions of the First Lectures Dedicated to the Development of the Scientific Heritage of K.E. Tsiolkovskiy, translation of "Trudy Pervykh Chteniy, Posvyashchennykh Razrabotke Nauchnogo Naslediya i Razvitiyu Idey K. E. Tsiolkovskogo," Academy of Sciences, USSR, Commission for the Development of the Scientific Legacy of K.E. Tsiolkovskiy, Moscow, 1967.* Washington, D.C.: NASA TT F-544, 1970.

Blaha, John. *NASA Oral History, ISS Phase I History Project.* Washington, D.C.: NASA, 1998. Available as of 8/13/02 at http://spaceflight.nasa.gov/history/shuttle-mir/people/oral-histories.htm.

Bogomokov, Valeri V. *Oral History, ISS Phase I History Project.* Washington, D.C.: NASA, 1998. Available as of 8/13/02 at http://spaceflight.nasa.gov/history/shuttle-mir/people/oral-histories.htm.

Bonting, Sjoerd L. *Advances in Space Biology and Medicine, a Research Annual, 1991–1997, volumes 1 through 5.* Greenwich, Connecticut: JAI Press, Inc., 1991–1997.

Branegan, John. *"Mir Communications in 1987,"* in *Spaceflight,* v30, 3/88, 108–112.

Brezhnev, Leonid. *Selected Speeches and Writings on Foreign Affairs.* Oxford: Pergamon Press, 1979.

———. *Peace, Detente, Cooperation.* New York: Consultants Bureau, 1981.

———. *Memoirs.* Oxford: Pergamon Press, 1982.

Brezhneva, Luba. *The World I Left Behind, Pieces of a Past.* New York: Random House, 1995.

Buergenthal, Thomas, ed. *Human Rights, International Law, and the Helsinki Accord.* Montclair, NJ: Allanheld, Osmun, & Co., Inc., 1977.

Burrough, Bryan. *Dragonfly, NASA and the Crisis Aboard* Mir. New York: HarperCollins Publishers, 1998.

Canby, Thomas Y. "Are the Soviets Ahead in Space?" in *National Geographic*, 10/86, 420–459.

Caprara, Giovanni. *Living in Space.* Willowdale, Ontario: Firefly Books, 2000.

Carlowicz, Michael J., and Ramon E. Lopez. *Storms from the Sun, the Emerging Science of Space Weather.* Washington, D.C.: Joseph Henry Press, 2002.

Cassutt, Michael. *Who's Who in Space, the International Space Year Edition.* New York: MacMillan Publishing Company, 1993.

Chafetz, Glenn R. *Gorbachev, Reform, and the Brezhnev Doctrine, Soviet Policy Toward Eastern Europe, 1985–1990.* Westport, Connecticut: Praeger Publishers, 1993.

Chaikin, Andrew. "The Loneliness of the Long-Distance Astronaut," in *Discover*, February 1985, 20–31.

Chernyak, Yuri B., and Joel L. Lebowitz. *Frontiers of Science, Reports from the Final International Session of the Moscow Refusnik Seminar.* New York: New York Academy of Sciences, 1992.

Churchill, Winston, *The Hinge of Fate, volume 4 of The Second World War.* Boston: Houghton Mifflin Company, 1950.

Clark, Phillip. *The Soviet Manned Space Program, an Illustrated History of the Men, the Missions, and the Spacecraft.* New York: Salamander Books Limited, 1988.

Clarke, Arthur C. *The Exploration of Space.* New York: Harper & Brothers, 1951.

———. *Islands in the Sky.* Philadelphia: John C. Winston Company, 1952.

———. *Man and Space.* New York: Time, Inc., 1964.

Collins, Guy. *Europe in Space.* New York: St. Martin's Press, 1990.

Collins, Joseph. *The Soviet Invasion of Afghanistan, a Study in the Use of Force in Soviet Foreign Policy.* Lexington, Massachusetts: Lexington Books, 1986.

Collins, Michael. *Carrying the Fire.* New York: Farrar, Straus, and Giroux, 1974.

Committee on Space Biology and Medicine, Space Studies Board, Commission on Physical Sciences, Mathematics, and Applications, National Research Council. *A Strategy for Research in Space Biology and Medicine in the New Century.* Washington, D.C.: National Academy Press, 1998.

Committee on the Space Station, National Research Council. *Report of the Committee on the Space Station.* Washington, D.C.: National Academy Press, 1987.

Compton, W. David, and Charles D. Benson. *Living and Working in Space, a History of* Skylab. Washington, D.C.: Scientific and Technical Information Branch, NASA, 1983.

Computer Sciences Corporation. International Space Station *Operations Architecture Study, Final Report.* Washington, D.C.: NASA, 2000.

Cooper, Henry S.F. *A House in Space.* New York: Holt, Rinehart, and Winston, 1976.

Crockett, James. *Crockett's Victory Garden.* Boston: Little, Brown, and Company, 1977.

Culbertson, Frank. *NASA Oral History, ISS Phase I History Project.* Washington, D.C.: NASA, 1998. Available as of 8/13/02 at http://spaceflight.nasa.gov/history/shuttle-mir/people/oral-histories.htm.

Dadykin, V.P., *Growing Plants in Space,* "Znaniye" Press, Moscow, 1968. Translation of *"Kosmicheskoye Rasteniyevodstvo,"* Wash-ington, D.C.: NASA TT F-704, 1972.

D'Agostino, Anthony. *Gorbachev's Revolution.* New York: New York University Press, 1998.

David, Leonard. *Space Station* Freedom, *A Foothold on the Future.* Washington, D.C.: NASA NP-107/10-88, 1988.

Dixon, Simon. *The Modernisation of Russia, 1676–1825.* Cambridge, UK: Cambridge University Press, 1999.

Donnelly, Alton. *The Orenburg Expedition: Russian Colonial Policies on the Southeastern Frontier, 1734–1740.*

Dornberger, Walter. *V2.* New York: Viking Press, 1958.

Dugan, William. *NOVA: The Russian Right Stuff: The Mission, Secrets of the Soviet Space Station,* Mir. 60 min. Boston: WGBH Educational Foundation, Windfall Films, Filmcentre, Toronto, and BBC-TV, 1991.

Drygas, Maciej. *State of Weightlessness.* 52 min. Poland: ADR Productions, distributed by First Run Icarus Films, New York, 1994.

Dzhanibekov, Vladimir. "Rescue in Outer Space," in *Parade Magazine,* 7/27/86, 12–14.

Eddy, John A. *A New Sun, the Solar Results from Skylab.* Washington, D.C.: Scientific and Technical Information Office, NASA, 1979.

Edmonds, Robin. *Soviet Foreign Policy, The Brezhnev Years.* Oxford: Oxford University Press, 1983.

Eisenhower, Susan. *Breaking Free, A Memoir of Love and Revolution.* New York: Farrar, Straus and Giroux, 1995.

Engle, Joe. *NASA Oral History, ISS Phase I History Project.* Washington, D.C.: NASA, 1998. Available as of 8/13/02 at http://spaceflight.nasa.gov/history/shuttle-mir/people/oral-histories.htm.

Felshman, Neil. *Gorbachev, Yeltsin, and the Last Days of the Soviet Empire.* New York: St. Martin's Press, 1992.

Feodoroff, Nicholas V. *History of the Cossacks.* Commack, New York: Nova Science Publishers, Inc.

Foale, Colin. *Waystation to the Stars, the Story of* Mir, *Michael, and Me.* London: Headline Book Publishing, 1999.

Foale, Michael. *NASA Oral History, ISS Phase I History Project.* Washington, D.C.: NASA, 1998. Available as of 8/13/02 at http://spaceflight.nasa.gov/history/shuttle-mir/people/oral-histories.htm.

Freeman, John. *Security and the CSCE Process, the Stockholm Conference and Beyond.* New York: St. Martin's Press, 1991.

Freeman, Marsha. *How We Got to the Moon, the Story of the German Space Pioneers.* Washington, D.C.: 21st Century Science Associates, 1993.

———. *Challenges of Human Space Exploration.* Chichester, UK: Praxis Publishing, 2000.

Gelman, Harry. *The Brezhnev Politburo and the Decline of Detente.* Ithaca, N.Y.: Cornell University Press, 1984.

General Accounting Office. *Space Station, NASA's Search for Design, Cost, and Schedule Stability Continues, Report to the Chairman, Committee on Science, Space, and Technology, House of Representatives.* Washington, D.C.: General Accounting Office, 1991.

————. *Space Station, Program Instability and Cost Growth Continue Pending Redesign, Report to the Honorable Tim Roemer, House of Representatives.* Washington, D.C.: General Accounting Office, 1993.

Geraci, Robert. *Window on the East, National and Imperial Identities in Late Tsarist Russia.* Ithaca and London: Cornell University Press, 2001.

Gerstenmaier, William H. *NASA Oral History, ISS Phase I History Project.* Washington, D.C.: NASA, 1998. Available as of 8/13/02 at http://spaceflight.nasa.gov/history/shuttle-mir/people/oral-histories.htm.

Geyer, Dietrich. *Russian Imperialism, the Interaction of Domestic and Foreign Policy, 1860–1914.* Leamington Spa, UK: Berg Publishers, Ltd., 1987.

Gilbert, Martin. *Shcharansky, a Hero of Our Time.* London: MacMillan, 1986.

Glazkov, Yuri N. *NASA Oral History, ISS Phase I History Project.* Washington, D.C.: NASA, 1998. Available as of 8/13/02 at http://spaceflight.nasa.gov/history/shuttle-mir/people/oral-histories.htm.

Glushko, V.P. *Rocket Engines GDL-OKB.* Moscow: Novosti Press Agency Publishing House, 1975.

Goldman, Marshall I. *What Went Wrong with Perestroika.* New York: W. W. Norton & Company, 1991.

Goldwater, Barry M. *With No Apologies, the Personal and Political Memoirs of United States Senator Barry M. Goldwater.* New York: William Morrow and Company, Inc., 1979.

Gorbachev, Mikhail. *Perestroika, New Thinking for Our Country and the World.* New York: Harper & Row, Publishers, 1987.

————. *Memoirs.* New York: Doubleday, 1996.

Grey, Jerry. *Enterprise.* New York: William Morrow and Company, Inc., 1979.

Gross, Roberta. *Letter to the Honorable F. James Sensenbrenner from NASA's Office of Inspector General, dated 9/12/97.* Text of letter available as of 8/26/02 at http://www.hq.nasa.gov/office/oig/hq/inspections/mir/sensenerrata.html.

Gruen, Adam L. *The Port Unknown, A History of the Space Station Freedom Program.* Typescript, located in the NASA History Office, Washington, D.C.

Hammond, Thomas T. *Red Flag Over Afghanistan, the Communist Coup, the Soviet Invasion, and the Consequences.* Boulder, Colorado: Westview Press, 1984.

Harford, James. *Korolev, How One Man Masterminded the Soviet Drive to Beat America to the Moon.* New York: John Wiley & Sons, Inc., 1997.

Harland, David. *The Mir Space Station, a Precursor to Space Colonization.* Chichester, UK: John Wiley & Sons, 1997.

————. *The Space Shuttle, Roles, Missions, and Accomplishments.* Chichester, UK: John Wiley & Sons, 1998.

Harrison, Albert A., Yvonne A. Clearwater, and Christopher P. McKay, eds. *From Antarctica to Outer Space: Life in Isolation and Confinement.* New York: Springer-Verlag, 1991.

Harvey, Brian. *The New Russian Space Programme, from Competition to Collaboration.* Chichester, UK: John Wiley and Sons, 1996.

————. *Russia in Space, The Failed Frontier.* Chichester, UK: Praxis Publishing, 2001.

Heinlein, Robert. *The Rolling Stones.* New York: Ace Books, 1952.

Heppenheimer, T.A. *The Space Shuttle Decision, NASA's Search for a Reusable Space Vehicle.* Washington, D.C.: NASA History Office, 1999.

Holder, William G., and Wiliam O. Siuru, Jr. Skylab, *Pioneer Space Station*. Chicago: Rand McNally & Company, 1974.

Hooper, Gordon R. *The Soviet Cosmonaut Team, a Comprehensive Guide to the Men and Women of the Soviet Manned Space Programme, volumes 1 and 2*. San Diego: GRH Publications, 1990.

———. "Missions to *Salyut 4*," "Missions to *Salyut 5*," "Missions to *Salyut 6*," articles in *Spaceflight*, 1975, 1976, 1977, 1978, and 1979.

Hutchings, Robert L. *Soviet-East European Relations, Consolidation and Conflict, 1968–1980*. Madison, Wisconsin: University of Wisconsin Press, 1983.

IMCE Task Force. *Report by the* International Space Station *(ISS) Management and Cost Evaluation (IMCE) Task Force to the NASA Advisory Council, November 1, 2001*. Washington, D.C.: NASA, 2001. Available as of 8/15/02 at ftp:// ftp.hq.nasa.gov/ pub/pao/reports/2001/imce.pdf.

Jenkins, Dennis R. *The History of Developing the National Space Transportation System, the Beginning Through STS-50*. Marceline, Missouri: Walsworth Publishing Co., 1992.

Johnson, Nicholas L. *Handbook of Soviet Manned Space Flight*. San Diego: Univelt, Inc., 1980.

Johnston, Richard S., and Lawrence F. Dietlein. *Biomedical Results from* Skylab. Washington, D.C.: Scientific and Technical Information Office, NASA, 1977.

Johnston, Richard S., Lawrence F. Dietlein, and Charles A. Berry, eds. *Biomedical Results from Apollo*. Washington, D.C.: Scientific and Technical Information Office, NASA, 1975.

Jones, Robert A. *The Soviet Concept of 'Limited Sovereignty' from Lenin to Gorbachev, the Brezhnev Doctrine*. New York: St. Mar-tin's Press, 1990.

Kalb, Marvin, and Bernard Kalb, *Kissinger*. Boston: Little, Brown, and Company, 1974.

Kamenskii, Aleksandr. *The Russian Empire in the Eighteenth Century: Searching for a Place in the World*. Armonk, New York: M.E. Sharpe, Inc., 1997.

Kargopolov, Yuri. *NASA Oral History, ISS Phase I History Project*. Washington, D.C.: NASA, 1998. Available as of 8/13/02 at http://spaceflight.nasa.gov/history/shuttle-mir/people/oral-histories.htm.

Kelley, Donald R., *Soviet Politics in the Brezhnev Era*. New York: Praeger Publishers, 1980.

Kerner, Robert J. *The Urge to the Sea, the Course of Russian History, the Role of Rivers, Portages, Ostrogs, Monasteries, and Furs*. New York: Russell & Russell, 1942.

Khrushchev, Nikita. *Khrushchev Remembers, the Last Testament*. Boston: Little, Brown, and Co., 1974.

Kidger, Neville. "*Salyut 6* Mission Reports," "*Salyut 7* Mission Reports," "*Mir* Mission Reports," "Above the Planet," articles in *Spaceflight*, 1980–1996, and the *Journal of the British Interplanetary Society*, 1983.

Kirkpatrick, Jeane. *The Withering Away of the Totalitarian State . . . and Other Surprises*. Washington, D.C.: AEI Press, 1990.

Kissinger, Henry. *White House Years*. Boston: Little, Brown, and Co., 1979.

———. *Years of Upheaval*. Boston: Little, Brown, and Co., 1982.

Kliuchevsky, Vasilii Osipovich. *A Course in Russian History, the Seventeenth Century*. Armonk, NY: M.E. Sharpe, Inc., 1994.

Kramer, Saunders B. "The Rescue of *Salyut 7*," in *Air & Space Smithsonian*. February/March 1990, v4, #6, 54–59.

La Fay, Howard. *The Vikings*. Washington, D.C.: Special Publications Division, National Geographic Society, 1972.

Lantzeff, George V., and Richard A. Pierce. *Eastward to Empire, Exploration and Conquest on the Russian Open Frontier, to 1750*. Montreal: McGill-Queen's University Press, 1973.

Lebedev, Valentin. *Diary of a Cosmonaut: 211 Days in Space*. New York: Bantam Books, 1990.

Lewis, Richard S. *The Voyages of* Columbia, *the First True Spaceship*. New York: Columbia University Press, 1984.

Ley, Willy. *The Conquest of Space*. New York: Viking Press, 1949.

———. *Engineers' Dreams*. New York: Viking Press, 1954.

———. *Space Stations*. Poughkeepsie, N.Y.: Guild Press, Inc., 1958.

———. *Rockets, Missiles, and Space Travel*. New York: Viking Press, 1961.

Ley, Willy, and Wernher von Braun. *The Exploration of Mars*. New York: Viking Press, 1956.

Lindroos, Marcus. "Space Station Freedom," available at *Encylopedia Astronautica* as of 1/21/03 at http://www.astronautix.com/craft/spaeedom.htm.

———. "Space Station Designs," available at *Encylopedia Astronautica* as of 1/21/03 at http://www.astronautix.com/craft/spas1982.htm.

———. "U.S. Space Stations," available at *Encylopedia Astronautica* as of 1/21/03 at http://www.astronautix.com/craftfam/usstions.htm.

———. "Dual Keel Space Station," available at *Encylopedia Astronautica* as of 1/21/03 at http://www.astronautix.com/craft/duan1985.htm.

Linenger, Jerry M. *Q&A with Astronaut Jerry Linenger*. Boston: WGBH & NOVA, 2002. Available as of 8/13/02 at http://www. pbs.org/wgbh/nova/mir/live.html.

———. *Off the Planet: Surviving Five Perilous Months Aboard the Space Station* Mir. New York: McGraw-Hill, 2000.

Logsdon, John M. *Space Station, A Policy History*. Washington, D.C.: NASA Johnson Space Center NAS9-16461, 1983.

———. *Together in Orbit, the Origins of International Participation in the Space Station, Monographs in Aerospace History Series, Number 11*. Washington, D.C.: NASA History Division, 1998.

Lomanov, Anatoli I. *NASA Oral History, ISS Phase I History Project*. Washington, D.C.: NASA, 1998. Available as of 8/13/02 at http://spaceflight.nasa.gov/history/shuttle-mir/people/oral-histories.htm.

Longley, David. *The Longman Companion to Imperial Russia, 1689–1917*. Essex, UK: Pearson Education Ltd., 2000.

Longworth, Philip. *The Cossacks*. New York: Holt, Rinehart, and Winston, 1969.

Lucid, Shannon. *NASA Oral History, ISS Phase I History Project*. Washington, D.C.: NASA, 1998. Available as of 8/13/02 at http://spaceflight.nasa.gov/history/shuttle-mir/people/oral-histories.htm.

———. "Six Months on *Mir*," in *Scientific American*, 5/98, 46–55.

Lutomski, Michael. *NASA Oral History, ISS Phase I History Project*. Washington, D.C.: NASA, 1998. Available as of 8/13/02 at http://spaceflight.nasa.gov/history/shuttle-mir/people/oral-histories.htm.

Maresca, John J. *To Helsinki, The Conference on Security and Cooperation in Europe, 1973–1975*. Durham, North Carolina: Duke University Press, 1985.

Marshall, Eliot. "Space Stations in Lobbyland," in *Air and Space Smithsonian*, 12/ 88–1/89, v3, #5, 54–61.

Mastny, Vojtech. *Helsinki, Human Rights, and European Security, Analysis and Documentation*. Durham, North Carolina: Duke University Press, 1986.

Matson, Wayne R., ed. *Cosmonautics, A Colorful History*. Washington, DC: Cosmo Books, 1994.

McCauley, Martin. *Gorbachev*. London: Addison Wesley Longman LTD, 1998.

McCurdy, Howard E. *The Space Station Decision, Incremental Politics and Technological Choice*. Baltimore: Johns Hopkins University Press, 1990.

McDonald, Sue. Mir *Mission Chronicle, November 1994–August 1996*. Washington, D.C.: NASA TP-98-207890, 1998.

McMichael, Scott R. *Stumbling Bear, Soviet Military Performance in Afghanistan*. London: Brassey's (UK), 1991.

Mel'kumov, T.M., ed. *Pioneers of Rocket Technology, Selected Works*. Translation of "*Pionery raketnoy tekhniki—Kibal'chich, Tsiolkovsky, Tsander, Kondratyuk—Izbrannyye trudy.*" Izdatel'stvo "Nauka," Moscow, 1964. Washington, D.C.: NASA, 1965.

Messerschmid, Ernst, and Reinhold Bertrand. *Space Stations, Systems and Utilization*. Berlin, Germany: Springer-Verlag, 1999.

Moore, Isaac W. "Caasi." *NASA Oral History*, ISS *Phase I History Project*. Washington, D.C.: NASA, 1998. Available as of 8/13/02 at http:// spaceflight.nasa.gov/history/shuttle-mir/people/oral-histories.htm.

Morgun, Valeri. *NASA Oral History, ISS Phase I History Project*. Washington, D.C.: NASA, 1998. Available as of 8/13/02 at http://spaceflight.nasa.gov/history/shuttle-mir/people/oral-histories.htm.

Morris, Edmund. *Dutch, A Memoir of Ronald Reagan*. New York: Random House, 1999.

Morrison, John. *Boris Yeltsin, From Bolshevik to Democrat*. New York: Dutton, 1991.

Murphy, Paul. *Brezhnev, Soviet Politician*. Jefferson, N.C.: McFarland & Company, Inc., 1981.

NASA. International Space Station *User's Guide*. Washington, D.C.: NASA, available as of 8/13/02 at http://spaceflight.nasa. gov/station/reference/index.html.

NASA Lyndon B. Johnson Space Center. Skylab *EREP Investigations Summary*. Washington, D.C.: Scientific and Technical Information Office, NASA, 1978.

NASA Mission Operations Directorate, Space Flight Training Division. International Space Station *Familiarization*. Houston, Texas: NASA, 1998.

NASA Public Affairs Office, *Space Station* Freedom *Media Handbook*. Washington, D.C.: NASA, 1989.

NASA *Skylab* transcripts, located in NASA History Office, NASA Headquarters, Washington, D.C.

NASA Space Station Engineering Division, *NASA Space Station* Freedom *Strategic Plan 1992*. Washington, D.C.: NASA, 1992.

National Commission on Space. *Pioneering the Space Frontier*. New York: Bantam Books, 1986.

Nechitailo, G.S., and A.L. Mashinsky. *Space Biology, Studies at Orbital Stations*. Moscow: Mir Publishers, 1993.

Newkirk, Dennis. *Almanac of Soviet Manned Space Flight*. Houston, Texas: Gulf Publishing Company, 1990.

———. "A Pioneer of Space: Georgi Grechko's Story," in *Quest*, summer 1993, 28–30.

Newkirk, Dennis, and Jim Plaxco. *Historical Reminiscences of Cosmonaut Georgi Grechko, Parts 1 and 2*. Originally published in *Spacewatch*, May 1993 and available as of 1/21/03 at http://www.astrodigital.org/space/intgrechko1.html.

Newkirk, Roland W., Ivan D. Ertel, and Courtney G. Brooks. Skylab, *a Chronology*. Washington, D.C.: Scientific and Technical Information Office, NASA, 1977.

Nicogossian, Arnauld, and James Parker. *Space Physiology and Medicine*. Washington, D.C.: Scientific and Technical Information Branch, NASA SP-447, 1982.

Nicogossian, Arnauld, Stanley Mohler, Oleg Gazenko, Anatoliy Grigoryev, eds. *Space Biology and Medicine, a Joint U.S./Russian Publication in Five Volumes*. Washington, D.C.: American Institute of Aeronautics and Astronautics, 1993.

Nixon, Richard. *The Memoirs of Richard Nixon*. New York: Grosset & Dunlap, 1978.

———. *The Real War*. New York: Warner Books, 1980.

———. *In the Arena, a Memoir of Victory, Defeat, and Renewal*. New York: Simon and Schuster, 1990.

Noordung, Hermann. *The Problem of Space Travel, the Rocket Motor*. Washington, D.C.: NASA SP-4026, 1995.

Norby, M.O. *Soviet Aerospace Handbook*. Washington, D.C.: Department of the Air Force, 1978.

Oberg, James E. *Red Star in Orbit*. New York: Random House, 1981.

———. *Uncovering Soviet Disasters: Exploring the Limits of Glasnost*. New York: Random House, 1988.

———. "Russia's Space Program: Running on Empty," in *IEEE Spectrum*, v32, #12, 12/95, 18–35.

———. *Star-Crossed Orbits, Inside the U.S.-Russian Space Alliance*. New York: McGraw-Hill, 2002.

Oberg, James, E., and Alcestis R. Oberg. *Pioneering Space, Living on the Next Frontier*. McGraw-Hill, 1986.

Oberth, Hermann. *Man into Space: New Projects for Rocket and Space Travel*. New York: Harper & Brothers, 1957.

———. *The Moon Car*. New York: Harper & Brothers, 1959.

Ordway, Frederick I. III. *Visions of Spaceflight, Images from the Ordway Collection*. New York: Four Walls Eight Windows, 2001.

Orlov, Yuri. *Dangerous Thoughts, Memoirs of a Russian Life*. New York: William Morrow and Company, Inc., 1991.

O'Rourke, Shane. *Warriors and Peasants, the Don Cossacks in Late Imperial Russia*. New York: St. Martin's Press, 2000.

Parker, John W. *Kremlin in Transition, volume I, From Brezhnev to Chernenko, 1978 to 1985*. Boston: Unwin Hyman, 1991.

———. *Kremlin in Transition, volume II, Gorbachev 1985 to 1989*. Boston: Unwin Hyman, 1991.

Patsayev, V.A., ed. Salyut *Space Station in Orbit*. Moscow: Mashinostroyeniye Press, 1973. Translation: Washington, D.C.: NASA TT F-15,450, 1974.

Pemberton, William E. *Exit with Honor, the Life and Presidency of Ronald Reagan.* Armonk, New York: M.E. Sharpe, 1997.

Pesavento, Peter. "From Aelita to the *International Space Station,* the Psychological and Social Effects of Isolation on Earth and in Space," in *Quest,* 8:2 (2000), 4–23.

Piszkiewicz, Dennis. *The Nazi Rocketeers, Dreams of Space and Crimes of War.* Westport, Connecticut: Praeger, 1995.

———. *Wernher von Braun, the Man Who Sold the Moon.* Westport, Connecticut: Praeger, 1998.

Pitts, John A. *The Human Factor, Biomedicine in the Manned Space Program to 1980.* Washington, D.C.: Scientific and Technical Information Branch, NASA, 1985.

Pogue, William R. *How Do You Go to the Bathroom in Space?* New York: Tom Doherty Associates, 1991.

Portree, David S.F. Mir *Hardware Heritage.* Washington, D.C.: NASA RP 1357, 1995.

———. *Humans to Mars, Fifty Years of Mission Planning, 1950–2000, Monographs in Aerospace History Series, Number 21.* Washington, D.C.: NASA SP-2001-4521, 2001.

Portree, David S.F., and Robert C. Trevino. *Walking to Olympus: An EVA Chronology, Monographs in Aerospace History Series, Number 7.* Washington, D.C.: NASA, 1997.

Precourt, Charles J. *NASA Oral History, ISS Phase I History Project.* Washington, D.C.: NASA, 1998. Available as of 8/13/02 at http://spaceflight.nasa.gov/history/shuttle-mir/people/oral-histories.htm.

Presidential Commission Report on the Space Shuttle Challenger *Accident,* volume 1. Dated June 6, 1986, Washington, D.C.

Raeff, Marc. *Imperial Russia, 1682–1825, the Coming of Age of Modern Russia.* New York: Alfred A. Knopf, 1971.

———. *Understanding Imperial Russia, State and Society in the Old Regime.* New York: Columbia University Press, 1984.

Reagan, Ronald. *An American Life.* New York: Simon and Schuster, 1990.

Riabchikov, Evgeny. *Russians in Space.* Moscow: Novosti Press Agency Publishing House, 1971.

Ryan, Cornelius, ed. *Across the Space Frontier.* New York: Viking Press, 1952.

Ryumin, Valeri Victorovitch. *NASA Oral History, ISS Phase I History Project.* Washington, D.C.: NASA, 1998. Available as of 8/13/02 at http://spaceflight.nasa.gov/history/shuttle-mir/people/oral-histories.htm.

———. *A Year Away From Earth: A Cosmonaut's Diary.* Moscow: Molodaya Gvardiya, 1987. Translation by JPRS-USP-90-002-L, 2/12/90.

Sagdeev, Roald Z. *The Making of a Soviet Scientist, My Adventures in Nuclear Fusion and Space from Stalin to Star Wars.* New York: John Wiley & Sons, 1994.

Sang, Anthony. *NASA Oral History, ISS Phase I History Project.* Washington, D.C.: NASA, 1998. Available as of 8/13/02 at http://spaceflight.nasa.gov/history/shuttle-mir/people/oral-histories.htm.

Savinykh, Viktor. Typescript of English translation of his diary from *Soyuz-T 13* mission.

Schauer, William H. *The Politics of Space, a Comparison of the Soviet and American Space Programs.* New York: Holmes & Meier, 1976.

Seaton, Albert. *The Horsemen of the Steppes, the Story of the Cossacks*. London: The Bodley Head, Ltd., 1985.

Sega, Ronald M. *NASA Oral History, ISS Phase I History Project*. Washington, D.C.: NASA, 1998. Available as of 8/13/02 at http://spaceflight.nasa.gov/history/shuttle-mir/people/oral-histories.htm.

Semyachkin, Vladimir. *NASA Oral History, ISS Phase I History Project*. Washington, D.C.: NASA, 1998. Available as of 8/13/02 at http://spaceflight.nasa.gov/history/shuttle-mir/people/oral-histories.htm.

Sharansky, Natan. *Fear No Evil*. New York: Random House, 1988.

Sharipov, Salihzan. *NASA Oral History, ISS Phase I History Project*. Washington, D.C.: NASA, 1998. Available as of 8/13/02 at http://spaceflight.nasa.gov/history/shuttle-mir/people/oral-histories.htm.

Sharman, Helen, and Christopher Priest. *Seize the Moment*. London: Victor Gollancz, 1993.

Shayler, David J. Skylab, *America's Space Station*. Chichester, UK: Springer Praxis Publishing, 2001.

Shcharansky, Avital. *Next Year in Jerusalem*. New York: William Morrow & Co., Inc., 1979.

Shepherd, Bill. *Expedition One Ship's Logs*. available in a censored edition as of 8/27/02 at http://spaceflight.nasa.gov/station/crew/exp1/ex1logs.html.

Shevchenko, Arkady N. *Breaking with Moscow*. New York: Alfred A. Knopf, 1985.

Siddiqi, Asif A. *Challenge to* Apollo*: The Soviet Union and the Space Race, 1945–1974*. Washington, D.C.: NASA History Division, Office of Policy and Plans, 2000.

Simpson, Theodore R., ed. *The Space Station, an Idea Whose Time Has Come*. New York: IEEE Press, 1985.

Skylab *On-Board Voice Transcriptions, including Ground-to-Capsule Communications, Voice Tape Dumps, Public Affairs Office Announcements, and Press Conferences*. Houston, Texas: NASA Johnson Space Center, 1973. Located in NASA History Office, NASA Headquarters, Washington, D.C.

Smith, Melvyn. *Space Shuttle*. Somerset, England: Haynes Publishing Group, 1986.

Soffen, Gerald A. *Visions of Tomorrow: A Focus on National Space Transportation Issues, 25th Goddard Memorial Symposium, Proceedings held March 18–20, 1987 at the Goddard Space Flight Center, Greenbelt, Maryland*. San Diego: American Astronautical Society, 1987.

Solovyov, Vladimir, and Elena Klepikova. *Boris Yeltsin, a Political Biography*. New York: G.P. Putnam's Sons, 1992.

Sotheby's. *Russian Space History, December 11, 1993 (Sale Number 6516), Auction Results*. New York: Sotheby's Holdings, Inc., 1993.

Steele, Jonathan. *World Power, Soviet Foreign Policy Under Brezhnev and Andropov*. London: Michael Joseph, 1983.

———. *Soviet Power, the Kremlin's Foreign Policy—Brezhnev to Chernenko*. New York: Simon & Schuster, Inc., 1984.

Stockton, William, and John Noble Wilford. *Spaceline, Report on the* Columbia's *Voyage into Tomorrow*. New York: Times Books, 1981.

Stoiko, Michael. *Soviet Rocketry: Past, Present, and Future*. New York: Holt, Rinehart, and Winston, 1970.

———. *Pioneers of Rocketry*. New York: Hawthorn Books, Inc., 1974.

Stuhlinger, Ernst, and Frederik I. Ordway III. *Wernher von Braun, Crusader for Space, a Biographical Memoir.* Malabar, Florida: Krieger Publishing Company, 1994.

Stuster, Jack. *Bold Endeavors, Lessons from Polar and Space Exploration.* Annapolis, Maryland: Naval Institute Press, 1996.

Summerlin, Lee B., ed. Skylab, *Classroom in Space.* Washington, D.C.: NASA SP-401, 1977.

Sumners, Carolyn. *Toys in Space, Exploring Science with the Astronauts.* Blue Ridge Summit, Pennsylvania: TAB Books, 1994.

Thagard, Norman E. *NASA Oral History, ISS Phase I History Project.* Washington, D.C.: NASA, 1998. Available as of 8/13/02 at http://spaceflight.nasa.gov/history/shuttle-mir/people/oral-histories.htm.

Thomson, Bob, with James Tabor. *The New Victory Garden.* Boston: Little, Brown, and Company, 1987.

Titov, Vladimir Georgievich. *NASA Oral History, ISS Phase I History Project.* Washington, D.C.: NASA, 1998. Available as of 8/13/02 at http://spaceflight.nasa.gov/history/shuttle-mir/people/ oral-histories.htm.

Tsygankov, Oleg S. *NASA Oral History, ISS Phase I History Project.* Washington, D.C.: NASA, 1998. Available as of 8/13/02 at http://spaceflight.nasa.gov/history/shuttle-mir/people/oral-histories.htm.

Ujica, Andrei. *Out of the Present.* 95 min. Germany: Bremer Institute Filmfernsehen/K Films, 1996.

United States Congress. 95-2, House, Committee on Science and Technology. *Space Shuttle, 1978 Status Report, Serial U, December 1978.* Washington, D.C.: U.S. Government Printing Office, 1978.

———. *Space Shuttle, 1979, Status Report, Serial ZZ, January 1978.* Washington, D.C.: U.S. Government Printing Office, 1978.

———. 96-1, House, Committee on Science and Technology. *Oversight: Space Shuttle Cost, Performance, and Schedule Review, Hearing, June 28, 1979.* Washington, D.C.: U.S. Government Printing Office, 1979.

———. *Oversight: Space Shuttle Program Cost, Performance, and Schedule Review, Serial U, August 1979.* Washington, D.C.: U.S. Government Printing Office, 1979.

———. 101-1, House, Committee on Science, Space, and Technology. *Hearing, October 31, 1989, No. 82.* Washington, D.C.: U.S. Government Printing Office, 1990.

———. 101-2, Senate, Subcommittee on Science, Technology, and Space. *NASA's Plan to Restructure the Space Station* Freedom, *Hearing, April 16, 1991.* Washington, D.C.: U.S. Government Printing Office, 1991.

———. 103-1, House, "Departments of Veterans Affairs and Housing and Urban Development and Independent Agencies Appropriations for Fiscal Year 1995: Hearings Before a Subcommittee of the Committee on Appropriations, April 23, 1993." Washington, D.C.: U.S. Government Printing Office, 1993.

———. 103-1, House, "Departments of Veterans Affairs and Housing and Urban Development and Independent Agencies Appropriations for Fiscal Year 1995: Hearings Before a Subcommittee of the Committee on Appropriations, June 17, 1993." Washington, D.C.: U.S. Government Printing Office, 1993.

———. 103-2, House, "Departments of Veterans Affairs and Housing and Urban Development and Independent Agencies Appropriations for Fiscal Year 1995: Hearings Before a Subcommittee of the Committee on Appropriations, May 17, 1994." Washington, D.C.: U.S. Government Printing Office, 1994.

———. 103-2, Senate, "Departments of Veterans Affairs and Housing and Urban Development and Independent Agencies Appropriations for Fiscal Year 1995: Hearings Before a Subcommittee of the Committee on Appropriations, June 7, 1994." Washington, D.C.: U.S. Government Printing Office, 1994.

Urban, Mark. *War in Afghanistan, 2nd Edition*. New York: St. Mar-tin's Press, 1990.

Van den Berg, Chris. *MirNews, 1988 to 2001*. Published on the web. 1997 to 1999 available as of 8/13/02 at http://infothuis.nl/muurkrant/mirmain.html.

Van der Linden, Frank. *Nixon's Quest for Peace*. Washington, D.C.: Robert B. Luce, Inc., 1972.

Vaughan, Diane. *The* Challenger *Launch Decision, Risky Technology, Culture, and Deviance at NASA*. Chicago: The University of Chicago Press, 1996.

von Braun, Wernher, Fred Whipple, Willy Ley. *Conquest of the Moon*. New York: Viking Press, 1953.

———. *The Mars Project*. Urbana, Illinois: University of Illinois, 1953.

———. *First Men to the Moon*. New York: Holt, Rinehart, and Winston, 1958.

———. *Space Flight, Past, Present, and Future, the Fifth Wings Club "Sight" Lecture*. New York: The Wings Club, 1968.

———. *Space Frontier*. New York: Holt, Rinehart, and Winston, 1971.

Vorobiev, Pavel Mikhailovich. *NASA Oral History, ISS Phase I History Project*. Washington, D.C.: NASA, 1998. Available as of 8/13/02 at http://spaceflight.nasa.gov/history/shuttle-mir/people/oral-histories.htm.

Wade, Mark. *Encylopedia Astronautica*. Available as of 8/23/02 at http://www.astronautix.com/.

Walker, Rachel. *Six Years That Shook the World, Perestroika—the Impossible Project*. Manchester, UK: Manchester University Press, 1993.

Ward, Ingaret, ed. *The Greatest Adventure*. Sydney, Australia: C. Pierson, Publishers, 1994.

Weeks, Albert L. *The Troubled Detente*. New York: New York University Press, 1976.

Westwood, J.N. Endurance *and* Endeavour, *Russian History 1812–1992, Fourth Edition*. Oxford: Oxford University Press, 1993.

Wetherbee, James D. *NASA Oral History, ISS Phase I History Project*. Washington, D.C.: NASA, 1998. Available as of 8/13/02 at http://spaceflight.nasa.gov/history/shuttle-mir/people/oral-histories.htm.

Wilford, John Noble. *We Reach the Moon*. New York: Bantam Books, 1969.

Wolfe, Tom. *The Right Stuff*. New York: Bantam Press, 1980.

Yeltsin, Boris. *Against the Grain, an Autobiography*. New York: Summit Books, 1990.

———. *The Struggle for Russia*. London: HarperCollins Publishers, 1994.

———. *Midnight Diaries*. New York: Pereus Books Group, 2000.

Zaehringer, Alfred J. *Soviet Space Technology*. New York: Harper & Brothers, 1961.

Ziman, John, Paul Sieghart, and John Humphrey. *The World of Science and the Rule of Law, a Study of the Observation and Violations of the Human Rights of Scientists in the Participating States of the Helsinki Accords.* Oxford: Oxford University Press, 1986.

Zimmerman, Keith. *NASA Oral History, ISS Phase I History Project.* Washington, D.C.: NASA, 1998. Available as of 8/13/02 at http://spaceflight.nasa.gov/history/shuttle-mir/people/oral-histories.htm.

Zimmerman, Robert. *The Chronological Encyclopedia of Discoveries in Space.* Phoenix, Arizona: Oryx Press, 2000.

Notes

Preface

1. Goldwater, 301.
2. Churchill, 475.
3. Kamenskii, 61.
4. Kamenskii, 245.
5. Churchill, 582.
6. Ley and von Braun, 107.
7. Harford, 129.

Chapter 1—Skyscrapers in the Sky

1. Khrushchev, 46.
2. Harford, 156, 166; Siddiqi, 117.
3. Harford, 93–94, 213–215.
4. Harford, 95, 16–167.
5. Harford, 33.
6. Siddiqi, 205–210; Harford, 247–248, 253–254.
7. Harford, 38; Siddiqi, 6–7.
8. Sagdeev, 181.
9. Siddiqi, 319–321; Harford, 113–115, 273.
10. Harford, 257–259; Siddiqi, 21–22.
11. Harford, 273.
12. Siddiqi, 227–236.
13. For just a sampling of the more practical engineering proposals, see for example Oberth (1957), 60–96; Noordung, 72–75, 93–95, 108–109, 128–129; Stoiko, 17–24.
14. *Wireless World*, October 1945, 305–308.

15. For a sampling only, see Clarke (1952), 25–26, 34–35, 48–49.
16. Clarke (1951), 58–59.
17. Clarke (1951), 150–162.
18. Freeman, Marsha (1993), 33; Ley (1961), 119–154.
19. Ley (1961), 117.
20. Ley (1961), 160–162; Freeman, Marsha (1993), 67–70.
21. Freeman, Marsha (1993), 216–249; *New York Times*, 6/25/69.
22. Ley (1949), 31.
23. *Colliers*, 4/30/54, 30; see also *Amazing Stories*, v14, #2, 2/40, 122–124.
24. Ley (1958), 23.
25. Ley (1949), 31–32.
26. Ley (1958), 16–17.
27. Stuhlinger & Ordway, 16–17; Piszkiewicz (1995), 13; Piszkiewicz (1998), 23–24.
28. Stuhlinger & Ordway, 23; Dornberger, 26, 139–140; Piszkiewicz (1998), 21.
29. Dornberger, 104; Piszkiewicz (1995), 197.
30. Piszkiewicz (1998), 91.
31. von Braun (1953), 65–66.
32. *Colliers*, 4/30/54, 23.
33. von Braun (1953), 6; *Colliers*, 4/30/54, 24.
34. *Colliers*, 4/30/54, 25–26.
35. *Colliers*, 4/30/54, 23.
36. Ley & Braun (1956), 88.
37. Siddiqi, 207.
38. Ley (1958), 44.
39. Dornberger, 62.
40. Stuster, 15–18.

Chapter 2—*Salyut*: "I Wanted Him to Come Home."

1. Nechitailo and Mashinsky, 22, 153–154.
2. Patsayev, 127.
3. Patsayev, 144.
4. *New York Times*, 10/23/69, 20; Siddiqi, 712.
5. Murphy, 5–20, 15–19; Brezhnev (1978), 23–24; Brezhnev (1982), 18–21.
6. Murphy, 20–35, 45–59, 59–82; Brezhnev (1978), 24–25.
7. Edmonds, 36–43; Murphy, 237–244; Brezhneva, 140–162.
8. Yeltsin (1990), 156–164.
9. Edmonds, 38–43, Brezhnev (1981), 3–8.
10. Siddiqi, 712.
11. Riabchikov, 278.
12. Siddiqi, 590–591.
13. Johnson (1980), 220; Siddiqi in *Quest*, v5, n3, 28.
14. Nechitailo interview; Patsayev interview.
15. Siddiqi, 766–769; Johnson (1980), 220–233; Clark, 56–60.
16. Harland (1997), 16; Johnson (1980), 108.
17. Patsayev, 11.
18. Johnson (1980), 229.

19. Patsayev, 20. Though the actual chemicals are not mentioned in this source, Patsayev describes the identical chemical process that occurs when lithium perchlorate is used. David Harland, in a personal communication, says that potassium superoxide was another chemical option.
20. Leonov interview; Siddiqi in *Quest*, v5, #3, 28.
21. Hooper (1990, v1), 69; Siddiqi in *Quest*, v5, n3, p30.
22. Leonov interview; Hooper (1990, v2), 131; *Quest*, v5, n3, 30.
23. Leonov interview; Siddiqi in *Quest*, v5, n3, 27.
24. Patsayeva interview; Cassutt, 278–279, 315; Hooper (1990, v2), 326.
25. Marina Dobrovolsky interview; Cassutt, 226.
26. Leonov interview; TASS press release, Feoktistov interview, June 8, 1971, L3.
27. Patsayev, 104.
28. Zimmerman, Robert, 46; Johnston, Dietlein, & Berry, 303–304, 315–316.
29. Patsayeva, 37.
30. Patsayeva, 109, 112, 116.
31. Patsayeva, 114.
32. Siddiqi in *Quest*, v5, #3, 31.
33. Harland (1997), says this occurred on June 18.
34. Leonov interview; Patsayeva interview; Sadiqqi in *Quest*, v5, n3, 31.
35. Patsayev, 122.
36. Serebrov interview; Leonov interview; Patsayev interview; Patsayev, 18–19, 22; Hooper in *Spaceflight*, v17, #6, 6/75, 225–6.
37. TASS press release, 6/19/71; *New York Times*, 6/20/71, 29.
38. Patsayeva, 130.
39. Vasilyev, et. al., 137, slightly reworded for grammar and clarity.
40. Patsayeva interview.
41. Patsayev, 141.
42. Siddiqi in *Quest*, v5, #3, 31.
43. Leonov interview; Note: Leonov's explanation does not match what others have written. See Sadiqqi in *Quest*, v5, n3, p31–32.
44. Patsayev inteview.
45. Dobrovolsky interview; Patsayev interview.

Chapter 3–*Skylab*: A Glorious Forgotten Triumph

1. Wilford, 136; Clarke (1964), 46–47; Wade, available as of 12/4/02 at http://www.astronautix.com/lvs/satint21.htm.
2. *Life*, 2/28/69, 22.
3. Compton & Benson, 10–11.
4. Wade, available as of 6/17/02 at http://www.astronautix.com/lvs/saturnib.htm .
5. Compton & Benson, 114–118.
6. Cooper, 81–84.
7. Belew, 25, 44–48; Compton & Benson, 172–174; Shayler, 200; Holder and Siuru, 31.
8. Compton & Benson, 255.
9. Cooper, 68; Belew, 23.
10. Belew, 80; Shayler, 50.
11. Compton & Benson, 152–158.

12. Belew, 67–70; Shayler, 70; Baker, 456.
13. Belew, 71.
14. Compton & Benson, 263; Shayler, 173.
15. Collins, Michael, 61; Chaikin, 5; Wolfe, 82–83.
16. Chaikin, 555.
17. *Skylab* transcripts, file sl2tec1.pdf, tape 145–03; Shayler, 176.
18. *Skylab* transcripts, file sl2tec1.pdf, day 145, 45; Belew, 61.
19. *Skylab* transcripts, file SL2146.pdf, tape 145–0015.
20. *Skylab* transcripts, file SL2146.pdf, tapes 146–01, 146–02, 146–003; file sl2pao1.pdf, 5/25/73.
21. Baker, 486.
22. *Skylab* transcripts, file SL2146.pdf, tapes 146–01, 146–02, 146–003; file sl2tec1.pdf, tapes 146–16, 147–01, 147–02.
23. Shayler, 184.
24. Johnston & Dietlein, 27.
25. *Skylab* transcripts, file sl2tec1.pdf, tape 158–08; Shayler, 183.
26. *Skylab* transcripts, file sl2tec1.pdf, tape 158–08; Baker, 490; Compton & Benson, 275.
27. *Skylab* transcripts, file sl2tec1.pdf, tape 158–11; Cooper, 137–143; Baker, 490; Belew, 75.
28. *Skylab* transcripts, file sl2tec1.pdf, tape 158–11; Cooper, 143.
29. Cooper, 139.
30. Baker, 490.
31. Compton & Benson, 290–291.
32. Cooper, 23, 73–74.
33. Cooper, 95.
34. Shayler, 293.
35. Cooper, 41.
36. Johnston & Dietlein, 333–335.
37. Johnston & Dietlein, 115.
38. Cooper, 169–173; Shayler, 293; Johnston & Dietlein, 119.
39. Johnston & Dietlein, 412.
40. Johnston & Dietlein, 187; Committee on Space Biology and Medicine, 88.
41. Baker, 494.
42. Compton & Benson, 297.
43. Baker 495.
44. Baker, 495; Shayler, 212–213.
45. *Skylab* transcripts, file sl2tec1.pdf, tapes 218–09/T117, 7–8 218–10/T118, 1–2.
46. Cooper, 137–140.
47. Shayler, 216; Compton & Benson, 310–311.
48. Cooper, 57–61.
49. Shayler, 215–216.
50. NASA Lyndon B. Johnson Space Center, 4; Compton & Benson, 303; Carlowicz & Lopez, 13.
51. Compton & Benson, 304; Shayler, 282–283; Cooper, 82.
52. Summerlin, 41–49; Holder & Siuru, 112–113.
53. Cooper, 70–71; Shayler, 203–204.
54. Shayler, 212; Belew, 115–117.
55. Belew, 117–118; Shayler, 221–222.

56. Johnston & Dietlein, 196; Shayler, 280.

57. Johnston & Dietlein, 173; Committee on Space Biology and Medicine, 88.

58. Compton & Benson, 298.

59. Shayler, 299–303.

60. *Skylab* transcripts, file sl4dump1.pdf, tapes 320–12, 320–17; file sl4dse1.pdf; file sl4tec1.pdf, tapes 320–06/T6, 4–6, 320–07/T7, 3–4; Compton & Benson, 314.

61. *Skylab* transcripts, file sl4dse1.pdf, 52–54, 75, 80–81.

62. *Skylab* transcripts, file SL4dump1.pdf, tapes 320–15, 320–17, 321–01; file sl4dse1.pdf, 95–99; Pogue, 79; Cassutt, 101; Shayler, 347; Compton & Benson, 315.

63. *Skylab* transcripts, file sl4dump1.pdf, tape 321–01, 321–02; file sl4tec1.pdf, tape 321–03/T10, 2; Compton & Benson, 315.

64. *Skylab* transcripts, file sl4tec1.pdf, tape 322–01/T20, 1–3; Shayler, 233.

65. Compton & Benson, 316; Shayler, 284–285; Cooper, 37–38.

66. Holder & Siuru, 120; Shayler, 227, 229.

67. Compton & Benson, 317.

68. Cooper, 85–87.

69. Cooper, 62; Compton & Benson, 317.

70. Baker, 502.

71. Compton & Benson, 317, 319.

72. Cooper, 128; *New York Times*, 11/25/73, 80; *New York Times*, 12/12/73, 14.

73. *Skylab* transcripts, file sl4dump4,pdf, tape 349–08, p2–3; file sl4dump5.pdf, tapes 354–03, 3–6, 354–04, 2–5.

74. *Skylab* transcripts, file sl4dump6.pdf, tapes 362–01; Compton & Benson, 327.

75. *Skylab* transcripts, file sl4tec4.pdf, tapes 362–12/T509, 1–2, 364–02/T523, 5–7, 365–01/T534, 365–02/T535.

76. Cooper, 130–131; Compton & Benson, 327–329; Baker, 504.

77. Pogue, 75–77.

78. Pogue, 133–134.

79. Johnston & Dietlein, 195–197.

80. Johnston & Dietlein, 173; Committee on Space Biology and Medicine, 88.

81. Baker, 506.

Chapter 4–The Early *Salyuts*: "The Prize of All People"

1. Van der Linden, 232; Brezhnev (1979), 225.

2. Nixon (1978), 609–621; Kissinger (1979) 1202–1275; Kalb, 312–335.

3. Kalb, 332.

4. Kissinger (1979), 1256.

5. *New York Times*, 5/30/72, 18.

6. Siddiqi, 805, 812–813; Harland (1997), 25; Wade, available as of 8/23/02 at http://www.astronautix.com/articles/almpart1.htm.

7. Siddiqi, 830–832; *Quest* v6, #1, Spring 1998, 20.

8. Sagdeev, 181–184; Sadiqqi, 832–838; Wade, available as of 7/18/02 at http://www.astronautix.com/craft/buran.htm; Wade, available as of 7/18/02 at http://www.astronautix.com/lvs/energia.htm.

9. Sagdeev, 205–206.

10. Harland (1997), 28, 32.

11. Siddiqi, 593–594; Wade, available as of 8/23/02 at http://www.astronautix.com/articles/almpart2.htm.
12. Portree (1995), 69; Siddiqi, 593–596; Newkirk, 125; Johnson, 242.
13. Harland (1997), 30.
14. Siddiqi, 593; Harland (1997), 31; Newkirk (1990), 125; Grechko interview; Note: though all these sources say that *Salyut 3*'s water-recycling system was the first installed on a *Salyut* station, in my interview with Patsayev she indicated that *Salyut 1* had a water-condensation system that pulled water into a tank for later use.
15. Siddiqi, 594.
16. Harland (1997), 29; Johnson, 242; Portree (1995), 69.
17. Harland (1997), 32.
18. *New York Times*, 7/5/74, 8; 8/28/74, 1; 8/29/74, 1; 8/30/74, 8; 8/31/74, 22; 9/6/74, 12.
19. Newkirk in *Quest*, summer 1993, 30.
20. Grechko interview; Newkirk & Plaxco.
21. Patsayev interview; Hooper in *Spaceflight*, v17, #6, 6/75, 222; Harland (1997), 28, 34.
22. Clarke, 81.
23. Grechko inteview; Makarov interview; Hooper in *Spaceflight*, v17, #6, 6/75, 223, 225; Hooper (1990, v2), 103.
24. Newkirk & Plaxco.
25. Nechitailo and Mashinsky, 27–28, 154; *Nature*, v253, 2/27/75, 675; Hooper in *Spaceflight*, v17, 4/75, 144–145; *Life Sciences and Space Research*, v15, 267–272; .
26. Grechko interview; Nechitailo & Mashinsky, 84.
27. Nechitailo interview; Grechko interview.
28. Grechko interview; Nechitailo interview; Hooper in *Spaceflight*, v17, #6, June 1975, 222, 224; Nechitailo and Mashinsky, 84–85, 181, 294–299, 394; Vinnikov, Ya. A., et al, *Arkhiv Anatomii, Gistologii i Embriologii*, v70, #1, 1/76, 11–17, abstract translated by NASA; *Life Sciences and Space Research*, v15, 267–272.
29. Grechko interview; Drygas.
30. *Trud*, 2/15/75, 4, translated by Library of Congress, Federal Research Division.
31. *Washington Post*, 2/10/75, A1; *New York Times*, 2/16/75, IV, 6.
32. Makarov interview; Ward, 59–65; Harvey (1996), 291. Note: Though some sources say that Lazarev suffered broken ribs, internal bleeding, and a concussion, in my interview with Makarov he claimed that neither man had been seriously injured. Lazerev's own account in Ward also agrees with this description.
33. *Aviation Week & Space Technology*, 6/2/75, 25–26.
34. Maresca, 143.
35. Hooper in *Spaceflight*, v18, #1, 1/76, 14.
36. Hooper (1990, v2), 128–129.
37. Hooper (1990, v2), 263–266.
38. Hooper in *Spaceflight*, v18, #1, 1/76, 14. The translation has been edited slightly for clarity.
39. Harland (1997), 41–44.

40. Nechitailo and Mashinsky, 27–28, 298–306, 394; Hooper in *Spaceflight*, v18, #1, 1/76, 14.
41. Hooper in *Spaceflight*, v18, #1, 1/76, 17.
42. *Life Sciences and Space Research*, v15, 267–272; Nechitailo and Mashinsky, 23, 85; Hooper in *Spaceflight*, v18, #1, 1/76, 14.
43. Clark, 82; Harvey (1960), 292.
44. Brezhnev (1975), 93, 95.
45. Hooper in *Spaceflight*, v19, #4, 4/77, 140–141, 143.
46. Nechitailo and Mashinsky, 29–30, 209–210.
47. Harvey, (1996) 294.
48. Harvey (1996), 295; Hooper (1990, v2), 330, 346; Drygas; Pesavento in *Quest*, v8, #2, 2000, 9. Interviews with two cosmonauts (Savinykh and Polyakov) confirmed the emotional and psychological problems that caused this flight to be cut short.
49. Hooper in *Spaceflight*, v19, #4, 4/77, 144.
50. Harland (1997), 47–48; Harvey (1996), 296–297; Hooper in *Spaceflight*, v19, #4, 4/77, 145; Hooper (1990, v2), 348.
51. Hooper in *Spaceflight*, v19, #4, 4/77, p145.
52. Portree (1995), 74.
53. Hooper in *Spaceflight*, v19, #7–8, 7–8/77, 267; Harvey (1996), 298.
54. Maresca, 150–152; Mastny, 65.
55. Buergenthal, 164, 166, 172–180; Edmonds, 148.
56. Edmonds, 148–149.
57. *New York Times*, 10/10/75, 1.
58. Orlov, 187–192, 203–209; Mastny, 102–103; Gilbert, 85–160.

Chapter 5—*Salyut 6*: The End of Isolation

1. Hooper (1990, v1), 79.
2. Hooper (1990, v1), 79; v2, 97–98.
3. Hooper (1990, v2), 98.
4. Clark, 106.
5. Harvey (1996) 305; Harland (1997), 56.
6. Cassutt, 257–258; Hooper (1990, v2), 229–230.
7. Portree & Trevino, 38.
8. Grechko interview; Dugan; Harland (1997), 56; Hooper in *Spaceflight*, v20 #8. 8/78, 294; Kidger in *Spaceflight*, v31, 2/89, 48–49.
9. Hooper (1990, v2), 230; Kidger in *Spaceflight*, v31, 2/89, 48–49.
10. Kidger in *Spaceflight*, v31, 2/89, 49; Harland (1997), 56; Grechko interview. Though in his interview Grechko did not remember anything about the failed sensor, it is obvious from the other sources that mission control withheld the details of the problem from the two cosmonauts.
11. Hooper in *Spaceflight*, v20, #6, 6/78, 230.
12. Hooper in *Spaceflight*, v20, #6, 6/78, 233.
13. Dzhanibekov interview; Hooper in *Spaceflight*, v20, #11, 11/78, 371.
14. Makarov interview.
15. Dzhanibekov interview; Leonov interview; Hooper (1990, v2), 132.
16. Grechko interview; Oberg (1981), 178.
17. Hooper in *Spaceflight*, v20, #11, 11/78, 372–373.

18. Grechko interview; Hooper in *Spaceflight*, v20, #11, 11/78, 378.
19. Portree (1995), 36–37.
20. Oberg (1981), 179; Portree (1995), 37, 78; Harland (1997), 62.
21. Nechitailo inteview.
22. Grechko interview.
23. Hooper (1990, v2), 231; Oberg (1981), 176–178.
24. Hooper (1990, v2), 97.
25. Grechko interview; Hooper (1990, v2), 231–232.
26. Hooper in *Spaceflight*, v20, #8, 8/79, 294.
27. Gilbert, 85–160; Sharansky, 19, 163–170, 364; *Nature*, v253, 2/27/75, 678; Orlov, 210–233.
28. Sagdeev, 228.
29. Hooper (1990, v1), 97–99; *New York Times*, 9/19/76, IV, 6; Edmonds, 159.
30. Edmonds, 61–64, 72–73; Mastny, 46; Sagdeev, 229–230; Hooper (1990, v1), 103–104.
31. Sagdeev, 228–229.
32. Grechko interview.
33. Hooper in *Spaceflight*, v20, #12, 12/78, 431.
34. Bluth, *Acta Astronomica*, v11, #2, 150; Oberg (1981), 183–184.
35. Hooper in *Spaceflight*, v19, #3, 3/77, 101.
36. Hooper (1990, v2), 227.
37. Drygas.
38. Grechko interview; Hooper in *Spaceflight*, v20, #8, 8/78, 294–295; TASS press release, 4/11/78.
39. Ryumin (1987), 8; *Aviation Week and Space Technology*, 10/10/83, 27.
40. Hooper (1990, v2), 238–239; Harland (1997), 76–77.
41. Ryumin (1987), 8; Chaikin, 30.
42. Harland (1997), 77; Hooper (1990, v2), 240–241; Kidger in *Spaceflight*, v22, #2, 2/80, 59.
43. Polyakov interview.
44. Ryumin (1987), 19; Hooper (1990, v2), 241–242.
45. Ryumin (1987), 3, 6–7; Chaikin, 30.
46. Ryumin (1987), 3–4; Cassutt, 248–249; Hooper (1990, v2), 173.
47. Kidger in *Spaceflight*, v22, #2, 2/80, 56; Ryumin (1987), 6.
48. Ryumin (1987), 13; Chaikin, 30.
49. Nechitailo & Mashinsky, 151–152.
50. Nechitailo & Mashinsky, 144–145.
51. Ryumin (1987), 4, 5, 9; Nechitailo & Mashinsky, 31, 44–47, 324–328; Harland (1997), 73.
52. Ryumin (1987), 11.
53. Ryumin (1987), 9.
54. Nechitailo & Mashinsky, 47–49.
55. Ryumin (1987), 13; Kidger in *Spaceflight* v22, #3, 3/80, 110.
56. Ryumin (1987), 13; TASS Press Release, May 14, 1979; Halstead & Dutcher, "Plants in Space," in *Annual Review of Plant Physiology*, v38, 1987, 320; Nechitailo & Mashinsky, 443–444; Nechitailo interview.
57. Nechitailo inteview; Ryumin (1987), 4, 13; Kidger in *Spaceflight*, v22, #4, 4/80, 151–152; Nechitailo & Mashinsky, 322, 439–444.
58. Ryumin (1987), 12–13.

59. Ryumin (1987), 19–20; Chaikin, 30; Stuster, 127, 240; Harrison, Clearwater & McKay, 332.
60. Ryumin (1987), 15.
61. Ryumin (1987), 14; Chaikin, 30.
62. Kidger in *Spaceflight*, v23, #3, 3/81, 76–77; Krikalev interview.
63. Kidger in *Spaceflight*, v22, #3, 3/80, 112; Oberg (1981), 217.
64. Chaikin, 30.
65. Ryumin (1987), 20.
66. Ryumin (1987), 57–58; Kidger in *Spaceflight*, v22, #4, 4/80, 146–154; *Advances in Space Research*, v12, #1, 1992, 324; DeCampli, William. "Medical Problems Associated with Long-Duration Space Flights," in *The Human Quest in Space, 24th Goddard Memorial Symposium*, ed. Gerald L. Burdett and Gerald A. Soffen, 206–210.
67. Nechitailo & Mashinsky, 440–444; Kidger in *Spaceflight*, v22, #4, 4/80, 151.
68. Oberg (1981), 182.
69. McMichael, 2–8; Hammond, 97–102; Urban, 38–48.
70. Hammond, 68–82; Urban, 7–38.
71. Hammond, 132–140; Collins, Joseph, 99–136; *Washington Post*, 10/17/80, A24; Hammond, 135; McMichael, 8.
72. Collins, Joseph, 85–89; Arnold, 112–142; Hammond, 105–124; *Washington Post*, 12/39/79, A1; 1/5/80, A1, A6–A7.
73. Ryumin (1987), 19.
74. Ryumin (1987), 30–33; Hooper (1990, v2), 159.
75. Ryumin (1987), 42–43.
76. Kidger in *Spaceflight*, v22, #11–12, 11–12/80, 345.
77. Nechitailo interview; Ryumin (1987), 34.
78. Nechitailo interview; Ryumin (1987), 44, 45; Nechitailo & Mashinsky, 85–86.
79. Ryumin (1987), 49–50.
80. Ryumin (1987), 39, 49; Caprara, 97.
81. Ryumin (1987), 35.
82. Nechitailo & Mashinsky, 49–54, 430–431; *Advances in Space Research*, v3, #8 (1983), 129–133.
83. Ryumin (1987), 45–46, 48; Harvey (1996), 327–328; Nechitailo & Mashinsky, 324–326; *Advances in Space Research*, v3 #8 (1983), 129–133; *Advances in Space Research*, v3, #9 (1983), 211–219.
84. Ryumin (1987), 57.
85. Ryumin (1987), 57–58.
86. Nechitailo & Mashinsky, 54–57.
87. Dzhanibekov interview.
88. Kidger in *Spaceflight*, v36, 3/94, 86–89; *Spaceflight*. v36. 10/94, 342–344.
89. Walker, 38.

Chapter 6–*Salyut 7*: Phoenix in Space

1. *New York Times*, 11/13/82, 4; 11/16/82, 1.
2. Lebedev, 259–260.
3. Lebedev, 262.

4. Ryumin (1987), 62; Harland (1997), 97–98; Kidger in *Journal of the British Interplanetary Society*, v36 (1983), 463–467; Serebrov interview.

5. Serebrov interview; Savinykh interview; Kidger in *Spaceflight*, v36, 3/94, 86–89; *Spaceflight*. v36. 10/94, 342–344.

6. Portree (1995), 162.

7. Lebedev, 181–185; Cassutt, 299–300; Hooper (1990, v2), 156–157.

8. Cassutt, 316.

9. Nechitailo interview.

10. Hooper (1990, v2), 41–44; Cassutt, 221–222; Berezovoi, 64; Oberg & Oberg, 151.

11. Lebedev, 33, 38, 40, 61–62, 78.

12. Nechitailo & Mashinsky, 35–37.

13. *Advances in Space Research* (1984) v4, #10, 56, 66–67; Nechitailo & Mashinsky 154–155.

14. Lebedev, 42, 46, 51–52.

15. Lebedev, 58, 62, 69; Nechitailo & Mashinsky, 58–60.

16. *New York Times*, 4/28/79, 7; 4/29/79, 11; Hooper (1990, v1), 108–109; (1990, v2), 57; *Daily Telegraph* (London), 4/28/79, 6.

17. *Times* (London), 12/10/80, 7; Hooper (1990, v1), 112; (1990, v2), 268; Kidger in *Spaceflight*, v26, Sept/Oct 1984, 371; *New York Times*, 2/23/80, 9.

18. Kidger in *Spaceflight*, v25, n3, March 1983, 122–123.

19. Dzhanibekov interview; Kidger in *Spaceflight*, v25, #3, 3/83, 122–123.

20. Dzhanibekov interview; Harland (1997), 100. A number of sources, including Oberg & Oberg (p. 20) and Hooper (1990, v2, pp. 58, 69), state that Malyshev was replaced because he couldn't get along with Chretien. In my interview with Dzhanibekov, however, he denied this charge, stating categorically that it was Malyshev's heart condition that grounded him.

21. Nechitailo & Mashinsky, 59; Lebedev, 110–111.

22. Lebedev, 114–116.

23. Lebedev, 149–157; Harland (1997) 102.

24. Lebedev, 123, 126, 129–130, 147, 186, 187.

25. Nechitailo & Mashinsky, 62–63; Lebedev, 186.

26. Lutomsky, 8–9.

27. Sagdeev, 226–227; Savinykh interview; Serebrov interview.

28. Lebedev, 195; Serebrov interview.

29. Lebedev, 191.

30. Nechitailo & Mashinsky, 63; Lebedev, 201.

31. Lebedev, 208, 241–242, 248, 249, 250, 251–252.

32. Lebedev, 208–9, 238, 245; Kidger in the *Journal of the British Interplanetary Society* (1983), v36, p275.

33. Lebedev, 243.

34. Lebedev, 262.

35. Ryumin (1987), 61–62.

36. Nechitailo inteview.

37. "Basic Results of Medical Studies During Prolonged Manned Flights On-Board the *Salyut-7-Soyuz-T* Orbital Complex," Washington, D.C.: NASA TT-20217, March 1988, 14–15, 16–17, 31; *Advances in Space Research*, v12, #1, 1992, 324.

38. Lebedev, 266.

39. Nechitailo interview.
40. *Advances in Space Research*, v4, #10 (1984), 55–63.
41. Moscow *News*, #28, 7/17-24/83, 1; #30, 7/31-8/7/83, 1; #31, 8/7-14/83, 1, 11; supplement 1.
42. Alexandrov interview; Kidger in *Spaceflight*, v26, 5/84, 230; Harland (1997) 109–110; *New Scientist*, 11/24/83, 597–600. Many sources say the leak was discovered during refueling from *Progress* to *Salyut 7*. This is not true according to Alexandrov. The refueling had already been completed.
43. Serebrov interview; Hooper (1990, v2), 302.
44. Alexandrov interview; Hooper (1990, v2), 19.
45. Harland (1997), 107; Kidger in *Spaceflight* v26, 3/84, 138.
46. Ryumin (1987), 65.
47. Alexandrov interview.
48. Wade, available as of 8/23/02 at http://www.astronautix.com/lvs/soya511u.htm.
49. *Aviation Week & Space Technology*, 12/12/83, 24.
50. Harland (1997), 110; Kidger in *Spaceflight*, v26, 5/84, 231.
51. Ryumin (1987), 65.
52. Alexandrov interview. While all other sources describe the discovery and repair of the spacesuit as occurring before their first space walk, Alexandrov in his interview told me it occurred *between* the two space walks. Because so little time passed between the two space walks, I have made the assumption that his memory was mistaken and that the repair occurred when they were preparing their spacesuits before going outside the first time.
53. Harvey (1996), 341; Kidger in *Spaceflight*, v26, 5/84, 233.
54. Alexandrov inteview; Harland (1997), 111; Kidger in *Spaceflight*, v26, 5/84, 233.
55. "Basic Results of Medical Studies During Prolonged Manned Flights On-Board the *Salyut 7-Soyuz-T* Orbital Complex," Washington, D.C.: NASA TT-20217, 1988, 31.
56. Cassutt, 237–238; Hooper (1990, v2), 124–125.
57. Ryumin (1987), 66; Hooper (1990, v2), 283.
58. Heinlein, 111.
59. Drygas.
60. "Basic Results of Medical Studies During Prolonged Manned Flights On-Board the *Salyut-7-Soyuz-T* Orbital Complex," Washington, D.C.: NASA TT-20217, March 1988, 36–37.
61. Portree & Trevino, 54–55; Kidger in *Spaceflight*, v31, 4/89, 140.
62. Savinykh interview.
63. Harland (1997, 116) states that the original plan was for Dzhanibekov to do the pinching. Dzhanibekov categorically denies this. "I was an instructor only."
64. Grechko interview.
65. *Advances in Space Research*, v12, #1, 1992, 343–345; *Spaceflight*, v27, #3, 3/85, 133–134; Thagard interview.
66. *Parade*, 7/27/86, 12.
67. Savinykh interview; Savinykh, diary typescript; Canby, 430–434; Kramer, 54–59.
68. Savinykh, diary typescript.
69. *Parade*, 7/27/86, 13; Savinykh interview.

70. Savinykh, diary typescript; Hooper (1990, v2), 83.
71. Savinykh interview; Cassutt, 309; Hooper (1990, v2), 250.
72. Ryumin (1987), 77.
73. Savinykh interview.
74. Savinykh, diary typescript.
75. Dzhanibekov interview; Savinykh inteview; Savinykh, diary typescript; Nechitailo & Mashinsky, 38.
76. Nechitailo interview; Nechitailo & Mashinsky, 64–67.
77. Hooper (1990, v2), 99; Cassutt, 292; Dugan.
78. Nechitailo & Mashinsky, 69.
79. Savinykh, diary typescript.
80. Savinykh, diary typescript.
81. Polyakov interview; Savinykh interview; Savinykh, diary typescript; Cassutt, 269; Harland (1997), 121–123.
82. Gorbachev (1996), 18–34; Felshman, 57–67.
83. Gorbachev (1987), 23.
84. *Washington Post*, 2/24/84, A2; 5/23/84, A23; 3/30/85, B7; 5/2/85, C3; 6/16/85, B3; Orlov, 275–277.

Chapter 7—*Freedom*:
"You've Got to Put on Your Management Hat . . . "

1. *Washington Post*, 2/23/83, F6; 4/21/83, A1; 5/20/83, E1; 6/22/83, A1; 9/22/83, C12; 10/21/83, A1; 12/11/83, A1; 12/22/83, D1, 12/23/83, A1; 1/5/84, A1, A2; *New York Times*, 2/6/83, IV2; 5/18/83, A20; 6/24/83, I1, IV6; 8/20/83, A34; 11/18/83 A1.
2. *New York Times*, 1/26/84, B8.
3. Jenkins, 49–71, 128; McCurdy, 22–30.
4. McCurdy, 31; Jenkins, 64.
5. Jenkins, 64.
6. Jenkins, 66; Grey, 66–68.
7. McCurdy, 31–32.
8. McCurdy, 36.
9. Stockton & Wilford, 46–47.
10. Jenkins, 138–145; Smith, 133–149.
11. Grey, 106–107.
12. Grey, 107.
13. Lewis, 62–64; Stockton & Wilford, 56–57.
14. Jenkins, 153–154; Stockton & Wilford, 57–58; Grey, 129–132.
15. Lewis, 84–85.
16. Lewis, 88–90; Stockton & Wilford, 59–61.
17. Lewis, 95–96.
18. U.S. Congress, 96:1, Hearing June 28, 1979, 14–15.
19. Smith, 159.
20. Marshall, 55.
21. Jenkins, 232–238.

22. Space Shuttle Overview, available as of 7/08/02 at http://science.ksc.nasa.gov/shuttle/technology/sts-newsref/stsover-prep.html; Jenkins, 290; Lewis, 191.

23. Lindroos, "Space Station Designs," available as of 1/21/03 at http://www.astronautix.com/craft/spas1982.htm.

24. McCurdy, 118–123, 127–129, 165–168, 183–184.

25. Reagan, 29.

26. Reagan, 33, 44–83; Pemberton, 3–24.

27. McCurdy, 177–185; *New York Times*, 1/26/84, B8.

28. See for example the essays in Bekey & Herman, 85–202 and Simpson, 107–204; also, see McCurdy, vii, 108, 148–156.

29. McCurdy, 86–87.

30. McCurdy, 75–80; Lindroos, "U.S. Space Stations," available as of 1/21/03 at http://www.astronautix.com/craftfam/usstions.htm.

31. *New York Times*, July 7, 1988, B5.

32. Gruen, Chapter 5, 2–9; Chapter 7, 1–8; Lindroos, "Dual Keel Space Stations," available as of 1/21/03 at http://www.astronautix.com/craft/duan1985.htm; McCurdy, 210, 212.

33. McCurdy, 209–212.

34. Lindroos, "Space Station *Freedom*," available on 1/21/03 at http://www.astronautix.com/craft/spaeedom.htm.

35. McCurdy, 224.

36. David, 25; GAO Report to the Chairman, Committee on Science, Space, and Technology, House of Representatives, March 1991, "Space Station, NASA's Search for Design, Cost, and Schedule Stability Continues." GAO/NSLAD-91-125, page 19.

37. *Johnson Space Center Space News Roundup*, 10/25/85, 1; Gruen, Chapter 5, 16–17; Space Station *Freedom* Media Handbook, 7–8, 65–66; David, 26; Lindroos, "Dual Keel Space Stations," available as of 1/21/03 at http://www.astronautix.com/craft/duan1985.htm; GAO Report, 3/91, 19–21.

38. Gruen, Chapter 6, 16–17; David, 26; Lindroos, available on 7/8/02 at http://www.astronautix.com/craft/spaeedom.htm.

39. Statement by Marlin Fitzwater, Assistant to the President for Press Relations, White House, Office of the Press Secretary, 7/18/88, in Space Station General 1988 file at NASA History Office; *Washington Post*, 7/18/88, A4.

40. McCurdy, 213–222.

41. Marshall, 54–61.

42. NASA Space Station *Freedom* Strategic Plan, 1992, 9–19; GAO Report, 3/91, 24–25; Lindroos, "Space Station *Freedom*," available on 1/21/03 at http://www.astronautix.com/craft/spaeedom.htm.

43. NASA Pocket Statistics, 1997, available as of 7/26/02 at http://history.nasa.gov/pocketstats/ "R&D Funding by Program." Note: Determining how much was spent on any government program can at times be guesswork, depending on what government document you wish to use. For example, 1/91 NASA Pocket Statistics, page C14, combined with the 3/91 House Report, 41–56, gives a figure of $4.7 billion for the same years. Other reports give other numbers, depending on what NASA programs they feel should be included in the station program.

44. Lindroos, "Space Station *Freedom*," available on 1/21/03 at http://www.astronautix.com/craft/spaeedom.htm.
45. McCurdy, 25, 78, 216; Grey, 74–76.
46. Vaughan, 153, 175, 383; *Presidential Commission*, 84–85.
47. Vaughan, 318–319.
48. *Presidential Commission* (1986), 8.
49. *Presidential Commission*, 19–39.

Chapter 8–*Mir*: A Year in Space

1. Manarov interview; Hooper (1990, v2), 189–191, 233.
2. Serebrov interview; Harland (1997), 129.
3. Harland (1997), 130, 133–135; Clark, 148–149.
4. Edmonds, 208–212; Felshman, 102–103, 216–222; Walker, 20, 27–38.
5. Yeltsin (1990), 142, 156–166; Walker, 51–55; Shevchenko, 62–63, 87–92, 125–126, 174–175.
6. Gorbachev (1996), 185.
7. Felshman, 141.
8. Felshman, 140–141; Walker, 106–110; Goldman, 89–90; Harvey (1996), 163–1644.
9. Yeltsin (1990), 115, 118–119; Aron, 155–156; *New York Times*, 1/1/87, A4; 3/23/87, A11.
10. *Spaceflight*, v31, 2/89, 38; Kidger in *Spaceflight*, v33, 8/91, 267.
11. Harland (1997), 132; Burrough, 66.
12. Portree (1995), 53.
13. Nechitailo & Mashinsky, 39, 73.
14. Nechitailo interview; Nechitailo & Mashinsky, 72.
15. *Spaceflight*, v32, 1/80, 7.
16. Hooper (1990, v2), 233.
17. Serebrov interview; Hooper (1990, v2), 260.
18. Serebrov interview; Kidger in *Spaceflight*, v29, 8/87, 282.
19. Harland (1997), 145, 370, 419–420.
20. Harland (1997), 143.
21. *Spaceflight*, v31, 9/89, 294.
22. Manarov interview; Alexandrov interview.
23. Portree & Trevino, 68–69.
24. *Spaceflight*, v29, 9/87, 318–319; Hooper (1990, v2), 152.
25. Kidger in *Spaceflight*, v29, 11/87, 377; Hooper (1990, v2), 152; Polyakov interview.
26. Alexandrov interview.
27. Alexandrov interview; Branegan in *Spaceflight*, v30, 3/88, 111–113.
28. Nechitailo interview; Nechitailo & Mashinsky, 73–75.
29. Kidger in *Spaceflight*, v30, 3/88, 114; *Aerospace Medicine*, v24, #4, July-August 1990, translated in JPRS-ULS-91-010, May 6, 1991.
30. Kidger in *Spaceflight*, v30, 3/88, 113.
31. *New York Times*, 11/8/87, A14; 11/3/87, A11–A13.
32. Walker, 73–78.
33. Yeltsin (1990), 212–25; Aron, 4–6.
34. Aron, 43.

35. Yeltsin (1990), 122–123, 126–130, 153–155, 177–183.
36. Aron, 205–212; Yeltsin (1990), 184–197; Gorbachev (1996), 242–248.
37. *New York Times*, 11/13/87, A1, A12; *Washington Times*, 11/12/87, A7; 11/13/87, A9.
38. Morrison, 72; Yeltsin (1990), 202.
39. Manarov interview.
40. Cassutt, 266–267; Hooper (1990, v2), 302–305; Manarov inteview.
41. Serebrov interview.
42. Manarov interview.
43. Some sources (Harland 1997, 152, and Hooper 1990, v2, 89) say that Levchenko was a late replacement for Valeri Polyakov. In my interviews, both Polyakov and Manarov denied this. Polyakov was never scheduled as the third crewman on this flight.
44. Manarov interview; *Spaceflight*, v30, 6/88, 115.
45. Manarov interview; Alexandrov interview.
46. Alexandrov interview.
47. Nechitailo interview; Nechitailo & Mashinsky, 75–76; Kidger in *Spaceflight*, v30, 3/88, 117; Siddiqi, 851.
48. Kidger in *Spaceflight*, v30, 3/88, 117; Harland (1997), 153.
49. *Spaceflight*, v31, 9/89, 295.
50. Bonting, v1, 1991, 25–26; *Aerospace Medicine*, v24, #4, July-August 1990, translated in JPRS-ULS-91-010, May 6, 1991, 9.
51. Manarov inteview.
52. Portree & Trevino, 69–70; Portree (1995), 105; Harland (1997), 154; Kidger in *Spaceflight*, v30, 6/88, 228–229.
53. Kidger in *Spaceflight*, v30, 6/88, 229.
54. Manarov inteview; Harland (1997), 161–162; Portree and Trevino, 70.
55. Manarov interview; Hooper (1990, v2), 191.
56. Collins, Joseph, 143; McMichael, 126–133; Urban, 237, 242–243, 316–317.
57. Hooper, (1990, v1), 105.
58. Kidger in *Spaceflight*, v30, 6/88, 229; Kidger in *Spaceflight*, v30, 12/88, 454–455; Hooper (1990, v1), 105.
59. Hooper (1990, v2), 209.
60. Polyakov interview; Hooper (1990, v2), 210.
61. Kidger in *Spaceflight*, v31, 6/89, 192.
62. Kidger in *Spaceflight*, v30, 12/88, 55–56.
63. Kidger in *Spaceflight*, v30, 12/88, 456–457.
64. Polyakov interview; Manarov interview; *Acta Astronautica*, v23 (1991), 149–151; *Advances in Space Research*, v12, #1, 1992, 339–341.
65. *Acta Astronautica*, v36 (1995), #1, 1–12; Portree & Trevino, 71–72.
66. Manarov interview; *Spaceflight*, v31 2/89, 65; Portree & Trevino, 71–72.
67. Kidger in *Spaceflight*, v30, 12/88, 455; Polyakov interview.
68. Harvey (2001), 36; Harvey (1996), 255.
69. Harvey (1996), 254–255; Harvey (2001), 23–24; Wade, available as of 1/22/03 at http://www.astronautix.com/lvs/energia.htm.
70. Manarov interview; Kidger in *Spaceflight*, v31, 3/89, 80; *Acta Astronautica*, v23 (1991), 3; *Aviation Week and Space Technology*, 1/2/89, 38–39.
71. *Acta Astronautica*, v23 (1991), 1–8; Kidger in *Spaceflight*, v31, 3/89, 81.
72. Walker, 102–110.
73. Kidger in *Spaceflight*, v31, 3/89, 77; spelling from Clark, 131.

74. Polyakov interview; Krikalev interview; Kidger in *Spaceflight*, v31, 6/89, 192; v31, 4/89, 116.
75. *New York Times*, 4/13/89, A24; Siddiqi, 851.
76. Kidger in *Spaceflight*, v31, 4/89, 116; Krikalev interview; Polyakov interview.
77. Aron, 230.
78. *New York Times*, March 27, 1989, A1, A6.
79. Gorbachev (1996), 258.
80. Aron, 230–249.
81. Aron, 242; Yeltsin (1990), 234–235.
82. Aron, 249–250; Yeltsin (1990), 238–239.
83. Hooper (1990, v2), 99–100.
84. Yeltsin (1990), 58.
85. Grechko interview; Yeltsin (1990), 57–60, 83–85.
86. Morrison, 88–90.
87. *New York Times*, March 29, 1989, A12.
88. Kidger in *Spaceflight*, v31, 6/89, 193.
89. Krikalev interview; Polyakov interview.
90. Sagdeev, 317–318.
91. Kidger in *Spaceflight*, v31, 6/89, 193–194.

Chapter 9—*Mir*: The Road to Capitalism

1. Dugan; Cassutt, 262–263; Hooper (1990, v2), 281.
2. Dugan.
3. Dugan.
4. Balandin interview; Dugan.
5. Dugan.
6. Harland (1997, 179) says the loose blankets were first noticed by the crew because the blankets partly blocked their field of view. This is incorrect. The blankets could not do this, located as they were *behind* their field of view. Moreover, Balandin in his interview said that they found out about the problem only when ground controllers told them.
7. *New York Times*, 9/7/89, A5.
8. Portree (1995), 43–44; Harvey (1996), 360; Kidger in *Spaceflight*, v31, 10/89, 332–333.
9. Harland (1997), 173, 418–419.
10. Portree (1995), 163–165; *Spaceflight*, v32, 1/90, 10, 12.
11. New York Times, 4/13/89, A24; Kidger in *Spaceflight*, v31, 10/89, 403; Portree (1995), 120; *Spaceflight*, v32, 1/80, 9.
12. Portree & Trevino (page 120) say that the docking was aborted when the module was only 20 meters from the station. This is incorrect. In my interview with Serebrov he explained that the docking was aborted much sooner, as described by Harland (1990, page 173) and in *Spaceflight* (v32, 1/80, 9–10).
13. Serebrov interview.
14. Portree & Trevino, 73–76; *Spaceflight*, v32, 3/90, 83; Kidger in *Spaceflight* v32, 7/90, 229–230; Harland (1997), 177–178.
15. Portree & Trevino, 76.
16. Yeltsin (1990), 245.
17. Aron, 305–307.
18. Gorbachev (1996), 249; Goldman, 118–120, 139–144.

19. *Spaceflight*, v31, 8/89, 273; Morrison, 180.
20. Kidger in *Spaceflight*, v32, 6/90, 192; Kidger in *Spaceflight*, v32, 10/90, 349, 357.
21. *Spaceflight*, v32, 8/90, 254.
22. Kidger in *Spaceflight*, v32, 6/90, 192; Kidger in *Spaceflight*, v32, 7/90, 232; Harland (1997), 180–181.
23. Kidger in *Spaceflight*, v32, 6/90, 192; 7/90, 232.
24. Nechitailo interview; Balandin interview; Harland (1997), 180; Kidger in *Spaceflight*, v32, 7/90, 232; *Aviakosmicheskaia i Ekologicheskaia Meditsina*, v26 (1992), #2, 65–66, abstract translated by NASA.
25. *Washington Times*, May 2, 1990, A1, A8; *New York Times*, May 2, 1990, A1, A11; Aron, 372.
26. *Spaceflight*, v32, 7/90, 222; Harland (1997), 181.
27. Kidger in *Spaceflight*, v31, 4/89, 116; Portree (1995), 104, 163; Harland (1997), 182.
28. Nechitailo interview; Balandin interview; *Acta Astronautica*, v42, 1998, 11–23; *Advances in Space Research*, v14, #11, 1994, 343–346.
29. *Acta Astronautica*, v42, #1–8, 1998, 11–23.
30. Portree (1995), 166.
31. Balandin interview; Kidger in *Spaceflight*, v32, 9/90, 311.
32. Balandin interview; Portree & Trevino, 77; Kidger in *Spaceflight*, v32, 10/90, 349.
33. Balandin interview; Kidger in *Spaceflight*, v32, 10/90, 349.
34. Kidger in *Spaceflight*, v32, 10/90, 349; 11/90, 357.
35. Harland (1997), 182.
36. Aron, 395–396; Morrison, 121; Gorbachev (1996), 370–371.
37. *Spaceflight*, v31, 7/89, 218; 8/89, 265–268, 273.
38. Sagdeev, 231; *Spaceflight*, v31, 8/89, 267; v32, 2/90, 49; v32, 6/90, 195; v32, 8/90, 255.
39. Kidger in *Spaceflight*, v31, 6/89, 194; v33, 3/91, 91, 96.
40. Harland (1997), 186; Kidger in *Spaceflight*, v33, 1/91, 14–15; 3/91, 91–94.
41. Kidger in *Spaceflight*, v33, 3/91, 92.
42. Cassutt, 216.
43. Dugan.
44. Hooper (1990, v2), 14.
45. Kidger in *Spaceflight*, v33, 3/91, 91–94.
46. Kidger in *Spaceflight* v33, 3/91, 94; Burrough, 223–224.
47. Manarov interview.
48. Manarov interview; Harland (1997), 188.
49. Kidger in *Spaceflight*, v33, 7/91, 228.
50. Sharman, 70.
51. Sharman, 29–30, 104; Ujica.
52. Sharman, 105–106; Krikalev interview.
53. *Spaceflight*, v33, 4/91, 138–139; 7/91, 231.

Chapter 10—*Mir*: The Joys of Freedom

1. Gorbachev (1996), 381–387; Aron, 404–406.
2. Solovyov and Klepikova, 211–212; Morrison, 194–206; Eisenhower, 255.
3. Orlov, 328–330.

4. Solovyov and Klepikova, 218–219; Morrison, 210.
5. Morrison, 213–227; Aron, 408–419.
6. Aron, 412.
7. *New York Times*, 6/13/91, A18.
8. Morrison, 266–267.
9. Aron, 434; Morrison, 267; *New York Times*, 6/14/91, A1; *Washington Times*, 6/13/91, A1.
10. Aron, 442.
11. Hooper (1990, v2), p143.
12. Cherniavsky interview; Portree & Trevino, 84; Harland (1997), 202.
13. Cherniavsky interview; Kidger in *Spaceflight*, v33, 10/91, 359; Harland (1997), 191.
14. *Acta Astronautica*, v36 (1995), #1, 1–12; Artsebarski interview.
15. Portree and Trevino, 85; Krikalev interview; Kidger in *Spaceflight*, v33, 10/91, 359.
16. Portree & Trevino, 86.
17. Aron, 440, 443.
18. Gorbachev (1995), 631–640.
19. Yeltsin (1994), 54–57.
20. Yeltsin (1994), 68–69; Aron, 445; Morrison, 283–284.
21. Aron, 450.
22. Yeltsin (1994), 76; Yeltsin (2000), 197, 260, 272.
23. Aron, 250.
24. Gorbachev (1996), 641.
25. Kidger in *Spaceflight*, v34, 1/92, 12.
26. Krikalev interview; Harland (1997), 193.
27. Kidger in *Spaceflight*, v33, 9/91, 293; v33, 10/91, 359; Aron, 475–477.
28. Artsebarski interview; Ujica.
29. Harland (1997), 194.
30. Aron, 479; Gorbachev (1996), xxvi–xxix, 670–672.
31. Portree (1995), 133; Harland (1997), 209.
32. Kidger in *Spaceflight*, v34, 4/92, 117; Portree (1995), 133.
33. Burrough, 78; Kidger in *Spaceflight*, v34, 8/92, 266.
34. Harland (1997), 197; Harvey (1996), 163–164; Aron, 473–474.
35. Polyakov interview; Krikalev interview; Kidger in *Spaceflight*, v34, 4/92, 119.
36. Ryumin's comments in the 9/12/97 STS-86 NASA Flight Readiness Review.
37. Van den Berg, *MirNews*, #221, July 25, 1994.
38. Harland (1997), 197.
39. Krikalev interview.
40. Drygas.
41. Krikalev interview; Van den Berg, *MirNews*, #122, 2/21/91; Portree & Trevino, 87.
42. Artsebarski interview; Krikalev interview; Ujica.

Chapter 11—*Mir*: Almost Touching

1. Hooper (1990, v1), 143; Siddiqi, 804; also available as of 1/22/03 at http://howe.iki.rssi.ru/GCTC/chiefs.htm.

2. Sotheby's, lots 16, 22, 40, 68a, 87–90, 94, 122, 123, 124, 175; *The Scientist*, 4/15/96, 10[8]; *New York Times*, 12/12/93, I47; *Quest*, Winter, 1993, 28–29.

3. Harland (1997), 201.

4. *Spaceflight*, v34, 8/92, 266; v35, 1/93, 20; *Aviation Week & Space Technology*, 8/23/93, 24.

5. Siddiqi, 844, 904–905; *Quest* v6, #1, Spring 1998, 21; Harvey (1996), 171.

6. *Spaceflight*, v36, 5/94, 159; v36, 6/94, 184.

7. Aron, 481–483; Walker, 257–260.

8. Walker, 262; Aron, 486–493.

9. *New York Times*, 10/19/92, A1.

10. *New York Times*, 4/7/93, 1; 6/24/93, A1; *Washington Post*, 6/24/93, A21; *Chemical And Engineering News*, 11/22/93, 29.

11. Burrough, 245–258; *Aviation Week & Space Technology*, 8/5/91, 18–19; *New York Times*, 10/6/92, C2; *Washington Post*, 6/18/92, A3; 10/6/92, A4; *Space News*, 8/24–30/92, 3; 5/3–9/93, 1; *Washington Times*, 10/6/92, A7.

12. Available as of 1/22/03 at http://clinton1.nara.gov/White_House/EOP/OSTP/other/apllocln.html.

13. Advisory Committee on the Redesign of the Space Station, 22–24.

14. *Aviation Week & Space Technology*, 6/21/93, 20–21.

15. Polyakov interview.

16. *New York Times*, 10/6/92, C2.

17. Portree (1995), 46, 146; Burrough, 66–67, 164–165.

18. Serebrov interview.

19. Burrough, 68–69; Cassutt, 268.

20. *Aviation Week & Space Technology*, 10/25/93, 68.

21. Burrough, 246–258; *New York Times*, 9/3/93, A1; *Washington Post*, 7/26/93, A6; 9/9/93, A7.

22. Serebrov interview; Van den Berg, *MirNews*, #184, August 11, 1993); #185, August 13, 1993.

23. Portree (1995), 141; Harland (1997), 208.

24. Van den Berg, *MirNews*, #185, August 13, 1993.

25. Cherniavsky interview; Serebrov interview; Krikalev interview.

26. Serebrov interview.

27. Van den Berg, *MirNews*, #191, October 8, 1993; Kidger in *Spaceflight*, v36, 5/94, 154.

28. Serebrov interview; Van den Berg, *MirNews*, #191, October 8, 1993; Note that Van den Berg misunderstood the radio communications, not realizing that it was Semenov himself who got on the radio to inform Serebrov and Tsibliev of the postponement.

29. Aron, 495–514.

30. Yeltsin (1994), 248; Aron, 556.

31. Aron, 518.

32. Aron, 538.

33. Aron, 534–538, 552; Morrison, 283–284.

34. Serebrov interview.

35. Kidger in *Spaceflight*, v36, 5/94, 154.

36. Harland (1997), 208–209.

37. Polyakov interview; Kidger in *Spaceflight*, v36, 5/94, 155.

38. *New York Times*, 12/12/93, I47.

39. *Spaceflight*, v36, 10/94, 328; Harvey (2001), 288–290.
40. Zimmerman, Robert, 276.
41. For a summary of later profits, see Oberg (2002), 255–256.
42. Polyakov interview; Drygas.
43. Serebrov interview; Polyakov interview.
44. Krikalev interview; *Spaceflight*, v36, 2/94, 38; *Spaceflight*, v36, 5/94, 146; *Johnson Space Center Space News Roundup*, 11/13/92, 1.
45. *Washington Post*, 2/3/94, A15.
46. Krikalev interview.
47. *Washington Times*, 3/10/93, A2.
48. *Washington Post*, 2/3/94, A15.
49. *Spaceflight*, v36, 4/94, 131–134; *Baltimore Sun*, 2/12/94, 9.
50. *New York Times*, 2/9/94, A19.
51. Transcript available as of 8/26/02 at http://spacelink.nasa.gov/NASA.Projects/ Human.Exploration.and.Development.of.Space/Human.Space.Flight/Shuttle/ Shuttle.Missions/Flight.060.STS-60/Russian.Premier.Calls.
52. Van den Berg, *MirNews*, January 25, 1994.
53. Van den Berg, *MirNews*, c. 3/25/94; c. 5/9/94.
54. *Spaceflight*, v36, 5/94, 155.
55. Drygas.
56. *Spaceflight*, v36, 10/94, 328.
57. *Spaceflight*, v36, 5/94, 159; v36, 6/94, 184.
58. Kidger in *Spaceflight*, v36, 11/94, 388–389.
59. Cassutt, 287; Kidger in *Spaceflight*, v37, 1/95, 7; Van den Berg, *MirNews*, c. 9/ 5/94.
60. Kidger in *Spaceflight*, v37, 1/95, 8.
61. Portree (1995), 46.
62. Kidger in *Spaceflight*, v37, 1/95, 7–8.
63. Polyakov interview; Van den Berg, *MirNews*, #226, August 31, 1994.
64. Van den Berg, *MirNews*, #227, September 2, 1994.
65. Polyakov interview; Harland (1997), 214; Portree & Trevino, 106; Kidger in *Spaceflight*, v37, 1/95, 9.
66. Gross, Appendix G; Van den Berg, *MirNews*, #232, 10/6/94; Kidger in *Space-flight*, v37, 2/95, 54.
67. *Spaceflight*, v38, 6/96, 190.
68. Cassutt, 296.
69. *Detroit News*, 5/11/97, available as of 6/27/01 at http://detnews. com/1997/ nation/9705/11/05110004/htm; Kidger in *Spaceflight*, v37, 1/85, 10.
70. Polyakov interview; *Spaceflight*, v38, 6/96, 190; Harland (1997), 218.
71. Kidger in *Spaceflight*, v37, 2/95, 55; Van den Berg, *MirNews*, #233, 10/13/94.
72. Portree (1995), 147–148; Van den Berg, *MirNews*, #233, 10/13/94; Kidger in *Spaceflight*, v37, 2/95, 55–56; Harland (1997), 216.
73. Polyakov interview; Gross, Appendix C, 4; Oberg (2002), 105–6; Harland (1997), 216.
74. Van den Berg, *MirNews*, #240, 12/20/94.
75. Kidger in *Spaceflight*, v37, 2/95, 57.
76. Portree (1995), 149.
77. Cassutt, 269; Hooper (1990, v2), 315–316.
78. *Quest*, 8:2 (2000), 14.
79. *Quest*, 8:2 (2000), 13–14; *Space News*, July, 15–21, 1996.

80. Van den Berg, *MirNews*, #242 1/26/95.
81. Kidger in *Spaceflight*, v37, 2/95, 58; Harland (1997), 218.
82. Van den Berg, *MirNews*, #239, 11/13/94.
83. Polyakov interview.
84. Van den Berg, *MirNews*, #241, 1/11/95; Harland (1997), 218.
85. Wetherbee, *Oral History*, 7; Foale, Colin, 43–44.
86. Foale, Michael, *Oral History*, session 3, 8; Titov, *Oral History*, 2; Polyakov interview.
87. *Aviation Week & Space Technology*, 2/13/95, 69.
88. Weatherbee, *Oral History*, 8.
89. *Aviation Week & Space Technology*, 2/13/95, 69; *Spaceflight*, v37 5/95, 155.
90. *Spaceflight*, v37, 7/95, 242; *Johnson Space Center's Space News Roundup*, 3/24/95, 1.
91. Thagard interview; Thagard, *Oral History*, 22–23.
92. Thagard interview; Thagard, *Oral History*, 25.
93. Polyakov interview; *Quest*, 8:2 (2000), 18; *Aviation Week & Space Technolgy*, 3/27/95, 23; Kidger in *Spaceflight*, v37, 6/95, 191.
94. Polyakov interview; Thagard interview.
95. Kidger in *Spaceflight*, v37, 6/95, 191.
96. *Space News*, 7/15–21/96.

Chapter 12—*Mir*: Culture Shock

1. Hooper (1990, v2), 291.
2. Cassutt, 225; see also NASA biography available as of 8/26/02 at http://www.jsc.nasa.gov/Bios/htmlbios/dezhurov.html.
3. Thagard interview; Thagard, *Oral History*, 23–24.
4. Harland (1997), 163–164.
5. Yeltsin (1990), 82.
6. Gorbachev (1996), 93.
7. Hooper (1990, v2), 67, 128, 173, 291, 296.
8. Savinykh, typescript of English translation of his diary from *Soyuz-T 13* mission.
9. Linenger (2000), 83; Shepherd, ship log for 11/19/00, 15.
10. Burrough, 71, 252, 305; Krikalev interview; Serebrov interview.
11. Thagard, *Oral History*, 33–35; *Washington Post*, 7/4/95, A7; *Aviation Week & Space Technology*, 7/24/95, 61.
12. Thagard interview; Thagard, *Oral History*, 3–4.
13. Thagard interview; Thagard, *Oral History*, 36.
14. Oberg (2002), 252–255; Linenger (2000), 34.
15. Linenger (2000), 45.
16. Linenger (2000), 36.
17. *Spaceflight*, v37, 1995.
18. *Spaceflight*, v38, 4/96, 140.
19. Thagard interview; Burrough, 314–318.
20. Thagard, *Oral History*, 12–13; *Space Times*, 9–10/95, 9.
21. Thagard interview; Burrough, 258–260; see also NASA biography available as of 8/26/02 at http://www.jsc.nasa.gov/Bios/htmlbios/thagard.html.
22. Kidger in *Spaceflight*, v37, 7/95, 223.

23. Thagard interview.
24. Thagard interview; Van den Berg, *MirNews*, #255, 5/17/95; Portree & Trevino, 110.
25. McDonald, 14–15; Kidger in *Spaceflight*, v37, 8/95, 274.
26. *Spaceflight*, v37, 11/95, 368–369.
27. Thagard interview; Van den Berg, *MirNews*, #261, 6/5/95.
28. Oberg in *IEEE Spectrum*, v32, #12, 12/95, 33.
29. Thagard interview; Thagard, *Oral History*, 35.
30. Thagard interview; *New York Times*, 6/7/95, D23.
31. Burrough, 322–323; Portree & Trevino, 112; *Houston Chronicle Interactive*, 8/13/97.
32. Thagard interview; Kidger in *Spaceflight*, v37, 9/95, 313; Stuster, 205–207; Thagard interview; Burrough, 323–324.
33. Zimmerman, Robert, 15, 295.
34. Cassutt, 58; Zimmerman, Robert, 268.
35. Thagard interview; Burrough, 21–27.
36. Thagard interview; Burrough, 25, 261–262, 275–278.
37. Burrough, 27.
38. Thagard, *Oral History*, 42.
39. Thagard interview; Kidger in *Spaceflight*, v37, 8/95, 276.
40. *Washington Post*, 7/4/95, A7.
41. Gross, Appendix I, 10–11.
42. *New York Times*, 7/1/95, 5.
43. Burrough, 324.
44. Thagard interview.
45. McDonald, 23; Harland (1997), 233–234; *Spaceflight*, v37, 10/95, 333–334.
46. Thagard interview; Thagard, *Oral History*, 43; *New York Times*, 7/11/95, C1, C5; 7/15/95, 22.
47. Thagard, *Oral History*, 48.
48. Thagard, *Oral History*, 37; *Countdown*, 7/8/95, 16.
49. Burrough, 324–327.
50. Thagard interview; Thagard, *Oral History*, 51.
51. *Washington Times*, 7/9/95, 1, 11; *New York Times*, 7/8/95, 6.
52. *Spaceflight*, v37, 8/95, 273; Kidger in *Spaceflight*, v37, 9/95, 311; *Aviation Week & Space Technology*, 7/24/95, 62–63; Portree & Trevino, 112–114. Note: Kidger in *Spaceflight* is the only source that describes the hatch problems. That is also the only source that mentions problems with an umbilical cord. Based on other sources, the hatch problem was far less worrisome than Kidger makes it sound. Unfortunately, because I was unable to arrange an interview with Solovyov, I could not get this issue clarified.
53. Van den Berg, *MirNews*, #269, 8/26/95.
54. Oberg in *IEEE Spectrum*, v32, #12, 12/95, 29.
55. *Spaceflight*, v38, 9/96, 299; Thagard interview.
56. Kidger in *Spaceflight*, v38, 1/96, 6; v38, 4/96, 140; Harland (1997), 237.
57. Oberg (2002), 167–169; Burrough, 211.
58. McDonald, 30–31.
59. Oberg in *IEEE Spectrum*, v32, 12/95, #12, 22.
60. Harland (1997), 353–354, 401.
61. Kidger in *Spaceflight*, v38, 9/96, 294–295.

62. Lucid in *Scientific American*, 5/98, 50; Burrough, 329–334; Van den Berg, *MirNews*, #327, 9/12/96.
63. Lucid, *Oral History*, recorded June 17, 1998, 7–8; *Quest*, v5, #3, 1996, 12–13; Gross, 10.
64. Burrough, 334.
65. Lucid, *Oral History*, recorded June 17, 1998, 18, 20.
66. Portree & Trevino, 118–122.
67. Portree & Trevino, 121; also, see Lucid's "Letter on EVAs," dated July 26, 1996, available as of 8/26/02 at http://spaceflight.nasa.gov/history/shuttle-mir/history/h-flights.htm.
68. Kidger in *Spaceflight*, v38, 9/96, 296; Harland (1997), 253, 349; NASA Memo, dated 6/7/93, "Government Standards of Ethical Conduct," located in "Impact: Ethics File" at the NASA History Office, Washington, D.C.; *Space News*, 2/4/02, 10.
69. Lucid in *Scientific American*, 5/98, 52.
70. *Acta Astronautica*, v42, 1998, 11–23.
71. See "Lucid's Postflight Conference," available as of 8/26/02 at http://spaceflight.nasa.gov/history/shuttle-mir/history/h-flights.htm.
72. *Newsweek*, 10/7/96, 31–32.
73. Van den Berg, *MirNews* #321, 8/13/96; Burrough, 105.
74. Burrough, 106.
75. Van den Berg, *MirNews* #337, 12/3/96; Van den Berg, *MirNews* #338, 12/10/96; Portree & Trevino, 122–123.
76. Van den Berg, *MirNews* #330, 10/7/96; Burrough, 92–114; *New York Times*, 1/22/97, A15.
77. *Orlando Sentinel*, 1/20/97, NASA Press Release, 12/12/96.
78. *Acta Astronautic*, v32, 1998, 11; Lucid in *Scientific American*, 5/98, 52; Freeman, Marsha (2000), 64–66.
79. Thagard interview; NASA Pocket Statistics, 1997, available as of 7/26/02 at http://history.nasa.gov/pocketstats/ "R&D Funding by Program".

Chapter 13—*Mir*: Spin City

1. MSNBC Report, February, 1997; *Space News*, 2/24–3/2/97, 1.
2. Van den Berg, *MirNews* #351, 3/21/97.
3. Linenger (2000), 217.
4. Van den Berg, *MirNews* #348, 3/2/97; Linenger (2000), 99–117; Burrough, 123–150.
5. Burrough, 145–146.
6. Available as of 1/22/03 at: http://spaceflight.nasa.gov/history/shuttle-mir/history/h-f-linenger-fire.htm; available as of 1/22/03 at http://spaceflight.nasa.gov/history/shuttle-mir/history/h-f-linenger-fire-cul.htm.
7. Burrough, 149.
8. Personal communication, summer 1995.
9. *Wall Street Journal*, 4/18/97, c15.
10. Van den Berg, *MirNews* #351, 3/21/97; #352, 4/3/97.
11. Van den Berg, *MirNews* #352, 4/3/97; *Space News*, 3/17-23/97, 8.
12. Linenger (2000), 160–172; Burrough, 157–164.
13. Linenger (2000), 173.

14. Van den Berg, *MirNews* #353, 4/7/97; *Space News*, 4/7–13/97, 4; Burrough, 188.
15. *New York Times*, 4/12/97; Linenger (2000), 190–191.
16. Burrough, 193.
17. Burrough, 205–206, 233–234; Linenger interview, 4/7/03. Though Linenger does not deny that there were some crew disagreements, he takes issue with the suggestion that there was any tension between him and his Russian crewmates. To quote him, "The most unbelievable thing to me was that you could live under such pressure cooker, claustrophobic conditions for months together with former Cold War enemies and get along." (Personal communication, April 5, 2003) He also claims that he was present during Tsibliev's comments to ground control, even though he also admits that he did not attend many communication sessions with the ground.
18. Linenger (2000), 130; Linenger interview, 4/7/03.
19. Burrough, 168–170; Linenger interview, 4/7/03.
20. Van den Berg, *MirNews* #361, 6/16/97; Burrough, 202–214.
21. Van den Berg, *MirNews* #358, 5/11/97.
22. Opening Statement, F. James Sensenbrenner, Jr., Chairman, Committee on Science, Hearing: FY1998 NASA Authorization: The International Space Station, Subcommittee on Space and Aeronautics, April 9, 1997.
23. Burrough, 212; *Space News*, 4/21–27/97, 3.
24. *Washington Post*, 4/10/97, A18.
25. Linenger interview, 4/7/03; Burrough, 222–223; Portree & Trevino, 127.
26. Linenger (2000), 205–207; Burrough, 460.
27. Linenger (2002), available as of 1/22/03 at http://www.pbs.org/wgbh/nova/mir/live.html.
28. Foale, Colin, 19–36.
29. Foale, Michael, *Oral History*, Session 1, 6–8, 11.
30. Foale, Colin, 191.
31. Burrough, 358.
32. Burrough, 371.
33. Burrough, 363–391; Foale, Michael, *Oral History*, Session 1, 9–16; NASA News Release: 97–214a, dated September 25, 1997.
34. Foale, Michael, *Oral History*, Session 1, 12–13.
35. Foale, Colin, 140.
36. Foale, Michael, *Oral History*, Session 1,13–15; Foale, Colin, 134.
37. Foale, Michael, *Oral History*, Session 1, 16.
38. Van den Berg, *MirNews* #366, 7/1/97; #369, 7/11/97.
39. Burrough, 417.
40. Foale, Colin, 165.
41. Van den Berg, *MirNews* #371, 7/20/97.
42. Engelauf, Philip L., NASA Memo dated July 28, 1997. Subject: MCC–M Response. Available as of 1/22/03 at http://www. reston.com/nasa/jsc/07.28.97.engleauf.html.
43. Foale, Michael, *Oral History*, Session 2, 4.
44. Van den Berg, *MirNews* #373, 7/28/97.
45. Van den Berg, *MirNews* #378, 8/17/97.
46. Foale, Michael, *Oral History*, Session 2, 1.
47. Bingham interview; Musgrave interview; Freeman, Marsha (2000), 71.
48. Foale, Colin, 141; Freeman, Marsha (2000), 71.

49. Foale, Colin, 249; Freeman, Marsha (2000), 74; Abstracts 17 and 87 presented at the 15th annual meeting of the American Society for Gravitational and Space Biology, 1999; Abstracts 51 and 112 presented at the 17th annual meeting of the American Society for Gravitational and Space Biology, 2001.

50. Gross, Letter, September 12, 1997), 2.

51. Burrough, 463; Oberg (2002), 141–148.

52. NASA News Release: 97–214a, dated September 25, 1997.

53. Keith Zimmerman, *Oral History*, 9.

54. Burrough, 444.

55. Foale, Colin, 211–212.

56. Ryumin, *Oral History* 6-7; *Washington Post*, 1/21/98; Yahoo, 1/21/98.

57. Valeri Ryumin's comments at the 9/12/97 Flight Readiness Review for STS-86; Precourt, *Oral History*, 7; Van den Berg, *MirNews* #427, 6/9/98.

58. Serebrov interview.

59. Foale, Michael, *Oral History*, Session 3, 14–15.

60. Burrough, 344–345.

Chapter 14–*International Space Station*: Ships Passing in the Night

1. MSNBC and Wire Reports, 5/6/01, available as of 8/13/02 at http://www.msnbc.com/news/564507.asp.

2. *Space News*, 3/12/01, 1, 20.

3. *Space News*, 2/5/01, 1, 18.

4. Spaceflightnow.com, available as of 8/14/02 at http://www.spaceflightnow.com/news/n0103/20tito/.

5. NASA Press Release, 3/19/01, available as of 8/13/02 at http://www.spaceref.com/news/viewpr.html?pid=4151; CNN reports, 3/20/01, available as of 8/13/02 at http://www.cnn.com/2001/TECH/space/03/20/space.tourist.flap/ and http://www.cnn.com/2001/TECH/space/03/20/alpha.tourist/; *USA Today*, 3/20/01, available as of 8/13/02 at http://www.usatoday.com/news/nation/2001-03-19-cosmonauts.htm.

6. *Space News*, 4/30/01, 1, 19; Spaceflightnow.com, available as of 8/14/02 at http://www.spaceflightnow.com/news/n0103/20tito/.

7. "Photovoltaic Array Assembly," available as of 1/22/03 at http://www.shuttlepresskit.com/STS-97/payload81.htm; "Multi-Purpose Logistics Modules," available as of 1/22/03 at http://spaceflight.nasa.gov/station/assembly/elements/mplm/.

8. NASA ISS Status Reports, #02-7, 2/4/02, #02-24 (5/17/02), #02-32 (7/19/02), #02-34 (8/2/02), available as of 8/13/02 at http://spaceflight.nasa.gov/spacenews/reports/issreports/2002/index.html.

9. "ISS to Date," available as of 12/3/02 at http://spaceflight.nasa.gov/station/isstodate.html.

10. NASA Fact Sheets: FS-2001-11-187-MSFC, FS-2001-03-47-MSFC, FS-2002-03-75-MSCF, available as of 8/12/02 at http://www.scipoc.msfc.nasa.gov/factchron.html.

11. Krikalev interview; Oberg (2002), 313–314.

12. NASA ISS Status Reports, #02-11 and #02-12; available as of 1/22/03 at http://spaceflight.nasa.gov/spacenews/reports/issreports/2002/index.html.

13. Krikalev interview.
14. Krikalev interview; McCurdy, 114–115; Messerschmid & Bertrand, 118–136.
15. IMCE Task Force, 1.
16. *Space News*, 3/5/01, 1; IMCE Task Force, 3, 17.
17. *Washington Post*, 7/31/86, A1; *New York Times*, 3/12/86, A1; Lewis, 190–192.
18. IMCE Task Force, 5.
19. Miller, James Grier. "Living Systems Theory Applied to Life in Space," in Harrison, Clearwater, & McKay. 195.
20. Manarov interview; Krikalev interview.
21. Burrough, 494.
22. Krikalev interview; *Space News*, 4/14–20/97, 4, 18.
23. Kerner, 1–11, 13–15, 103–104; La Fay, 11–15, 36–56; Lantzeff & Pierce, 21–23; Westwood, 133–134; Raeff (1971), 62–64.
24. Shepherd, ship log for 1/6/01.
25. Ley (1949), 71.

Index

A

Abbey, George, 395-396, 400
Academy of Sciences of USSR, 43, 268
    Institute of Atmospheric Physics, 266
    Institute of Chemical Physics, 266
Afanasyev, Viktor, 293, 294, 295, 298-300, 304, 336, 347, 348, 353-354, 357, 428
Afghanistan
    civil war, 258-259
    Communist Party, 152
    cosmonauts, 256, 258-259, 311
    Soviet invasion of, 151-153, 155, 161, 172, 173, 190, 246, 255-256
Akiyama, Toyohiro, 293, 294, 295, 302, 326
Alexandrov, Alexander, 181, 183-188, 189, 193, 243, 244-245, 250, 251, 252, 272, 457
Almaz station, 26, 27, 84, 85-86, 125, 139, 160, 327
Alpha space station, 332, 338, 457
Altair 1, 235
American-Russian Shuttle-*Mir* program, 203

Atlantis missions, 392-402, 404-405, 406, 410, 411, 415, 427, 428, 437
cosmonaut/astronaut exchange program, 330-331, 332-333, 339, 349-353, 406
costs and funding problems, 334, 342, 407, 441
crew interaction and cultural differences, 374, 376-377, 378-379, 380-381, 384, 385-386, 400-401, 409, 411-412, 424-425, 429-430, 441, 459
crises and disasters, 418-425, 430-437
Discovery mission, 367-370, 442
docking test, 367-370
equipment, 365–366
injuries, 385-386
Mir system failures, 422-426
mission control, 420, 441
NASA's management of, 379-380, 381-383, 392, 401, 407-408, 409-410, 411-414, 420-421, 423, 425, 426, 439-441
negotiations, 337-338, 346, 383
Pepsi commercial, 409
politics and, 330, 331, 339-341, 381-383

public relations gestures, 353, 392, 398-399

refueling and resupplying *Mir*, 403, 405, 406

schedule and phases, 338-339, 346

visa waiver for Russians, 400-401

American space program. *See also* National Aeronautics and Space Administration; *individual programs*

accomplishments and records, 110, 117, 127, 329, 374, 391-392, 395, 414, 449

privatization, 454

recovery after *Challenger* accident, 329, 454

Amin, Hafizullah, 152

Andre-Deshays, Claudie, 410

Andropov, Yuri, 163, 181, 190, 205

Apollo program, 25, 49, 50, 82
    (*1*), 50, 97
    (*11*), 392
    (*12*), 57, 58, 66

Apollo-Soyuz Test Project, 82, 90, 98, 99-100, 101, 104-105, 106, 110, 111, 120, 125, 162, 172, 174, 283, 375, 440

Apollo spacecraft
    design, 52, 58, 59, 105
    problems with, 66-68, 69, 74

Armstrong, Neil, 50-51

"Artificial settlements," 3, 15

Artsebarski, Anatoli, 301, 302, 306-307, 308-309, 310-313, 317, 318, 320, 323-324, 325, 335, 370

Artyukhin, Yuri, 89

Arzamazov, Gherman, 345, 357

Asimov, Isaac, 6

Astronauts. *See also* Cosmonauts; *specific individuals*

cosmonaut exchange program, 330-331, 332-333, 339, 349-353, 357, 396

crew assignments, 395-396

culture, 379

deaths, 50, 97, 224-226

endurance records, 110, 117, 127, 374, 391-392, 411

female, 176-177, 194, 214, 224, 360, 367, 396, 406-410, 411

in mission control, 457

NASA's treatment of, 381-382, 401, 445

spacesuits, 260

training, 352, 370

Western (non-American), 171

Atkov, Oleg, 188-189, 191, 193, 195, 197-198, 199, 244, 256, 258, 345

Attitude control systems, 16, 29, 53, 222. *See also individual spacecraft*

Aubakirov, Toktar, 319, 320

Austria, 279, 319, 320, 327, 331

Avdeyev, Sergei, 335, 443, 445

B

Baikonur Cosmodrome, 104, 136-137, 160, 180, 188, 235, 251, 270, 317, 319, 322, 324, 328, 338, 341, 347, 356, 359, 445

Balandin, Alexander, 270, 271-272, 279, 280-281, 285, 286, 287-290, 294, 295, 388, 402

Bean, Alan, 66-73, 74, 77

Beggs, James, 216, 219

Berezovoi, Anatoli, 164, 165, 166-167, 168-170, 174, 175, 177, 178-180, 188, 189, 195, 214, 236-237

Beria, 316

Berlin Wall, 100, 161, 279, 360

Big Sky, The (Guthrie), 463-464

Biogravistat, 142-143, 150, 169

Blagov, Viktor, 245, 246, 249

Blaha, John, 410, 411-413, 414, 415, 425, 428-429

Boeing, 216, 219, 455

Boisjoly, Roger, 224, 225

Boland, Edward, 221-222

Bolshevik (October) Revolution, 22, 117, 246, 248

Botany experiments in space
American experiments, 329-330, 370, 410
Arabidopsis, 142, 158, 159-160, 170, 175-176, 178, 180-181, 200, 201, 236-237, 245, 285, 330, 437-438, 439, 450-451

Astroculture facility, 370
Biogravistat, 142-143, 150, 169
brassica rapa, 437-439
Chromex greenhouse, 370
cotton, 200, 201
Fiton greenhouse, 142-143, 169,
    175-176, 178, 245
flax, 19, 21, 47
flowers, 143, 145, 146, 156, 158,
    200, 285
hawksbeard, 19, 21, 47, 107, 109,
    158
on *International Space Station*,
    450-451
light requirements, 413
on *Mir* space station, 236, 245, 251,
    284-285, 410, 413
mushrooms and fungi, 109
Oasis greenhouse, 27, 29, 42, 43,
    47, 103, 105, 142-143, 145,
    169-170, 174-175, 178
practical joke, 156-157
psychological benefits of, 142, 169-
    170, 174-175, 456
on *Salyut* stations, 19-21, 27, 29,
    42, 43, 47, 95-96, 103, 105,
    107, 109, 142-143, 145, 146,
    150, 156, 158, 159-160, 169-
    170, 174-176, 178, 180-181,
    200, 201, 202, 214, 236-237
on Space Shuttle, 370
Svet greenhouse, 284-285, 410,
    437-438
Svetoblok greenhouse, 170, 178
Vazon containers, 143, 145, 169-
    170
vegetables, 19, 21, 95-96, 103, 105,
    142, 143, 150, 156, 158, 169-
    170, 175-176, 200, 202, 285
viability of seeds, 156, 170, 176,
    180-181, 285, 410, 413, 438-
    439, 451
watering systems, 19-20, 42, 47, 95-
    96, 103, 150, 245, 330, 410,
    413, 437-438
wheat, 142, 150, 175-176, 200, 285,
    370, 410, 413
Bradbury, Ray, 6
Brakov, Evgeniy, 266, 268
Brand, Vance, 104

Brezhnev, Galina, 163
Brezhnev, Leonid Ilyich, 14, 22-24, 26,
    47, 51, 81, 82-84, 85, 99, 100-
    101, 105, 106, 109, 110, 112-
    113, 114, 125, 131, 134, 148,
    152-153, 155, 161-162, 163,
    171, 172, 176, 177, 178, 189-
    190, 203, 205, 206, 216, 217,
    218, 235, 254, 258, 259, 306,
    316, 328, 357, 367
Brezhnev, Viktoriya, 163
Brezhnev doctrine, 112
Budarin, Nikolai, 396, 398, 399, 401-
    402, 442
Bulgaria, 100, 136-137, 139, 141, 246,
    252, 254, 285
Buran (Soviet space shuttle), 261, 263,
    264, 292, 293
Bush, George H., 220, 317, 330, 331,
    332-333, 357
Bush, George W., 454, 455
Bush-Yeltsin agreement, 330-331, 332-
    333, 339, 350, 357

C

Cabana, Bob, 448, 451
Canada, 171, 329, 333-334
Cape Canaveral, 212, 218
Carr, Jerry, 74-79, 391
Carter, Jimmy, 153, 207
Central Aerological Observatory
    (Soviet), 35
Chaffee, Roger, 50, 97
Chazov, Yevgeni, 257-258
Chelomey, Vladimir, 4-6, 12, 14, 26,
    84, 85-86, 110, 160, 222, 239,
    273
Chernenko, Konstantin, 163, 205, 399
Chernigov Higher Air Force College,
    120, 249, 271
Chernomyrdin, Viktor, 338, 342, 344,
    353, 367, 417, 418, 422, 426
Chretien, Jean-Loup, 171, 173, 174,
    189, 198, 204, 261, 263, 363
Clarke, Arthur C., 6-7, 9, 12, 13, 14, 17
Clinton, Bill, 330, 331, 338, 356, 358,
    367, 381-382, 421, 427, 440,
    444, 454, 457

CNN, 408
Coca-Cola, 317-318, 370
Cold War, 1, 24, 82, 204
Collins, Eileen, 367
Collins, Mike, 57
Commonwealth of Independent
    States, 319, 320
Communication satellites, 6-7, 235,
    323, 329, 346, 452
Compton Gamma Ray Observatory,
    329
Conquest of Space (Ley), 462-463
Conrad, Pete, 57-59, 60-63, 64, 65-66,
    67, 69, 70, 74, 147, 196, 446
Coronal mass ejections, 70
Cosmonauts (Soviet), 2, 33-35. *See
    also specific individuals*
    astronaut exchange program, 330-
        331, 332-333, 339, 349-353,
        357, 370, 382
    bonus for flying in space, 99
    civilian, 90, 92, 102, 168
    crew selection, 117, 125, 227, 228-
        229, 357, 376, 411
    cultural challenges in America,
        350-352
    deaths, 45-46, 97, 376
    Eastern European, 131, 132-135,
        136, 138, 139, 141-142, 148,
        155, 159, 161-162, 171-172,
        213, 246, 252, 254, 441
    education and literacy, 378
    election to public office, 268, 293-
        294
    female, 176, 177, 193-194, 198,
        203, 301, 359, 360
    fine for refusal of orders from
        mission control, 393-394
    medical screening, 238-239, 249,
        308, 337
    military, 92, 168, 376
    mission commander, 376
    from non-Communist countries,
        171-172, 173, 189-190, 204
    physicians as, 188-189, 191, 193,
        195, 197-198, 199, 244, 256-
        258, 344-345
    personality conflicts, 40, 166-167,
        171, 174-175, 195, 243, 345
    training program, 134-135, 238,
        293, 310, 352
    uniforms, 168
Cosmos 557, 84
Cosmos 1267 (transport-support
    module), 160, 165, 273, 283
Cosmos 1443, 183-184
Cosmos 1686, 201-202
Cosmos spy satellites, 107
Crippen, Robert, 72, 214
Cronkite, Walter, 48
Cruise missiles, 4, 5
Cuba, 133, 155, 238
Culbertson, Frank, 382, 421, 441
Czechoslovakia, 24, 100, 132, 133,
    134, 161, 171, 213, 278-279,
    384

D

Dagastan, 227, 293-294
Davis, Jan, 353
Demin, Lev, 89
Desai, Morarji, 171
Detente, 24, 25-27, 51, 81-83, 90, 100-
    101, 106, 109, 110-111, 113,
    152, 153, 162, 173, 190
Dezhurov, Vladimir, 371, 374, 375,
    376-377, 379, 384-391, 393,
    394, 397, 400-401, 437
Dobrovolsky, Georgi, 20, 21, 33-34, 36,
    39, 40-41, 42, 43, 44, 45, 46,
    92, 97, 98, 139-140, 151, 376,
    378, 439
Dobrovolsky, Marina, 43, 46, 47
Dunbar, Bonnie, 396, 397
Dzhanibekov, Valdimir ("Johnny"),
    114, 123, 124, 125, 126-127,
    132, 171, 173-174, 193, 194,
    196-197, 198, 199, 200-201,
    203, 204, 205, 236, 308

E

East Germany, 100, 133, 161, 213, 279,
    324, 360; *see also* Germany
Edwards Air Force Base, 74, 211, 214
Egypt, 161

Elektron oxygen electrolysis system, 239, 274, 322, 354, 361, 387, 388, 408-409, 422, 428, 432, 449-450

Empire Test Pilot School, 74

Energia Scientific-Production Association
administration, 251, 261, 264, 294, 328, 333, 338, 345, 357
biology team and experiments, 95, 158, 167-168, 170, 237, 285
and cosmonaut program, 168, 177, 228-229, 251, 257, 272, 302, 307, 347, 350, 360
funding problems and solutions, 302, 318-319, 322, 333, 336, 341, 346, 405, 409, 422, 447, 453
and Glavkosmos, 234, 295-296
internal politics, 295-296
origin and structure, 307, 328
reorganization and privatization, 328, 356
rocket program, 84-85, 261
and space shuttle program, 261, 264, 272
space station program, 110-111, 165, 294, 295-296, 302, 318-319, 322, 328; *see also* Mir; *Salyut*

Engle, Joe, 210-211, 440

Estonia, 268, 305

Europe, cooperative missions with U.S., 329, 333-334

European Space Agency (ESA), 357, 360, 361, 402, 410-411, 440-441, 447, 455

Eward, Reinhold, 415, 418, 419, 422

Exploration of Space, The (Clarke), 7

F

Fairchild Hiller Corporation, 54

Fitzwater, Martin, 221

Flade, Klaus-Dietrich, 324-325, 354

Fletcher, James, 209, 210, 219

Foale, Michael, 368-369, 370, 428, 429, 431, 433, 434, 436, 437-439, 441, 444, 458-459

Ford, Gerald, 101

France, 171-172, 173, 189, 198, 204, 261, 263, 292, 327, 331, 335, 357, 410-411, 440-441

Freedom station, 262, 381
attitude control thrusters, 222
budget and funding, 208, 215-216, 220, 221, 222-223, 330, 331, 444, 454
bureaucratic battles, 218-220
collapse of project, 222-223
Congress and, 220, 221-222, 330
design proposals, 216, 219, 220-223, 224, 330, 331-332, 339, 414, 454
dual-keel design, 220-221
European module, 221
"Fred" redesign, 222
Japanese module, 221
life support system, 222
mission plans, 333
name, 457
opposition to, 216
power-tower design, 220
solar power systems, 219, 220
Soviet rocket offer, 292
task force, 216, 217-218

Freedom-7 capsule, 395

Fullerton, Gordon, 210-211

Functional Auxiliary Block (space tug), 240

G

Gagarin, Valentina, 326

Gagarin, Yuri, 3, 168, 306, 326, 337

Gagarin Air Force Academy, 135, 189, 238

Gagarin Cosmonaut Training Center, 32, 148, 238, 272, 293, 295, 326, 360, 370, 376

Gallai, Mark, 2

Gandhi, Indira, 172, 190

Garriott, Helen, 72

Garriott, Owen, 66-73, 77, 246, 456

Gemini capsule, 86

Gemini program, 26, 37, 49, 70, 430

Georgia (republic), 320

Germany. *See also* East Germany;
  West Germany
  cosmonauts, 346, 357, 360, 402-
    403, 415, 418
  Nazis, 8, 10, 34, 91, 140, 189, 205,
    378
  Peenemünde launch facility, 10, 16
  Society for Space Travel, 7, 10
  U.S. space endeavors with, 329,
    352-353
Gerstenmaier, Bill, 408
Gibson, Bob "Hoot," 395-396, 397,
  398, 399, 440
Gibson, Ed, 74-79, 391, 392
Giscard-d'Estaing, Valéry, 171-172, 173
Glavkosmos, 234, 279, 280, 292, 295-
  296, 301, 322
Glenn, John, 207
Glushko, Valentin, 3-4, 6, 23, 84, 85,
  177, 193-194, 229, 251, 258,
  261, 264, 294, 328, 345
Goddard, Robert, 464
Goldin, Dan, 332, 337-338, 346, 382,
  383, 401, 412, 427, 440, 444-
  445
Good Morning America, 353
Gorbachev, Mikhail, 161, 203, 205,
  229, 232, 233-235, 238, 246,
  247, 248, 256, 259, 263, 264,
  265, 277, 278, 279, 282, 283,
  291, 303-304, 305, 306, 314,
  316, 317, 320, 328, 331, 367,
  377-378
Gore, Al, 338, 342, 399, 417, 418, 426
Gravity, artificial, 17. *See also*
  Weightlessness
Great Britain, 292, 301, 327
Great Depression, 8, 217
Grechko, Andrei, 24
Grechko, Georgi, 90-92, 94-96, 103,
  114, 117-122, 123, 124, 125-
  131, 133-134, 135-136, 138,
  140, 147, 151, 158, 169, 177-
  178, 195, 201, 205, 238, 241,
  252, 266-268, 269, 327, 378,
  420
Green, Bill, 221-222
Grissom, Gus, 50, 97
Gromyko, Andrei, 24
Gross, Roberta, 439-440

Grumman, 216
Gubarev, Alexei, 92, 94-95, 103, 131,
  135
Gurovski, Nikolai, 358
Guthrie, A.B., 463-464
Gyroscopes, 16, 53, 71

H

Haise, Fred, 210-211
Harris, Bernard, 368, 370
Heinlein, Robert, 6, 191
Helsinki Accords, 83, 100, 101, 106,
  111-113, 131, 430
Helsinki Watch, 131, 206
Hitler, Adolph, 10
Hubble Space Telescope, 329, 330
Hungary, 100, 133, 139, 155, 161, 278,
  279, 283
Husak, Gustav, 134
Hutchinson, Neil, 77
Hydrazine fuel, 4, 138-139, 222
Hydrogen bomb, 2
Hydrogen-breathing bacteria, 167

I

Icarus jetpack, 274-275, 276, 277, 284,
  336, 340, 362, 364
Igla radar docking system, 89, 108,
  116-117, 118, 119, 123-124,
  136, 173, 236, 240-241
India, 171, 172, 189-190, 204
Industrial Space Facility, 222
Institute of Medical-Biological
  Problems (IMBP), 257, 332-333
Intercontinental ballistic missiles, 6,
  388
Intercosmos program, 131, 132-135,
  136, 138, 139, 141-142, 148,
  155, 159, 161-162, 171-172,
  189-190, 198-199, 213, 238,
  244, 252, 254, 255-256, 263,
  319
International Launch Services, 346
International space conferences, 246
International Space Station, 307
  airlocks and docking modules, 449,
  452

American expeditions to, 379, 452
assembly, 448, 450, 451
attitude control system, 449
communications, 452
computerization, 451-452
construction delays and funding
    problems, 346, 416-418, 421-
    422, 426-427, 439, 440-441,
    453-454
design, 339, 416-417, 449-450, 451
Destiny laboratory, 449, 450, 452,
    454
first expedition, 451, 459
goals, 453, 460-462
life support systems, 449-450, 452
maintenance and repair, 449-450
mission control, 451
naming, 457-458
paid berths on, 446-448, 453
participants, 339, 453-454
permanent occupation, 449, 451
Pirs module, 449
plant experiments, 450-451
political issues, 427, 440, 441, 444,
    454-455
refueling and resupply, 449
robot arms, 448, 449
self-sufficiency issues, 451-452, 453
solar panel arrays, 449, 454
Unity (Node 1) module, 416, 448,
    449, 451, 454
volume and mass, 450
Zarya (Functional Cargo Block)
    module, 416, 417, 426, 445,
    446, 448, 451
Zvezda (service module), 416, 417,
    421-422, 426, 440, 441, 448-
    449, 452
Interplanetary spacecraft
attitude control systems, 16
Kosmoplan (Space Glider), 5
life-support problems, 14-15, 36
power supplies, 16, 46
propulsion systems, 16
space stations as, 3, 14, 232, 294
systems reliability, 17
temperature extremes, 16
Iran, 161
Israel, 132
Italy, 346

ITAR-TASS news agency, 356
Ivanchenkov, Alexander, 171, 174
Ivanov, Georgi, 136, 137, 141, 246

J

Jahn, Sigmund, 325
Japan, 292-293, 326, 327, 329, 333-334,
    376
Jet Propulsion Laboratory, 446
Joanneum, 319
Johnson Space Center [Manned Space
    Center], 57, 218, 219, 395-396,
    448
Journey to the Cosmos (Ley), 7
Jupiter probe, 329

K

Kaleri, Alexander (Sasha), 319, 324,
    410, 411, 415, 419, 422
Karmal, Barak, 152, 153
Kazakhstan, 319, 320, 328, 356, 360,
    363
Kennedy, John F., 13, 22, 25, 50, 111,
    208, 217, 331
Kennedy Space Center, 218, 395
Kerwin, Joe, 57, 59, 60-64, 65-66, 69,
    147, 196, 286-287
Kharkov Higher Air Force School, 140,
    376
Khrunichev State Scientific-
    Production Center, 328, 346,
    417
Khrushchev, Nikita, 2, 5-6, 13, 23, 26,
    82, 100, 205, 222, 316
Khrushchev, Sergei, 5
Kiev Aviation Institute, 4
Kiev Institute of Civil Engineering,
    249
Kikuchi, Ryoko, 293
Kinzer, Jack, 57
Kiselev, Anatoli, 417
Kizim, Leonid, 159, 188, 189, 190-193,
    194-195, 197-198, 204, 205,
    214, 235, 236, 237, 244, 257,
    310, 378
Klimuk, Pyotr, 101-103, 104-105, 111,
    127, 168, 430

Kobzey, Yeugeny, 169
Kolodin, Petr, 32, 33, 125, 239, 253
Komorev, Vladimir, 97
Kondakova, Elena, 359, 360, 362-363,
    364-365, 366-367, 368, 371-
    372, 376, 411
Koptev, Yuri, 328, 337-338, 346, 357,
    383, 422
Korean War, 74, 92
Korolev, Sergei Pavlovich, 2-3, 4, 5, 10,
    12, 14, 16, 23, 25, 27, 33, 84,
    92, 110-111, 183, 222, 367,
    460, 463
Korzun, Valeri, 410, 411-412, 415, 418-
    419, 420, 422-426
Kosmoplan (Space Glider), 5, 14, 26
Kosygin, Aleksei, 23, 24
Kovalev, Anatoly, 111-112
Kovalyonok, Vladimir, 116-117, 123,
    159-160, 198-199
Krikalev, Elena, 351
Krikalev, Sergei, 261, 263, 264, 268-
    269, 275, 301, 302, 306, 307-
    309, 311-313, 317, 318, 319-
    325, 326-327 n., 332, 333, 335,
    350-353, 364, 368, 370, 379,
    402, 434, 451, 452, 453, 457,
    459, 461
Kubasov, Valeri, 32, 104
Kurs radar rendezvous system, 236,
    240 n., 273, 276, 299-301, 320,
    335, 358, 359, 363, 366-367,
    422-423, 430
Kuznetsov, Major-General, 32
Kvant transport-support module, 166,
    203, 239-241

L

Lakets people, 227
Latvia, 305
Launch vehicles, 5-6
Laveikin, Alexander, 238, 239, 241-
    244, 253, 434-435
Lazarev, Vasili, 98-99, 101
Lazutkin, Alexander, 415, 418-419,
    422, 423, 425-426, 429-432,
    434-436, 442

Lebedev, Valentin, 154, 164, 165, 165,
    166-168, 169-170, 174-176,
    177-180, 188, 189, 195, 214,
    236-237, 261, 345, 410, 413,
    439
Lenin, 328
Leningrad [St. Petersburg] Institute of
    Mechanics, 92, 325
Leonardo reusable cargo module, 449
Leonov, Alexei, 31, 32, 33, 45, 104,
    125, 172, 301, 326
Levchenko, Anatoly, 250, 251, 252
Lewis, Jerry, 426
Ley, Willy, 7-9, 10, 12, 13, 14, 17, 51,
    367, 460, 462-463
Life-support systems, 14-15, 36
    Rodnik water system, 116, 155,
        169, 185, 200, 230-231, 274,
        354
    Vozdukh carbon dioxide scrubber,
        231, 239-240
Lindbergh, Anne Morrow, 49
Lindroos, Martin, 222
Lineger, Jerry, 379, 382, 412, 415,
    418, 419, 420-421, 423, 425,
    427-429, 445
Liquid oxygen propellant, 4
Lithuania, 268, 304-305
Lockheed, 216, 219, 346
Lockheed-Martin, 455
Loral, 346
Lousma, Jack, 66-73, 77
Low, George, 73
Lucid, Shannon, 406, 407-410, 411,
    412, 413, 461
Lunar landing, 50, 331
Lund, Bob, 225
Lyakhov, Vladimir, 136, 139, 140-142,
    147-150, 151, 171, 181, 182,
    183-188, 189, 193, 195, 205,
    256, 259, 280, 378

M

Makarov, Oleg, 98-99, 101, 114, 123,
    124, 126-127, 132, 159, 326
Malenchenko, Yuri, 357-359, 361, 363,
    367

Malenkov, Georgi, 5
Malyshev, Yuri, 174
Manakov, Gennadi, 290, 295, 376, 411, 428
Manarov, Musa, 227-229, 232, 235-236, 237, 243, 249, 250, 252-254, 258, 259-262, 263, 264, 271, 272, 293-294, 295, 297-300, 304, 308, 309, 311, 336, 347, 366, 368, 393, 457
Manned Maneuvering Unit, 214, 263
Manned Orbital Laboratory (MOL), 26, 86
Mariner Mars probes, 446
Mars exploration, 5, 11-12, 14, 17, 79, 239, 363, 372-373, 374, 461
Mars Project, The (von Braun), 11
Marshall Space Flight Center, 79, 218, 219, 224
Martin-Marietta, 219
Martsinovski Institute of Medical Parasitology and Tropical Medicine, 257
Mashinsky, Alexander, 43, 237
Masum, Mohammad Dauran Ghulam, 256
McAuliffe, Christa, 224
McCandless, Bruce, 214, 277
McDonnell-Douglas, 216
Merbold, Ulf, 359-360, 362-363, 367, 419
Mercury mission, 49
Mercury capsule, 86, 395
Merkur capsule, 86, 160, 165, 166, 184, 240
Meteoroids, hazards to spacecraft, 17, 56
Mir space station, 96. *See also* American-Russian Shuttle-*Mir* program
air leak, 390-391, 392, 393, 402, 431-432, 442
animal experiments, 274, 282, 289
antifreeze pollution, 413
assembly, 203, 226, 236, 275-276, 291, 309-313, 340, 376, 386-387, 406-407
astrophysical facility, 239, 253-255, 260, 321

attitude control system, 230, 242, 276, 309, 321, 349, 361-362, 365, 397-398, 399-400, 425, 432-433, 434, 435-436, 441, 443
collisions with spacecraft, 348-349, 353-354, 357-359, 397, 422-423, 430-437, 439
communications, 235, 322-323, 325, 380, 425
cooling system, 403-404, 408, 423-424, 427
core module, 229-230, 361
crews and missions, *see Soyuz-T* mission *15, Soyuz-TM* missions *1-25*
design and construction, 51, 203, 229-232, 236, 237, 263, 277, 283-284, 298-299, 311-313, 332, 336, 384-385, 389, 406-407
deterioration, 321, 335, 361-362, 365, 388-389, 402-404, 422-426, 439-441
docking systems and problems, 231, 235, 236, 240-241, 272, 276, 283, 299-301, 309, 321, 324, 325, 335, 348, 358-359, 363, 366-367, 384, 389-390, 391, 397-398, 402, 404, 412, 422-423, 430-431, 434
endurance records, 239, 244-245, 252, 261, 262, 323, 325, 334-335, 341, 361, 366, 368, 443
fires, 362, 410, 418-421, 441
first manned mission, 238-246, 248-249
flag, 310, 312-313, 324, 335
foreign payloads, 354
goals, 263-264
international missions, 244, 252, 254, 255-256, 258-259, 261, 292-293, 319, 320, 324-325, 331, 357-358, 361, 364, 384
journalist on, 292-293
Kristall module, 279-280, 281-282, 283, 284-285, 287, 289, 290, 296, 297, 299, 336, 348, 349, 359, 361, 365, 368, 376, 384-385, 386-387, 389-390, 391, 397, 402, 404, 410, 416, 431, 434, 437-439

Kvant module, 203, 239-241, 243, 254, 260, 281, 284, 296, 297, 298-300, 311, 321, 336, 340, 362, 365, 368, 386, 387, 402, 409, 418-420, 425, 436-437

Kvant-2 module, 273-276, 277, 281, 283, 284, 286, 289-290, 297, 299, 304, 318, 335, 336, 340, 358, 361, 376, 383, 384, 388, 392, 402, 412, 416, 425, 428, 432, 434, 442

life support systems, 230-231, 235, 239-240, 255, 260, 274, 361, 362, 365, 387, 393, 403-404, 408-409, 412, 418, 422-424, 434, 436-437, 440-441, 443

living quarters, 274

main engines, 230, 242

maintenance and repair activities, 253-255, 260, 296-298, 301, 304, 322, 335, 336, 354, 376, 383, 399-400, 402, 412, 422-424, 442

medical research, 258, 354-355

meteorite damage, 339-340, 344

mission control and, 243, 253, 271, 318, 321-322, 323, 325, 377, 393-394, 429-430, 432, 436, 442

mission statistics, 443

mothballing and abandonment, 263, 269, 273, 277

name, 229

orbital decay and burnup, 443

orientation system, 275, 301-302, 309

paid berths on, 292-293, 295, 301, 302, 319, 324-325, 327, 331, 333-334, 335, 336, 357, 447

political pressures, 234-235, 269

permanent occupancy, 232

plant experiments, 236, 245, 251, 284-285, 410, 413, 437-439

press coverage, 234-235, 238, 244, 259, 270-271, 361, 383

Priroda module, 291, 328, 406-407, 409, 419, 434

profit-making ventures, 280-281, 285, 290-292, 296, 299, 302, 317-318, 319, 327, 357-358

psychological adjustments of crews, 244-245, 252-253, 260-261, 364-365

reactivation, 272, 273, 336-337

refueling and resupply system, 235, 239, 245, 252, 253, 321, 322, 354, 357-359, 365-366, 370, 398, 403-404, 405

replacement station (*Mir-2*) , 328, 338, 339, 340, 417

robot arm, 389, 391, 404, 406

shift system, 294-295, 332

Sofora construction and installation, 309-313, 318, 324, 335, 340, 344, 353-354, 369, 385, 442

solar arrays and electrical system, 230, 232, 242-244, 253, 276, 284, 296, 298, 321, 324, 336, 339, 344, 361, 362, 365, 385, 386, 387, 388-389, 392, 393, 397, 398, 402-403, 404-405, 409, 412, 413, 434, 435-436

space walks, 194, 232, 242-244, 252, 253, 254-255, 260, 263, 276-277, 281, 285-289, 296-298, 300, 309, 310-313, 323-324, 335, 336, 340, 343-344, 353-354, 359, 365, 385, 386-387, 389-391, 393, 402, 409, 427-428, 434, 436-437, 442

Spektr module, 281, 328, 365, 383, 384-385, 386, 388-389, 390, 391-392, 398, 402, 406, 418, 429, 431-432, 436, 437, 442

stowage and inventory problems, 442

Strela cargo cranes, 296-298, 304, 311, 312, 324, 336, 386, 387, 409, 428

television link-ups, 245-246, 254, 408

test girder, 340

thermal shielding, 344

volume and mass, 406, 443

waste disposal, 242

MirCorp, 447

Mishin, Vasily, 26, 30, 31, 33, 41, 84, 85

Mohmand, Abdul Ahad, 256, 258-259

Moldavia, 268
Mondale, Walter, 207
Mongolia, 133, 159, 198-199
Morton Thiokol, 224, 225
Moscow Aviation Institute, 33, 228
Moscow Geographic Institute, 35
Moscow International Film Festival, 181
Mukai, Chiaki, 360
Murashev, Arkady, 268
Musabayev, Talgat, 357-359, 361, 363, 367, 442, 446

N

NASA Lewis Research Center, 219
National Aeronautics and Space Administration (NASA), 25, 447. *See also specific programs*
   administrators, 332, 337, 395-396
   Astronaut Office, 395-396
   budgets and funding, 208, 209, 210, 212, 215-216, 294
   lobbying for programs, 216, 217-218
   lunar mission, 111, 209
   management problems, 209-210, 216, 218-220, 223, 224-225, 330, 381-382, 392, 395-396, 400, 401, 407-408, 409-410, 420-421, 426-428, 444-445, 455-458
   press coverage of missions, 104, 369, 392, 401
   scheduling of astronaut time, 352, 455-456
   Stafford Commission, 440
   and Shuttle-*Mir* missions, 379-380, 381-383, 392, 401, 407-408, 409-410, 411-414, 420-421, 423, 425, 426, 439-441
   space stations, 208, 209, 215-223; *see also Freedom; Skylab*
   structure, 218-219
National Air & Space Museum, 73
Nazarbayev, Nursultan, 356
Nechitailo, Galina, 42-43, 47, 96, 103, 128, 145, 156, 157, 168, 176, 201, 237, 245, 251, 285

Nikolayev, Andrian, 364
Nitric acid propellant, 4
Nitrogen tetroxide fuel, 182
Nixon, Richard, 14, 51, 81, 82, 83, 100, 153, 215, 331, 367
North American Rockwell, 210
North Atlantic Treaty Organization (NATO), 106
NOVA documentary, 270-271

O

O'Connor, Bryan, 380
Olympic Games, boycotts, 153, 155, 190, 206
Onufrienko, Yuri, 405-406, 408, 410, 413
Orbital Sciences, 455
Orlov, Yuri, 113, 131
Oswald, Steve, 371
Oxygen
   life support systems, 15, 30, 41
   liquid rocket propellant, 4

P

Paine, Thomas, 50-51
Pakistan, 255, 256
Patsayev, Svetlana, 33, 43-44, 95
Patsayev, Viktor, 19, 20, 33, 34-35, 36, 38, 39, 40-41, 42, 43-44, 46, 97, 98, 140, 151, 169, 376, 378, 439
Penza Industrial College, 35
Pepsi, 409
Perestroika (Gorbachev), 205-206
Perestroika and glasnost, 205-206, 233-234, 238, 244, 246, 247, 248, 259, 269, 282, 304, 317
Perot, Ross, 327, 331
Perseid meteorite shower, 339-340
Persian Gulf War, 297, 302
Pirs docking and airlock module, 449
Pogue, Bill, 74-79, 391
Poland, 100, 133, 161, 171, 173, 190, 206, 213, 278
Politics and political forces
   and American-Russian Shuttle-*Mir* program, 330, 331, 381-382

cost issues, 18, 56, 80
detente, 24, 25-27, 51, 81-83, 90,
    100-101, 106, 109, 110-111,
    113, 152, 153, 162, 173, 190
economic conditions, 207-208, 232
in *Freedom* space station, 216, 217-
    218, 220, 221-222, 381
gender politics, 204
Helsinki Accords, 83, 100, 101,
    106, 111-113, 153
Moscow Summit, 81-83, 90
Olympic Game boycotts, 153, 155
in *Salyut* missions, 24-35, 81-84,
    90, 100, 101, 111-113, 132-
    133, 148, 159
in *Skylab* mission, 56
in Space Shuttle program, 80, 224-
    226, 381
and space race, 13-14, 25, 47, 83-84
Polyakov, Valeri, 256-257, 258, 259,
    260, 261, 263-264, 269, 275,
    332, 333, 334-335, 337, 341,
    345, 346, 347-348, 349, 353,
    357, 358, 360-361, 362, 364-
    369, 371-373, 400, 402, 406,
    418, 435, 461
Popov, Leonid, 154, 155-156, 157, 158,
    168, 177, 180
Popovich, Pavel, 89
Precourt, Charles, 442
Progress freighters, 116, 123, 127, 138,
    141, 144, 147, 154, 166, 175,
    181, 184, 190, 193, 200, 214,
    230, 235, 239, 245, 252, 253,
    254
Progress-M freighters, 273, 281, 298,
    299, 312, 317, 321, 335, 344,
    354, 357-359, 365-366, 367,
    404, 405, 422-423
Proxmire, William, 220
Psychological stresses of space
    missions
    bad news from home, 394
    behavioral patterns on long-term
        missions, 150-151
    delusions, 107-108, 109
    early concerns, 12, 17-18
    greenhouse experiments and, 142,
        169-170, 174-175
    isolation, loneliness, and
        claustrophobia, 129-130, 142,
        150-151, 157-158, 174-175,
        180, 408
    mitigation of, 78, 332, 140, 180,
        195, 204, 243
    NASA's approach, 456-457
    outside communication and, 144,
        157-158, 401, 408
    personality conflicts, 40, 166-167,
        171, 174-175, 195, 243, 250-
        251, 345
    physician as crewmember and, 188-
        189
    workload and, 77-78, 144, 204, 412-
        413, 425-426, 429-430

Q

Quest airlock, 449, 452

R

Radiation hazards, 17, 20, 361
Radio antenna array, 147-149, 151
Raduga capsule, 273, 336, 344
Rafaello reusable cargo module, 449
Reagan, Ronald, 161, 207, 208, 216-
    217, 221, 222, 223, 235, 331,
    346, 457
Reiter, Thomas, 402-403
Remek, Vladimir, 131, 133, 134-135
Reustel-Buecht, Jorg E., 447
Rocketdyne Corporation, 212, 213
Rockets
    advertisements on, 273
    Ariane, 455
    Atlas, 346
    early experiments, 8, 464
    Energia, 261, 262, 263, 292
    military boosters, 403
    N1, 25, 84-85, 167
    propellants, 4, 6
    Proton, 6, 26, 49, 229, 328, 346, 417
    R7, 2, 3, 92
    Saturn IB, 49, 50, 67
    Saturn 5, 25, 48-49, 50, 51, 53, 79,
        80, 84-85, 167, 216

Soviet marketing of, 292
Space Shuttle boosters, 223-224
Titan, 346
U.S. industry, 455
V2, 5, 10, 11, 16, 91
Rodnik water system, 116, 155, 169,
    185, 200, 230-231, 274, 354,
    425-426
Romanenko, Roman, 252
Romanenko, Yuri, 114, 117-121, 123,
    124, 125-132, 133-135, 138,
    140, 151, 195, 198, 238, 239,
    241-246, 248-252, 253, 258,
    260-261, 327, 332, 400, 420,
    425, 457
Romania, 100, 132-133, 159, 199
Rozdestvensky, Valery, 108
Rukavishnikov, Nikolai, 30, 136-137,
    141
Russia. *See* Soviet Union/Russia
Russian Space Agency, 322
    administrators, 337, 357
    foreign satellite launches, 346
    and *International Space Station*,
        422, 453
    structure and role, 327-328
Ryumin, Valeri, 116-117, 136, 137,
    139-142, 143-151, 154-157,
    158, 169, 171, 177-178, 179,
    182, 186, 191, 192, 195, 202,
    203, 227, 228, 229, 241, 249,
    250, 268, 280, 282, 299, 302,
    319, 320, 337, 350, 360, 362,
    365 n., 376, 379-380, 441-442,
    451, 456

S

Sakharov, Andrei, 2, 113, 131
Salyut stations
    (*1*), 19-21, 26-32, 35-45, 47, 85, 86,
        87, 88, 92, 93, 94, 95, 96, 97,
        106, 110, 120, 125, 128, 139,
        154, 169, 261, 361, 411, 441;
        see also Soyuz mission *11*
    (*2*), 84
    (*3*), 85-90, 92, 93, 94, 98, 106, 107,
        110, 116; *see also Soyuz*
        missions *14, 15*

(*4*), 91, 92-96, 99, 101, 102-106,
    107, 110, 111, 115, 122, 131,
    169, 430; *see also Soyuz*
    missions *17, 18*
(*5*), 86, 106-111, 116, 191, 327; *see
    also Soyuz* missions *22-24*
(*6*), 114-151, 154-162, 169, 170,
    173, 182, 184, 196, 198, 201,
    213, 214, 236, 238, 241, 280,
    282, 308, 319, 325, 327, 336,
    357, 376, 441, 442, 451; *see
    also Soyuz* missions *26, 27,
    28, 32, 33, 35; Soyuz-T*
    missions *3-7*
(*7*), 164-206, 214, 229, 235, 236-
    237, 238, 239, 249, 257, 271,
    308, 310, 336, 337, 364, 376,
    378-379, 430-431; *see also
    Soyuz-T* missions *8-15*
animal experiments, 95, 96, 103,
    105, 107, 122, 126, 128, 145-
    146
attitude control system, 29, 86, 88,
    115, 141, 143
communication systems, 146-147
crews, *see individual Soyuz
    missions*
crises and failures, 74, 110, 128-129,
    137-139,   173-174,   181-188,
    202-203, 236, 238, 249, 376
design, 26-29, 30, 41-42, 51, 52, 85-
    88, 92-93, 108, 110, 115-117,
    124-125, 138, 144, 164-166,
    216, 336
docking system and problems, 28,
    31, 36, 89, 93, 108, 114, 116-
    119, 121-122, 123-124, 128,
    136-137, 139, 144, 165, 173-
    174, 182, 196-197, 237, 238,
    241, 308
engineering problems, 39-40, 106,
    128-129
flight program workload, 95, 122,
    169, 204
guidance system, 93-94
hospitality rituals, 124, 174, 197
Intercosmos missions, 131, 132-
    135, 136, 138, 139, 141-142,
    148, 155, 159, 161-162, 171-
    174, 189-190, 198-199, 213

life support systems, 30, 31, 41-42, 87-88, 94, 103-104, 115, 116, 155, 167, 169, 185, 200
living quarters, 38, 44, 87, 88, 146, 158, 165
maintenance and repair, 39-40, 95, 102, 117, 119, 138, 139, 141, 155, 159, 182, 186-187, 189-194, 201, 235, 257, 308
military surveillance, 29 n., 88, 89, 98, 106, 107, 109, 165, 327
mission code names, 21 n.
orbital deterioration and burnup, 47, 89-90, 92, 105, 138, 161, 183-184, 188, 237
plant experiments, 19, 20, 21, 27, 29, 42, 43, 47, 95-96, 103, 105, 107, 109, 142-143, 145, 146, 150, 156, 158, 159-160, 169-170, 174-176, 178, 180-181, 200, 201, 202, 214, 236-237, 439
political pressures and, 24-35, 81-84, 90, 100, 101, 111-113, 132-133, 148, 159
practical jokes, 156-158
press coverage and conferences, 39, 42, 107, 109, 126-127, 162, 172
records and successfulness, 127, 130, 135, 138, 139, 164, 188, 196, 202, 203, 208-209, 244
refueling and resupply, 107, 110, 116, 123, 127, 144, 166, 175, 200, 214
research laboratory and equipment, 25, 88, 89, 93, 94, 102, 143-145, 147
rescue mission to repower station, 196-204
solar panels, 86, 94, 115, 125, 127, 143, 165, 166, 183, 184-185, 187-188, 189, 193, 197, 200, 201
space walks, 117, 118-121, 147-149, 165, 175, 185, 186-187, 189-194, 198, 201, 202, 235, 236, 241, 310
transport-support modules, 160, 165-166, 183-184, 201-202, 239

waste disposal system, 41-42, 95
weapons systems, 86
Sarafanov, Gennady, 89
Satellites, 6-7, 35, 209
recovery of, 370
U.S., on Soviet rockets, 346
Savinykh, Viktor, 159-160, 196, 197-199, 200-201, 202, 204, 205, 228, 268, 272, 378-379
Savitskaya, Svetlana, 177, 180, 193-194, 203, 204, 214, 236, 238, 268, 336, 360
Schweickart, Rusty, 60-61
Science fiction and science writers, 6-13
Sechenov First Moscow Medical School, 256
Semashko Institute of Social Hygiene and Public Health, 257
Semenov, Yuri, 264, 272, 294, 296, 302, 333, 334, 338, 346, 357, 362
Sensenbrenner, F. James, 416, 417-418, 420, 422, 426-427, 439-440
Serebrov, Alexander, 177, 180, 182-183, 238-239, 249, 272-273, 276, 295, 336-337, 339-341, 342, 343-344, 348, 349, 364, 380, 429, 433, 443
Sevastynov, Vitali, 101, 102-103, 104, 111, 122, 127
Sharansky, Anatoly, 131
Sharma, Rakesh, 189-190, 204
Sharman, Helen, 301, 302, 305, 327, 361
Shatalov, Vladimir, 30, 31, 134, 238, 295, 320, 326
Shenin, Oleg, 343
Shepard, Alan, 75, 395
Shepherd, Bill, 379, 448, 459, 461
Shevardnadze, Eduard, 304
Shuttleworth, Mark, 453
Simak, Clifford, 6
Six-Day War, 132
Skylab, 25, 210, 231, 392
crew dynamics, 72, 195
design, 49-50, 51-57, 108, 160, 244, 285, 332, 389, 449
docking, 51-52, 59, 74
endurance record, 110, 117, 127, 374, 391-392

experiments, 56, 59-60, 63, 69, 71, 75

first manned mission (*Skylab 2*), 57-66, 74, 95, 196

flight program workload, 69-71, 75-78, 412, 425

follow-up station mission, 73

launch vehicle, 48-51

life support systems, 53, 55-56, 239

living quarters, 54-55, 64-65, 146

maintenance and repair, 68, 71, 79, 147, 204

meteoroid shield, 58

orbital deterioration and burnup, 80, 212

practical jokes, 72

research equipment/facilities, 53, 54

second manned mission, 66-73, 74, 77

sleeping arrangements, 55, 63, 65

solar panels, 58, 59, 60-63, 65-66, 284

space walks, 60, 62-63, 68, 69, 70-71, 76, 147, 286-287

television broadcasts, 70, 71

third manned mission, 73-79

thermal heat shielding, 55-57, 59, 67, 68-69, 280

waste disposal, 54-55

wet-tank concept, 49-50

Slayton, Deke, 104

Solar Max research satellite, 214

Solar sails, 16

Solar power, 16, 28, 46, 53, 56, 60, 214-215

Solar telescope, 53, 63, 70, 93

Solidarity Union movement, 206, 278

Solovyov, Anatoly, 270-271, 272, 279, 280-281, 283, 286, 287-290, 294, 295, 335, 388, 395, 396, 399-400, 401-402, 436-437, 439, 459

Solovyov, Vladimir, 188, 189, 190-193, 194-195, 197-198, 199, 204, 214, 235, 236, 237, 244, 249, 257, 271, 285, 297-298, 299, 310, 362, 378-379, 387-388, 432, 436

Soskovets, Oleg, 426

South Korean airliner incident, 206

Soviet-Russian space program. *See also* Cosmonauts; Energia Scientific-Production Association; *Mir* space station; *Salyut* space stations; *Soyuz* missions

breakup of Soviet Union and, 314-325

Chelomey's design bureau, 6, 26, 84, 110, 219, 328

democracy and, 263-269

funding problems, 85, 110, 263-264, 272, 275-276, 279, 294, 301-302, 318-319, 321-322, 327, 403, 407

Glavkosmos, 234, 279, 280, 292, 295-296, 301, 322

Intercosmos program, 131, 132-135, 136, 138, 139, 141-142, 148, 155, 159, 161-162, 171-172, 189-190, 198-199, 213, 238, 244, 252, 254, 255-256, 263, 319

internal politics, 295-296

Korolev's design bureau, 2, 25, 27, 34, 35, 43, 47, 139, 167, 183, 189, 198, 219, 307; *see also* Energia

lunar base, 85, 102, 139

May Day parade, 282

military vs civilian crews, 92

mission control, 179, 228, 235-236, 237, 238, 245, 271, 321-322, 327, 377

press coverage, 293 n.

privatization, 327-328; *see also* Russian Space Agency

profit-making schemes and profitability, 273, 279-281, 292-293, 326-327, 345-347, 407

propaganda, 19-35, 101, 131, 133, 155, 161-162, 164, 204, 258-259, 277, 328

records and accomplishments, 21, 127, 130, 135, 138, 139, 164, 188, 196, 202, 203, 244, 256-257, 262, 306, 326, 332, 333, 359, 366, 405, 414

relationships within, 156, 218
Sotheby's auction of memorobilia,
    326, 327, 345, 346
space shuttle, 26, 84, 132-133, 193,
    227, 250, 261, 262, 263, 272,
    283, 293, 294, 306, 336
visionaries of, 1-6
Soviet Union
    Afghanistan invasion, 151-153,
        155, 161, 172, 173, 190, 246,
        255-256
    All-Union Treaty, 306, 319
    anti-alcohol campaign, 233-234,
        263
    authoritarian culture, 377-378, 380,
        393-394
    and Berlin Wall, 100, 161, 279
    Bolshevik (October) Revolution, 22,
        117, 246, 248
    breakup of, 277-278, 304-305, 314-
        325, 328, 335
    Brezhnev doctrine, 112
    Congress of People's Deputies, 264-
        265, 277-278, 291, 308
    constitution, 341
    corruption in government, 232-233,
        234, 246, 263
    coup attempt, 314-316, 317, 341,
        342-343
    democracy and elections, 246, 264-
        269, 277-278, 303, 304, 305,
        308
    demonstrations, 282-283
    detente, 24, 25-27, 51, 81-83, 90,
        100-101, 106, 109, 110-111,
        113, 152, 153, 162, 173, 190
    economic conditions and reforms,
        232, 234, 256, 263, 279, 303,
        328-329, 345-346, 355-356,
        448-449
    Enterprise Law, 279
    ethnic bigotry in, 228-229
    grain embargo, 153
    Helsinki Accords, 83, 100, 101,
        106, 111-113, 131
    Helsinki Watch, 131, 206
    miners' strike, 278
    Ministry of General Machine
        Building, 322
    national sovereignty declaration,
        291

    Nazi invasion of, 8, 10, 34, 91, 140,
        189, 205, 378
    Olympic Game boycotts, 153, 155,
        190, 206
    open fairs, 234
    perestroika and glasnost, 205-206,
        233-234, 238, 244, 246, 247,
        248, 259, 269, 282, 304, 317
    State Committee for the State of
        Emergency, 314-315
    Supreme Soviet of Russia, 265, 320-
        321, 341-343
    "War of Laws," 341-342
    "White House," 314, 322, 343
Soyuz missions
    (1), 97
    (9), 36, 37, 102
    (10), 30-32, 136
    (11), 32-47
    (12), 98
    (13), 101, 167, 168
    (14), 88-89
    (15), 89
    (17), 91-92, 94-97, 102
    (18-1), 98-99, 100, 114, 125, 326
    (18-2), 100 n., 101-104, 105, 111
    (19), see Apollo-Soyuz Test project
    (20), 105
    (22), 106-108
    (23), 108-109
    (24), 109
    (25), 116-117
    (26), 118-124, 125-126
    (27), 123-124, 125-126, 135, 173
    (28), 131
    (32), 139, 141, 143, 144, 145
    (33), 136-137, 141, 142, 143, 145
    (34), 141, 144
    accidents/emergencies, 45-47, 97,
        98-99, 108-109, 110, 136-137,
        139-140, 141, 144, 145, 326
    triple, 33
Soyuz spacecraft
    attitude control for Mir, 433, 434,
        441
    auctioned items, 326
    design and performance, 26, 27, 28,
        31, 38, 46, 98, 127, 131-132,
        144
    docking system, 299-300, 312, 326

heat shield problems, 280
repairs in space, 286-289
Soyuz-T redesign and testing, 151,
   153-154, 257
Soyuz-TM redesign, 235-236, 272,
   280, 299-300, 312, 406, 433
Soyuz-T missions
   (3), 159
   (4), 159
   (6), 171, 173-174
   (7), 176-177
   (8), 182-183, 185, 237, 238, 337, 376
   (9), 182
   (12), 193
   (13), 196-197, 199
   (14), 201
   (15), 235
   accidents/emergencies, 185-186,
      238, 285-286, 337, 376
   launch-escape system, 185-186
   test flight, 189
Soyuz-TM missions
   (1), 235-236
   (2), 238-239
   (7), 261
   (8), 285, 430
   (9), 270, 272, 279, 289, 290
   (10), 326
   (13), 320
   (14), 324
   (17), 348-349
   (18), 347
   (19), 356-357, 363
   (20), 359-360, 367
   (25), 415
   accidents/emergencies, 259, 272,
      290, 347-348, 430
   civilian passengers, 453
Space flight
   longest, 36
   physical and emotional stresses,
      12, 17-18, 36-38, 78, 107-108;
      see also Weightlessness
Space race, 13-14, 25, 47, 83-84, 194,
   307
Space shuttle (U.S.), 80, 105, 132, 262.
   See also American-Russian
   Shuttle-Mir program
   accidents and disasters, 219, 223-
      226, 229, 329, 381, 444, 454

Atlantis, 215, 392-402, 404-405,
   406, 410, 411, 415, 427, 428,
   437
   budget/cost problems, 212-213,
      215, 426
Challenger, 214, 219, 223-226, 229,
   318, 329, 381, 444, 454
Columbia, 212-214, 331-332, 392
   contractor, 210
   design and development, 209-210,
      211, 224, 395
Discovery, 214-215, 352-353, 367-
   370, 397, 451
   engine problems, 211, 212, 213
Endeavour, 371, 395
Enterprise (prototype), 210-211
   foreign astronauts in, 171, 352-353
   foreign payloads, 352-353
   female crew, 176-177, 194, 224, 360
   funding and costs, 209, 210
   heat shield for reentry, 211, 212-
      213, 215
   inaugural launch, 212, 213
   landing, 210, 214, 395
   management and budget problems,
      209, 212-213, 224-226
   military involvement, 209-210
   missions, 214-215, 236, 352-353,
      371
   planned uses, 215, 454
   plant experiments, 370
   private commercial use, 454
   refueling system, 214
   robot arm, 353, 370
   rocket booster system, 223-224,
      329
   testing, 212-213, 214
   space walks, 194, 214, 277, 370
   volume and mass, 406
   weather delays, 394-395, 411
Space sickness, 37, 60, 66, 67, 73, 74-
   75, 150, 295
Space stations. *See also Salyut*
   as interplanetary spacecraft, 3, 460-
      462
   as research laboratory, 24-25
   Soviet visionaries, 1-6
   Western visionaries, 6-13, 208, 216-
      217
   wheel-shaped, 9, 14, 49, 51

Space treaties, 83-84
Space walks, 238
    Europeans, 261
    first, 32
    internal, 390-391, 434
    jetpacks/manned maneuvering
        units, 70-71, 214, 273, 274-
        275, 276, 277
    records, 324, 388, 402, 437
    Salyut maintenance and repair,
        117, 118-121, 147-149, 165,
        175, 185, 186-187, 189-194,
        198, 201, 202, 235, 236, 241
    Skylab repairs, 60, 62-63, 68-69,
        70-71, 76, 147
    Soyuz repairs, 285-289
    simulation, 189, 286, 310
    spacesuit failures, 343-344, 387, 402
    women, 214
Spacelab missions, 329
Spacesuits
    American design, 260
    gloves, 370
    memorobilia auctioned, 326, 327
    Orlan-D spacesuit, 118-119, 148
    Orlan-DMA, 260, 277, 287, 288,
        387, 402
    problems and solutions, 186-187,
        241-242, 289, 311, 313, 323-
        324, 326-327 n., 340, 343-344,
        387, 402
Sputnik, 1, 2, 3, 5, 25, 35, 90, 92, 246,
    376, 446
Stafford, Tom, 104, 105, 440
Star Trek, 210
Stalin, 4, 5, 35, 83, 100, 163, 246, 316,
    378
Stewart, Robert, 214, 277
Stockman, Dave, 216
Strategic Arms Limitation Talks
        (SALT agreements), 82, 83, 153
Strekalov, Gennady, 159, 182, 183,
        185-186, 290, 295, 357, 371,
        374, 375-377, 378, 384, 385-
        391, 393-394, 397, 400-401,
        437
Sullivan, Kathryn, 194, 214
Surveyor probe, 58
Suslov, Mikhail, 23, 24
Syria, 161, 244, 364

T

Taraki, Nur Mohammad, 152
Temperature control systems, 16
Tereshkova, Valentina, 176, 203
Thagard, Norman, 370, 371, 374, 375-
        377, 379, 380-381, 382, 383-
        384, 386, 387, 389, 391-392,
        393, 396, 397, 399, 400, 401,
        407, 408, 411, 412, 414, 428-
        429, 431-432, 437
Thermal heat shielding, 55-56
Thunderbirds (U.S. Air Force), 74
Tito, Dennis, 446-448, 449, 453, 458
Titov, Gherman, 120
Titov, Vladimir, 182-183, 185-186,
        238-239, 249, 250, 252-254,
        258, 259-262, 263, 264, 271,
        308, 311, 324, 325, 350, 366,
        368, 369, 370, 376, 430-431
Tokyo Broadcasting System (TBS),
        292-293, 295
TORU docking system, 335, 358, 422-
        423, 430-431
Trafton, Wilbur, 426
Truly, Richard, 210-211
TRW, 216
Tsibliev, Vasili, 336, 337, 339-341, 342,
        344, 348-349, 415, 418, 422-
        423, 425-436, 442
Tsiolkovsky, Konstantin, 3
TTM X-ray telescope, 254-255

U

Ukraine, 320, 335, 346, 421-422
United Nations, 320
    Outer Space Treaty, 458 n.
University of Bremen, 352-353
University of Utah, 438
Usachev, Yuri, 347, 348, 353-354, 357,
        406, 408, 410, 446
Ustinov, Dimitri, 24

V

van Laak, Jim, 445
Vasyutin, Vladimir, 197-198, 201, 202,
        203, 235, 249, 308, 337

Vazon containers, 143, 145
Venus probe, 329
Verne, Jules, 3
Viehbock, Franz, 319, 320
Vietnam, 133, 155, 156
Vietnam War, 82, 100, 380
Vikings, 460, 464
Viktorenko, Alexander, 272-273, 276, 295, 324, 336-337, 359, 362, 363-364, 365, 366-367, 368, 369-370, 372, 433
Vinogradov, Pavel, 411, 436, 437, 439
Volk, Igor, 193
Volkov, Alexander, 197-198, 201, 202, 261, 263, 264, 269, 275, 319, 320, 321, 322, 323-324, 325, 326-327 n.8, 333
Volkov, Vladislav, 20, 21, 33, 34, 35, 36, 39, 40-41, 42, 44, 46, 97, 98, 140, 151, 261, 376, 439
Volynov, Boris, 106, 107-108, 195, 345
von Braun, Wernher, 10-13, 14, 16, 17, 37, 49, 51, 78, 79, 180, 218, 367, 460, 463, 465
Voskhod missions, 25, 32
    (1), 256-257
Vostok capsule, 25, 183
Vostok 6 mission, 176
Vozdukh carbon dioxide scrubber, 231, 239-240, 403-403, 408, 423-424

W

Walker, Rachel, 161
Water recycling systems, 15, 41
Watergate scandal, 100
Weather satellites, 341, 346
Webb, Jim, 50
Weightlessness. *See also* Space sickness
    adaptation to, 14, 17, 38, 64, 67, 75-76, 88, 94-95, 140, 244, 364
    animal experiments, 95, 96, 103, 105, 107, 122, 126, 128, 145-146, 274
    botany experiments, 19-20, 21, 27, 28, 42, 43, 47, 95-96, 103, 105, 107, 109, 142-143, 145, 146,
        150, 156, 158, 159-160, 169-170, 174-176, 177, 180-181, 200, 201, 202, 214, 236-237
    Chibis system, 38, 39, 59, 94, 103, 131, 366
    design issues, 64-85, 336
    exercise in, 37-38, 39-40, 60, 102, 103, 108, 122, 124-125, 146, 149-150, 157-158, 159, 180, 190, 195, 245, 252, 260, 366, 373, 403
    experience of, 63-64, 69, 259, 287
    and fire propagation, 410
    fluid behavior experiments, 274, 317-318
    medical experiments, 49, 59-60, 94, 103, 107, 258, 302
    "Penguin" suits, 38, 39, 89, 94, 103, 245, 362, 366
    physiological effects of, 12, 15, 36-38, 39-40, 43, 49, 65-66, 72-73, 79, 95, 97-98, 135, 150, 159, 180, 188, 195, 252, 258, 262, 354-355, 372-373, 400, 403
    readjustment to gravity, 36, 66, 72-73, 78, 89, 108, 135-136, 149-150, 158-159, 180, 188, 195, 251-252, 262, 372-373, 411, 415, 428-429
    relaxation in, 190, 318
    tools and equipment designed for, 54-55, 59, 190-191, 194, 236, 296
    yoga and, 190
Weinberger, Caspar, 216
Weitz, Paul, 57, 58-59, 60, 65-66, 67, 196
West Germany, 100, 132, 171, 292, 319, 327, 331
Western space programs. *See also* National Aeronautics and Space Administration
    research-oriented goals, 51
    visionaries, 6-13
Wetherbee, James, 367, 368, 369-370
White, Ed, 50, 97
Whitson, Peggy, 380
Wilshire Associates, 447
Wireless World, 6-7
Wolf, David, 440

World War I, 22
World War II, 10, 23, 34, 35, 140, 189,
    205, 328, 378

Y

Yanaev, Gennadiy, 314
Yegorov, Boris, 256
Yeliseyev, Alexei, 30, 44, 154, 179
Yeltsin, Boris, 233, 247-248, 263, 264,
    265-268, 269, 278, 290, 303,
    304-306, 308, 314-315, 316-
    317, 319, 322, 327-328, 328,
    332-333, 334, 341-342, 343,
    356, 377, 421-422, 441
Young, John, 214

Z

Zero-gravity scale, 59
Zholobov, Vitaly, 106, 107-108, 151,
    191, 195, 345
Zhuchenko, Aleksandr, 268
Zimmerman, Keith, 440
Zloynikova, Antonina, 2
Zond spacecraft, 139, 154
Zudov, Vyacheslav, 108